Generis

PUBLISHING

INTERMEDIATE STATISTICS

VUKENKENG ANDREW WUJUNG

University of Bamenda

vukenkengwujung@yahoo.come

MBOKA NYAMSI GEORGES BIENVENUE

University of Bamenda

georgesmboka@gmail.com

Title: INTERMEDIATE STATISTICS

ISBN: 978-1-63902-264-9

Authors: VUKENKENG ANDREW WUJUNG
          MBOKA NYAMSI GEORGES BIENVENUE

Cover image: www.pixabay.com

Generis Publishing
Online orders: www.generis-publishing.com
Orders by email: info@generis-publishing.com

PREFACE

This book explores the frames of the practice and science of collecting and analysing numerical data in large quantities especially for the purpose of **inferring** in a whole from those in a representative sample from an intermediate perspective. Statistical concepts are applied with the help of numerical examples to issues and problem of all walks of life. More specifically this book starts with the presentation of basic algebra and followed by some cone concepts of statistics. It ends with derivatives and integrals of functions.

This work is based on a number of years of study and field experience of teaching statistics by the author at the intermediate level.

ACKNOWLEDGEMENT

This piece of work could not be completed without the support of some individual person, their contributions to this work is either direct or indirect.

We are sincerely grateful to the contribution of Mr. Ngalim Valentine (PhD) who out of his busy schedule and tied office duties, made critical comments which were considered in the realization of this work.

We are indebted to the following persons: Mr. Banseka Siyu Zephaniah, Ms. Azongwa Doris Ngeche, Chi Brandone, Nfor Amalia, Mr. Derick Viban, Saidu Sali, Fai Sairatu Liybarfeh and Tanfen Marinette, who took out time from the precious time to edit the work

Our particular thanks goes to the following persons: Mr. Wingo Kilian, Mrs. Ngoran Enerstine, Mrs. Fondzenyuy nee Anester Bonjeh, Mr. Anthony Lupiya. Mr. Tati Daniel for the moral contribution to the work

Special thanks goes to the following friends and colleague: Mr. Mbinka Benoit Shalanyuy, Mrs Aghen Laura Endah epse Banseka, Mr. Sewoyehbaa Marcel, Mr. Shu Elvis, Mr. Eric Njong Seka, Mr. Akumbom Paul, Mr. Penanje Christain, Miss Emeniguene Balindi Laure, Mr. Wanda Njinkeu Bonaventure and Mr. Bonou Nkeh Blaise for their moral and motivational support.

DEDICATION

TO

Readers of Intermediate Statistics

Table of Contents

PART ONE

ALGEBRA

CHAPTER ONE

NUMBERS

1.1 Meaning of Numbers

A numbers is an arithmetical value expressed by a word, symbol or figure representing a particular quantity and used in counting and making calculations and conclusions. It is also seen as a mathematical object used to count, measure and label things. The different types of numbers are based on their mathematical signs, expressions and combinations.

1) Natural Numbers($\mathbb{N}$): Natural numbers are a set of numbers used to count something. They are also called counting numbers or positive integer.

$$\mathbb{N} = \{1, 2, 3, 4, 5, 6, 7, 8, 9, 10, \cdots \cdots + \infty\} \ldots\ldots\ldots\ldots\ldots(1)$$

2) Whole Numbers($\mathbb{W}$): Whole numbers consist of a set of natural numbers with zero included. It is worth knowing that fractions, decimals and negative are not included in a whole number.

$$\mathbb{W} = \{0, 1, 2, 3, 4, 5, 6, 7, 8, 9, 10 \cdots + \infty\} \ldots\ldots\ldots\ldots(2)$$

Pythagoras

Hippasus

3) Integers($\mathbb{Z}$): An integer consists of a set of whole numbers and their image (opposite value). That is it is a set of positive numbers, zero and negative numbers.

$$\mathbb{Z} = \{-\infty \cdots - 5, -4, -3, -2, -1, 0, 1, 2, 3, 4, 5 \cdots + \infty\}\ldots(3)$$

4) Rational Numbers($\mathbb{Q}$): Rational numbers are numbers expressed as a fraction $\left(\frac{a}{b}\right)$ and where (a) and (b) are integers with (b) not equal to zero. That is, $\left[\mathbb{Q} = \frac{a}{b}, \text{ where}, a \in \mathbb{Z}, b \in \mathbb{Z} \text{ and } b \neq 0\right]$.

$$\mathbb{Q} = \left\{\frac{2}{3}, \frac{1}{5}, \frac{7}{1}, \frac{-5}{9}, \frac{17}{11}, \frac{153}{100}, \frac{2}{9} \ etc\right\} \ldots\ldots\ldots\ldots\ldots\ldots\ldots\ldots\ldots\ldots\ldots\ldots(4)$$

5) Irrational Numbers: Irrational numbers are numbers that cannot be expressed in rational number form (fraction).

$$\text{Irrational Number} = \left\{\sqrt{2}, \sqrt{5}, 6 - \sqrt{3}, 3 + \sqrt{7}, \pi \ etc\right\} \ldots\ldots\ldots\ldots\ldots\ldots\ldots(5)$$

6) Real Numbers($\mathbb{R}$): Real numbers consists of all the rational numbers and irrational numbers.

$$\mathbb{R} = \{\text{Rational Numbers and Irrational Numbers}\} \ldots\ldots\ldots\ldots\ldots\ldots(6)$$

7) Prime Numbers: They are numbers greater than one (> 1) and that have only one (1) and itself as factor.

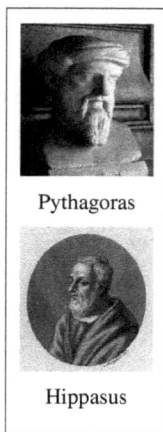

$$\text{Prime Numbers} = \{2, 3, 7, 11, 13, 17, 19, 23, 31 \text{ etc}\} \ldots\ldots\ldots\ldots\ldots\ldots\ldots\ldots\ldots(7)$$

1.2 Mathematical Approximation

Measurement results can be exact (=) or non-exact to a given unit of measurement. When non-exact measurements occur, the result is being approximated according to the unit of measurement in question. To approximate a value means to express a value within a specified degree of accuracy and approximation is represented by the sign ($\approx$). Approximation methods are of different types but our analysis will be limited to rounding and significance figures.

1.2.1 Rounding

Rounding can be rounding up (overestimation) and rounding down (underestimation). Rounding can be expressed to a specific decimal place (decimal numbers) or to a ten, hundred, thousand, ten of thousand, millions (whole numbers) etc. Decimal numbers are numbers having a decimal point (.) amongst its digits.

1) **Round to the Nearest Tenth:** A ten is any two digits number (10, 20, 30, 40, 50, 60, 70, 80 and 90). Any number rounded up is changed to 10 and any number rounded down is changed to 0.

Example 1: Round the following to the nearest tenth. 54 and 98

<u>**Solution**</u>

	54
Transformation Part	**[54]** This is because a ten is made up of two digits. The first digit of the transformation part 5 is changed to 50.
Rounding Decision	The last digit of the transformation part [4] is less than 5. We change it to 0 and add to the existing 50.
Rounded Number	$54 = [54] = [50 + 0] = \mathbf{50}$

	98
Transformation Part	**[98]** This is because a ten is made up of two digits. The first digit of the transformation part 9 is changed to 90.
Rounding Decision	The last digit of the transformation part [8] is greater than 5. We change it to 10 and add to the existing 90.
Rounded Number	$98 = [98] = [90 + 10] = \mathbf{100}$

Example 2: Round the following to the nearest ten

    1) 857

    2) 13,912

<u>**Solution**</u>

1) 857

	857
Transformation Part	8[**57**] This is because a ten is made up of two digits. The first digit of the transformation part 5 is changed to 50.
Rounding Decision	The last digit of the transformation part [7] is greater than 5. We change it to 10 and add to the existing 50.
Rounded Number	$857 = 8[57] = 8[50 + 10] = \mathbf{860}$

2) 13,912

	13,912
Transformation Part	13,9[**12**] This is because a ten is made up of two digits. The first digit of the transformation part 1 is changed to 10.
Rounding Decision	The last digit of the transformation part [2] is less than 5. We change it to 0 and add to the existing 10.
Rounded Number	$13,912 = 13,9[12] = 13,9[10 + 0] = \mathbf{13,910}$

<u>**Test Your Understanding**</u>

Exercise 1: Round the following to the nearest ten

    1) 74...**[Answer: 70]**

    2) 37...**[Answer: 40]**

    3) 2,058..**[Answer: 2,060]**

    4) 178,133...**[Answer: 178,130]**

2) **Round to the Nearest Hundred:** A hundred is any three digits number (100, 200, 300, 400, 500, 600, 700, 800 and 900). Any value rounded up is changed to 100 and any number rounded down is changed to 0.

Example 1: Round the following numbers to the nearest hundred; 1,685 and 13,912.

4

	1,685
Transformation Part	1,[685]
	This is because a hundred is made up of three digits. The first digit of the transformation part 6 is changed to 600.
Rounding Decision	The two last digit of the transformation part [85] is greater than 50. We change them to 100 and add to the existing 600.
Rounded Number	$1,685 = 1,[685] = 1,[600 + 100] = 1,[700] = \mathbf{1,700}$

	13,912
Transformation Part	13,[912]
	This is because a hundred is made up of three digits. The first digit of the transformation part 9 is changed to 900.
Rounding Decision	The two last digit of the transformation part [812] is less than 50. We change them to 0 and add to the existing 900.
Rounded Number	$13,912 = 13,[912] = 13,[900 + 0] = 13,[900] = \mathbf{13,900}$

Example 2: Round 236 to the following

1) Nearest ten
2) Nearest hundred

1) **Nearest ten**

	236
Transformation Part	2[36]
	This is because a ten is made up of two digits. The first digit of the transformation part 3 is changed to 30.
Rounding Decision	The last digit of the transformation part [6] is greater than 5. We change them to 10 and add to the existing 30.
Rounded Number	$236 = 2[36] = 2[30 + 10] = \mathbf{240}$

2) Nearest hundred

	236
Transformation Part	[236] This is because a hundred is made up of three digits. The first digit of the transformation part 2 is changed to 200.
Rounding Decision	The two last digits of the transformation part [36] is less than 50. We change them to 0 and add to the existing 200.
Rounded Number	$236 = [236] = [200 + 0] = \mathbf{200}$

Test Your Understanding

Exercise 2: Round the following to the nearest hundred

1) 781...[Answer: 800]

2) 916...[Answer: 900]

3) 555...[Answer: 600]

4) 89...[Answer: 100]

Exercise 3: Round 860 to the following

1) Nearest ten...[Answer: 860]

2) Nearest hundred...[Answer: 900]

---

3) **Round to the Nearest Thousand:** A thousand is any four digits number (1000, 2000, 3000, 4000, 5000, 6000, 7000, 8000 and 9000). Any value rounded up is changed to 1,000 and any number rounded down is changed to 0.

Example 1: Round the following numbers to the nearest thousand; 6,567 and 6,281

Solution

	6,567
Transformation Part	[6,567] This is because a thousand is made up of four digits. The first digit of the transformation part 6 is changed to 6,000.
Rounding Decision	The three last digit of the transformation part [**567**] is greater than 500. We change them to 1,000 and add to the existing 6,000.
Rounded Number	$6,567 = [6,567] = [6,000 + 1,000] = \mathbf{7,000}$

	6,281
Transformation Part	[6,281] This is because a thousand is made up of four digits. The first digit of the transformation part 6 is charged to 6,000.
Rounding Decision	The three last digit of the transformation part [281] is less than 500. We change them to 0 and add to the existing 6,000.
Rounded Number	$6,281 = [6,281] = [6,000 + 0] = \mathbf{6,000}$

Example 2: Round 34,519 to the following

1) Nearest ten
2) Nearest hundred
3) Nearest thousand

Solution

1) Nearest ten

	34,519
Transformation Part	34,5[19] This is because a ten is made up of two digits. The first digit of the transformation part 1 is changed to 10.
Rounding Decision	The last digit of the transformation part [9] is greater than 5. We change it to 10 and add to the existing 10.
Rounded Number	$34,519 = 34,5[19] = 34,5[10 + 10] = \mathbf{34,520}$

2) Nearest hundred

	34,519
Transformation Part	34,[519] This is because a hundred is made up of three digits. The first digit of the transformation part 5 is changed to 500.
Rounding Decision	The two last digits of the transformation part [19] is less than 50. We change them to 0 and add to the existing 500.
Rounded Number	$34,519 = 34,[519] = 34[500 + 0] = \mathbf{34,500}$

3) Nearest thousand

	34,519
Transformation Part	3[4,**519**] This is because a thousand is made up of four digits. The first digit of the transformation part 4 is changed to 4,000.
Rounding Decision	The last three digits of the transformation part [**519**] is greater than 500. We change them to 1,000 and add to the existing 4,000.
Rounded Number	$34,519 = 3[4,519] = 3[4,000 + 1,000] = \mathbf{35,000}$

Test Your Understanding

Exercise 4: Round the following numbers to the nearest thousand

1) 34,344...[Answer: **34,000**]

2) 78,753...[Answer: **79,000**]

3) 899..[Answer: **1,000**]

Exercise 5: Round 18,765 to the following

1) Nearest ten...[Answer: **18,770**]

2) Nearest hundred..[Answer: **18,800**]

3) Nearest thousand...[Answer: **19,000**]

---

4) **Round to the Nearest Million:** A million is any seven digits number (1,000,000.... and 9,000,000). Any value rounded up is changed to 1,000,000 and any number rounded down is changed to 0.

Example 1: Round the following numbers to the nearest million; 78,168,951 and 8,534,576

Solution

	78,168,951
Transformation Part	7[8,**168,951**] This is because a million is made up of seven digits. The first digit of the transformation part 8 is changed to 8,000,000.
Rounding Decision	The six last digits of the transformation part [**168,951**] is less than 500,000. We change them to 0 and add to the existing 8,000,000.
Rounded Number	$78,168,951 = 7[8,168,951] = 7[8,000,000 + 0] = \mathbf{78,000,000}$

	8,534,576
Transformation Part	**[8,534,576]** This is because a million is made up of seven digits. The first digit of the transformation part 8 is changed to 8,000,000.
Rounding Decision	The six last digits of the transformation part **[534,576]** is greater than 500,000. We change them to 1,000,000 and add to the existing 8,000,000.
Rounded Number	$8,534,576 = [8,534,576] = [8,000,000 + 1,000,000] = \mathbf{9,000,000}$

Example 2: Round the following numbers to the nearest millions

1) 2,100,384
2) 56,518,720

<u>Solution</u>

1) 2,100,384

	2,100,384
Transformation Part	**[2,100,384]** This is because a million is made up of seven digits. The first digit of the transformation part 2 is changed to 2,000,000.
Rounding Decision	The six last digits of the transformation part **[100,384]** is less than 500,000. We change them to 0 and add to the existing 2,000,000.
Rounded Number	$2,100,384 = [2,100,384] = [2,000,000 + 0] = \mathbf{2,000,000}$

2) 56,518,720

	56,518,720
Transformation Part	**5[6,518,720]** This is because a million is made up of seven digits. The first digit of the transformation part 6 is changed to 6,000,000.
Rounding Decision	The six last digits of the transformation part **[518,720]** is greater than 500,000. We change them to 1,000,000 and add to the existing 6,000,000.
Rounded Number	$56,518,720 = 5[6,518,720] = 5[6,000,000 + 1,000,000] = \mathbf{57,000,000}$

Test Your understanding

Exercise 6: Round the following numbers to the nearest millions

1) 1,121,565...[**Answer: 1,000,000**]
2) 6,039,510...[**Answer: 6,000,000**]
3) 6,125,189 ....[**Answer: 6,000,000**]

Exercise 7: Round 355,504,257 to the following
1) Nearest ten ....[Answer: **355,504,260**]
2) Nearest hundred...[Answer: **355,504,300**]
3) Nearest thousand...[Answer: **355,504,000**]
4) Nearest millions..[Answer: **356,000,000**]

Exercise 8: Round the following
1) 10.01 to the nearest ten...[Answer: **10.00**]
2) 45.76 to the nearest ten...[Answer: **45.80**]
3) 256 to the nearest hundred..[Answer: **300**]
4) 13,912 to the nearest hundred.......................................[Answer: **13,900**]
5) 185,165 to the nearest ten-thousand..............................[Answer: **190,000**]

5) Round to a Specific Decimal Place: When rounding to a specific decimal place, the following principles are worth respecting,

1) If the value is less than 5{0, 1, 2, 3, and 4}, we underestimate the value (drop the value) and maintain the preceding value.

2) If the value is equal to 5, to change the preceding value to an even number (if it is an odd number) or maintain the preceding number (if it is already an even number). It is worth knowing that zero is considered an even number.

3) If the number is greater than 5{6, 7, 8, and 9}, we overestimate and add 1 to the preceding value.

Example 1: Express the following in two decimal places

1) 67.973
2) 398.8892

Solution

Hint: Two decimal places means there should be two digits after the decimal point.

67.973	398.8892
N/B: The preceding value here is 7. The value to be changed to 1 is 3. Since 3 is less than 5, we change it to zero or drop the value (3). The new preceding value is $(7 + 0 = 7)$	**N/B:** The preceding value here is 8. The value to be changed to 1 is 9. Since 9 is greater than 5, we change it to 1 and add to the preceding value. $(8 + 1 = 9)$
67.97	**398.89**

Example 2: Express the following in three decimal places

1) 785.9739
2) 970.0971

Hint: Three decimal places means there should be three digits after the decimal point.

785.9739	970.0971
N/B: The preceding value here is 3. The value to be drops or changed to 1 is 9. Since 9 is greater than 5, we change it to 1. The new preceding value is $(3 + 1 = 4)$	**N/B:** The preceding value here is 7. The value to be drops or changed to 1 is 1. Since 1 is less than 5, we change it to 0 and add to the preceding value. $(7 + 0 = 7)$
785.974	**970.097**

Test Your Understanding

Exercise 9: Express the following,

1) 5.976 to one decimal place................................... ....[Answer: 6.0]

2) 13.872 to two decimal places............................... ....[Answer: 13.87]

3) 22.5638 to three decimal places............................. ....[Answer: 22.564]

1.2.2 Significance Figure

The significance figure of a number refers to those digits that have meaning in reference to a measured or specified value. All non-zero integer numbers are significance. The complication comes with the decision of zero being significant or non-significant. The following rules help in deciding if a zero or zeros are significant or non-significant

Rule 1: Zeros that are between non-zero integers are always significant.

Example 1: How many significance figures do the following numbers contain?

1) 26.38

2) 2,552

Solution

26.38	2,552
Four significance figure	Four significance figures
[4 significance figure]	[4 significance figure]
It is four significance figures because all digits are non-zero integers	It is 4 significance figures because all numbers are non-zero integers

Example 2: How many significance figures do the following numbers contain?

1) 406

2) 103.4

406	103.4
Three significance figure	Four significance figure
[3 significance figure]	[4 significance figure]
This is because the number is made up of two non-zero integer digits and the zero is between non-zero integer digits making it significance This sum up to three significance figure.	This is because the number is made up of three non-zero integer digits and the zero is between non-zero integer digits making it significance This sum up to three significance figure.

Test Your Understanding

Exercise 10: How many significance figures do the following numbers contain?

1) 3,457...[Answer: **4 significance figures**]

2) 258,997...[Answer: **6 significance figures**]

3) 2,0576.6..[Answer: **6 significance figures**]

4) 67,609...[Answer: **5 significance figures**]

Rule 2: Zeros that comes before non-zero integers are never significant (leading zeros)

Example 1: How many significance figures do the following numbers contain?

1) 0.3465

2) 0.986

0.3465	0.986
Four significance figure	Three significance figure
[4 significance figure]	[3 significance figure]
This is because the zero is before a non-zero integer digit, making it not significance and the remaining digits are non-zero integers giving us a total four significance figure.	This is because the zero is before a non-zero integer digit, making it not significance and the remaining digits are non-zero integers giving us a total four significance figure.

Example 2: How many significance figures do the following numbers contain?

1) 09

2) 901

Solution

09	901
One significance figure [1 significance figure]	Three significance figure [3 significance figure]
The zero is before a non-zero integer digit, making it not significance and the other digit (9) is a non-zero integer, giving us a one significance figure.	There are two non-zero integer digits and the zero is between non-zero integer digits making it significance and giving us a total of three significance figures

---

Test Your Understanding

Exercise 11: How many significance figures do the following numbers contain?

1) 45,678...[Answer: 5 Significance Figures]
2) 401,354..[Answer: 6 Significance Figures]
3) 0.897...[Answer: 3 Significance Figures]
4) 05..[Answer: 1 Significance Figure]

Rule 3: Any zero followed by a decimal point is significant

Example 1: How many significance figures do the following numbers contain?

1) 400.000
2) 7,097

Solution

400.000	7,097
Three significance figures [3 significance figures]	
The number is made up a non-zero integer digit. The next two zeros are followed by a decimal point making them significant and giving a total of 3 significance figures. The three last zeros are not significant since they are not before a non-zero integer or followed by a decimal point.	Four significance figures [4 significance figures]

Example 2: How many significance figures do the following numbers contain?

1) 0.00456
2) 130.

0.00456	130.
Three significance figures [3 significance figure]	Three significance figures [3 significance figure]
The first zero is not significant because it is not after a non-zero integer. The two zeros after the decimal are not significant because they are not between or after non-zero integer digits. The three last numbers are significant since they are non-zero integer digits.	The first two numbers are non-zero integer digits making them significant. The zero is significant since it is after a non-zero integer digit and is followed by a decimal point. This gives a total of three significance figures.

Test Your Understanding

Exercise 12: How many significance figures do the following numbers contain?

1) 79,564……………………………………...………[Answer: 5 significance figures]

2) 23.04…………………………………...…………[Answer: 4 significance figures]

3) 970.00………………………………………………[Answer: 3 significance figures]

Rule 4: Any zero that comes after a non-zero integer but not followed by a decimal point is not significant.

Example 1: How many significance figures do the following numbers contain?

1) 700

2) 700.

700	700.
One significance figure [1 significance figure]	Three significance figure [3 significance figure]
The number is made up of a non-zero integer digit. The two zeros are after a non-zero integer digit but are not followed by a decimal point making them non-significant. This gives a total of one significance figure	The number is made up of a non-zero integer digit. The two zeros are after a non-zero integer digit and are followed by a decimal point making them significant. This gives a total of three significance figures.

Example 2: How many significance figures do the following numbers contain?

1) 70,400.00
2) 0.80900

<u>Solution</u>

70,400.00	0.80900
Five significance figures [5 significance figure]	Three significance figure [3 significance figure]
The first zero is significant since it is between non-zero integer digits. The next two zeros are significant since they are after a non-zero integer digit and are followed by a decimal point	The leading zero is not significant since it is not after a non-zero integer. The next zero is significance since it is between non-zero integer digits and the trailing zeros are not significance since they are not followed by a decimal point although they comes after a non-zero integer digit.

<div align="center">**Test Your Understanding**</div>
Exercise 13: How many significance figures contain the following numbers? 1) 34,987...[Answer: **5 significance figures**] 2) 5,010.00..[Answer: **4 significance figures**] 3) 600,200..[Answer: **4 significance figures**] 4) 9,0000..[Answer: **1 significance figure**]

Rule 5: Any zero that comes after a non-zero integer and comes after the decimal point is significant.

Example 1: How many significance figures do the following numbers contain?

1) 100,045
2) 130.00

<u>Solution</u>

100,045	130.00
Six significance figures [6 significance figures]	Five significance figures [5 significance figures]
The three zeros are significant because they are between non-zero integer digits. With the three non-zero integer digits, we have a total of 6 significance figures.	The first zero is significance because it is after a non-zero integer digit and is followed by a decimal point. The last two zeros (trailing zeros) are significance because they come after a decimal.

Example 2: How many significance figures do the following numbers contain?

1) 12.2300

2) 0.0560

Solution

12.2300	0.0560
Six significance figure	Three significance figures
[6 significance figure]	[3 significance figures]
The number is made up of four non-zero integer digits. The trailing zeros are significant since they are after a non-zero integer digit and after a decimal point.	The leading zero is not significant as it is not after a non-zero integer digit and not after a decimal point. The next zero is not significant as it is after a decimal point but not after a non-zero integer digit. The last zero is significant as it is both after a non-zero integer digit and decimal point.

Test Your Understanding
Exercise 14: How many significance figures do the following numbers contain?

1) 36.509...[**Answer: 5 significance figures**]

2) 0.6096...[**Answer: 4 significance figures**]

3) 708.000..[**Answer: 6 significance figures**]

4) 0.100..[**Answer: 3 significance figures**]

It is worth knowing that all the above principles must be considered to determine if a given digit is significance or not significance.

Note: The mathematical operation of decimal points makes use of the limiting term. The limiting term is the factor or term having the smallest decimal places or having the smallest significance figure. The mathematical operation respects the following principles.

1) When adding or subtracting, the result should have the same number of decimal place as the limiting terms.

2) When multiplying or dividing, the result should have the same significance figure as the limiting terms.

1.3 The Four Rules of Number Manipulation

The four rules of number manipulation include; addition rule, subtraction rule, multiplication rule and division rule. The summary table of these rules is given below.

1.3.1 Multiplication Rule

Multiplication is represented by the cross sign ($\times$) or asterisk ($*$) or dot (.). In multiplication, the first value is called multiplier, the second value after the multiplication sign is called multiplicand and the result is called product.

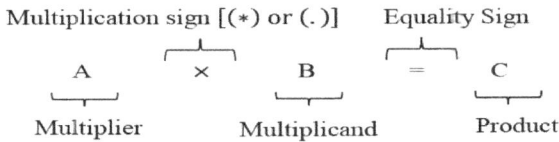

Multiplication sign [($*$) or (.)]   Equality Sign

A $\times$ B $=$ C

Multiplier   Multiplicand   Product

N/B: The multiplication of integers follows specific rules as illustrated below.

$(+a) \times (+b) = a \times b = ab$ ....[Result (+)]

$(+a) \times (-b) = a \times -b = -ab$ ....[Result (−)]

$(-a) \times (+b) = -a \times b = -ab$ ....[Result (−)]

$(-a) \times (-b) = -a \times -b = ab$ ....[Result (+)]

Note: Apart from the above rules, the following multiplication properties such as property of zero, identity property of one, inverse property, commutative property and associative property are worth knowing.

$A \times 0 = 0$ ....[Property of zero]

$A \times 1 = A$ ....[Identify property of one]

$A \times \frac{1}{A} = 1$ ....[Inverse property]

$A \times B = B \times A$ ....[Commutative property]

$A \times (B \times C) = (A \times B) \times C$ ....[Associative property]

Example 1: Solve the following equations

1) -6×4
2) $4 \times (-10)$
3) $-8 \times (-2)$
4) 2×7

17

	1) −6 × 4	2) 4 × (−10)	3) −8 × (−2)	4) 2 × 7
Formulas	[(−a) × (+b) = −a × b = −ab]	[(+a) × (−b) = a × −b = −ab]	[(−a) × (−b) = −a × −b = ab]	[(+a) × (+b) = a × b = ab]
Working	−6 × 4 = **−24**	4 × (−10) = **−40**	8 × 2 = **16**	2 × 7 = **14**

Example 2: The price of a bag of rice is 5,000 FCFA. What is the total revenue of the seller given that he sold 25 bags?

Step 1: Formula. Total Revenue = Total Number of Bags of Rice × Price per Bag of Rice

Where; Total Number of Bags of Rice = 25 and Price per Bag of Rice = 5,000

Step 2: Data substitution and solving

Total Revenue = 25 × 5,000 ....[Solve equation]

Total Revenue = **125, 000 FCFA**

Example 3: The unit price of a bag of cement is 5,500 FCFA. The quantity purchased by business man from Monday to Thursday is given below.

Days	Monday	Tuesday	Wednesday	Thursday
Quantity Purchased	25	15	30	8

Determine his total expenditure using the table above.

1) Method 1

Step 1: Formula. Total Expenditure = Total Number of Bags of Cement ×

Price per Bag of Cement

Where; Price per Bag of Cement = 5,500 and Total Number of Bags of Cement = 78

Step 2: Data substitution and solving

Total Expenditure = 78 × 5,500 ....[Solve equation]

Total Expenditure = **429, 000 FCFA**

2) Method 2

Step 1: Formula. Total Expenditure = Sum of Daily Expenditure

Step 2: Data substitution and solving

Days	Monday	Tuesday	Wednesday	Thursday	Total
Quantity Purchased	25	15	30	8	Expenditure
Daily Expenditure	$25 \times 5,500 = 137,500$	$15 \times 5,500 = 82,500$	$30 \times 5,500 = 165,000$	$8 \times 5,500 = 44,000$	**429,000 FCFA**

Example 4: Fill the empty squares in the digits multiplication table below.

Multiplication ($\times$)	5	0	-5
-5			
0			
5			

Solution

($\times$)	5		0		-5	
-5	$-5 \times 5 = -25$	$5 \times (-5) = -25$	$-5 \times 0 = 0$	$0 \times (-5) = 0$	$-5 \times (-5) = 25$	$-5 \times (-5) = 25$
0	$0 \times 5 = 0$	$5 \times 0 = 0$	$0 \times 0 = 0$	$0 \times 0 = 0$	$0 \times (-5) = 0$	$-5 \times 0 = 0$
5	$5 \times 5 = 25$	$5 \times 5 = 25$	$5 \times 0 = 0$	$0 \times 5 = 0$	$5 \times (-5) = -25$	$-5 \times 5 = -25$

Test Your Understanding

Exercise 15: Given that A = 15, B = 20 and C = 5, show that multiplication respect the following,

1) Commutative property…………………………...………….[Answer: See appendix]

2) Associative property……………………………….………..[Answer: See appendix]

3) Inverse property…………………………………………….[Answer: See appendix]

4) Identity property of one……………………………….…...[Answer: See appendix]

Exercise 16: Given that a bag of cement is 50kg. What will be the weight of 25 bags?

[Answer: 1,250kg]

Exercise 17: Fill the empty squares in the digits multiplication table below.

Table 1						Table 2				

Table 1

Multiplication	-1	-2	0	1	2
1					
2					
0					
-1					
-2					

Table 2

($\times$)	1	2		4
4	4		-12	16
		-6	18	
2	2		-6	8

[Answer: ……………………………………………… …………..……………...See Appendix]

19

1.3.2 Division Rule

Division means splitting into equal parts or group and it is represented by a division sign $(\div)$ or moins $(-)$ or $(/)$. In division, the first value is called dividend, the second value after the division sign is called divisor and the result is called quotient.

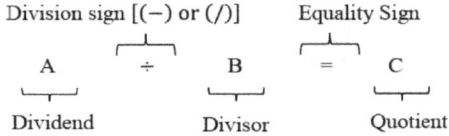

$$\text{Division sign } [(-) \text{ or } (/)] \qquad \text{Equality Sign}$$

$$\underbrace{A}_{\text{Dividend}} \quad \overbrace{\div} \quad \underbrace{B}_{\text{Divisor}} \quad \overbrace{=} \quad \underbrace{C}_{\text{Quotient}}$$

N/B: The Division of integers follows specific rules as illustrated below.

$$(+a) \div (+b) = \frac{+a}{+b} = +\frac{a}{b} \dots\dots\dots\dots\dots\dots\dots\dots\dots\dots\text{[Result } (+)]$$

$$(+a) \div (-b) = \frac{+a}{-b} = -\frac{a}{b} \dots\dots\dots\dots\dots\dots\dots\dots\dots\dots\text{[Result } (-)]$$

$$(-a) \div (+b) = \frac{-a}{+b} = -\frac{a}{b} \dots\dots\dots\dots\dots\dots\dots\dots\dots\dots\text{[Result } (-)]$$

$$(-a) \div (-b) = \frac{-a}{-b} = +\frac{a}{b} \dots\dots\dots\dots\dots\dots\dots\dots\dots\dots\text{[Result } (+)]$$

Note: Apart from the above rules, the following division properties such as; property of zero, property of one and identity property of one are worth knowing. It is worth knowing that division does not respect the inverse property, commutative property and associative property.

$$\frac{0}{A} = 0, \text{ when A} \neq 0 \dots\dots\dots\dots\dots\dots\dots\dots\dots\dots\text{[Property of zero]}$$

$$\frac{A}{A} = 1, \text{ when A} \neq 0 \dots\dots\dots\dots\dots\dots\dots\dots\dots\dots\text{[Property of one]}$$

$$\frac{A}{1} = A \dots\dots\dots\dots\dots\dots\dots\dots\dots\dots\text{[Identify property of one]}$$

Example 1: Compute the following

1) $-50 \div 25$
2) $100 \div (-50)$
3) $-15 \div (-5)$
4) $30 \div 10$

Solution

	1) $-50 \div 25$	2) $100 \div (-50)$	3) $-15 \div (-5)$	4) $30 \div 10$
Formulas	$\left[(-a) \div (+b) = \frac{-a}{+b} = -\frac{a}{b}\right]$	$\left[(+a) \div (-b) = \frac{+a}{-b} = -\frac{a}{b}\right]$	$\left[(-a) \div (-b) = \frac{-a}{-b} = +\frac{a}{b}\right]$	$\left[(+a) \div (+b) = \frac{+a}{+b} = +\frac{a}{b}\right]$
Working	$-50 \div 25 = -\frac{50}{25} = -2$	$100 \div (-50) = -\frac{100}{50} = -2$	$-15 \div -5 = \frac{-15}{-5} = \frac{15}{5} = 3$	$30 \div 10 = \frac{30}{10} = 3$

Example 2: Given that A = 30, B = 10 and C = 2. Show that the division rule does not respect the inverse property $\left[A \div \frac{1}{A} = 1\right]$, commutative property $[A \div 3 = B \div A]$ and associative property $[A \div (B \div C) = (A \div B) \div C]$.

Solution

	Inverse property	Commutative property	Associative property
Formula	$\left[A \div \frac{1}{A} = 1\right]$	$[A \div B = B \div A]$	$[A \div (B \div C) = (A \div B) \div C]$
Working	$30 \div \frac{1}{30} = 1$ ......[Transform] $30 \times \frac{30}{1} = 1$ ....[Solve] $60 \neq 1$	$30 \div 10 = 10 \div 30$ .....[Solve] $3 \neq 1/3$	$30 \div (10 - 2) = (30 \div 10) \div 2$ ...[Simplify bracket] $30 \div 5 = 3 \div 2$ ....[Solve] $6 = 1.5$
Conclusion	$A \div \frac{1}{A} \neq 1$	$A \div B \neq B \div A$	$A \div (B \div C) \neq (A \div B) \div C$

Example 3: How many 4 (divisor) are in a 12 (dividend)?

Solution

Step 1: Formula. Number of Four $= \frac{\text{Dividend}}{\text{Divisor}}$, where; Dividend = 12 and Divisor = 4

Step 2: Data substitution and solving

Number of Four $= \frac{12}{4}$ ....[Solve equation]

Number of Four $= 3$ ....[There are three fours in 12]

Test Your Understanding

Exercise 18: Solve the following

1) $-75 \div 5$..[Answer: -15]

2) $250 \div (-25)$...[Answer: -10]

3) $-5 \div (-5)$...[Answer: 1]

4) $1,000 \div 250$...[Answer: 4]

Exercise 19: Given that A = 100, B = 50 and C = 25, show that the division rule does not respect,

1) The commutative law $[A \div B = B \div A]$...............................[Answer: $2 \neq 1/2$]

2) Associative law $[A \div (B \div C) = (A \div B) \div C]$...................[Answer: $50 \neq 2/25$]

Exercise 20: The amount 125,000 FCFA is shared between 5 children. How much will each child receive?...[Answer: 25,000 FCFA]

1.3.3 Addition Rule

Addition is the total or summation of two or more numbers. The number before the addition sign is called addend and after the addition sign is also called addend. Addition is represented by the sign (+)

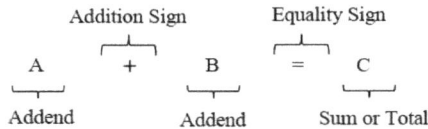

$$\underbrace{\text{Addition Sign}}$$

Addition Sign   Equality Sign

A   +   B   =   C

Addend   Addend   Sum or Total

N/B: The addition of integers follows specific rules as illustrated below.

$(+a) + (+b) = a + b$ ....[Result (+)]

$(+a) + (-b) = a - b$ .... $\begin{bmatrix} \text{Result (+) if a} > b \\ \text{Result (-) if b} > a \end{bmatrix}$

$(-a) + (+b) = -a + b$ .... $\begin{bmatrix} \text{Result (+) if b} > a \\ \text{Result (-) if a} > b \end{bmatrix}$

$(-a) + (-b) = -a - b$ ....[Result (-)]

Note: Just as multiplication and unlike division, the addition rule respects the inverse property, commutative property and associative property.

$A + 0 = A$ ....[Identity property of zero]

$A + (-A) = 0$ ....[Inverse property]

$A + B = B + A$ ....[Commutative property]

$A + (B + C) = (A + B) + C$ ....[Associative property]

Example 1: Evaluate the following
  1) $5 + 17$
  2) $10 + (-5)$
  3) $-20 + (-2)$
  4) $-9 + 4$

Solution

	1) $5 + 17$	2) $10 + (-5)$	3) $-20 + (-2)$	4) $-9 + 4$
Formulas	$[(+a) + (+b) = a + b]$	$[(+a) + (-b) = a - b]$	$[(-a) + (-b) = -a - b]$	$[(-a) + (+b) = -a + b]$
Working	$5 + 17 = 22$	$10 - 5 = 5$	$-20 - 2 = -22$	$-9 + 4 = -5$

Example 2: Add 2516 and 255

Solution

Hint: When adding, we start from the right to the left. When the sum of values is in ten, hundred and/or thousand, we considered the last digit and take the others to the next stage.

	Stage 1	Stage 2	Stage 3	Stage 4
	2 5 1 **6** +2 5 **5** **1**	2 5 **1** 6 +2 **5** 5 **7** 1	2 **5** 1 6 +2 **5** 5 **7** 7 1	**2** 5 1 6 +**2** 5 5 **2** 7 7 1
Working	$6 + 5 = 11$	$1 + 5 + 1 = 6$	$5 + 2 = 7$	$2 + 0 = 2$

Example 3: During Christmas, your father gave you 1,000 FCFA (addend) and your mother added 500 FCFA (addend). How much is your total money?

Solution

Step 1: Formula. Total Money = First Addend + Second Addend

Where; First Addend = 1,000 and Second Addend = 500

Step 2: Data substitution and solving

Total Money = 1,000 + 500 ....[Add values]

Total Money = **1, 500 FCFA**

Test Your Understanding

Exercise 21: Compute the following

1) $20 + 50$...**[Answer: 70]**

2) $25 + (-30)$...**[Answer: −5]**

3) $-9 + (-12)$...**[Answer: −21]**

4) $-5 + 30$..**[Answer: 25]**

Exercise 22: Miss Doris feed sales from Monday to Friday are given below.

Days	Monday	Tuesday	Wednesday	Thursday	Friday
Sales	200	1,000	450	500	95

Compute his total sales...**[Answer: 2,245 Units]**

Exercise 23: Answer true or false

1) $3 + 5 \neq 5 + 3$...**[Answer: False]**

2) $10 + 0 = 0$...**[Answer: False]**

3) $(2 + 3) + 5 = 2 + (3 + 5)$...**[Answer: True]**

1.3.4 Subtraction Rule

Subtraction means to take from a group or number or finding the difference between two numbers or quantities. The first number is called minuend and second number is called subtrahend. The result is called difference.

Subtraction Sign   Equality Sign

A   −   B   =   C

Minuend   Subtrahend   Difference

N/B: The subtraction of integers follows specific rules as illustrated below.

$(+a) - (+b) = a - b$ .... $\begin{bmatrix} \text{Result } (+) \text{ if a} > b \\ \text{Result}(-) \text{ if b} > a \end{bmatrix}$

$(+a) - (-b) = a + b$ ....[Result $(+)$]

$(-a) - (+b) = -a - b$ ....[Result $(-)$]

$(-a) - (-b) = -a + b$ .... $\begin{bmatrix} \text{Result } (+) \text{ if b} > a \\ \text{Result}(\ ) \text{ if a} > b \end{bmatrix}$

N/B: It is worth knowing that subtraction is not commutative $[A - B \neq B - A]$ and associative$[(A - B) - C \neq A - (B - C)]$.

Example 1: Solve the following
1) $51 - 17$
2) $11 - (-5)$
3) $-4 - (+2)$
4) $-4 - (-2)$

<u>Solution</u>

	1) $51 - 17$	2) $11 - (-5)$	3) $-4 - (+2)$	4) $-4 - (-2)$
Formulas	$[(+a) - (+b) = a - b]$	$[(+a) - (-b) = a + b]$	$[(-a) - (+b) = -a - b]$	$[(-a) - (-b) = -a + b]$
Working	$51 - 17 = 34$	$11 + 5 = 16$	$-4 - 2 = -6$	$-4 + 2 = -2$

Example 2: Mr. John was given 10,000 FCFA and later taken 3,000 FCFA to purchase goods and service. How much is remaining (balance) with Mr. John?

<u>Solution</u>

Step 1: Formula. Balance = Minuend − Subtrahend

Where; Minuend = 10,000 and Subtrahend = 3,000

Step 2: Data substitution and solving

  Balance $= 10,000 - 3,000$ ....[Solve equation]

  Balance $= \mathbf{7,000\ FCFA}$

Test Your Understanding

Exercise 24: Solve the following

  1) $13 - 33$.. ....…..[Answer: -20]

  2) $4 - (-6)$...[Answer: 10]

  3) $-6 - (+15)$...[Answer: -21]

  4) $-90 - (-70)$..[Answer: -20]

Exercise 25: Given that; A = 14, B = 8 and C = 4, proof that subtraction is not commutative $[A - B \neq B - A]$ and   associative $[(A - B) - C \neq A - (B - C)]$.   ....[Answer: See Appendix]

Exercise 26: Solve the following

  1) $-5 \times (-4)$...[Answer: 20]

  2) $-4 \div (2)$..[Answer: -2]

  3) $-4 + (-16)$...[Answer: $-\frac{1}{4}$]

  4) $9 - (-7)$ ....[Answer: 16]

1.4 The concept of Power

The expression of power is made up of two components; the base (A) and the exponent or index or power(n). Putting them together, we obtain the power expression,

  $\text{Base}^{\text{Index}} = A^n$ ....(1)

N/B: The exponent (n) is the value that specifies how many times the base will be multiplied by itself. The mathematical operation and property of power are given below

  $A^m \times A^n = A^{n+n}$ ....[Multiplication with same Base Rule]

  $A^m \div A^n = A^{m-n}$ ....[Division with same Base Rule]

  $(A^m)^n = A^{m \times n}$ ....[The Power or Exponential Rule]

  $A^m \times B^m = (A \times B)^m$ ....[Multiplication with same Power Rule]

  $\left(\frac{A}{B}\right)^m = \frac{A^m}{B^m}$ ....[The power of a Fraction Rule]

  $A^{-m} = \frac{1}{A^m}$ ....[The negative power or Inverse Rule]

25

$A^0 = 1$, if $A \neq 0$ ....[The zero Power Rule]

$A^1 = A$ ....[The One Power Rule]

Note: We may be required to determine the value of the base when the power and result are given or to determine the value of the power when the base and result are given. To do this, we apply the rules below.

If $A = B^m$, therefore $B = \sqrt[m]{A}$ ....(2)

If $A^m = B$, therefore $m = \dfrac{\log A}{\log B}$ ....(3)

Example 1: Determine the following; 10^2 and 5^4

<u>Solution</u>

10^2	5^4
10^2 ....[Multiply 10 by itself 2 times] 10×10 ....[Solve equation] $10^2 = \mathbf{100}$	5^4 ....[Multiply 5 by itself 4 times] $5 \times 5 \times 5 \times 5$ ....[Solve equation] $5^4 = \mathbf{625}$
N/B: If using scientific calculator, press 10 and the button (x^2) to obtain the square root of 10.	**N/B:** If using a scientific calculator, press 5, press the button(y^x), press 4 and finally press the equal to button $(=)$ to obtain the value 625.

Example 2: Solve the following,

    1) $5^2 \times 5^3$

    2) $10^5 \div 10^3$

<u>Solution</u>

Step 1: Formula. $A^m \times A^n = A^{m+n}$ and $A^m \div A^n = A^{m-n}$

Step 2: Data substitution and solving

$5^2 \times 5^3$	$10^5 \div 10^3$
$5^2 \times 5^3$.[Maintain base and add powers] 5^{2+3}.........................[Simplify power] $5^5 = 5 \times 5 \times 5 \times 5 \times 5$ ....[Solve] $5^2 \times 5^3 = \mathbf{3,125}$	$10^5 \div 10^3$.[Maintain base and subtract powers] 10^{5-3} ....[Simplify power] $10^2 = 10 \times 10$ ....[Solve] $10^5 \div 10^3 = \mathbf{100}$

Example 3: Determine the following,

    1) $2^2 \times 2^{-3}$

    2) $4^7 \div 4^8$

Solution

1) $2^2 \times 2^{-3}$

Step 1: Formula. $A^m \times A^n = A^{m+n}$ and $A^{-m} = \frac{1}{A^m}$

Step 2: Data substitution and solving

$2^2 \times 2^{-3}$ ....……....…[Maintain base and add power]

$2^{2+(-3)}$ ....…..............[Simplify power]

2^{-1} ....…..…… ….[Apply the inverse rule]

$2^2 \times 2^{-3} = \frac{1}{2^1} = \frac{1}{2}$

2) $4^7 \div 4^8$

Step 1: Formula. $A^m \div A^n = A^{m-n}$ and $A^{-m} = \frac{1}{A^m}$

Step 2: Data substitution and solving

$4^7 \div 4^8$ ....….… ...[Maintain base and subtract powers]

4^{7-8} ....…..…….[Simplify power]

4^{-1} ....…............. ....[Apply inverse rule]

$4^7 \div 4^8 = \frac{1}{4^1} = \frac{1}{4}$

Example 4: Compute the following

  1) $(4^2)^2$

  2) $3^5 \times 2^5$

Solution

Step 1: Formula. $(A^m)^n = A^{m \times n}$ and $A^m \times B^m = (A \times B)^m$

Step 2: Data substitution and solving

$(4^2)^2$	$3^5 \times 2^5$
$(4^2)^2$ ....[Multiply powers]	$3^5 \times 2^5$ ...[Multiply bases and maintain power]
$4^{2\times2} = 4^4$ ....[Solve]	$(3 \times 2)^5 = 6^5$ ....…...[Solve]
$(4^2)^2 = \mathbf{256}$	$3^5 \times 2^5 = \mathbf{7,776}$

Example 5: Find the value of the missing elements

  1) $6^m = 216$

  2) $A^2 = 25$

Solution

Step 1: Formula. If $A = B^m$, therefore $B = \sqrt[m]{A}$ and If $A^m = B$, therefore $m = \frac{\log A}{\log B}$

Step 2: Data substitution and solving

$6^m = 216$	$A^2 = 25$
$6^m = 216$ ....[Log both sides of equation]	$A^2 = 25$ ...[Square root both sides of equation]
m Log 6 = Log 216 ....[Make m subject]	$\sqrt{A^2} = \sqrt{25} \rightarrow \left(A^{\frac{1}{2}}\right)^2 = \sqrt{25}$ ..[Simplify power]
Power(m) $= \frac{Log\ 216}{Log\ 6}$ ....[Work log]	
Power(m) $= \frac{2.334453751}{0.77815125}$ ...[Solve equation]	$A^{\frac{2}{2}} = \sqrt{25} \rightarrow A = \sqrt{25}$ ....[Work square root]
Power(m) $= 3$	Base(A) $= 5$

Test Your Understanding

Exercise 27: Solve the following

1) $6^3 \times 6^1$..[Answer: 1,296]

2) $4^4 \div 4^2$...[Answer: 16]

3) $5^6 \div 5^7$..[Answer: 1/5]

4) $2,000^0$..[Answer: 1]

5) $(3^2)^{-2}$..[Answer: 1/81]

Exercise 28: Determine the missing elements

6) $X^2 = 81$...[Answer: 9]

7) $2^X = 8$...[Answer: 3]

8) $Y^{-2} = 1/4$..[Answer: 2]

1.5 The Concept of Root

A root expression is made up of three components, the radical sign $(\sqrt{})$, the radicand (A) and the index or power (n). Putting the three components together, we obtain the root expression below.

$$\sqrt[Power]{Radicand} = \sqrt[n]{A} \dots\dots\dots\dots\dots\dots\dots\dots\dots\dots\dots\dots\dots\dots\dots\dots\dots\dots(1)$$

When the power component is equal to two, it is called square root and when (n) it is equal tp three it is called cube root. It is worth knowing that the value of (n) can be greater than three but our analysis will be limited to square root and cube root.

$$\sqrt{A} = \sqrt[2]{A} \dots\dots\dots\dots\dots\dots\dots\dots\dots\dots\dots\dots\dots\dots\dots\dots\dots\dots[Square\ Root]$$

$$\sqrt[3]{A} \dots\dots\dots\dots\dots\dots\dots\dots\dots\dots\dots\dots\dots\dots\dots\dots\dots\dots\dots\dots[Cube\ Root]$$

Note: The mathematical operation (multiplication and division) is given below.

$$\sqrt[n]{A} \times \sqrt[n]{B} = \sqrt[n]{A \times B} = \sqrt[n]{AB} \dots\dots\dots\dots\dots\dots\dots\dots\dots\dots\text{[Multiplication Rule]}$$

$$\sqrt[n]{A} \div \sqrt[n]{B} = \frac{\sqrt[n]{A}}{\sqrt[n]{B}} = \sqrt[n]{\frac{A}{B}} \dots\dots\dots\dots\dots\dots\dots\dots\dots\dots\dots\dots\text{[Division Rule]}$$

N/B: It is worth knowing that the above rule does not hold for addition and subtraction. That is, $\left[\sqrt[n]{A} + \sqrt[n]{B} \neq \sqrt[n]{A + B}\right]$ and $\left[\sqrt[n]{A} - \sqrt[n]{B} \neq \sqrt[n]{A - B}\right]$. It is also worth knowing that $\left[A^{\frac{m}{n}} = \sqrt[n]{(A)^m}\right]$

Example 1: Solve the following; $\sqrt{144}$ and $\sqrt[3]{125}$

<u>Solution</u>

$\sqrt{144}$		$\sqrt[3]{125}$	
	Explanation		**Explanation**
$\sqrt{144} = 12$	If using a scientific calculator, press 144 and $(\sqrt{})$ button to get the square root of 144.	$\sqrt[3]{125} = 5$	If using a scientific calculator, press second function button (2ndf), press the button (y^x), press 3 and finally press equal to button $(=)$.

Example 2: Compute the following

1) $\sqrt{4} \times \sqrt{16}$
2) $\sqrt[3]{8} \times \sqrt[3]{125}$

<u>Solution</u>

Step 1: Formula. $\sqrt[n]{A} \times \sqrt[n]{B} = \sqrt[n]{A \times B} = \sqrt[n]{AB}$

Step 2: Data substitution and solving

$\sqrt{4} \times \sqrt{16}$	$\sqrt[3]{8} \times \sqrt[3]{125}$
$\sqrt{4} \times \sqrt{16}$ ....[Multiply radicands]	$\sqrt[3]{8} \times \sqrt[3]{125}$ ....[Multiply radicands]
$\sqrt{4 \times 16} = \sqrt{64}$ ....[Work square root]	$\sqrt[3]{8 \times 125} = \sqrt[3]{1,000}$ ....[Work cube root]
$\sqrt{4} \times \sqrt{16} = \sqrt{64} = 8$	$\sqrt[3]{8} \times \sqrt[3]{125} = \sqrt[3]{1,000} = 10$

Example 3: Determine the following

1) $\sqrt[2]{400} \div \sqrt[2]{100}$
2) $\sqrt[3]{1,728} \div \sqrt[3]{64}$

Solution

Step 1: Formula. $\sqrt[n]{A} \div \sqrt[n]{B} = \dfrac{\sqrt[n]{A}}{\sqrt[n]{B}} = \sqrt[n]{\dfrac{A}{B}}$

Step 2: Data substitution and solving

$\sqrt[2]{400} \div \sqrt[2]{100}$	$\sqrt[3]{1,728} \div \sqrt[3]{64}$
$\sqrt[2]{400} \div \sqrt[2]{100}$ ....[Divide radicands]	$\sqrt[3]{1,728} \div \sqrt[3]{64}$ ....[Divide radicands]
$\sqrt[2]{\dfrac{400}{100}} = \sqrt[2]{4}$ ....[Work square root]	$\sqrt[3]{\dfrac{1,728}{64}} = \sqrt[3]{27}$ ....[Work cube root]
$\sqrt[2]{400} \div \sqrt[2]{100} = \sqrt[2]{4} = \mathbf{2}$	$\sqrt[3]{1,728} \div \sqrt[3]{64} = \sqrt[3]{27} = \mathbf{3}$

Example 4: Given that A = 25 and B = 16, using the square root idea, show that $\left[\sqrt[n]{A} + \sqrt[n]{B} \neq \sqrt[n]{A + B}\right]$ and $\left[\sqrt[n]{A} - \sqrt[n]{B} \neq \sqrt[n]{A - B}\right]$.

Solution

$\left[\sqrt[n]{A} + \sqrt[n]{B} \neq \sqrt[n]{A + B}\right]$	$\left[\sqrt[n]{A} - \sqrt[n]{B} \neq \sqrt[n]{A - B}\right]$
$\sqrt[2]{25} + \sqrt[2]{16} \neq \sqrt[2]{25 + 16}$...[Work sides]	$\sqrt[2]{25} - \sqrt[2]{16} \neq \sqrt[2]{25 - 16}$ ......[Work sides]
$5 + 4 \neq \sqrt[2]{41}$ ....[Solve]	$5 - 4 \neq \sqrt[2]{9}$ ....[Solve]
$9 \neq \mathbf{6.403124237}$	$\mathbf{1 \neq 3}$

Test Your Understanding

Exercise 29: Compute the following

1) $\sqrt[2]{225}$...[Answer: 15]

2) $\sqrt[3]{64}$...[Answer: 4]

Exercise 30: Determine the following

1) $\sqrt[2]{36} \times \sqrt{9}$..[Answer: 18]

2) $\sqrt[3]{512} \times \sqrt[3]{8}$ ....[Answer: 16]

3) $\sqrt[3]{2,744} \div \sqrt[3]{343}$ ....[Answer: 2]

4) $\sqrt{25} \div \sqrt{225}$..[Answer: 1/3]

Exercise 31: Solve the following

1) $\sqrt{4} - \sqrt{9}$..[Answer: -1]

2) $\sqrt[3]{27} + \sqrt[3]{1,000}$ ....[Answer: 13]

3) $\sqrt{9} - \sqrt[3]{8}$...[Answer: 1]

4) $\sqrt[3]{343} + \sqrt{9}$..[Answer: 10]

1.6 Combination of Four Rules Operations

In situation where there are multiple operations in a given problem, we make use of order of operation. The rule used is called BIMDAS (Brackets, Indices, Multiplication and Division, Addition and Subtraction). It can also be called PEMDAS (Parenthesis, Exponent, Multiplication or Division from left to right and Addition or Subtraction from left to right) or BODMAS (Bracket of Division, Multiplication, Addition and Subtraction) or BEDMAS (E representing exponents).

Example 1: Compute the following

1) $6 \times 6 + 4$
2) $12 + 6 \times 2$
3) $9 \div 3 - 1$

Solution

1) $6 \times 6 + 4$

Procedure	Working	New equation
Original Equation		$6 \times 6 + 4$
Multiplication	$6 \times 6 = 36$	$36 + 4$
Addition	$36 + 4 = 40$	40

2) $12 + 6 \times 2$

Procedure	Working	New equation
Original Equation		$12 + 6 \times 2$
Multiplication	$6 \times 2 = 12$	$12 + 12$
Addition	$12 + 12 = 24$	24

3) $9 \div 3 - 1$

Procedure	Working	New equation
Original Equation		$9 \div 3 - 1$
Division	$9 \div 3 = 3$	$3 - 1$
Subtraction	$3 - 1 = 2$	2

Example 2: Determine the following

1) $(5 \times 3) \times 6$
2) $45 \div (4 + 5)$

31

3) $(9 \div 3) + (2 \times 4)$

Solution

1) $(5 \times 3) \times 6$

Procedure	Working	New equation
Original Equation		$(5 \times 3) \times 6$
Work Bracket	$(5 \times 3) = 15$	$\mathbf{15} \times 6$
Multiplication	$15 \times 6 = 90$	**90**

2) $45 \div (4 + 5)$

Procedure	Working	New equation
Original Equation		$45 \div (4 + 5)$
Work Bracket	$(4 + 5) = 9$	$45 \div \mathbf{9}$
Division	$45 \div 9 = 5$	**5**

3) $(9 \div 3) + (2 \times 4)$

Procedure	Working	New equation
Original Equation		$(9 \div 3) + (2 \times 4)$
Work Bracket	$(9 \div 3) = 3$ and $(2 \times 4) = 8$	$3 + 8$
Addition	$3 + 8 += 11$	**11**

Example 3: Solve the following

1) $21 \div 3 + (3 \times 9) \times 9 + 5$

2) $4 \times 5 + (14 + 8) - 36 \div 9$

3) $(12 \div 3) + 3 + (16 - 7) \times 4$

4) $3 - 2 + 5 \times 9 + 56 \div 7$

Solution

1) $21 \div 3 + (3 \times 9) \times 9 + 5$

Procedure	Working	New equation
Original Equation		$21 \div 3 + (3 \times 9) \times 9 + 5$
Work Bracket	$(3 \times 9) = 27$	$21 \div 3 + \mathbf{27} \times 9 + 5$
Multiplication	$27 \times 9 = 243$	$21 \div 3 + \mathbf{243} + 5$
Division	$21 \div 3 = 7$	$\mathbf{7} + 243 + 5$
Addition	$7 + 243 + 5 = 255$	**255**

2) $4 \times 5 + (14 + 8) - 36 \div 9$

Procedure	Working	New equation
Original Equation		$4 \times 5 + (14 + 8) - 36 \div 9$
Work Bracket	$(14 + 8) = 22$	$4 \times 5 + 22 - 36 \div 9$
Multiplication	$4 \times 5 = 20$	$20 + 22 - 36 \div 9$
Division	$36 \div 9 = 4$	$20 + 22 - 4$
Addition	$20 + 22 = 42$	$42 - 4$
Subtraction	$42 - 4 = 38$	38

3) $(12 \div 3) + 3 + (16 - 7) \times 4$

Procedure	Working	New equation
Original Equation		$(12 \div 3) + 3 + (16 - 7) \times 4$
Work Bracket	$(12 \div 3) = 4$ and $(16 - 7) = 9$	$4 + 3 + 9 \times 4$
Multiplication	$9 \times 4 = 36$	$4 + 3 + 36$
Addition	$4 + 3 + 36 = 43$	43

4) $3 - 2 + 5 \times 9 + 56 \div 7$

Procedure	Working	New equation
Original Equation		$3 - 2 + 5 \times 9 + 56 \div 7$
Multiplication	$5 \times 9 = 45$	$3 - 2 + 45 + 56 \div 7$
Division	$56 \div 7 = 8$	$3 - 2 + 45 + 8$
Subtraction	$3 - 2 = 1$	$1 + 45 + 8$
Addition	$1 + 45 + 8 = 54$	54

Example 4: Solve the following

1) $17 - 2^3 + 4 \times 5$

2) $6 + (36 \div 9)^3 \div 2 - 1$

3) $10 \times (0.4 + 0.3) - 2^2 \div 5^0$

4) $7^2 - 4^2 \div \sqrt{4} + \sqrt{25}$

1) $17 - 2^3 + 4 \times 5$

Procedure	Working	New Equation
Original Equation		$17 - 2^3 + 4 \times 5$
Exponent or power	$2^3 = 8$	$17 - \mathbf{8} + 4 \times 5$
Multiplication	$4 \times 5 = 20$	$17 - \mathbf{8} + \mathbf{20}$
Subtraction	$17 - 8 = 9$	$\mathbf{9 + 20}$
Addition	$9 + 20 = 29$	$\mathbf{29}$

2) $6 + (36 \div 9)^3 \div 2 - 1$

Procedure	Working	New Equation
Original Equation		$6 + (36 \div 9)^3 \div 2 - 1$
Bracket	$(36 \div 9) = 4$	$6 + \mathbf{4}^3 \div 2 - 1$
Exponent or power	$4^3 = 64$	$6 + \mathbf{64} \div 2 - 1$
Division	$64 \div 2 = 32$	$6 + \mathbf{32} - 1$
Addition	$6 + 32 = 38$	$\mathbf{38} - 1$
Subtraction	$38 - 1 = 37$	$\mathbf{37}$

3) $10 \times (0.4 + 0.3) - 2^2 \div 5^0$

Procedure	Working	New Equation
Original Equation		$10 \times (0.4 + 0.3) - 2^2 \div 5^0$
Bracket	$(0.4 + 0.3) = 0.7$	$10 \times \mathbf{0.7} - 2^2 \div 5^0$
Exponents	$2^2 = 4$ and $5^0 = 1$	$10 \times \mathbf{0.7} - \mathbf{4} \div \mathbf{1}$
Multiplication	$10 \times 0.7 = 7$	$\mathbf{7} - 4 \div 1$
Division	$4 \div 1 = 4$	$\mathbf{7} - \mathbf{4}$
Subtraction	$7 - 4 = 3$	$\mathbf{3}$

4) $7^2 - 4^2 \div \sqrt{4} + \sqrt{25}$

Procedure	Working	New Equation
Original Equation		$7^2 - 4^2 \div \sqrt{4} + \sqrt{25}$
Root	$\sqrt{4} = 2$ and $\sqrt{25} = 5$	$7^2 - 4^2 \div \mathbf{2} + \mathbf{5}$
Exponent or power	$7^2 = 49$ and $4^2 = 16$	$\mathbf{49} - \mathbf{16} \div \mathbf{2} + \mathbf{5}$

Division	$16 \div 2 = 8$	$\mathbf{49 - 8 + 5}$
Addition	$8 + 5 = 13$	$\mathbf{49 - 13}$
Subtraction	$49 - 13 = 36$	$\mathbf{36}$

Test Your Understanding

Exercise 32: Determine the following

1) $20 - 6 \times 2$..[Answer: 8]
2) $8 + 6 - 3$..[Answer: 11]
3) $12 \div 3 + 5$..[Answer: 9]

Exercise 33: Compute the following

1) $63 \div (17 - 8)$..[Answer: 7]
2) $(4 + 7) \times 3$..[Answer: 33]
3) $(18 + 15) - (13 + 12)$..[Answer: 8]

Exercise 34: Solve the following equations

1) $48 \div 6 + (5 \times 6) \times 13 + 6$..[Answer: 404]
2) $(11 - 8) \times 3 + 7 + 27 - 3$..[Answer: 40]
3) $(18 \div 3) + 6 + (14 - 8) \times 5$..[Answer: 42]
4) $56 \div 8 + 30 - 18 \div 3$..[Answer: 31]

Exercise 35: Solve the following

1) $30 - (5 \times 2^3 - 15)$..[Answer: 5]
2) $(95 \div 19)^2 + 3$..[Answer: 28]
3) $(3^3 - 2 \times 7) + (5 + 3 - 2^2)$..[Answer: 17]
4) $(3 + 2)^2 - 5 \times 3 + 2^3$..[Answer: 2]
5) $\sqrt{36} \times 3^2 - \sqrt{16} \div 2^2$..[Answer: 53]

Note: It is worth knowing that the application of multiplication and division vary with objective to avoid decimal points and the application of addition and subtraction vary with objective to avoid negative results.

1.7 Directed Numbers

Directed numbers have either positive or negative sign in front of them. Directed numbers are represented on a line called number line. The center of a number line is zero; right from the zero point is made up of positive numbers and left of negative numbers.

Number Line

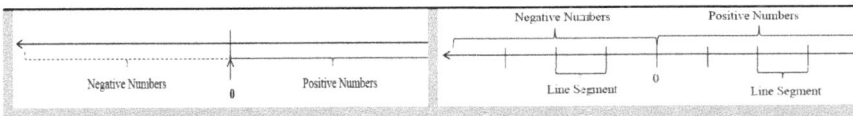

N/B: A number line is expressed horizontally (X-number line) and vertically (Y-number line).

1.7.1 Establishment of Number Line

To establish a number line, we make use of the concept of interval. Interval means the length between lines segments must be the same.

Example 1: Establish a number for the coordinate $(X = -5)$ and $(X = 8)$ using the following.

1) Interval of 2
2) Interval of 3

<u>Solution</u>

1) Interval of 2

Hint: Interval of 2 means the first value after the zero point should be(± 2). From the(± 2), we add 2 (positive sides) or subtract 2 (negative side) successively to obtain the next value. We end when the value on the number line is equal or more than the given values.

Negative Numbers (0 to ≥ -5)   Positive Numbers (0 to ≥ 8)

2) Interval of 3

Hint: Interval of 3 means the first value after the zero point should be(± 3). From the(± 3), we add 3 (positive sides) or subtract 3 (negative side) successively to obtain the next value. We end when the value on the number line is equal or more than the given values.

Negative Numbers (0 to ≥ -5)   Positive Numbers (0 to ≥ 8)

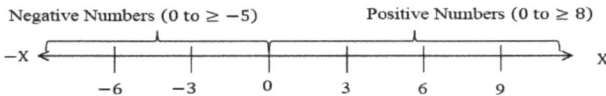

N/B: From the two number lines, we realized that the bigger the interval, the shorter the number line and the smaller the interval, the longer the number line. Therefore, the length of a number line is inversely related to the interval chosen.

Example 2: Establish a number for the coordinate $(Y = -25)$ and $(Y = 16)$ using the following.

1) Interval of 4
2) Interval of 5

Solution

1) Interval of 4	2) Interval of 5
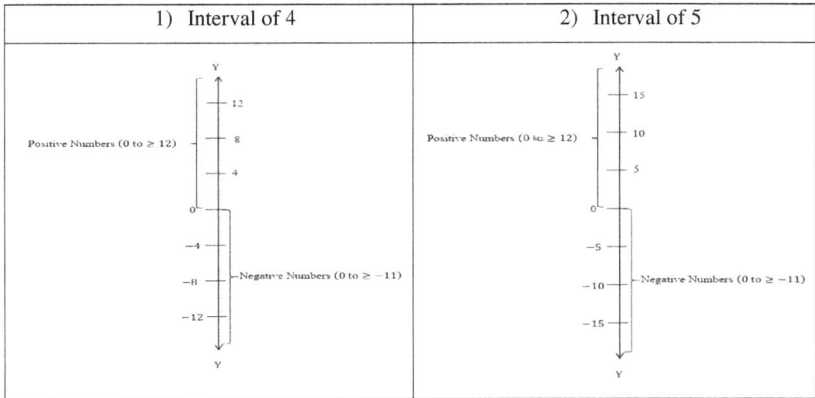	

1.7.2 Addition and Subtraction of Directed Number using the Number Line

N/B: To add or subtract directed numbers using the number line, the following steps are recommended.

a) Start Point: The start point value is the first value in the equation which can be a positive number (+a) or a negative number(−a).

b) Movement

    i) Addition: From the start point, move right if the second value is a positive value (+b) and left if the second value is a negative value(−b).

    ii) Subtraction: From the start point, move right if the second value is a negative value (−b) and left if the second value is a positive value(+b).

c) End Point: Move till the second value of time from the number line direction.

Example 1: Evaluate the following using the number line.

1) $(-5) + (+3)$

2) $(+6) + (-2)$

Solution

$(-5) + (+3)$	$(+6) + (-2)$
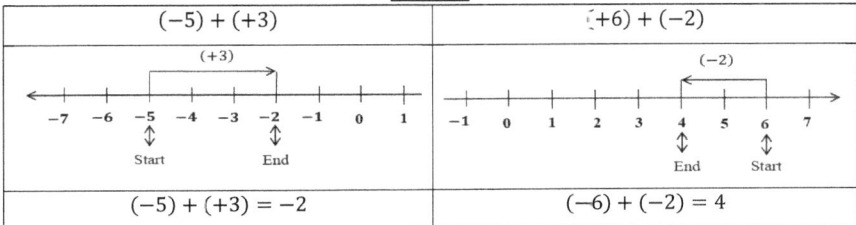	
$(-5) + (+3) = -2$	$(-6) + (-2) = 4$

Example 2: Solve the following using the number line.

   1) $(-3) + (+6)$

   2) $(+7) + (-9)$

Solution

	Illustration	Result
(-3) $+ (+6)$		**3**
$(+6)$ $+ (-9)$		**−2**

Example 3: Evaluate the following using the number line.

   1) $(-1) + (-5)$

   2) $(+3) + (+3)$

Solution

$(-1) + (-5)$	$(+3) + (+3)$
$(-1) + (-5) = -6$	$(+3) + (+3) = 6$

Example 4: Evaluate the following using the number line.

   1) $(-4) - (-3)$

   2) $(-2) - (+3)$

Solution

$(-4) - (-3)$	$(-2) - (+3)$
$(-4) - (-3) = -1$	$(-2) - (+3) = -5$

38

Example 5: Solve the following using the number line.

1) $(+3) - (+5)$

2) $(+4) - (-2)$

Solution

$(+3) - (+5)$	$(+4) - (-2)$
$(+3) - (+5) = -2$	$(+4) - (-2) = 6$

Test Your Understanding
Exercise 36: Using the number line, calculate the following. 1) $(-5) + (+5)$ ....[See Appendix] 2) $(+5) + (-7)$..[See Appendix] 3) $(-1) - (+4)$ ....[See Appendix] 4) $(+2) + (-2)$..[See Appendix] 5) $(-3) + (-7)$..[See Appendix] 6) $(0) - (-5)$..[See Appendix]

1.7.3 Coordinate Plane

A coordinate plane is a combination of two number lines used to locate points using ordered pairs. A coordinate plane is made up of four quadrants as seen below.

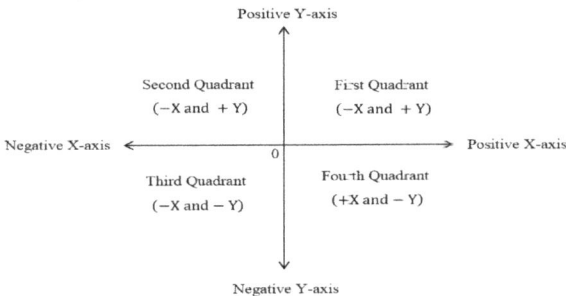

An ordered pair is a pair coordinates representing the x-axis and y-axis and in which their intersection gives a point.

Ordered Pair $= (x, y)$ ....(1)

The distance between two points on a coordinate plane and the midpoint of the distance between two points are calculated using the formula below.

$$\text{Distance(d)} = \sqrt{(X_2 - X_1)^2 + (Y_2 - Y_1)^2} \quad \text{..(2)}$$

$$\text{Midpoint} = \left(\frac{X_1 + X_2}{2}, \frac{Y_1 + Y_2}{2}\right) \quad \text{..(3)}$$

N/B: Two or more points are connected by a line and the slope of the line is called a gradient. The gradient of a straight line measures the slope or steepness of the line and can be positive (line sloping from right to left) or negative (line sloping from left to right) in nature. The gradient is calculated as,

$$\text{Gradient} = \frac{Y_2 - Y_1}{X_2 - X_1} \quad \text{..(4)}$$

Example 1: Plot the points of the following pair coordinates; $A(2, -4)$, $B(4, 3)$, $C(-5, 5)$, $D(-3, -5)$, $E(-3, 0)$ and $F(0, 6)$.

<u>**Solution**</u>

Hint: To derive a point. We draw a line from the values and mark the point at which they intersect. The intersection point is named after the coordinate pairs.

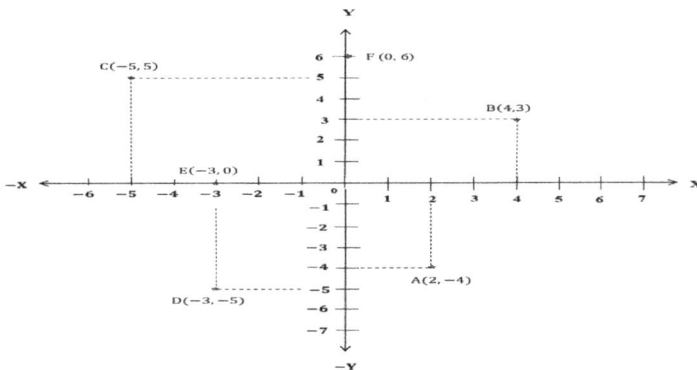

Example 2: Calculate the distance and midpoint between point $A(1,2)$ and $B(-4, 3)$, assuming that distance is measured in meters.

<u>**Solution**</u>

1) Calculation of distance

Step 1: Formula. $\text{Distance(d)} = \sqrt{(X_2 - X_1)^2 + (Y_2 - Y_1)^2}$

Step 2: Data substitution and solving

Moving from Point A to B	Moving from point B to A
Where; $X_2 = -4, X_1 = 1, Y_2 = 3, Y_1 = 2$	Where; $X_2 = 1, X_- = -4, Y_2 = 2, Y_1 = 3$
$d = \sqrt{(-4-1)^2 + (3-2)^2}$.[Work bracket] $d = \sqrt{(-5)^2 + (1)^2}$ ....[Work power] $d = \sqrt{25+1}$ ....[Sum and work root] Distance(d) $\approx$ **5m**	$d = \sqrt{(1-(-4))^2 + (2-3)^2}$.[Work bracket] $d = \sqrt{(5)^2 + (-1)^2}$ ....[Work power] $d = \sqrt{25+1}$ ....[Sum and work root] Distance(d) $\approx$ **5m**

2) Calculation of midpoint

Step 1: Formula. Midpoint $= \left(\frac{X_1+X_2}{2}, \frac{Y_1+Y_2}{2}\right)$

Step 2: Data substitution and solving

Moving from Point A to B	Moving from point B to A
Where; $X_2 = -4, X_1 = 1, Y_2 = 3, Y_1 = 2$	Where; $X_2 = 1, X_1 = -4, Y_2 = 2, Y_1 = 3$
Midpoint $= \left(\frac{1+(-4)}{2}, \frac{2+3}{2}\right)$ ....[Simplify] Midpoint $= \left(\frac{-3}{2}, \frac{5}{2}\right)$	Midpoint $= \left(\frac{-4+1}{2}, \frac{3+2}{2}\right)$ ....[Simplify] Midpoint $= \left(\frac{-3}{2}, \frac{5}{2}\right)$

Example 3: Locate the point and calculate the distance and midpoint of the following coordinates pairs.

1) $A(-3, 1)$ and $B(1, -4)$.
2) $C(5, 0)$ and $D(3, 3)$.

Solution

Step 1: Location of points

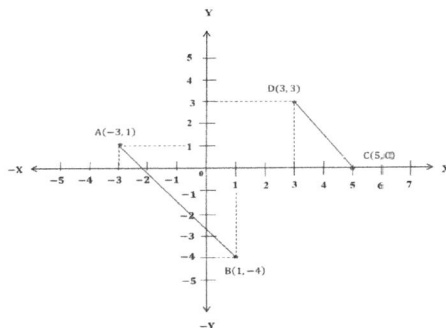

Step 2: Calculation of distance

1) Formula. Distance$(d) = \sqrt{(X_2 - X_1)^2 + (Y_2 - Y_1)^2}$

2) Data substitution and solving

A$(-3, 1)$ and B $(1, -4)$...............[A → B]	C$(5, 0)$ and D$(3, 3)$ ....[C → D]
Where; $X_2 = 1$, $X_1 = -3$, $Y_2 = -4$, $Y_1 = 1$	Where; $X_2 = 3$, $X_1 = 5$, $Y_2 = 3$, $Y_1 = 0$
$d = \sqrt{(1 - (-3))^2 + (-4 - 1)^2}$ [Work bracket] $d = \sqrt{(4)^2 + (-5)^2}$ ....[Work power] $d = \sqrt{16 + 25}$ ....[Sum and work root] Distance$(d) \approx$ **6m**	$d = \sqrt{(3 - 5)^2 + (3 - 0)^2}$..[Work bracket] $d = \sqrt{(-2)^2 + (3)^2}$ ....[Work power] $d = \sqrt{4 + 9}$ ....[Sum and work root] Distance$(d) \approx$ **4m**

Step 3: Calculation of Midpoints

1) Formula. Midpoint $= \left(\frac{X_1 + X_2}{2}, \frac{Y_1 + Y_2}{2}\right)$

2) Data substitution and solving

A$(-3, 1)$ and B $(1, -4)$.........[A → B]	C$(5, 0)$ and D$(3, 3)$ ....[C → D]
Where; $X_2 = 1$, $X_1 = -3$, $Y_2 = -4$, $Y_1 = 1$	Where; $X_2 = 3$, $X_1 = 5$, $Y_2 = 3$, $Y_1 = 0$
Midpoint $= \left(\frac{-3+1}{2}, \frac{1+(-4)}{2}\right)$ Midpoint $= \left(\frac{-2}{2}, \frac{-3}{2}\right) = (-1, -1.5)$	Midpoint $= \left(\frac{5+3}{2}, \frac{0+3}{2}\right)$ Midpoint $= \left(\frac{8}{2}, \frac{3}{2}\right) = (4, 1.5)$

N/B: The above calculations and location can be put together as seen below

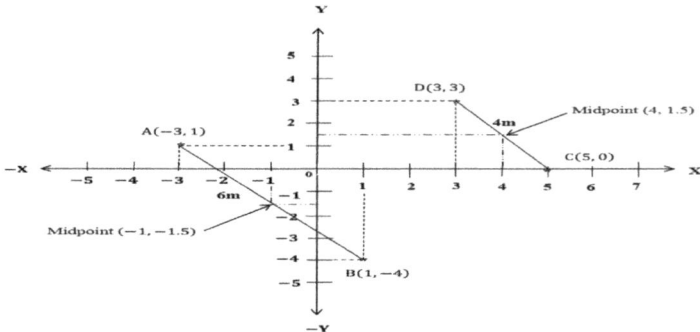

Example 4: Plot the coordinate A $(0, 0)$ and B $(3, 4)$ and calculate the distance and midpoint coordinate. Assume that distance is measured in meters.

Solution

Step 1: Formula. $\text{Distance}(d) = \sqrt{(X_2 - X_1)^2 + (Y_2 - Y_1)^2}$ and $\text{Midpoint} = \left(\frac{X_1 + X_2}{2}, \frac{Y_1 + Y_2}{2}\right)$

Step 2: Plotting, data substitution and solving

Graphical Illustration	Calculation of distance and midpoint	
	Distance	Midpoint
	$d =$ $\sqrt{(3-0)^2 + (4-0)^2}$ $d = \sqrt{(3)^2 + (4)^2}$ $d = \sqrt{9 + 16}$ $d = \sqrt{25}$ $\text{Distance}(d) = \mathbf{5m}$	$\text{Midpoint} =$ $\left(\frac{0+3}{2}, \frac{0+4}{2}\right)$ $\text{Midpoint} = \left(\frac{3}{2}, \frac{4}{2}\right)$ $\text{Midpoint} =$ $(\mathbf{1.5, 2})$

Test Your Understanding

Exercise 37: Plot the points of the following pair coordinates; $A(-3, 2)$, $B(2, 3)$ and $C(1, -4)$. Use an interval of one..[**Answer: See Appendix**]

Exercise 38: Compute the distance and midpoint between the point $A(-5, -3)$ and $B(9, 3)$. Assume distance is measured in meters.....[**Answer: Distance $\approx$ 16m and Midpoint $(2, 0)$**]

Exercise 39: Compute the gradient of the following coordinate pairs.

1) $A(1, 4)$ and $B(3, 3)$...............................[**Answer: -0.5 (Negative gradient)**]

2) $A(1, 2)$ and $B(3, 4)$...................................[**Answer: 1(Positive gradient)**]

Exercise 30: Plot and calculate the distance and midpoint of the following coordinate pairs.

1) $A(-4, 4)$ and $B(-3, -2)$...[**Answer: See Appendix**]

2) $C(3, -3)$ and $D(1, 4)$.............................. ....[**Answer: See Appendix**]

1.8 The Concept of Fraction

A fraction (fractus in Latin) is a number that represents a part of a whole. The major parts of a fraction are given below.

$$\frac{A}{B} = A/_B = A/B = A \div B \dots\dots\dots\dots\dots\dots\dots\dots\dots\dots\dots(1)$$

Where; (A) and (B) are called terms

(A) represents the divident or numerator

(B) represents the divisor or denominator

($-$ or/) represents the fraction bar (solidus or vinculum)

N/B: The following are worth knowing; $\left(\frac{A}{A} = 1\right)$, $\left(\frac{A}{1} = A\right)$, $\left(\frac{0}{A} = 0\right)$ and $\left(\frac{A}{A} = 1\right) =$ Undefined

1.8.1 Types of Fraction

There are three types fractions; proper fraction, improper fraction and mixed fraction.

1) **Proper Fraction:** It is a fraction in which the numerator (A) is smaller than the denominator (B) or in which the dividend (A) is smaller than the divisor (B). That is, $\left(\frac{A}{B}, \text{where } A < B\right)$.

2) **Improper Fraction:** It is a fraction in which the numerator (A) is greater than the denominator (B) or in which the dividend is greater than the divisor. That is, $\left(\frac{A}{B}, \text{where } A > B\right)$.

3) **Mixed Fraction:** It is a fraction showing a combination of a whole number (W) and proper fraction$\left(\frac{A}{B}\right)$. That is, $\left(W\frac{A}{B}\right)$ and $\left(\frac{A}{B}\right)$ is the fractional section.

1.8.2 Conversion analysis

It is possible to change from a proper fraction or improper fraction to mixed fraction and from mixed fraction to improper fraction.

1) **Changing from improper fraction to mixed fraction:** To change a mixed fraction to an improper fraction, we apply the following formula.

$$W\frac{A}{B} = \frac{(W \times B) + A}{B} \quad \dots\dots\dots\dots\dots\dots\dots\dots\dots\dots\dots\dots\dots\dots\dots\dots(1)$$

Example 1: Change the following mixed fraction into improper fractions; $\left(2\frac{3}{4}\right)$ and $\left(36\frac{6}{11}\right)$

<u>**Solution**</u>

Step 1: Formula. $W\frac{A}{B} = \frac{(W \times B) + A}{B}$

Step 2: Data substitution and solving

First Mixed Fraction $\left(2\frac{3}{4}\right)$	Second Mixed Fraction $\left(36\frac{6}{11}\right)$
Where; W = 2, A = 3 and B = 4	Where; W = 36, A = 6 and B = 11
$2\frac{3}{4} = \frac{(2 \times 4) + 3}{4}$ ....[Simplify bracket]	$36\frac{6}{11} = \frac{(36 \times 11) + 6}{11}$ ....[Simplify bracket]
$2\frac{3}{4} = \frac{8+3}{4}$ ....[Simplify numerator]	$36\frac{6}{11} = \frac{396+6}{11}$ ....[Simplify numerator]
$2\frac{3}{4} = \frac{11}{4}$	$36\frac{6}{11} = \frac{402}{11}$

Example 2: Change the following mixed fraction into improper fraction; $\left(1\frac{5}{12}\right)$ and $\left(3\frac{1}{2}\right)$

Solution

Step 1: Formula. $W\frac{A}{B} = \frac{(W \times B) + A}{B}$

Step 2: Data substitution and solving

First Mixed Fraction $\left(1\frac{5}{12}\right)$	Second Mixed Fraction $\left(3\frac{1}{2}\right)$
Where; W = 1, A = 5 and B = 12	Where; W = 3, A = 1 and B = 2
$1\frac{5}{12} = \frac{(1\times12)+5}{12}$ ....[Simplify bracket]	$3\frac{1}{2} = \frac{(3\times2)+1}{2}$ ....[Simplify bracket]
$1\frac{5}{12} = \frac{12+5}{12}$ ....[Simplify numerator]	$3\frac{1}{2} = \frac{6+1}{2}$ ....[Simplify numerator]
$1\frac{5}{12} = \frac{17}{12}$	$3\frac{1}{2} = \frac{7}{2}$

Test Your Understanding
Exercise 41: Change the following mixed fraction into improper fraction.

1) $8\frac{2}{3}$..[Answer: 26/3]

2) $15\frac{3}{8}$..[Answer: 123/8]

3) $5\frac{3}{4}$..[Answer: 23/4]

4) $1\frac{4}{9}$..[Answer: 13/9]

5) $2\frac{1}{5}$..[Answer: 11/5]

2) **Changing mixed fraction to improper fraction:** To change an improper fraction to a mixed fraction, we use the following formula.

$$\frac{A}{B} = \text{Quotient}\frac{\text{Remainder}}{\text{Divisor(or Original Denominator)}} \quad ....(2)$$

N/B: To perform this, we apply the concept of long division and short division to obtain the components. We consider the numbers before the decimal as the quotient number. We multiply the quotient with the denominator to obtain a product. Finally, we subtract the product from the numerator to obtain the remainder.

Example 1: Change $\left(\frac{51}{8}\right)$ into mixed fraction.

Solution

Step 1: Formula. $\frac{A}{B} = \text{Quotient}\frac{\text{Remainder}}{\text{Divisor(or Original Denominator)}}$, where; Divisor = 8

Step 2: Determination of components.................................[Quotient and remainder]

$\frac{51}{8} = 6.375$ ....[**Quotient = 6**. Multiply 6 by denominator (8)]

$6 \times 8 = \mathbf{48}$ ....[Product = 48. Subtract 48 from the numerator (51)]

$51 - 48 = 3$ ....[**Remainder = 3**]

Step 3: Data substitution and solving

$\frac{51}{8} = \mathbf{6\frac{3}{8}}$ ....[Proof: $6\frac{3}{8} = \frac{(6\times8)+3}{8} = \frac{51}{8}$]

Example 2: Change the following improper fractions into mixed fractions; $\left(\frac{14}{8}\right)$ and $\left(\frac{36}{7}\right)$

Solution

Step 1: Formula. $\frac{A}{B} = \text{Quodient}\frac{\text{Remainder}}{\text{Divisor(or Original Denominator)}}$

Step 2: Data substitution and solving

First Fraction $\left(\frac{14}{8}\right)$, where; Divisor = 8	Second Fraction$\left(\frac{36}{7}\right)$, where; Divisor = 7
$\frac{14}{8}$ = 1 Remainder 6, where; Quodient = 1	$\frac{36}{7}$ = 5 Remainder 1, where; Quodient = 5
$\frac{14}{8} = 1\frac{6}{8}$ ....[Proof: $1\frac{6}{8} = \frac{(1\times8)+6}{8} = \frac{14}{8}$]	$\frac{36}{7} = 5\frac{1}{7}$ ....[Proof: $5\frac{1}{7} = \frac{(5\times7)+1}{7} = \frac{36}{7}$]

Test Your Understanding

Exercise 42: Change the following improper fractions into mixed fractions

1) $\frac{13}{2}$...[**Answer: $6\frac{1}{2}$. Proof:** $6\frac{1}{2} = \frac{(6\times2)+1}{2} = \frac{13}{2}$]

2) $\frac{19}{8}$...[**Answer: $2\frac{3}{8}$. Proof:** $2\frac{3}{8} = \frac{(2\times8)+3}{8} = \frac{19}{8}$]

3) $\frac{67}{9}$...[**Answer: $7\frac{4}{9}$. Proof:** $7\frac{4}{9} = \frac{(7\times9)+4}{9} = \frac{67}{9}$]

1.8.3 Mathematical Operation of Fraction

It's worth knowing that for simplicity reasons, all mixed fractions should be changed to improper fraction for any mathematical operation application.

1) **Addition of Fraction:** The formula or procedure of addition fraction depends on the nature of the denominator of both fractions. That is denominator can be the same or different.

46

$$\frac{A}{B} + \frac{C}{B} = \frac{A+C}{B} \dots\dots\dots\dots\dots\dots\dots\dots\dots\dots\text{[Situation of same denominator]}$$

$$\frac{A}{B} + \frac{C}{D} = \frac{A(D)}{B(D)} + \frac{B(C)}{B(D)} = \frac{A(D)+B(C)}{B(D)} \dots\dots\dots\dots\text{[Situation of different denominator]}$$

Example 1: Evaluate the following fractions

1) $\frac{5}{2} + \frac{3}{2}$

2) $\frac{2}{7} + \frac{4}{3}$

Solution

First Fractions $\left(\frac{5}{2} + \frac{3}{2}\right)$	Second Fractions $\left(\frac{2}{7} + \frac{4}{3}\right)$
Step 1: Formula. $\frac{A}{B} + \frac{C}{B} = \frac{A+C}{B}$ Where; A = 5, B = 2 and C = 3	Step 1: Formula. $\frac{A}{B} + \frac{C}{D} = \frac{A(D)+B(C)}{B(D)}$ Where; A = 2, B = 7, C = 4 and D = 3
Step 2: Data substitution and solving $\frac{5}{2} + \frac{3}{2} = \frac{5+3}{2}$ .....[Simplify numerator] $\frac{5}{2} + \frac{3}{2} = \frac{8}{2} = 4$	Step 2: Data substitution and solving $\frac{2}{7} + \frac{4}{3} = \frac{2(3)+7(4)}{7(3)}$ ....[Open brackets] $\frac{2}{7} + \frac{4}{3} = \frac{6+28}{21}$ ....[Simplify numerator] $\frac{2}{7} + \frac{4}{3} = \frac{34}{21} \dots\dots\frac{34}{21} = 1\frac{13}{21} = \frac{(1\times21)+13}{21} = \frac{34}{21}]$

Example 2: Solve the following fractions

1) $8\frac{1}{2} + \frac{5}{2}$

2) $3\frac{1}{7} + 4\frac{2}{6}$

Solution

First Fractions $\left(8\frac{1}{2} + \frac{5}{2} = \frac{17}{2} + \frac{5}{2}\right)$	Second Fractions $\left(3\frac{1}{7} + 4\frac{2}{6} = \frac{22}{7} + \frac{26}{6}\right)$
Step 1: Formula. $\frac{A}{B} + \frac{C}{B} = \frac{A+C}{B}$ Where; A = 17, B = 2 and C = 5	Step 1: Formula. $\frac{A}{B} + \frac{C}{D} = \frac{A(D)+B(C)}{B(D)}$ Where; A = 22, B = 7, C = 26 and D = 6
Step 2: Data substitution and solving $\frac{17}{2} + \frac{5}{2} = \frac{17+5}{2}$ ...[Simplify numerator] $8\frac{1}{2} + \frac{5}{2} = \frac{17}{2} + \frac{5}{2} = \frac{22}{2} = 11$	Step 2: Data substitution and solving $\frac{22}{7} + \frac{26}{6} = \frac{22(6)+7(26)}{7(6)}$ ....[Open brackets] $\frac{22}{7} + \frac{26}{6} = \frac{132+182}{42}$ ......[Simplify numerator] $\frac{22}{7} + \frac{26}{6} = \frac{314}{42} \dots\frac{314}{42} = 7\frac{20}{42} = \frac{(7\times42)+20}{42} = \frac{314}{42}]$

Exercise 34: Evaluate the following fractions

1) $\frac{8}{5} + \frac{17}{5}$...[Answer: $\frac{25}{5} = 5 = \frac{5}{1} = 5\frac{0}{5} = \frac{(5\times5)+0}{5} = \frac{25}{5}$]

2) $\frac{9}{4} + \frac{15}{7}$...[Answer: $\frac{123}{28} = 4\frac{11}{28} = \frac{(4\times28)+11}{28} = \frac{123}{28}$]

3) $3\frac{9}{10} + \frac{51}{8}$...[Answer: $\frac{822}{80} = \frac{411}{40} = 10\frac{11}{40} = \frac{(10\times40)+11}{40} = \frac{411}{40}$]

4) $2\frac{3}{8} + 1\frac{13}{8}$...[Answer: $\frac{32}{8} = 4 = \frac{4}{1} = 4\frac{0}{1} = \frac{(4\times1)+0}{1} = \frac{4}{1} = 4$]

2) **Subtraction of Fraction:** The formula or procedure of addition fraction depends on the nature of the denominator of both fractions. That is denominator can be the same or different.

$$\frac{A}{B} - \frac{C}{B} = \frac{A-C}{B}$$ ....[Situation of same denominator]

$$\frac{A}{B} - \frac{C}{B} = \frac{A(D)}{B(D)} - \frac{B(C)}{B(D)} = \frac{A(D)-B(C)}{B(D)}$$ ....[Situation of different denominator]

Example 1: Compute the following fractions

1) $\frac{3}{5} - \frac{6}{5}$

2) $\frac{7}{12} - \frac{3}{8}$

Solution

First Fractions $\left(\frac{3}{5} - \frac{6}{5}\right)$	Second Fractions $\left(\frac{7}{12} - \frac{3}{8}\right)$
Step 1: Formula. $\frac{A}{B} - \frac{C}{B} = \frac{A-C}{B}$ Where; A = 3, B = 5 and C = 6	Step 1: Formula. $\frac{A}{B} - \frac{C}{B} = \frac{A(D)-B(C)}{B(D)}$ Where; A = 7, B = 12, C = 3 and D = 8
Step 2: Data substitution and solving $\frac{3}{5} - \frac{6}{5} = \frac{3-6}{5}$ .....[Simplify numerator] $\frac{3}{5} - \frac{6}{5} = \frac{-3}{5} = -\frac{3}{5}$	Step 2: Data substitution and solving $\frac{7}{12} - \frac{3}{8} = \frac{7(8)-12(3)}{12(8)}$ ....[Open brackets] $\frac{7}{12} - \frac{3}{8} = \frac{56-36}{96}$ ....[Simplify numerator] $\frac{7}{12} - \frac{3}{8} = \frac{20}{96} = \frac{5}{24}$ ...$[\frac{5}{24} = 0\frac{5}{24} = \frac{(0\times24)+5}{24} = \frac{5}{24}]$

Example 2: Compute the following fractions

1) $4\frac{3}{2} - \frac{4}{2}$

2) $5\frac{2}{5} - 7\frac{1}{2}$

First Fractions $\left(4\frac{3}{2} - \frac{4}{2} = \frac{11}{2} - \frac{4}{2}\right)$	Second Fractions $\left(5\frac{2}{5} - 7\frac{1}{2} = \frac{27}{5} - \frac{15}{2}\right)$
Step 1: Formula. $\frac{A}{B} - \frac{C}{B} = \frac{A-C}{B}$ Where; $A = 11$, $B = 2$ and $C = 4$	Step 1: Formula. $\frac{A}{B} - \frac{C}{E} = \frac{A(D)-B(C)}{B(D)}$ Where; $A = 27$, $B = 5$, $C = 15$ and $D = 2$
Step 2: Data substitution and solving $\frac{11}{2} - \frac{4}{2} = \frac{11-4}{2}$ ....[Simplify numerator] $\frac{11}{2} - \frac{4}{2} = \frac{7}{2}$ ....$\left[\frac{7}{2} = 3\frac{1}{2} = \frac{(3\times2)+1}{2} = \frac{7}{2}\right]$	Step 2: Data substitution and solving $\frac{27}{5} - \frac{15}{2} = \frac{27(2)-5(15)}{5(2)}$ ....[Open brackets] $\frac{27}{5} - \frac{15}{2} = \frac{54-75}{10}$ ....[Simplify numerator] $\frac{27}{5} - \frac{15}{2} = \frac{-21}{10}$.$\left[-\frac{21}{10} = -2\frac{-1}{10} = \frac{(-2\times10)+(-1)}{10} = \frac{-21}{10}\right]$

Exercise 44: Compute the following fractions

1) $\frac{7}{4} - \frac{13}{4}$.........................[**Answer:** $\frac{-6}{4} = -\frac{6}{4} = -1\frac{-2}{4} = \frac{(-1\times4)+(-2)}{4} = \frac{-6}{4} = -\frac{6}{4}$]

2) $\frac{9}{5} - \frac{10}{3}$.....................[**Answer:** $\frac{-23}{15} = -\frac{23}{15} = -1\frac{-8}{15} = \frac{(-1\times15)+(-8)}{15} = \frac{-23}{15} = -\frac{23}{15}$]

3) $5\frac{1}{3} - \frac{12}{7}$..................................[**Answer:** $\frac{76}{21} = 3\frac{13}{21} = \frac{(3\times21)+13}{21} = \frac{76}{21}$]

4) $1\frac{2}{3} - 2\frac{9}{3}$....................[**Answer:** $\frac{-10}{3} = -\frac{10}{3} = -3\frac{-1}{3} = \frac{(-3\times3)+(-1)}{3} = \frac{-10}{3} = -\frac{10}{3}$]

3) **Multiplication and Division of Fraction:** The multiplication and division of fraction is done using the formula below.

$$\frac{A}{B} \times \frac{C}{D} = \frac{A(C)}{B(D)} \quad ....[\text{Multiplication of Fraction}]$$

$$\frac{A}{B} \div \frac{C}{D} = \frac{A}{B} \times \frac{D}{C} = \frac{A(D)}{B(C)} \quad ....[\text{Division of Fraction}]$$

N/B: It is worth knowing that $\left[\frac{A}{B} \times \frac{C}{B} \neq \frac{A\times C}{B}\right]$ and $\left[\frac{A}{B} \div \frac{C}{B} \neq \frac{A-C}{B}\right]$. The equations above are applied for both the same and different denominator situation.

Example 1: Evaluate the following fractions

1) $\frac{3}{4} \times \frac{5}{6}$

2) $\frac{7}{12} \div \frac{2}{3}$

Solution

First Fractions $\left(\frac{3}{4} \times \frac{5}{6}\right)$	Second Fractions $\left(\frac{7}{12} \div \frac{2}{3}\right)$
Step 1: Formula. $\frac{A}{B} \times \frac{C}{D} = \frac{A(C)}{B(D)}$	Step 1: Formula. $\frac{A}{B} \div \frac{C}{D} = \frac{A(D)}{B(C)}$
Where; A = 3, B = 4, C = 5 and D = 6	Where; A = 7, B = 12, C = 2 and D = 3
Step 2: Data substitution and solving	Step 2: Data substitution and solving
$\frac{3}{4} \times \frac{5}{6} = \frac{3(5)}{4(6)}$ ....[Open bracket]	$\frac{7}{12} \div \frac{2}{3} = \frac{7(3)}{12(2)}$ ....[Open brackets]
$\frac{3}{4} \times \frac{5}{6} = \frac{15}{24} . \left[\frac{15}{24} = 0\frac{15}{24} = \frac{(0\times24)+15}{24} = \frac{15}{24}\right]$	$\frac{7}{12} \div \frac{2}{3} = \frac{21}{24} .....\left[\frac{21}{24} = 0\frac{21}{24} = \frac{(0\times24)+(21)}{24} = \frac{21}{24}\right]$

Example 2: Compute the following fractions

1) $3\frac{1}{5} \times 2\frac{3}{4}$

2) $5\frac{1}{4} \div 4\frac{2}{3}$

Solution

First Fractions $\left(3\frac{1}{5} \times 2\frac{3}{4} = \frac{16}{5} \times \frac{11}{4}\right)$	Second Fractions $\left(5\frac{1}{4} \div 4\frac{2}{3} = \frac{21}{4} \div \frac{14}{3}\right)$
Step 1: Formula. $\frac{A}{B} \times \frac{C}{D} = \frac{A(C)}{3(D)}$	Step 1: Formula. $\frac{A}{B} \div \frac{C}{D} = \frac{A(D)}{B(C)}$
Where; A = 16, B = 5, C = 11 and D = 4	Where; A = 21, B = 4, C = 14 and D = 3
Step 2: Data substitution and solving	Step 2: Data substitution and solving
$\frac{16}{5} \times \frac{11}{4} = \frac{16(11)}{5(4)}$ ....[Open bracket]	$\frac{21}{4} \div \frac{14}{3} = \frac{21(3)}{4(14)}$ ....[Open brackets]
$\frac{16}{5} \times \frac{11}{4} = \frac{176}{20}$	$\frac{21}{4} \div \frac{14}{3} = \frac{63}{56}$

Test Your Understanding
Exercise 45: Solve the following
1) $\frac{4}{2} \times \frac{5}{2}$...[Answer: $\frac{20}{4} = 5$]
2) $\frac{11}{13} \times \frac{150}{13}$...[Answer: $\frac{143}{1,950}$]
3) $2\frac{4}{7} \times 1\frac{3}{8}$...[Answer: $\frac{128}{77}$]
4) $3\frac{3}{8} \div 4\frac{4}{5}$...[Answer: $\frac{81}{5}$]

1.9 Highest Common Factor (HCF) and Lowest Common Multiple (LCM)

Lowest common multiple (LCM) or least common denominator (LCD) is the smallest number which is exactly divisible by all the given numbers That is, the lowest common multiple (LCM) is the smallest number that if divided by all the given numbers gives no remainder. Highest common factor (HCF) or greatest common divisor (GCD) or greatest common factor (GCF) is the greatest number which exactly divides all given numbers. That is, highest common factor (HCF) is the largest number that can divide all given numbers without a remainder.

N/B: To determine the lowest common multiple (LCM) and highest common factor (HCF), we can use the listing multiple (LCM) or factor (HCF) method or prime factor or multiple method. Our analysis will use the prime factor method. It is worth knowing that the HCF of same value is the value, that is; $[HCF(A, A) = A]$.

1) **Whole Number Analysis:** Here, we are going to limit our analysis to non-fraction and non-decimal numbers.

Example 1: Find the lowest common multiple of 6, 7 and 9

Solution

Step 1: Tabula determination of multiplication values

Factors	Values		
2	6	7	9
3	$6/2 = 3$	7	9
	$3/3 = 1$	7	$9/3 = 3$

Step 2: Multiplication of values

Lowest Common Multiple(LCM) $= 2 \times 3 \times 1 \times 7 \times 3$ ....(Solve)

Lowest Common Multiple(LCM) $= \mathbf{126}$

Step 3: Justification. The value 126 if divided by the given numbers should not have a remainder.

$126/6 = \mathbf{21}$	$126/7 = \mathbf{18}$	$126/9 = \mathbf{14}$

Example 2: Fine the lowest common multiple (LCM) of 10, 12 and 11

Solution

Step 1: Tabula determination of multiplication values

Factors	Values		
2	10	11	12
2	$10/2 = 5$	11	$12/2 = 6$
3	5	11	$6/2 = 3$
	5	**11**	$3/3 = 1$

Step 2: Multiplication of values

Lowest Common Multiple(LCM) $= 2 \times 2 \times 3 \times 5 \times 11 \times 1$ ....[Solve]

Lowest Common Multiple(LCM) $=$ **660**

Step 3: Justification. The value 660 if divided by the given numbers should not have a remainder.

$660/10 = 66$	$660/11 = 60$	$660/12 = 55$

Example 3: Fine the highest common factor (HCF) of 24 and 56

Solution

Step 1: Tabula determination of multiplication values

24		56	
Factors	Division	Factors	Division
2	$24/2 = 12$	2	$56/2 = 28$
2	$12/2 = 6$	2	$28/2 = 14$
2	$6/2 = 3$	2	$14/2 = 7$
3	$3/3 = 1$	7	$7/7 = 1$
$2 \times 2 \times 2 \times 3 = 2^3 \times 3$		$2 \times 2 \times 2 \times 7 = 2^3 \times 7$	

Step 2: Multiplication of values

N/B: The common prime number is 2 with smallest powers are 3. That is $[2^3]$.

Highest Common Factor(HCF) $= 2^3 = 2 \times 2 \times 2$ ....[Solve]

Highest Common Factor(HCF) $=$ **8**

Step 3: Justification. The number 8 should divide all the given numbers without remainder.

$24/8 = 3$	$56/8 = 7$

Example 4: Determine the highest common factor (HCF) of 12, 36 and 48

Solution

Step 1: Tabula determination of multiplication values

12		36		48	
Factors	Division	Factors	Division	Factors	Division
2	$12/2 = 6$	2	$36/2 = 18$	2	$48/2 = 24$
2	$6/2 = 3$	2	$18/2 = 9$	2	$24/2 = 12$
3	$3/3 = 1$	3	$9/3 = 3$	2	$12/2 = 6$
$2 \times 2 \times 3 = 2^2 \times 3$		3	$3/3 = 1$	2	$6/2 = 3$
		$2 \times 2 \times 3 \times 3 = 2^2 \times 3^2$		3	$3/3 = 1$
				$2 \times 2 \times 2 \times 2 \times 3 = 2^4 \times 3$	

Step 2: Multiplication of values

N/B: The common prime numbers are 2 and 3 and their least or smallest powers are 2 and 1 respectively. That is $[2^2]$ and $[3 = 3^1]$.

Highest Common Factor(HCF) $= 2^2 \times 3$ ....[Work exponent]

Highest Common Factor(HCF) $= 4 \times 3$ ....[Solve]

Highest Common Factor(HCF) $= \mathbf{12}$

Step 3: Justification. The number 12 should divide all the given numbers without remainder.

$12/12 = 1$	$26/12 = 3$	$48/12 = 4$

Test Your Understanding

Exercise 46: Find the lowest common multiple (LCM) of the following numbers,

1) 2, 5 and 13...[Answer: 130]

2) 3, 7, 11 and 15..[Answer: 1,155]

3) 14, 36 and 42..[Answer: 252]

Exercise 47: Determine the highest common factor (HCF) of the following numbers,

1) 12 and 56..[Answer: 4]

2) 4, 16 and 12..[Answer: 4]

3) 34 and 60..[Answer: 2]

---

2) **Fraction Number Analysis:** The derivation of lowest common multiple (LCM) and highest common factor (HCF) is more complex than with the situation of whole numbers. They are calculated using the formulas below.

$$\text{Lowest Common Multiple(LCM)} = \frac{\text{Lowest Common Multiple(LCM) of Numerator}}{\text{Highest Common Factor(HCF) of Denominator}} \quad ..(1)$$

$$\text{Highest Common Factor(HCF)} = \frac{\text{Highest Common Factor(HCF) of Numerators}}{\text{Lowest Common Multiple (LCM) of Denominators}}$$
..(2)

N/B: The concept of lowest common multiple (LCM) is use for comparing and ordering of fractions with different denominators. The idea of equivalent fraction is also applied here.

Example 1: Determine the lowest common multiple (LCM) of; $\frac{2}{3}, \frac{7}{18}$ and $\frac{11}{12}$.

Solution

Step 1: Formula. $\text{Lowest Common Multiple(LCM)} = \frac{\text{Lowest Common Multiple(LCM) of Numerator}}{\text{Highest Common Factor(HCF) of Denominator}}$

Step 2: Determination of unknown components

Lowest Common Multiple(LCM) of Numerator					Highest Common Factor(HCF) of Denominator				
2	2	7	11		3		18		12
	1	7	11	3	3	2	18	2	12
LCM = 2 × 1 × 7 × 11 = 154					1	3	9	2	6
				3 = 3^1		3	3	3	3
						3	1		1
						2×3^3		$2^2 \times 3$	
						HCF = $3^1 = 3$			

Step 3: Data substitution and solving: $\text{Lowest Common Multiple(LCM)} = \frac{154}{3}$

Step 4: Justification.

$\frac{154}{3} \div \frac{2}{3} = \frac{154}{3} \times \frac{3}{2} = \frac{462}{6} = 77$	$\frac{154}{3} \div \frac{7}{18} = \frac{154}{3} \times \frac{18}{7} = \frac{2{,}772}{21} = 132$	$\frac{154}{3} \div \frac{11}{12} = \frac{154}{3} \times \frac{12}{11} = \frac{1{,}848}{33} = 56$

Example 2: Calculate the highest common factor (HCF) of; $\frac{4}{9}, \frac{16}{15}$ and $\frac{12}{21}$.

Solution

Step   1:   Formula. $\text{Highest Common Factor(HCF)} =$
$\frac{\text{Highest Common Factor(HCF) of Numerators}}{\text{Lowest Common Multiple (LCM) of Denominators}}$

Step 2: Determination of unknown component

Lowest Common Multiple (LCM) of Denominators				Highest Common Factor(HCF) of Numerators					
3	9	15	21	4		16		12	
3	3	5	7	2	4	2	16	2	12
	1	5	7	2	2	2	8	2	6
LCM = 3 × 3 × 1 × 5 × 7 = **315**					1	2	4	3	3
				$2 \times 2 = 2^2$		2	2		1
							1	$2 \times 2 \times 3 = 2^2 \times 3$	
						$2 \times 2 \times 2 \times 2 = 2^4$			
				HCF $= 2^2 = 2 \times 2 = 4$					

Step 3: Data substitution and solving: Highest Common Factor(HCF) $= \frac{4}{315}$

Step 4: Justification

$$\frac{4}{9} \div \frac{4}{315} = \frac{4}{9} \times \frac{315}{4} = \frac{1,260}{36} = 35 \qquad \frac{16}{15} \div \frac{4}{315} = \frac{16}{15} \times \frac{315}{4} = \frac{5,040}{60} = 84 \qquad \frac{12}{21} \div \frac{4}{315} = \frac{12}{21} \times \frac{315}{4} = \frac{3,780}{84} = 45$$

Test Your Understanding

Exercise 48: Find the lowest common multiple (LCM) of the following

1) $\frac{1}{3}, \frac{5}{6}, \frac{5}{9}$ and $\frac{10}{27}$ ....[Answer: $\frac{10}{3}$]

2) $\frac{108}{375}, 1\frac{17}{25}$ and $\frac{54}{55}$ ....[Answer: $\frac{756}{5}$]

Exercise 49: Find the highest common factor (HCF) of the following

1) $\frac{1}{2}$ and $\frac{3}{2}$ ....[Answer: $\frac{3}{2}$]

2) $\frac{54}{9}, 3\frac{9}{17}$ and $\frac{36}{51}$ ....[Answer: $\frac{6}{17}$]

Exercise 50: Determine the lowest common multiple (LCM) and highest common factor (HCF) of $3\frac{1}{4}$ and $8\frac{3}{4}$ ....[Answer: LCM $= \frac{455}{4}$ and HCF $= \frac{7}{4}$]

3) **Decimal Number Analysis:** To obtain the lowest common multiple and highest common factor of a decimal, we change the decimal to a whole number or fraction, derive the lowest common multiple and highest common factor and finally change the value to decimal.

Example 1: Determine the lowest common multiple (LCM) of 0.25, 0.5 and 0.75.

Solution

a) **Method 1: Using Whole Number**

Step 1: Formula. Lowest Common Multiple(LCM) $=$

$$\frac{\text{Whole Number Lowest Common Multiple(LCM)}}{\text{Decimal Number Transformation Multiplier}}$$

Where; Decimal Number Transformation Multiplier $= 100$

Step 2: Change decimal to whole number...............[0.25, 0.5 and 0.75] = [25, 50 and 75]

Step 3: Determination of lowest common multiple of whole numbers.

2	25	50	75
5	25	25	75
5	5	5	3
	1	1	3
$2 \times 5 \times 5 \times 1 \times 1 \times 3 = 2 \times 5^2 \times 1^2 \times 3 = \mathbf{150}$			

Step 4: Determination of lowest common multiple in decimal point

Lowest Common Multiple(LCM) $= \frac{150}{100}$ ....[Solve equation]

Lowest Common Multiple(LCM) $= \mathbf{1.5}$

Step 5: Justification

$1.5/0.25 = \mathbf{6}$	$1.5/0.5 = \mathbf{3}$	$1.5/0.75 = \mathbf{2}$

b) Method 2: Using Fraction Number

Step 1: Formula. Lowest Common Multiple(LCM) $= \frac{\text{Lowest Common Multiple(LCM) of Numerator}}{\text{Highest Common Factor(HCF) of Denominator}}$

Step 2: Change decimal to fraction.................$\left[(0.25, 0.5 \text{ and } 0.75) = \left(\frac{25}{100}, \frac{50}{100} \text{ and } \frac{75}{100}\right)\right]$

Step 3: Determination of unknown components

Lowest Common Multiple(LCM) of Numerator = 150 ......[Already calculated above]

Highest Common Factor(HCF) of Denominator = 100[Since all numbers are the same]

Step 4: Data substitution and solving

Lowest Common Multiple(LCM) $= \frac{150}{100}$ ....[Solve equation]

Lowest Common Multiple(LCM) $= \mathbf{1.5}$

Example 2: Compute the highest common factor of 0.3, 0.2, and 0.5

<u>**Solution**</u>

a) Method 1: Using Whole Number

Step 1: Formula. Highest Common Factor (HCF) $= \frac{\text{Whole Number Highest Common Factor (HCF)}}{\text{Decimal Number Transformation Multiplier}}$

Where; Decimal Number Transformation Multiplier = 100

Step 2: Change decimal to whole number.................[0.3, 0.2 and 0.5] = [30, 20 and 50]

Step 3: Determination of highest common factor of whole numbers

30		20		50	
2	30	2	20	2	50
3	15	2	10	5	25
5	5	5	5	5	5
	1		1		1
$2 \times 3 \times 5 = 2^1 \times 3^1 \times 5^1$		$2 \times 2 \times 5 = 2^2 \times 5^1$		$2 \times 5 \times 5 = 2^1 \times 5^2$	
Whole Number Highest Common Factor (HCF) $= 2 \times 5 = \mathbf{10}$					

Step 4: Determination of highest common factor in decimal point

Highest Common Factor (HCF) $= \frac{10}{100}$ ....[Solve equation]

Highest Common Factor (HCF) $= \mathbf{0.1}$

Step 5: Justification

$0.3/0.1 = \mathbf{3}$	$0.2/0.1 = \mathbf{2}$	$0.5/0.1 = \mathbf{5}$

b) **Method 2: Using Fraction Number**

Step 1: Formula. Highest Common Factor(HCF) $=$

$\frac{\text{Highest Common Factor(HCF) of Numerators}}{\text{Lowest Common Multiple (LCM) of Denominators}}$

Step 2: Change decimal to fraction...................$\left[(0.3, 0.2 \text{ and } 0.5) = \left(\frac{30}{100}, \frac{20}{100} \text{ and } \frac{50}{100}\right)\right]$

Step 3: Determination of unknown components

Highest Common Factor(HCF) of Denominator $= 10$ ....[Already calculated above]

Lowest Common Multiple (LCM) of Denominators $= 100$..[Since all numbers are the same]

Step 4: Data substitution and solving

Highest Common Factor(HCF) $= \frac{10}{100}$ ....[Solve equation]

Highest Common Factor(HCF) $= \mathbf{0.1}$

Example 3: Determine the lowest common multiple (LCM) and highest common factor (HCF) of 0.48, 0.72 and 0.108.

Solution

Step 1: Change decimal to whole number.....[$(0.48, 0.72 \text{ and } 0.108) = (480, 720 \text{ and } 108)$]

Step 2: Calculation of LCM and HCF

1) Determination of lowest common multiple

a) Formula.   Lowest Common Multiple(LCM) =

$\frac{\text{Whole Number Lowest Common Multiple(LCM)}}{\text{Decimal Number Transformation Multiplier}}$

Where; Decimal Number Transformation Multiplier = 1,000

b) Determination of unknown

Factors	Values		
2	480	720	108
2	240	360	54
2	120	180	27
2	60	90	27
2	30	45	27
3	15	45	27
3	5	15	9
3	5	5	3
	5	5	1
Whole Number Lowest Common Multiple(LCM) = $2^5 \times 3^3 \times 5^2 = \mathbf{21,600}$			

c) Data substitution and solving

Lowest Common Multiple(LCM) = $\frac{21,600}{1,000}$ ....[Solve equation]

Lowest Common Multiple(LCM) = $\mathbf{21.6}$

2) Determination of highest common factor

a) Formula. Highest Common Factor (HCF) = $\frac{\text{Whole Number Highest Common Factor (HCF)}}{\text{Decimal Number Transformation Multiplier}}$

Where; Decimal Number Transformation Multiplier = 1,000

b) Determination of unknown component

480		720		108	
Factors	Division	Factors	Division	Factors	Division
2	480	2	720	2	108
2	240	2	360	2	45
2	120	2	180	3	27
2	60	2	90	3	9
2	30	3	45	3	3
3	15	3	15		1

5	5	5	5	$2^2 \times 3^3$
	1		1	
$2^5 \times 3^1 \times 5^1$		$2^4 \times 3^2 \times 5^1$		
Whole Number Highest Common Factor (HCF) $= 2^2 \times 3 = 4 \times 3 = \mathbf{12}$				

c) Data substitution and solving

Highest Common Factor (HCF) $= \dfrac{12}{1{,}000}$ ....[Solve equation]

Highest Common Factor (HCF) $= \mathbf{0.012}$

Test Your Understanding

Exercise 51: Compute the lowest common multiple (LCM) of the following,

1) 0.06 and 0.08...[Answer: $\mathbf{0.24}$]

2) 0.1, 0.2 and 0.25...[Answer: $\mathbf{1.00}$]

Exercise 52: Determine the highest common factor of the following,

1) 0.18 and 0.24...[Answer: $\mathbf{0.06}$]

2) 0.2, 0.4 and 0.45...[Answer: **0.05**]

Exercise 53: Find the lowest common multiple and highest common factor of the following

1) 1.20 and 22.5............................[Answer: $\mathbf{LCM = 90.00}$ and $\mathbf{HCF = 0.30}$]

2) 1.7, 0.51 and 0.153....................[Answer: $\mathbf{LCM = 15.300}$ and $\mathbf{HCF = 0.017}$]

CHAPTER TWO

FINANCIAL, GEOMETRIC AND WEIGHT MEASUREMENT

Measurement is the process of determining the size, length, weight and equivalence of something. Our analysis will be limited to money, shapes and weight measurements.

2.1 Financial Measurement
Financial measurement deals with the exchange of currencies. That is to determine the equivalence of a unit of currency to another currency.

2.1.1 Meaning of Exchange Rate
International trade or exchange of goods and service has led to the utilization of different currencies. Since each country accepts her legal tender for trade, currencies are exchange to make trade possible. The price of a currency in terms of another is called exchange rate. The major currencies in the world include;

Country	Currency	Code	Sign
United State	US Dollar	USD	US$
United Kingdom	Pound Sterling	GBP	£
China	Yuan	CNY	Y
Japan	Yen	JPY	¥
Canada	Canadian Dollar	CAD	Can$
Russian Federation	Ruble	RUB	R
India	Repee	INR	Rs
France	Euro	EUR	€
Ghana	Cedi	GHC	¢
Cameroon	CFA Franc	XAF	CFAF

2.1.2 Method of currency Exchange
The exchange from one currency to another currency can be done using the direct exchange rate or indirect exchange rate.

1) Direct exchange Rate
Direct exchange rate is an exchange rate in which one currency is exchange to another without passing through any intermediary currency.

Currency A	→	Currency B
[Franc CFA]	←	[US Dollar]

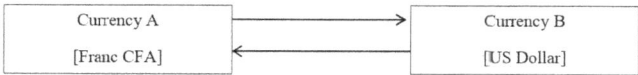

Example 1: The exchange rate between American dollar and Franc CFA is $1 = 500 FCFA. How much will you receive if you exchange $15,000?

Solution

Step 1: Exchange rate $1 = 500 FCFA

Step 2: Data substitution and solving

$$\left.\begin{array}{l} \$1 = 500 \text{ FCFA} \\ \$15,000 = Y \text{ FCFA} \end{array}\right\} \dots\dots\dots\dots[\text{Perform cross multiplication}]$$

$Y(\$1) = 7,500,000\$/FCFA \dots\dots\dots\dots$ [Divide both sides of equation by ($1)]

$X = \mathbf{7,500,000\ FCFA} \dots$[You will receive 7,500,000 FCFA if you exchange $15,000]

Example 2: The exchange rate between pound sterling and Franc CFA is £1 = 250 FCFA. How much will you receive if you exchange 10,000 FCFA?

Solution

Step 1: Exchange rate £1 = 250 FCFA

Step 2: Data substitution and solving

$$\left.\begin{array}{l} £1 = 250 \text{ FCFA} \\ X = 10,000 \text{ FCFA} \end{array}\right\} \dots\dots\dots\dots[\text{Perform cross multiplication}]$$

$X(250 \text{ FCFA}) = 10,000 \text{ FCFA}/£ \dots$[Divide both sides of equation by (250 FCFA)]

$X = \mathbf{£40} \dots\dots\dots\dots$[You will receive £40 if you exchange 10,000 FCFA]

Example 3: A Cameroonian importer is asked to choose between importing a product from America or United Kingdom. The cost is 1,000 dollar in America and 800 pound sterling in United Kingdom. The exchange rate of these currencies and Franc CFA are $1 = 1.5 FCFA and £1 = 1.9 FCFA. Which country will you advise him to import from?

Solution

Step 1: Exchange rates. $1 = 1.5 FCFA and £1 = 1.4 FCFA.

Step 2: Data substitution and solving

$1 = 1.5 FCFA	£1 = 1.9 FCFA
$\left.\begin{array}{l} \$1 = 1.5 \text{ FCFA} \\ \$1,000 = X \end{array}\right\}$ ....[Cross multiply]	$\left.\begin{array}{l} £1 = 1.9 \text{ FCFA} \\ £800 = X \end{array}\right\}$ ....[Cross multiply]
$X(\$1) = \$1,500/FCFA$ ....[Solve]	$X(£1) = £1,520/FCFA$ ....[Solve]
$X = \mathbf{1,500\ FCFA = \$1,000}$	$X = \mathbf{1,520\ FCFA = £800}$

N/B: Importing the product from America will cost him 1,500 FCFA and from united kingdom will cost him 1,520 FCFA. Therefore, he should import from America since the cost is lower (1,500 FCFA < 1,520 $FCFA$).

Test Your Understanding

Exercise 1: The exchange rate between American dollar ($) and Franc CFA is $1 = 500 FCFA. How much will a business man receive if he exchange the following?

1) 200 dollar..**[Answer: 100,000 Franc CFA]**
2) 1,000 dollar..**[Answer: 500,000 Franc CFA]**

Exercise 2: The exchange rate between pound sterling (£) and Franc CFA is £1 = 400 FCFA. How much pound sterling much will you receive if you exchange the following?

1) 10,000 FCFA...**[Answer: £25]**
2) 800,000 FCFA..**[Answer: £2,000]**

Exercise 3: A Cameroonian exporter is asked to accept $25,000 or Y30,000. The exchange rates are $1 = 500 FCFA and Y1 = 400 FCFA. Which offer should the man accept?

[Answer: $25,000]

---

2) Indirect Exchange Rate

Indirect exchange rate is an exchange rate in which one currency is exchange to another by passing through one or more intermediary currency.

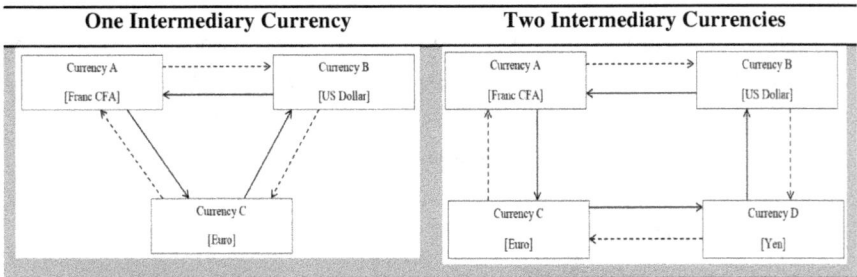

One Intermediary Currency	Two Intermediary Currencies

N/B: The number of intermediary currencies can be more than two. The exchange from one currency to another goes through a multiple stage.

Example 1: The exchange market shows that $1 = £1.95 and $1 = ¥2.5. How much will you receive if you exchange 200 pound starling for Yen?

Solution

Step 1: Movement Identification. Pound sterling and Yen are linked through American dollar. The amount to be exchanged is given in pound sterling. Hence, we move from American dollar and pound sterling to American dollar to Yen.

Step 2: First Movement

1) Exchange Rate. $1 = £1.95$

2) Data substitution and solving

$$\left.\begin{array}{l} \$1 = £1.95 \\ X = £200 \end{array}\right\} \text{..} \text{......[Cross multiplication]}$$

$X(£1.95) = £200/\$ \text{..}$[Solve equation for X]

$X \approx \$103 \text{.......................................} \text{............[That is £200} \approx \$103 \text{]}$

Step 3: Final Movement.....................................[We use $103 as the exchange amount]

1) Exchange Rate. $1 = ¥2.5$

2) Data substitution and solving

$$\left.\begin{array}{l} \$1 = ¥2.5 \\ \$103 = X \end{array}\right\} \text{..} \text{......[Cross multiplication]}$$

$X(\$1) \approx ¥258/\$ \text{..}$[Solve equation for X]

$X \approx ¥258 \text{.......................................} \text{............[That is £200} \approx ¥258 \text{]}$

Example 2: The exchange rate table of the exchange market is given as follows; ($1 = 1.5$ Rs), ($1 = 7.4$ FCFA) and (1 FCFA $= 0.94$ ¢). How much will you receive exchanging 150,000 Rs to Cedi?

Solution

Step 1: First Movement

1) Exchange Rate. $1 = 1.5$ Rs

2) Data substitution and solving

$$\left.\begin{array}{l} \$1 = 1.5 \text{ Rs} \\ X = 150,000 \text{ Rs} \end{array}\right\} \text{..} \text{......[Cross multiplication]}$$

$X(1.5 \text{ Rs}) = 150,000 \text{ Rs}/\$ \text{..}$[Solve equation for X]

$X = \$100,000 \text{..}$[That is 150,000 Rs $= \$100,000$]

Step 2: Second Movement.........................[Use $100,000 as the exchange amount]

$$\left.\begin{array}{l} \$1 = 7.4 \text{ FCFA} \\ \$100,000 = X \end{array}\right\} \text{..}$$[Cross multiplication]

$X(\$1) = 740,000 \text{ FCFA}/\$ \text{..}$[Solve equation for X]

$X = \mathbf{740,000 \text{ FCFA}} \text{..}$[That is $100,000 = 740,000$ FCFA]

Step 3: Final Movement...........................[Use 740,000 FCFA as the exchange amount]

$$\left.\begin{matrix} 1\ FCFA = 0.94 \text{ ¢} \\ 740{,}000\ FCFA = X \end{matrix}\right\}$$ ....[Cross multiplication]

X(1 FCFA) = 695,600 FCFA/¢ ....[Solve equation for X]

X = 695,600¢................................[That is 150,000 Rs = 695,600¢]

Test Your Understanding

Exercise 4: A man wishes to purchase a car in China worth ¥100,000. Given that the exchange rate between Euro and Yen is €1 = ¥0.5 and between Euro and Franc CFA is €1 = 150 FCFA. How much should he budget in FCFA to buy the car...**[Answer: 30,000,000 FCFA]**

Example 5: A business man wishes to exchange 1,000,000 Yen for Franc CFA and is proposed two exchange rate path.

First path: (1$ = ¥2.0), (1$ = £1.2) and (1FCFA = £0.5)

[¥1,000,000 = 1,200,000 FCFA]

Second path: (1€ = ¥0.5), (1€ = £0.5) and (1FCFA = £0.5)

[¥1,000,000 = 2,000,000 FCFA]

Which of the path will you advise him to take?........................[**Answer; Second path:**]

2.2 Geometric Measurement

Here, we measure the perimeter and area of geometric shapes. Geometric shapes take different dimensions, such as; one, two and three dimensions.

2.2.1 One Dimensional Shape

A one dimensional shape is a shape without width. For instance, a line is a one dimensional shape. A line (ray) is a straight line connecting points. The distance between the points is called a length (line segment or interval).

Horizontal Expression of a Line

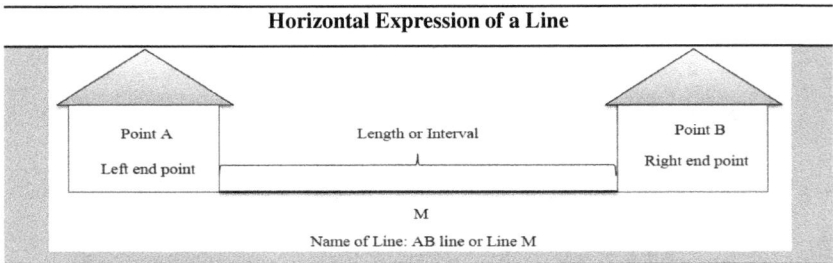

Point A — Left end point

Length or Interval

Point B — Right end point

M

Name of Line: AB line or Line M

N/B: Lines are of different types. Two lines are said to be congruent if they have the same length. Two lines are said to be perpendicular if they intersect to form 90^{0} angle. Two lines are said to be parallel if they are equidistant from one another and never intersect. A line that crosses two lines is called transversal line. Two or more points belonging to the same line are called collinear. Three or more lines that meet at the same intersecting point are called concurrent lines.

Types of Lines

Congruent Lines	Perpendicular Lines	Parallel Lines	Transversal Line (FE)	Concurrent Lines

Note: The distance between two or more points can be measured in length and/or time. The conversion table for unit of length and time is given below.

Conversional Table for Unit of Length	Conversional Table for Unit of Time
10 Millimeter(mm) = 1 Centimeter(cm)	60 Seconds(60sec) = 1 Munites(1 min)
10 Centimeter(cm) = 1 Meter(m)	60 Munites(60min) = 1 Hour(1 hr)
10 Meter(m) = 1 Kilometer(km)	24 Hour(24 hr) = 1 Day
	7 Days = 1 Week
	28 − 31 Days = 1 Month
	12 Months = 1Year

Exercise 1: Convert the following

1) 200 centimeter to meter
2) 5 days to hours

Solution

1) 200 centimeter to meter

Step 1: Formula. 10 Centimeter(cm) = 1 Meter(m)

Step 2: Data substitution and solving

$$\left. \begin{array}{l} 10 \text{ Centimeter(cm)} = 1 \text{ Meter(m)} \\ 200 \text{ Centimeter(cm)} = X \end{array} \right)$$ ....[Perform cross multiplication]

X(10 Centimeter) = 200 Centimeter(cm)/Meter ....[Solve for X]

X = 20 Meters = 200 Centimeter

2) 5 days to hours

Step 1: Formula. 24 Hour(24 hr) = 1 Day

Step 2: Data substitution and solving

$$\left.\begin{array}{l} 24 \text{ Hour} = 1 \text{ Day} \\ X = 5 \text{ Day} \end{array}\right)$$ ....[Perform cross multiplication]

X(1 Day) = 120 Hour/Days ....[Divide both sides by (1 Day)]

X = 120 Hour = 5Days

Example 2: The length and time distance information of Mr. John and Mr. Peter houses to the school compound is given below.

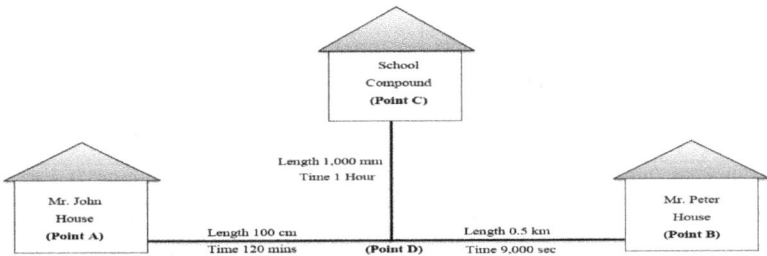

School Compound (Point C)

Length 1,000 mm
Time 1 Hour

Mr. John House (Point A)

Length 100 cm
Time 120 mins

(Point D)

Length 0.5 km
Time 9,000 sec

Mr. Peter House (Point B)

Given that transport cost per meter is 1,000 FCFA, who amongst them is having advantage in the following,

1) Distance if measured in meters
2) Time if measured in hours
3) Transport cost

Solution

1) Distance if measured in meters

Step 1: Formula. Total Distance = ∑ of distance covered in meter

Step 2: Data substitution and solving

Mr. John	Mr. Peter
Where: 100 cm = 10m and 1,000 mm = 1m	Where: 0.5 km = 5m and 1,000 mm = 1m
Total Distance = 10m + 1m ....[Solve]	Total Distance = 5m + 1m ....[Solve]
Total Distance = **11m**	Total Distance = **6m**
Note: Mr. Peter is having advantage over distance if measured in meter	

2) Time if measured in hours

Step 1: Formula. Total Time = ∑ of Time covered in hours

66

Step 2: Data substitution and solving

Mr. John	Mr. Peter
Where: 120 mins = 2hrs	Where: 9,000 sec = 2.5 hours
Total Time = 2hrs + 1hr ………..[Solve] Total Time = **3hours**	Total Time = 2.5 hrs + 1hr ……….[Solve] Total Time = **3.5 hours**
Note: Mr. John is having advantage over time if measured in hours	

3) Transport cost

Step 1: Formula. Transport Cost = Total Distance × Cost per unit distance

   Where; Cost per unit distance = 1,000 FCFA

Step 2: Data substitution and solving

Mr. John	Mr. Peter
Where: Total Distance = 11	Where: Total Distance = 6
Transport Cost = 11 × 1,000 …...[Solve] Transport Cost = **11,000 FCFA**	Transport Cost = 6 × 1,000 ……….[Solve] Transport Cost = **6,000 FCFA**
Note: Mr. Peter is having advantage over transport cost since he spend less	

Example 3: Use the information in the picture below and determine the following. Assume the transport cost per kilometer is 1,000 FCFA. [$1 = 500 FCFA]. It is worth knowing that point CE and point DC are congruent.

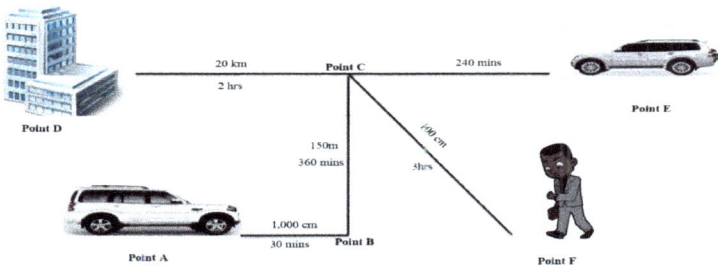

1) The time required for the man to arrive the company (Point D) in minutes
2) How much will the passengers in point A pay to reach the company in dollar
3) The total distance it will take the passengers in point B to arrive the company in meters

Solution

1) The time required for the man to arrive the company (Point D) in minutes

Step 1: Formula. Total Time $= \sum$ of time used to cover the distance

Step 2: Data substitution and solving

Total Time $= 3\text{hrs} + 2\text{hrs}$ ....[Solve equation]

Total Time $= \mathbf{5hrs} = \mathbf{300\ mins}$

2) How much will the passengers in point A pay to reach the company in dollar

Step 1: Formula. Transport Cost = Total Distance Covered × Cost per distance unit

Where; Cost per distance unit = 1,000 FCFA = \$2

Total Distance Covered = 1,000cm + 150m + 20km = 10km + 15km + 20km = 45km

Step 2: Data substitution and solving

Transport Cost $= 45 \times 2$ ....[Solve equation]

Transport Cost = \$**90**

3) The total distance it will take the passengers in point B to arrive the company in meters

Step 1: Formula. Total Distance $= \sum$ of distance used to cover the distance

Where; Point(DC) = Point(CE), therefore, Point(DC) = 20km and Point(CE) = 20km

Step 2: Data substitution and solving

Total Distance = 20km + 20km ....[Solve equation]

Total Distance $= \mathbf{40km} = \mathbf{400m}$

Example 4: The distance (measured in meters) covered by a driver from start to end is given by the following coordinates; $(-4,1)$, $(2,4)$, $(5,4)$ and $(8,2)$. Assume the transport cost per meter is 500 FCFA, how much is the drivers total revenue.

Solution

Step 1: Formula. Total Revenue = Total Distance × Price per unit distance

Where; Price per unit distance = 500 FCFA

Step 2: Derivation of total distance

Sketch without distance	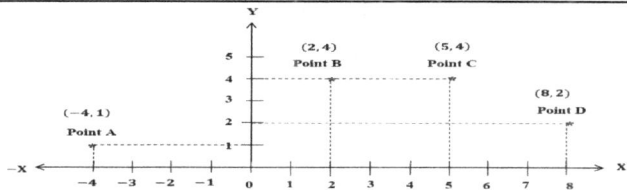
Sketch with distance	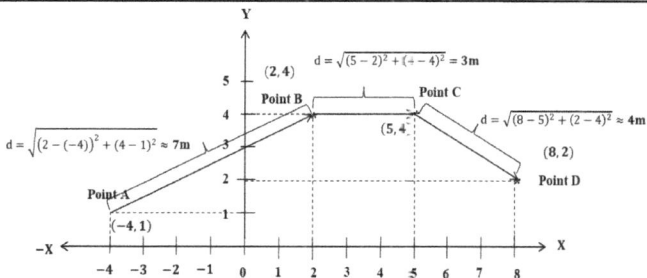
Total distance	Total Distance Covered $= 7m + 3m + 4m = \textbf{14m}$

Step 3: Data substitution and solving

    Total Revenue $= 14 \times 500$ ....[Solve equation]

    Total Revenue $= \textbf{7,000 FCFA}$

Test Your Understanding

Exercise 6: Convert the following,

  1) 1,000 centimeter (cm) to meters (m)..............................…......[Answer: 100 meters]

  2) 2 days to munites (mins) ....…......[Answer: 120 munites]

  3) 1 kilometer (km) to millimeters (mm)........................ [Answer: 1,000 milimeters]

Exercise 7: Use the information in the table below and answer the following questions. Assume the transport cost per kilometer is 10,000 FCFA and the exchange rate between dollar and FCFA is $1 = 500 FCFA.

  1) How much will the man pay in dollar to arrive point D using point B and C direction?

  2) How many hours will the man take to reach point F passing through point E

[Answer: 1) $ 410, 2) 58.5 hours]

Exercise 8: The coordinate point of Miss Mary house is (0,0). The coordinate point of the hospital building is (1,3) and the coordinate point of the church building is (2,6). Assuming that the distance is measured in kilometer and the transport cost per kilometer is 1,000 FCFA, how much will miss Mary need to arrive the,

1) Hospital building..[Answer: 3,000 FCFA]
2) Church building..[Answer: 6,000 FCFA]
3) Church building and back to the hospital building..............[Answer: 9,000 FCFA]
4) Church building and back home and back to the hospital building

[Answer: 15,000 FCFA]

2.2.2 Two Dimensional Shapes

Two dimensional shapes are shapes with either length and width or base and height. Major shapes of interest include; rectangle, right angled triangle, square and circle.

1) Rectangle: A rectangle is a plan shape with two pairs of side of equal distance and four right angled corners.

Horizontal Expression	Vertical Expression	Slanting Expression

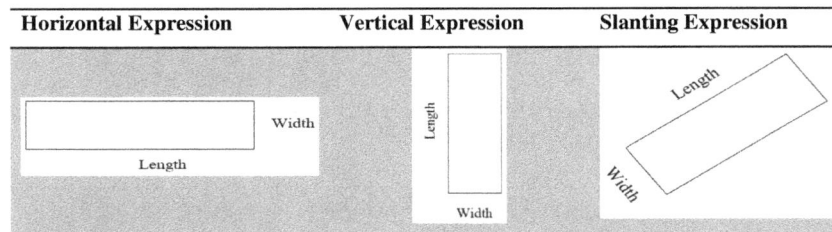

2) Triangle: A triangle is a plan enclosed by three straight segments (sides) and has three vertices. Triangles are of different types based on their sides and angles. Our analysis will be limited to right angled triangle and equilateral triangle. A right angled triangle has an angle of (90^0) and an equilateral triangle has three sides of equal length and three equal angles. The interior angle of any triangle is equal to (180^0) and each angle of an equilateral triangle is (60^0).

	Right angled Triangle	Equilateral Triangle

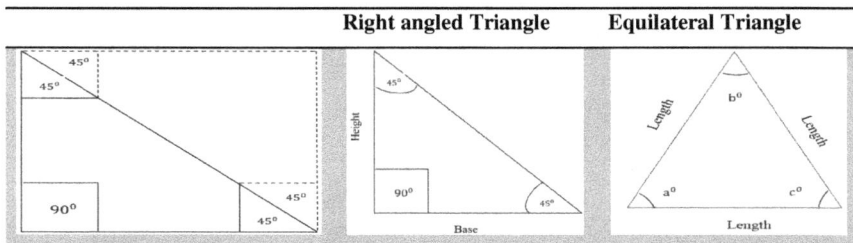

70

3) Circle: A circle is an enclosed shape where all points on the boundary are at a fixed distance from the center. The boundary of a circle is called circumference, the line from the center to any given point on the circumference is called radius and a line from one point through the center to another point on the circumference is called diameter (diameter produces semi-circles).

Circle	Semi-circle	Quadrant	Sector

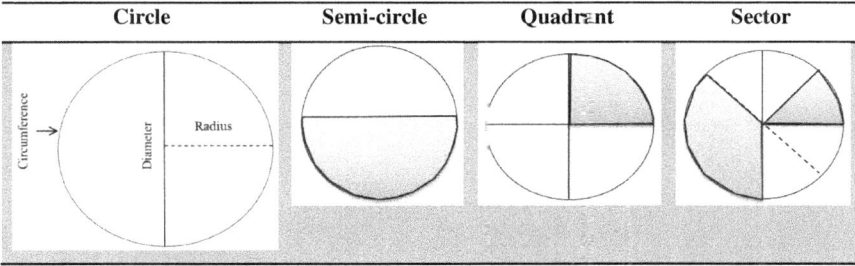

1) Perimeter of a Plan

A perimeter of a plan shape is the total length of the sides of the plan shape. That is, the distances around the outside of the plan shape.

a) **Perimeter of a rectangle:** The perimeter of a triangle is simply the sum of the two lengths plus the sum of the two widths. It is calculated using the formula below.

$$\text{Perimeter of a Rectangle} = 2(\text{Lenght} + \text{Width}) = 2(L + W) \quad \ldots\ldots\ldots\ldots\ldots(1)$$

Example 1: The length of a rectangle is 50 km and the width is 0.25 of the length. What is the perimeter of the rectangle?

Solution

Step 1: Formula. Perimeter of a Rectangle $= 2(\text{Lenght} + \text{Width})$

　　　　Where; $L = 50$ and $W = 12.5$

Step 2: Data substitution and solving

　　　　Perimeter of a Rectangle $= 2(50 + 12.5)$ $\ldots\ldots\ldots\ldots\ldots\ldots\ldots\ldots$[Simplify bracket]

　　　　Perimeter of a Rectangle $= 2(62.5)$ $\ldots\ldots\ldots\ldots\ldots\ldots\ldots\ldots\ldots\ldots$[Solve equation]

　　　　Perimeter of a Rectangle $= \mathbf{125km^2}$

Example 2: Compute the perimeter of the following rectangles using the coordinates below.

　　Rectangle A: $(-4,3)$, $(-1,3)$, $(-4,-2)$ and $(-1,-2)$

　　Rectangle B: $(2,2)$, $(6,2)$, $(2,1)$ and $(6,1)$

　　Rectangle C: $(1,-3)$, $(2,-2)$, $(3,-6)$ and $(4,-5)$

Solution

Step 1: Formula. Perimeter of a Rectangle $= 2(\text{Lenght} + \text{Width})$

Step 2: Derivation of rectangles shape

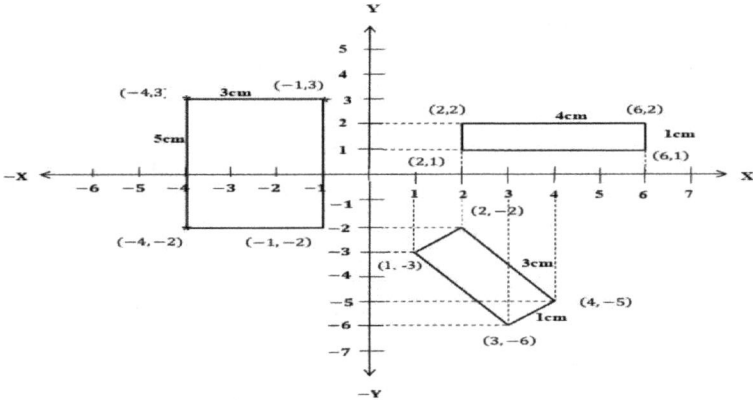

Step 3: Data substitution and solving

Rectangle A	Rectangle B	Rectangle C
Where; L = 5 and W = 3	Where; L = 4 and W = 1	Where; L = 3 and W = 1
Perimeter $= 2(5 + 3)$	Perimeter $= 2(4 + 1)$	Perimeter $= 2(3 + 1)$
Perimeter $= 2(8)$	Perimeter $= 2(5)$	Perimeter $= 2(4)$
Perimeter $= \textbf{16cm}$	Perimeter $= \textbf{10cm}$	Perimeter $= \textbf{8cm}$

Test Your Understanding

Exercise 9: Determine the perimeter of the following rectangles

   1) Length is 15m and width 5 m..[**Answer: 75m^2**]

   2) Length is 20km and the width is 0.5 of length..........................[**Answer: 200m^2**]

   3) Length is 150 and the width is 25% of length..........................[**Answer: 5,625m^2**]

   4) Length is 300 and the width is 35/100 of length..................[**Answer: 31,500m^2**]

Exercise 10: A car is required to go round the following rectangles.

Rectangle 1	**Rectangle 2**	**Rectangle 3**
5m (1.2m)	10m (2m)	8m, 4m

A square meter distance consumes 2 liters of petrol and a liter of petrol cost 500 FCFA. Assuming that the car goes round the rectangles, compute the following,

1) The shortest perimeter...[Answer: 6m^2]
2) The number of liters of petrol needed to go round the longest perimeter

[Answer: 64 Liters]

3) The cost required to go round rectangle 2.....................[Answer: 20,000 FCFA]
4) The rectangle to go round if objective is to minimize cost.....[Answer: Rectangle 1]

Exercise 11: Compute the distance of the width of a rectangle with area 28km^2 and the length 7 km. ....[Answer: 4m]

b) **Perimeter of a triangle:** The perimeter of a triangle is simply the sum of the side's length. It is calculated using the formula below.

Perimeter $=$ Side(a) Lenght $+$ Side(b)Lenght $+$ Side(c) Lengt...[All triangles]
Perimeter $= 3 \times$ One Side Length $= 3L$ ....[Equilateral triangle only]

N/B: In situation where all lengths are not given, we apply the Pythagoras theorem. The Pythagoras theorem states the relationship between the lengths of the three sides of a right angled triangle. It is expressed graphically and algebraically as seen below.

Graphical Expression	Algebraic Expression
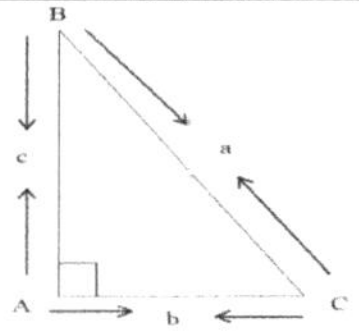	$a^2 = b^2 + c^2$ ....[BC relationship] $c^2 = a^2 - b^2$ ....[BA relationship] $b^2 = a^2 - c^2$ ....[AC relationship]

Example 1: The base of a right angled triangle is 6m and the height is 8m. Determine the perimeter of the triangle.

Solution

Step 1: Formula. Perimeter $= \sum$ of Sides Lenght

Step 2: Determination of unknown angle

First Expression	Working	Final Expression
Height (24m) [triangle: Height (24m), Base (18m)]	Step 1: Formula. $c^2 = a^2 + b^2$ Where; a = 18 and b = 24 Step 2: Data substitution and solving $c^2 = 18^2 + 24^2$.[Work power] $c^2 = 900$ ....[Work square root] c = **30m**	[triangle: Height (24m), Hypotenuse (30m), Base (18m)]

Step 3: Data and solving

Perimeter = 24m + 18m + 30m ………………………….…………[Solve equation]

Perimeter = **72m**

Example 2: Compute the perimeter of the triangle with coordinates; $(-1,1)$, $(2,1)$ and $(-4,3)$. Assume distance is measured in kilometer.

<u>**Solution**</u>

Step 1: Formula. Perimeter = $\sum$ of Sides Lenght

Step 2: Derivation of triangle shape

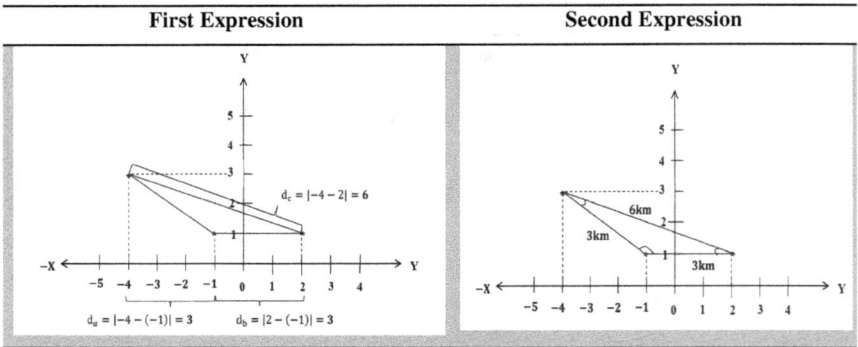

First Expression	Second Expression
$d_c = \lvert -4 - 2 \rvert = 6$ $d_a = \lvert -4 - (-1) \rvert = 3$ $d_b = \lvert 2 - (-1) \rvert = 3$	6km 3km 3km

Step 3: Data substitution and solving

$\quad$ Perimeter $= 3km + 3km + 6km$ ....[Solve equation]

$\quad$ Perimeter $= 12\mathbf{km}$

Example 3: A business man is presented two lands of different shapes as seen below.

Land A	Land B

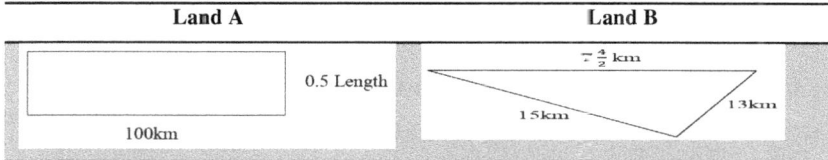

0.5 Length

100km

$7\frac{4}{2}$ km

15km

13km

One kilometer (1km) of land A is covered in 10 minutes and of land B is in 4.8 minutes and 5 liters of petrol is used to cover 1km. Assuming ($1 = 500$ FCFA), determine the following,

1) The time taken to move around the triangle in hours and rectangle in days

2) The time taken to move around both lands in seconds

3) The liters of petrol required to move around both lands

4) The total expenditure of petrol in FCFA if a liter is worth 0.25 dollar.

Solution

1) The time taken to move around the triangle in hours and rectangle in days

Step 1: Formula. Total Time = Perimeter × Time per unit distance

Step 2: Data substitution and solving

Land A (Rectangle)	Land B (Triangle)
Where; Perimeter $= 100 + 50 = 150$km	Where; Perimeter $= 15 + 13 + 14 = 42$km
Time per unit distance $= 4.8$	Time per unit distance $= 10$
Total Time $= 150 \times 4.8$ ....[Solve equation]	Total Time $= 42 \times 10$ ....[Solve equation]
Total Time $= \mathbf{720mins} = \mathbf{0.5days}$	Total Time $= \mathbf{420mins} = \mathbf{7hrs}$

2) The time taken to move around both lands in seconds

Step 1: Formula. Total Time = Triangle Time + Rectangle Time

$\quad$ Where; Triangle Time $= 7hrs = 420mins = 25,200sec$

$\quad\quad$ Rectangle Time $= 0.5days = 720mins = 43,200sec$

Step 2: Data substitution and solving

$\quad$ Total Time $= 25,200 + 43,200$ ....[Solve equation]

$\quad$ Total Time $= \mathbf{68,400sec}$

3) The liters of petrol required to move around both lands

Step 1: Formula. Total Liter Utilised = Total Perimeter × Liter per unit distance

      Where; Total Perimeter = 150km + 42km = 192km, Liter per unit distance = 5

Step 2: Data substitution and solving

      Total Liter Utilised = 192 × 5 ....[Solve equation]

      Total Liter Utilised = **960 Liters of Petrol**

4) The total expenditure of petrol in FCFA if a liter is worth 0.25 dollar.

Step 1: Formula. Total Expenditure (TE) = Total Liters Utilised × Price per Liter

Step 2: Data substitution and solving

Where; Price per Liter = 0.25 dollar	Where; Price per Liter = $0.25 dollar = 125 FCFA
Total Expenditure = 960 × 0.25	Total Expenditure = 960 × 125
Total Expenditure = $240	Total Expenditure = **120, 000 FCFA**
Total Expenditure = **120, 000 FCFA**	Total Expenditure = $240

Test Your Understanding

Exercise 12: The sides length of a scalene triangle are given as; 17cm, 10cm and 8cm. Determine the perimeter of the triangle in meter..................................…...…....…[**Answer: 3.5m**]

Exercise 13: Draw the triangles with the following coordinates and compute their respective perimeter.

1) Triangle 1: (1,1), (4,5) and (4,1)...…...…….[**See Appendix**]
2) Triangle 2: (1, −2), (1,5) and (5, −2)...................................…….[**See Appendix**]

Exercise 14: Compute the perimeter of the following triangles

Triangle 1	Triangle 2	Triangle 3	Triangle 4

[**Answer:** ....…...See Appendix]

c) **Perimeter of a circle:** The perimeter of a circle is also called circumference and the perimeter of a sector is called arc length. Generally, the circumference of a complete circle is 360^0

$$\text{Perimeter of a Circle} = 2 \times \pi \times \text{Radius} = 2\pi r \dots\dots\dots\dots\dots\dots\dots\dots\dots(1)$$

$$\text{Perimeter of a Circle} = \text{Diameter} \times \pi = d\pi \dots\dots\dots\dots\dots\dots\dots\dots\dots(2)$$

$$\text{Perimeter of a Sector} = \frac{\text{Angle}}{360} \times 2 \times \pi \times \text{Radius} = \frac{\theta}{360} \times 2\pi r \dots\dots\dots\dots(3)$$

Example 1: Find the circumference of a circle under the following situation,

1) When the radius is 21 cm
2) When the diameter is 14.2 m

Solution

1) When the radius is 21 cm

Step 1: Formula. Perimeter of a Circle $= 2 \times \pi \times$ Radius, where; Radius $= 21$ and $\pi = 22/7$

Step 2: Data substitution and solving

$$\text{Perimeter of a Circle} = 2 \times \frac{22}{7} \times 21 \dots\dots\dots\dots\dots\dots\dots\dots\text{[Solve equation]}$$

Perimeter of a Circle $= \mathbf{132cm}$

2) When the diameter is 14.2 m

Step 1: Formula. Perimeter of a Circle $= \text{Diameter} \times \pi$

Where; Diameter $= 14.2$ and $\pi = 22/7$

Step 2: Data substitution and solving

Method 1 (Using perimeter)	Method 2 (Using radius)
Perimeter of a Circle $= 14.2 \times \frac{22}{7}$..[Solve]	Perimeter of a Circle $= 2 \times \frac{22}{7} \times 7.1$.[Solve]
Perimeter of a Circle $\approx \mathbf{45m}$	Perimeter of a Circle $\approx \mathbf{45m}$

Example 2: A business man wishes to move around the land he purchased as seen below.

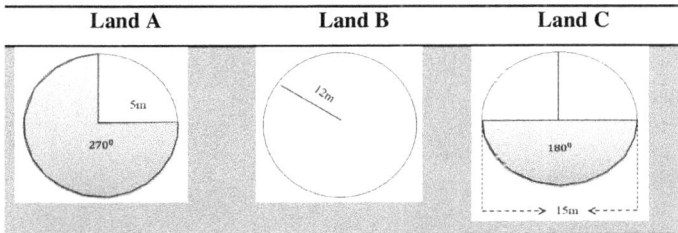

Assuming that the transport cost per km is 4,000 FCFA, how much should he budget to make this movement possible?

Solution

Step 1: Formula. Total Transport = Total Perimeter × Cost per unit distance

Where; Cost per unit distance = 4,000

Step 2: Determination of total perimeter

1) Formula.   Perimeter of a Circle = 2 × π × Radius and Perimeter of a Sector = $\frac{\theta}{360} \times 2\pi r$

2) Data substitution and solving

Land A	Land B	Land C
Perimeter $= \frac{270}{360} \times 2 \times \frac{22}{7} \times 5$ Perimeter $\approx$ **24m**	Perimeter $=$ $2 \times \frac{22}{7} \times 12$ Perimeter $\approx$ **75m**	Perimeter $= \frac{180}{360} \times 2 \times \frac{22}{7} \times 7.5$ Perimeter $\approx$ **24m**
Total Perimeter = 24m + 75m + 24m = **123m**		

Step 3: Data substitution and solving

Total Transport = 123 × 4,000 ....……..............[Solve equation]

Total Transport = **492,000 FCFA** ....[He should budget 492,000 FCFA]

Test Your Understanding

Exercise 15: Determine the perimeter of the following circle and sectors

Figure 1	Figure 2	Figure 3

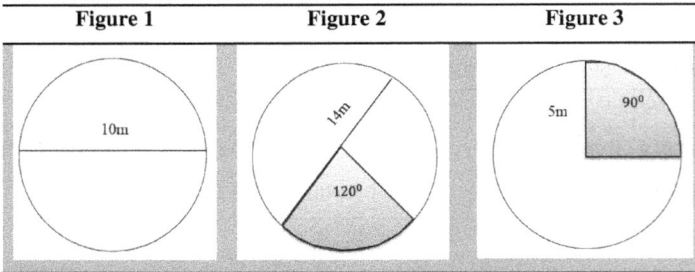

[Answer: Figure 1 ≈ 31m, Figure 2 ≈ 15 and Figure 3 ≈ 8m]

Exercise 16: An entrepreneur is having 100,000 FCFA to move around his plantains.

Rice Plantation	Banana Plantation	Coc a Plantation

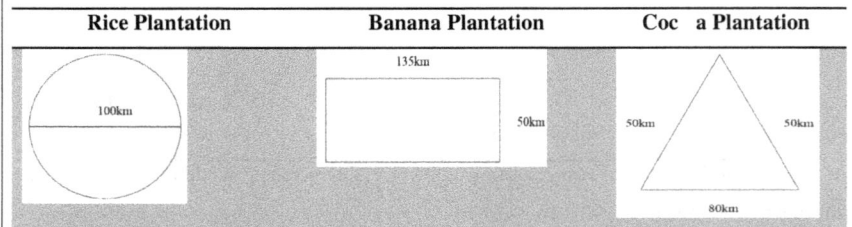

78

Assuming that the transport cost per km is 500 FCFA, which of the plantation will he successfully move around?..........................[**Answer: Cocoa Plantation (90,000 FCFA)**]

d) **Perimeter of a Square:** The perimeter of a square is simply the sum of the equal square sides length. Since all sides are equal, we simply multiply the length of a side by 4.

$$\text{Perimeter} = 4 \times \text{Lenght}(L) = 4L \dots\dots\dots\dots\dots(1)$$

Example 1: The side of a square is of length 25 km, what is the perimeter of the square?

<u>Solution</u>

Step 1: Formula. Perimeter = $4 \times$ Lenght(L) = 4L, where; Lenght(L) = 25

Step 2: Data substitution and solving

$$\text{Perimeter} = 4 \times 25 \dots\dots\dots\dots\dots[\text{Solve equation}]$$

Perimeter = **100km**

Example 2: Compute the perimeter of the square with coordinates; $(-5,6)$, $(0,8)$, $(-3,1)$ and $(2,3)$

<u>Solution</u>

Step 1: Formula. Perimeter = $4 \times$ Lenght(L) = 4L

Step 2: Determination of side distances..................................[Lenght]

Sketch without distance

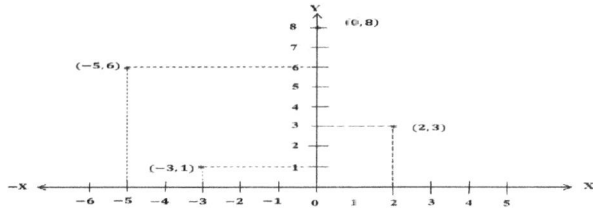

Sketch with distance

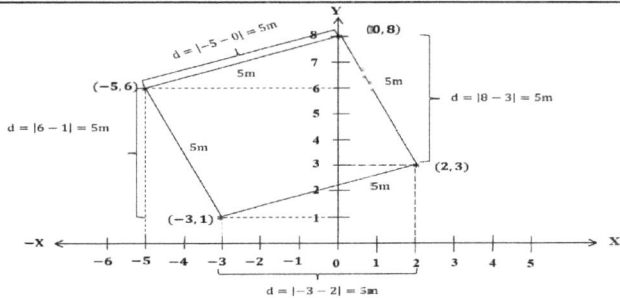

79

Step 3: Data substitution and solving...……..……[Lenght = 5]

Perimeter = 4 × 5 ………………………………...…………………..[Solve equation]

Perimeter = **20m**

Example 3: A drive covers 1km distance in 2 hours and consumes 1.5liters of petrol. The transport cost for 1km is 100 FCFA. Using the information of the square below, compute the following

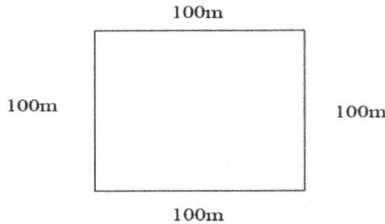

100m

100m   100m

100m

1) The number of hours required to go round the square
2) The transport cost required to go round the square
3) The number of liters of petrol required to go round the square.

<u>Solution</u>

1) The number of hours required to go round the square

Step 1: Formula. Time = Perimeter × Time per unit distance

Where; Time per unit distance = 2hrs and Perimeter = 4 × 100 = 400

Step 2: Data substitution and solving

Time = 400 × 2 ………………………………………………..……..[Solve equation]

Time = **800 hrs**

2) The transport cost required to go round the square

Step 1: Formula. Transport Cost = Perimeter × Cost per unit distance

Where; Cost per unit distance = 100 and Perimeter = 4 × 100 = 400

Step 2: Data substitution and solving

Transport Cost = 400 × 100 …………………....…..................……..[Solve equation]

Transport Coste = **40, 000 FCFA**

3) The number of liters of petrol required to go round the square.

Step 1: Formula. Number of liters of petrol = Perimeter × Liter per unit distance

Where;Perimeter = 400 and Liter per unit distance = 1.5

Step 2: Data substitution and solving

Number of liters of petrol = 400 × 1.5 …………..…….....……..[Solve equation]

Number of liters of petrol = **600 Liters of Petrol**

Test Your Understanding

Exercise 17: The perimeter of a square is 80cm. what is the length of a side of the square in the following,

1) Meter (m)...[**Answer: 2m**]

2) Millimeter (mm) ....[**Answer: 200mm**]

3) Kilometer (km)..[**Answer: 0.2km**]

Exercise 18: A builder wishes to lay the first line blocks for a square room. The length of a block is 5cm and the length of the room side is 1m. Assuming that the price per block is 250 FCFA, how much should he budget to achieve his objective?.............[**Answer: 2,000 FCFA**]

Exercise 19: Which of the following shape is having the largest perimeter

Rectangle	Triangle	Circles	Square

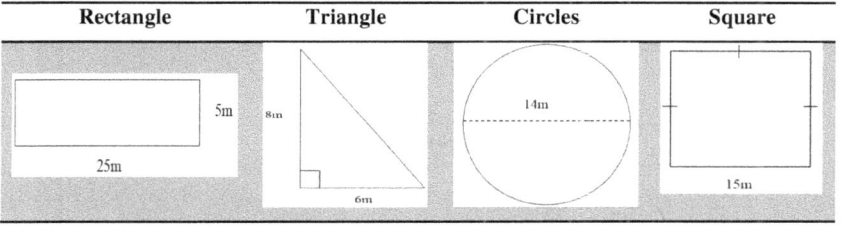

[**Answer: ....See Appendix**]

---

2) Area of a Plan

An area of a figure or plan describes the size of the total surface inside the perimeter of the figure. That is, area is the amount of space or surface occupied or covered by a figure or plan. Area is expressed in squared units such as; mm^2, cm^2, m^2 and km^2. The conversion table for units of area is given below.

100 square millimeters(mm^2) = 1 square centimeter (cm^2)

100 square centimeter (cm^2) = 1 square decimeter (dm^2)

100 square decimeter (dm^2) = 1 square meter(m^2)

100 square meter(m^2) = 1 are (a)

100 are (a) = 1 hectare (ha)

100 hectare (ha) = 1 square kilometer (km^2)

N/B: The conversion table for units of area helps in bringing different area units to common area units and to change from one area unit to another area unit.

a) **Area of a Rectangle:** The area of a rectangle is the product between the length and width of the rectangle. It is calculated using the formula below.

$$\text{Area} = \text{Length} \times \text{Width} \dots\dots\dots\dots\dots\dots\dots\dots\dots\dots\dots\dots\dots(1)$$

Example 1: The length of a rectangular land 12m and the width is 7m. Calculate the area of the land.

Solution

Step 1: Formula. Area = Length × Width, where; Length = 12m and Width = 7m

Step 2: Data substitution and solving

$$\text{Area} = 12\text{m} \times 7\text{m}\dots\dots\dots\dots\dots\dots\dots\dots\dots\dots\dots\dots\dots\dots\dots[\text{Solve equation}]$$

$$\text{Area} = \mathbf{84m^2}$$

Example 2: A business man is faced with three choices of land (A, B and C) and has as objective to minimize cost. Assuming a m^2 of land cost 1,000 FCFA, which land should he buy?

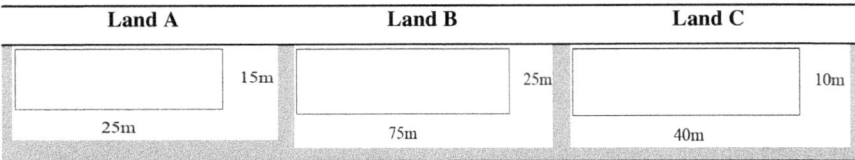

Land A	Land B	Land C
15m	25m	10m
25m	75m	40m

Solution

Step 1: Formula. Total Cost(TC) = Area × Price per unit Area

   Where; Price per unit Area = 1,000

Step 2: Determination of land areas

1) Formula. Area = Length(L) × Width(W), where; Length = 12m and Width = 7m
2) Data substitution and solving

Land A	Land B	Land C
L = 25m and W = 15m	L = 75m and W = 25m	L = 40m and W = 10m
Area = 25m × 15m	Area = 75m × 25m	Area = 40m × 10m
Area = 375m²	Area = 1,875m²	Area = 400m²
Land B$(1,875^2) > $ Land $C(400^2) > $ Land $A(375^2)$		

Step 3: Data substitution and solving

Land A	Land B	Land C
TC = 375 × 1,000	TC = 1,875 × 1,000	TC = 400 × 1,000
TC = 375,000 FCFA	TC = 1,875,000 FCFA	TC = 400,000 FCFA
N/B: Cost Minimization preference: Land A $>$ Land C and Land B. The business man should choose land A since it is having the lowest cost (375,000 FCFA).		

Example 3: An entrepreneur spends 1,000,000 FCFA to obtain a land with a rectangular shape. Given that the area of the land is 500m^2, calculate the cost per m^2.

Solution

Step 1: Formula. Total Cost(TC) = Area × Price per unit Area

Where; Total Cost(TC) = 1,000,000 FCFA and Area = 500m^2

Step 2: Data substitution and solving

1,000,000 FCFA = 500m^2 × Price per unit Area ......[Divide both sides by (500m^2)]

Price per unit Area = $\frac{1,000,000 \text{ FCFA}}{500 \text{m}^2}$ ....[Solve equation]

Price per unit Area = **2,000 FCFA/m^2**

Example 4: The area of a rectangular land is 0.16 hectare(ha). What is the area in square metre(m^2).

Solution

Step 1: Formula. Square Metre(m^2) = $\frac{10,000 \text{ m}^2 \times \text{Given hectare}}{1 \text{ hectare}}$

Where; Given hectare = 0.16 hectare

Step 2: Data substitution and solving

Square Metre(m^2) = $\frac{10,000 \text{ m}^2 \times 0.16 \text{ hectare}}{1 \text{ hectare}}$ ....[Simplify numerator]

Square Metre(m^2) = $\frac{1,600 \text{ m}^2/\text{hectare}}{1 \text{ hectare}}$ ....[Simplify equation]

Square Metre(m^2) = **1,600 m^2**

Example 4: A manager wishes to purchase a land with coordinates; (2,0),(0,2),(1,3), and (3,1). Assuming a square cost 6,000 FCFA, how much does he need to purchase the land?

Solution

Step 1: Derivation of shape using the coordinates...[x, y]

Illustration

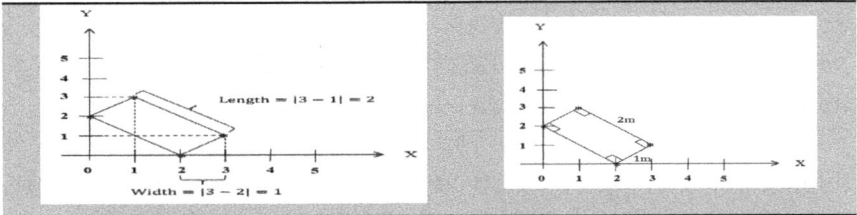

Step 2: Calculation of total cost..........................[Length(L) = 2m and Width(W) = 1]

1) Formula. Total Cost(TC) = Area × Price per unit Area

Where; Price per unit Area = 6,000 FCFA and Area = 2m × 1m = 2m²

2) Data substitution and solving

Total Cost(TC) = 2 × 6,000 ....[Solve equation]

Total Cost(TC) = **12,000 FCFA** .....[He needs 12,000 FCFA to purchase the land]

Test Your Understanding

Exercise 20: A farmer requires labour for farm clearing as seen below. Assuming a worker can clear 25m² and is to be paid 25,000 FCFA, calculate the following

50cm

10cm

1) The number of workers required by the farmer..................[**Answer: 20 Workers**]
2) The total labour cost...[**Answer: 500,000 FCFA**]

Exercise 21: A man is presented a rectangular land for purchase as seen below.

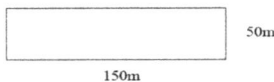

50m

150m

How much will he pay for the land if a hectare is worth 500,000 FCFA?

[Answer: 373,000 FCFA]

Exercise 22: The length of a rectangle is 100 km and the width is 50% of the length. Given that the cost price per square kilometer is 15,500 FCFA, compute the following,

1) The area of the rectangle...[**Answer: 5,000km²**]

2) The amount required to purchase the land..................[**Answer: 77,500,000 FCFA**]

b) **Area of a Triangle:** The formula for calculating the area of a triangle depends on the type of triangle in question. When all triangle lengths are known, we use the Heron's or Hero formula.

$\text{Area} = \frac{1}{2}\text{Base} \times \text{Height}$ .... [Right angled triangle only]

$\text{Area} = \sqrt{S(S-a)(S-b)(S-c)}$ ....[Triangles where all lengths are known]

$S = \frac{1}{2}(\text{Perimeter}) = \frac{1}{2}(a+b+c)$ ....(1)

Example 1: Find the area of the following triangles

Triangle 1	Triangle 2

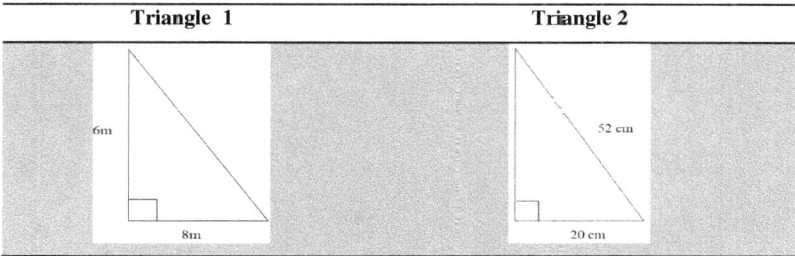

| Triangle 1 (6m, 8m) | Triangle 2 (52 cm, 20 cm) |

Solution

Step 1: Formula. $\text{Area} = \frac{1}{2}\text{Base} \times \text{Height}$

Step 2: Data substitution and solving

Triangle 1	Triangle 2
Where; Base = 8m and Height = 6m	Where; Base = 20m and Height = $\sqrt{2,304} = 48$m
$\text{Area} = \frac{1}{2}(8)\text{m} \times 6\text{m}$ ...[Open bracket]	$\text{Area} = \frac{1}{2}(20)\text{m} \times 48\text{m}$ ....[Open bracket]
$\text{Area} = 4\text{m} \times 6\text{m}$ ...[Solve equation]	$\text{Area} = 10\text{m} \times 48\text{m}$ ....[Solve equation]
$\text{Area} = \mathbf{24m^2}$	$\text{Area} = \mathbf{480m^2}$

Example 2: Determine the perimeter of the following triangles

Triangle 1	Triangle 2

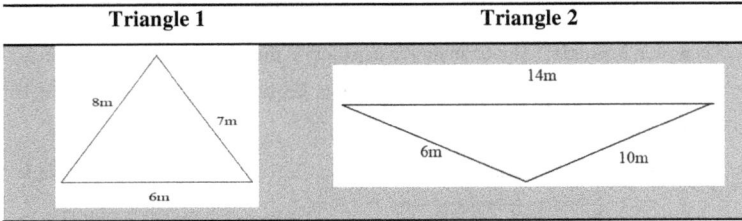

Solution

Step 1: Formula. Area $= \sqrt{S[(S-a)(S-b)(S-c)]}$ and $S = \frac{1}{2}(a+b+c)$

Step 2: Data substitution and solving

Triangle 1 (let a = 8, b = 6 and c = 7)	Triangle 2 (let a = 6, b = 10 and c = 14)
Where: $S = \frac{1}{2}(8+6+7) = 10.5$	Where: $S = \frac{1}{2}(6+10+14) = 15$
$A = \sqrt{10.5[(10.5-8)(10.5-6)(10.5-7)]}$	$A = \sqrt{15[(15-6)(15-10)(15-14)]}$
Area $= \sqrt{10.5[(2.5)(4.5)(3.5)]}$	Area $= \sqrt{15[(9)(5)(1)]}$
Area $= \sqrt{10.5[39.375]}$	Area $= \sqrt{15[45]}$
Area $= \sqrt{413.4375}$	Area $= \sqrt{675}$
Area $\approx$ **20m**	Area $\approx$ **26m**

Example 3: The price per meter square of a land in Bamenda is 5,500 FCFA. Which of the following land below is more expensive?

Land 1	Land 2

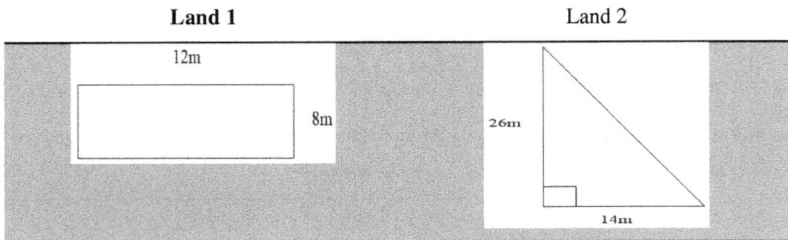

Solution

Step 1: Formula. Total Cost(TC) $=$ Area $\times$ Price per unit Area

Where; Price per unit Area $= 5,500$

Step 2: Determination of area

Land 1	Land 2
Formula. Area = Length × Width Where; L = 12 and W = 8	Formula. Area = $\frac{1}{2}$ Base × Height Where; Base = 14 and Height = 26
Area = 12m × 8m …[Solve equation] Area = $\mathbf{96m^2}$	Area = $\frac{1}{2}$(14) × 26 ………[Solve equation] Area = $\mathbf{182m^2}$

Step 3: Data substitution and solving

Land 1	Land 2
Total Cost(TC) = 96 × 5,500	Total Cost(TC) = 182 × 5,500
Total Cost(TC) = **528,000 FCFA**	Total Cost(TC) = **1,001,000 FCFA**
N/B: Land 2 (1,001,000 FCFA) is more expensive than land 1(528,000 FCFA)	

Example 4: A land coordinate of (0,0), (4,0) and (4,3) is presented for sale. Assuming that a square meter cost 5,500 FCFA, how much does the land cost?

Solution

Step 1: Derivation of shape using coordinates……………………………………………..[x,y]

Illustration

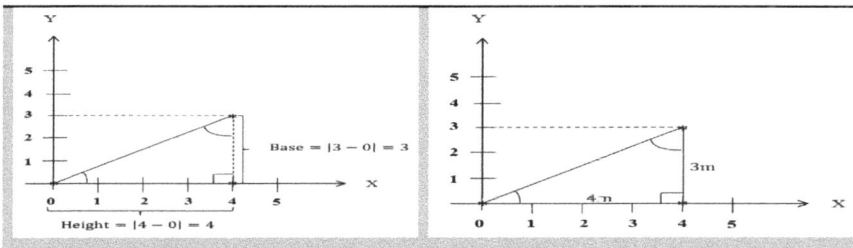

Step 2: Calculation of total cost…………………………..[Base(B) = 3m and Height(H) = 4]

1) Formula. Total Cost(TC) = Area × Price per unit Area

Where; Price per unit Area = 5,500 FCFA and Area = $\frac{1}{2}$(3) × 4m = 6m^2

2) Data substitution and solving

Total Cost(TC) = 6 × 5,500 …………… ………… …………[Solve equation]

Total Cost(TC) = **33,000 FCFA** ……… …….…..[The land cost 33,000 FCFA]

Exercise 23: Plantation agriculture is practiced in two lands with different shapes as seen below. Which of the plantation is the largest, using decimeter (dm^2) as the common measurement?

Land 1	Land 2

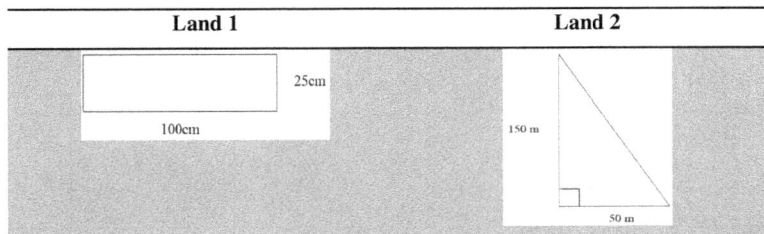

[**Answer: Triangle** $= 3,750m^2 = 375,000dm^2 > Rectangle = 2,500cm^2 = 25dm^2$]

Exercise 24: A farmer is wishing to purchase and work in a triangular shape farm as seen below.

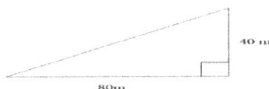

The price per square meter is 1,000 FCFA and the price per worker is 500 FCFA. Assuming a worker can work 20 square meters, computer the total cost of the farmer

[Answer: 1,640,000 FCFA]

Exercise 25: A dying father wishes to offer a piece of land to his son. The lands coordinates are given as; land 1: (2,1), (4,0) and (5,5) and land 2: (−2,1), (3,4) and (3,1). Assuming the son is having as objective area maximization, which of the land should hechoose

[Answer: See Appendix]

c) **Area of a Circle and Area of a Sector:** The area of a circle can be the area of a complete circle and/or the area of a section of a circle called sector. The formula for calculating the total area of a complete circle (area of a circle) and area of a section of a circle (area of sector) is given below.

Area of a Circle $= \pi r^2$ ....(1)

Area of a Sector $= \dfrac{\theta}{360} \pi r^2$ ....(2)

Example 1: A father offers three land of circular shape to his son to select one. The son wishes to create plantation agriculture and have as objective land size maximization. Which of the land will you advice the son to select?

Circle 1	Circle 2	Circle 3

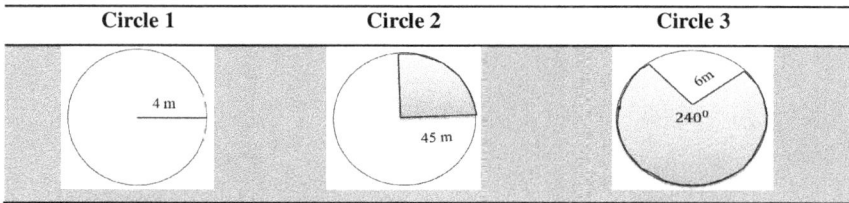

Solution

Step 1: Formula. Area of a Circle $= \pi r^2$ ....[For circle 1]

$$\text{Area of a Sector} = \frac{\theta}{360}\pi r^2 \text{ ....[For circle 2 and circle 3]}$$

Step 2: Data substitution and solving

Circle 1	Circle 2	Circle 3
$\pi = 22/7$ and r = 4	$\theta = 90$,$\pi = 22/7$ and r = 45	$\theta = 240$,$\pi = 22/7$ and r = 6
Area $= \frac{22}{7}(4^2)$	Area $= \frac{90}{360} \times \frac{22}{7} \times 45^2$	Area $= \frac{240}{360} \times \frac{22}{7} \times 6^2$
Area $= \frac{22}{7}(16)$	Area $= \frac{90}{360} \times \frac{22}{7} \times 2,025$	Area $= \frac{240}{360} \times \frac{22}{7} \times 36$
Area $= \mathbf{50m^2}$	Area $= \mathbf{1,591m^2}$	Area $= \mathbf{75m^2}$

N/B: The land maximization preference is given as; $Circle\ 2(1,591m^2) > Circle\ 3(75m^2) > Circle\ 1(50\ m^2)$. The son should select circle 2 land.

Example 2: A study shows that $5m^2$ of land yield a bag of cocoa and a bag of cocoa is worth 800,000 FCFA. Giving that the price of $1m^2$ of land is estimated to be 6,500 FCFA, which of the land below will yield the highest profit?

Land 1	Land 2	Land 3

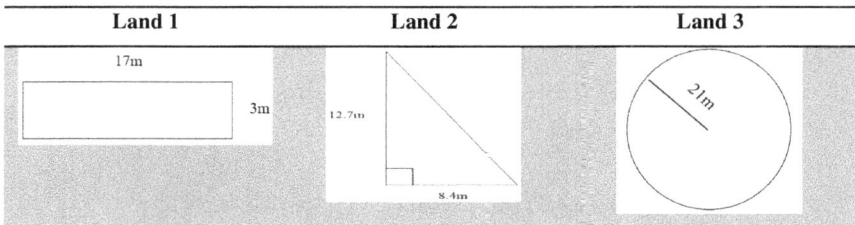

Solution

Step 1: Formula. Profit = Total Revenue(TR) − Total Cost(TC)

Step 2: Determination of total revenue (T)R and total cost (TC)

    1) Formula. Total Cost(TC) = Area × Price per unit Area

            Total Revenue(TR) = Total Ouput × Price per output

    2) Determination of area

Land 1	Land 2	Land 3
Area = Length × Width [L = 17 and W = 3]	$Area = \frac{1}{2} Base \times Height$ [Base = 8.4 and Height = 12.7]	$Area = \pi r^2$ [$\pi = 22/7$ and r = 21]
Area = 17 × 3 Area = **51m²**	$Area = \frac{1}{2}(8.4) \times 12.7$ Area = **53.34m²**	$Area = \frac{22}{7}(21^2)$ Area = **1,386m²**

    3) Data substitution and solving

	Land 1	Land 2	Land 3
Total Revenue (TR) [Total Cost(TC) = Area × Price per unit Area]	TC = 51 × 6,500 TC = **331,500 FCFA**	TC = 53.34 × 6,500 TC = **346,710 FCFA**	TC = 1,386 × 6,500 TC = **9,009,000 FCFA**
Total Cost (TC) [Total Revenue(TR) = Total Ouput × Price per output]	$TR = \frac{51}{5} \times 800,000$ TR = **8,160,000 FCFA**	$TR = \frac{53.34}{5} \times 800,000$ TR = **8,534,400 FCFA**	$TR = \frac{1,386}{5} \times 800,000$ TR = **221,760,000 FCFA**

Step 3: Data substitution and solving

Land 1	Land 2	Land 3
Profit = 8,160,000 − 331,500 Profit = **7,828,500 FCFA**	Profit = 8,534,400 − 346,710 Profit = **8,187,690 FCFA**	Profit = 221,760,000 − 9,009,000 Profit = **212,751,000 FCFA**

N/B: Land 3 (circle) yields the highest profit. This is logical because holding other factors constant, large surface area mean large output and large output mean large revenue and profit.

Example 3: A man presents two circular lands (Y and Z) to two entrepreneurs (A and B) with diameter 10 m for land Y and 8 m for land Z. Entrepreneur A is willing to purchase (3/4) of the land Y and entrepreneur B 25% of the land Z. Assuming the selling price per square meter is 5,000 FCFA and the man sells to the entrepreneurs, what is the man total revenue.?

Solution

Step 1: Formula. Total Revenue = Land Y Revenue + Land Z revenue

Step 2: Determination of lands revenue

    1) Formula. Land Revenue = Land Area × Selling Price per unit land

Where; Selling Price per unit land $= 5,000$

2) Determination of land area

a) Graphical illustration of land size purchased

Land Y $(3/4 = 75\% = 0.75)$	Land Z $(25\% = 1/4 = 0.25)$

b) Formula. Area of a Sector $= \frac{\theta}{360}\pi r^2$, where; $\pi = 22/7$

c) Data substitution and solving

Land Y (r = 5m)	Land Z (r = 4m)
Area $= \frac{270}{360} \times \frac{22}{7} \times 5^2$ .....[Work Power]	Area $= \frac{90}{360} \times \frac{22}{7} \times 4^2$ ....[Work Power]
Area $= \frac{270}{360} \times \frac{22}{7} \times 25$ ....[Solve]	Area $= \frac{90}{360} \times \frac{22}{7} \times 16$ ....…...[Solve]
Area $\approx$ **59m^2**	Area $\approx$ **13m^2**

3) Data substitution and solving

Land Y (r = 5m)	Land Z (r = 4m)
Land Y Revenue $= 59 \times 5,000$	Land Z Revenue $= 13 \times 5,000$
Land Y Revenue = **295, 000 FCFA**	Land Z Revenue = **65, 000 FCFA**

Step 3: Data substitution and solving

Total Revenue $= 295,000 + 65,000$ .... …..………[Solve equation]

Total Revenue = **360, 000 FCFA**

Test Your Understanding

Exercise 26: The size of a company is related to the land area of its construction as seen below.

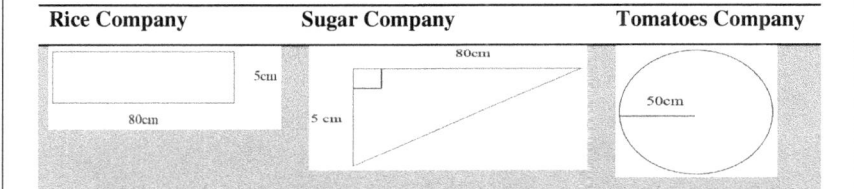

Rice Company	Sugar Company	Tomatoes Company

Which of the company above is the largest? ....[**Answer Rice Company (400cm^2)**]

Exercise 27: An entrepreneur wishes to purchase three lands with different shapes. The shaded sector in the circle below is the purchased area and is measured in square meter.

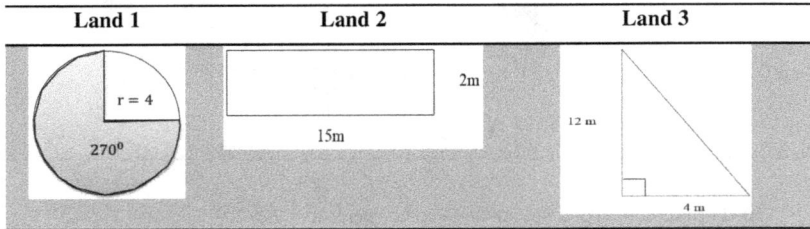

Land 1	Land 2	Land 3
r = 4 270⁰	2m 15m	12 m 4 m

Assuming a square meter of land is worth 3,500 FCFA, how much does he need to achieve his objective? ....[**Answer: 322,000 FCFA**]

Exercise 28: A business man is presented a circular land and he decides to purchase (3/4) of the land at a price of 2,000 FCFA per square meter. Determine the following assuming the radius is 7m.

1) The total area purchased..[**Answer: 115.5m^2**]
2) The total area not purchased...[**Answer:38.5m^2**]
3) Total area of the land...[**Answer:154m^2**]
4) The total amount spend by the man to purchase the land.....[**Answer: 231,000 FCFA**]

Note: An additional figure of interest is a square. A square is a figure with four sides of equal length and four right angled. To determine the area of a square, we simply square the value of one length amongst the four lengths.

Square Representation	Square Formula.
Length (2) Length (4) Length (3) Length (1)	Area of a Square $= L^2$ Where; (L) represents the value a lenght

2.2.3 Three Dimensional Shapes

A three dimensional shape is any shape having a length, width and height. Three dimensional shape have as main characteristics face, edge (line where two faces meet) and vertices (the point where the edges meet) and are also called solids. Major examples include; cube and cuboid.

Three dimensional shapes have surface area and volume and capacity. Capacity measures how much a container can hold and volume is the amount of three dimensional spaces occupied. Our analysis will be limited to the calculation of cuboid volume.

$$Volume(V) = Length(L) \times Width(W) \times Height(H) \dots\dots\dots\dots\dots\dots\dots\dots\dots\dots(1)$$

Label Cuboid

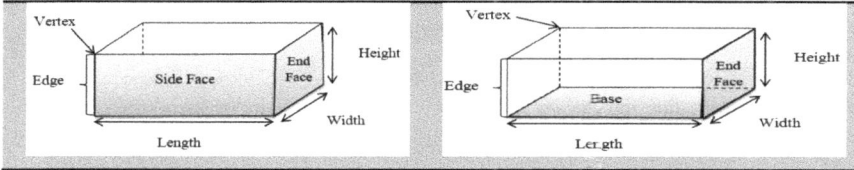

The volume of a figure is measured in cubic and the conversion table for units of volume is given below.

1,000 cubic millimeters(mm^3) = 1 square centimeter(cm^3)

1,000 square centimeter(cm^3) = 1 cubic decimeter(dm^3)

1,000 cubic decimeter(dm^3) = 1 cubic meter(m^3)

1,000 cubic meter(m^3) = 1 liter(l)

Example 1: A manager is presented two cuboids as seen below. Which of the cuboid should he buy if his objective is to maximize the volume?

Cuboid 1	Cuboid 2

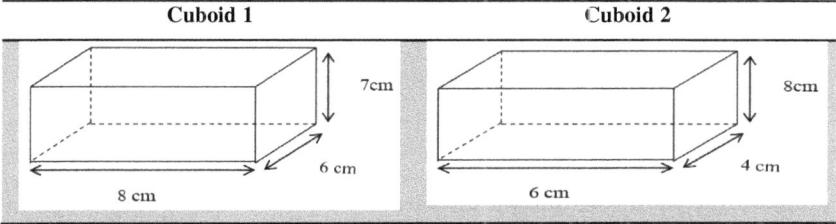

Solution

Step 1: Formula. Volume(V) = Length(L) × Width(W) × Height(H)

Step 2: Data substitution and solving

Cuboid 1	Cuboid 2
Where; L = 8, W = 6 and H = 7	Where; L = 6, W = 4 and H = 8
Volume(V) = 8 × 6 × 7 ….[Solve equation]	Volume(V) = 6 × 4 × 8 ….[Solve equation]
Volume(V) = $336cm^3$	Volume(V) = $192cm^3$

Example 2: Fine the missing values in the cuboids below.

Cuboid 1	Cuboid 2	Cuboid 3
$[V = 12,000m^3]$	$[V = 10,800m^3]$	$[V = 3,750m^3]$

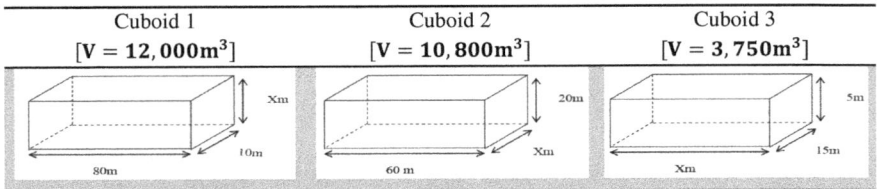

Solution

Step 1: Formula. $Volume(V) = Length(L) \times Width(W) \times Height(H)$

Step: Data substitution and solving

Cuboid 1	Cuboid 2	Cuboid 3
Where; L = 80 and W = 10	Where; L = 60 and H = 20	Where; W = 15 and H = 5
$12,000 = 80 \times 10 \times H$	$10,800 = 60 \times W \times 20$	$3,750 = L \times 15 \times 5$
$12,000 = 800H$	$10,800 = 1,200W$	$3,750 = 75L$
Height = **15m**	Width = **9m**	Length = **50m**

Example 3: A business man takes 2 litters of oil to the market. Which of the following cuboid should he buy to achieve his objective?

Cuboid 1	Cuboid 2

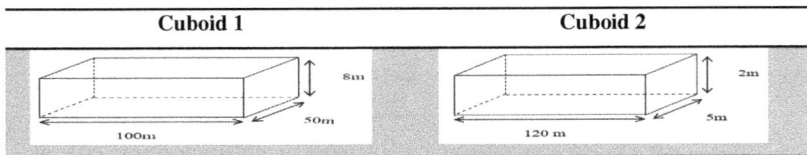

Solution

Step 1: Formula. $Volume(V) = Length(L) \times Width(W) \times Height(H)$

Step 2: Data substitution and solving

Cuboid 1	Cuboid 2
Where; L = 100, W = 50 and H = 8	Where; L = 120, W = 5 and H = 2
$Volume(V) = 100 \times 50 \times 8$ ....[Solve]	$Volume(V) = 120 \times 5 \times 2$ ....[Solve]
$Volume(V) = 40,000m^3$	$Volume(V) = 1,200m^3$
$Volume(V) = 40,000m^3 = 3$ Liters	$Volume(V) = 1,200m^3 = 1.2$ Liters
N/B: He should buy cuboid 1 since the volume is more than the volume needed to contain the 2 liters of oil.	
Test Your Understanding	
Exercise 29: Compute the volume of the following cuboids	
Cuboid 1	**Cuboid 2** **Cuboid 3**

[Answer: (Cuboid 1 = 345m³), (Cuboid 2 = 225m³) and (Cuboid 3 = 935m³)]

Exercise 30: Given that 30 cubic meters cost 10,000 FCFA, what is the cost of a cuboid with length 5 meters, width 3 meters and height 6 meters in dollar if 1 dollar ($) equals 500 FCFA?.. ....[Answer: $6]

Exercise 31: Compute the missing element of the following cuboids

Cuboid 1(Area = 400m³)	Cuboid 2(Area = 800m³)	Cuboid 3(Area = 875m³)

[Answer: Length = 20m, Height = 2m and Width = 5m]

2.3 Measurement of Weight

Weight is the mass of an object and our analysis will be limited to measurement of weight in metric. Conversion table for unit of weight/mass is given below.

1,000 Grams(g) = 1 kilogram(kg)

1,000 kilogram(kg) = 1 Tonne(t)

N/B: The measurement of weight varies with countries and the other conversion unit of weight includes; drams, ounce, pounds, stone, quarter etc.

Example 1: Convert 1,300g grams (g) into kilograms (kg)

Solution

Step 1: Formula. 1,000 Grams(g) = 1 kilogram(kg)

Step 2: Data substitution and solving

$$\left.\begin{array}{l} 1,000 \text{ Grams(g)} = 1 \text{ kilogram(kg)} \\ 1,300 \text{ Grams(g)} = X \end{array}\right\}$$ ....[Perform cross multiplication]

X(1,000 Grams) = 1,300 Grams/kilogram ....[Divide both sides by (1,000 Grams)]

X = **1.3 Kilograms** ....….. ..[That is, 1,300g = 1.3kg]

Example 2: Convert 1.2 kilograms (kg) into grams (g)

Solution

Step 1: Formula. 1,000 Grams(g) = 1 kilogram(kg)

Step 2: Data substitution and solving

$$\left.\begin{array}{l} 1{,}000 \text{ Grams(g)} = 1 \text{ kilogram(kg)} \\ X = 1.2\text{kg} \end{array}\right) \dots\dots\dots\dots\dots\text{[Perform cross multiplication]}$$

X(1 kilogram) = 1,200 Grams/kilogram ....[Divide both sides by (1 kilogram)]

X = **1,200 Grams** ....[That is, 1.2kg = 1,200g]

Example 3: Convert 100,000 grams (g) in to tonnes (t)

Solution

Step 1: Formula. 1,000 Grams(g) = 1 kilogram(kg) and 1,000 kilogram(kg) = 1 Tonnes(t)

Step 2: Data substitution and solving

First movement [1,000 g = 1 kg]	Final Movement
$\left.\begin{array}{l} 1{,}000\text{g} = 1\text{kg} \\ 100{,}000\text{g} = \text{X} \end{array}\right)$ ....[Cross multiply]	$\left.\begin{array}{l} 1{,}000\text{kg} = 1\text{t} \\ 100\text{kg} = \text{X} \end{array}\right)$ ....[Cross multiply]
X(1,000g) = 100,000g/kg ....[Solve]	X(1,000kg) = 100kg/t ....[Solve]
X = 100kg = 100,000g	X = **0.1t = 100kg = 100,000g**

Test Your Understanding

Exercise 32: Convert the following weights

1) 1,400 grams to Kilogram..[Answer: **1.4kg**]

2) 1.5 kilograms to grams..[Answer: **1,500g**]

3) 200,000 kilograms to tonnes......................................[Answer: **200 tonnes**]

4) 500 tonnes to kilograms..[Answer: **500,000kg**]

Exercise 33: To celebrate her birthday, she purchased 100kg of meat. Assuming a kg of meat is worth 1,000 FCFA and that $1 = 500 FCFA, what is the total cost of the meat in the following

1) Franc CFA...[Answer: **100,000 FCFA**]

2) American dollar...[Answer: **$200**]

CHAPTER THREE

MATHEMATICAL EQUATIONS

An equation is a statement that shows that two mathematical expressions have the same value. Major mathematical equations include; simple linear equation, polynomial equation (quadratic equation), simultaneous equation, inequality equation and absolute value equation.

3.1 Simple Linear Equation
The simple linear equation or linear equation is an equation with variables having an exponent equal to one(1).

3.1.1   Linear Equation with Single Variable
A linear equation with single variable is an equation with one independent variable and one dependent variable. The standard form of a linear equation with single variable is given as seen below.

$$y = mx + c \dots\dots\dots\dots\dots\dots\dots\dots\dots\dots\dots\dots\dots\dots\dots(1)$$

N/B: (y) represents the dependent variable, (x) represents the independent variable, (m) represents the coefficient of the independent variable and (c) represents the constant term. (m) is also known as the slope or gradient of the line and (c) represent (y) intercept (where the line cut the y-axis).

Example 1: Compute the value of y and x in the following equations
1) $3y + 1 = 22$
2) $3y - 9 = -27$
3) $-3x - 4 = 2$

<u>Solution</u>
1) **$3y + 1 = 22$**

   $3y + 1 = 22$ ....[Collect like terms]

   $3y = 21$ ....[Divide both sides of equation by (3)]

   Variable $(y) = $ **7**

2) **$3y - 9 = -27$**

   $3y - 9 = -27$ ....[Collect like terms]

   $3y = -18$ ....[Divide both sides of equation by (3)]

   Variable $(y) = $ **−6**

3) $-3x - 4 = 2$

$\quad -3x - 4 = 2$ ....…...……....[Collect like terms]

$\quad -3x = 6$ ....…..[Divide both sides of equation by (-3)]

$\quad$ Variable $(x) = -2$

Example 2: Find the value of variables x and y in the following equations

1) $2x + (-7) = -21$

2) $-3y - (+6) = -27$

Solution

1) $2x + (-7) = -21$

$\quad 2x + (-7) = -21$ ....…...…………...….………[Open bracket]

$\quad 2x - 7 = -21$ ....…......…………..…….[Collect like terms]

$\quad 2x = -14$ ....…...…......[Divide both sides by (2)]

$\quad$ Variable $(x) = -7$

2) $-3y - (+6) = -27$

$\quad -3y - (+6) = -27$ ....….......………...…..………[Open bracket]

$\quad -3y - 6 = -27$ ....…….[Collect like terms]

$\quad -3y = -21$ ....…….......…...……..[Divide both sides by (-3)]

$\quad$ Variable $(y) = 7$

Example 3: Calculate the value of x in the following equations

1) $6 + 3x = 5(x - 1) - 3(x - 2)$

2) $\frac{2x-1}{3} - \frac{3x}{4} = \frac{5}{6}$

Solution

1) $6 + 3x = 5(x - 1) - 3(x - 2)$

$\quad 6 + 3x = 5(x - 1) - 3(x - 2)$ ....….......…..…....……[Open bracket]

$\quad 6 + 3x = 5x - 5 - 3x + 6$ ....….......…..……..[Collect like terms]

$\quad$ **Variable $(x) = -5$**

2) $\frac{2x-1}{3} - \frac{3x}{4} = \frac{5}{6}$

a) Method 1

$\quad \frac{2x-1}{3} - \frac{3x}{4} = \frac{5}{6}$ ....…......[Multiply both sides of equation by (3)]

$\quad 2x - 1 - \frac{9x}{4} = \frac{15}{6}$ . ...….....…...[Multiply both sides of equation by (4)]

98

$8x - 4 - 9x = \frac{60}{6}$ ....….. [Multiply both sides of equation by (6)]

$48x - 24 - 54x = 60$ ....…....………[Collect like terms]

$-6x = 84$ ....…....…..[Divide both sides by (-6)]

Variable $(x) = -14$

b) **Method 2**

$\frac{2x-1}{3} - \frac{3x}{4} = \frac{5}{6}$ ....…..…… [Equate equation to zero (0)]

$\frac{2x-1}{3} - \frac{3x}{4} - \frac{5}{6} = 0$ ....…..……[Apply LCM]

$\frac{4(2x-1)-3(3x)-2(5)}{12} = 0$ ....…..........……..…[Open brackets]

$\frac{8x-4-9x-10}{12} = 0$ ....[Multiply both sides by (12)]

$8x - 4 - 9x - 10 = 0$ ....…….......[Collect like terms]

$-x = 14$ ....…..........…[Divide both sides by (-1)]

Variable $(x) = -14$

Test Your Understanding
Exercise 1: Solve the following equations

1) $2y - 7 = -1$..…..………..**[Answer: 3]**

2) $2x + 9 = 11$......................................…... ....**[Answer: 1]**

3) $2x + (-8) = -24$...…..……..**[Answer: -8]**

4) $-2y - (-9) = 9$...................................…..............................**[Answer: 0]**

5) $\frac{1}{2}(x + 7) = \frac{3x}{5} + 9$..…..………….**[Answer: -55]**

3.1.2 Linear Equation with Multiple Variables

A linear equation with multiple variables is an equation with one independent variable and two or more independent variables. The standard form of a linear equation with multiple variables is given as seen below.

$y = mx + by + c$ ....….............. .[Linear Equation with Two Variables]

$y = mx + by + \cdots nz + c$ ....[Linear Equation with Multiple (n) Variables]

N/B: (y) represents the dependent variable, (x, b, n) represents the independent variables, (m, y, z) represents the coefficient of the independent variables respectively and (c) represents the constant term.

The slope or gradient of a line is calculated using the coordinate pair values, two points on the line. The slope is calculated using the formula below.

$$\text{Slope(m)} = \frac{\Delta y}{\Delta x} = \frac{y_2 - y_1}{x_2 - x_1} \dots\dots\dots\dots\dots\dots\dots\dots\dots\dots\dots\dots\dots\dots\dots\dots\dots\dots\dots\dots\dots(1)$$

To calculate the (x) intercept, we equate the other variable (y) in the equation to zero and solve for variable (x) and to calculate the (y) intercept, we equate the other variable (x) in the equation to zero and solve for variable(y). To determine the table of value, we substitute the given range values into the equation and solve for the unknown.

Example 1: Find the (x) and (y) intercept of the following equations

1) $3x - y = 3$
2) $2x = 4y - 8$

Solution

1) $\mathbf{3x - y = 3}$

Variable(y) intercept .....[x = 0]	Variable(x) intercept ....[y = 0]	Illustration
$3x - y = 3$ ....[Substitute (x)] $3(0) - y = 3$ ...[Open bracket] $0 - y = 3$ ..[Collect like terms] $-y = 3$ ...[Divide sides by (-1)] Variable (y) Intercept = -3	$3x - y = 3$ ....[Substitute (y)] $3x - 0 = 3$ [Collect like terms] $3x = 3$ .....[Divide sides by (3)] Variable (x) Intercept = $\mathbf{1}$	

2) $\mathbf{2x = 4y - 8}$...$[2x = 4y - 8 \Rightarrow 2x - 4y = -8]$

Variable(y) intercept ....[x = 0]	Variable(x) intercept ....[y = 0]	Illustration
$2x - 4y = -8$ .....[Substitute (x)] $2(0) - 4y = -8$.[Open bracket] $0 - 4y = -8$.[Collect like terms] $-4y = -8$..............[Solve] Variable (y) Intercept = $\mathbf{2}$	$2x - 4y = -8$ .....[Substitute (y)] $2x - 4(0) = -8$..[Open bracket] $2x - 0 = -8$ [Collect like terms] $2x = -8$....[Divide sides by (2)] Variable (x) Intercept = $\mathbf{-4}$	

Example 2: Determine the values of the equations below

1) $y = 2x + 3$, when $x = -2, -1, 0, 1,$ and 2
2) $x = y - 1$, when $y = 0, 1, 2, 3$ and 4
3) $y = -2x$, when $x = 0, 2, 4, 6$ and 8

1) $y = 2x + 3$, when $x = -2, -1, 0, 1,$ and 2

x	-2	-1	0	1	2
y	$2(-2) + 3 = -1$	$2(-1) + 3 = 1$	$2(0) + 3 = 3$	$2(1) + 3 = 5$	$2(2) + 3 = 7$

2) $x = y - 1$, when $y = 0, 1, 2, 3$ and 4

y	0	1	2	3	4
x	$0 - 1 = -1$	$1 - 1 = 0$	$2 - 1 = 1$	$3 - 1 = 2$	$4 - 1 = 3$

3) $y = -2x$, when $x = 0, 2, 4, 6$ and 8

x	0	2	4	6	8
y	$-2(0) = 0$	$-2(2) = -4$	$-2(4) = -8$	$-2(6) = -12$	$-2(8) = -16$

Example 3: Plot the following simple linear equations

1) $y = x - 2$, when $x = 0, 1, 2, 3, 4$ and 5

2) $y = 2x$, when $x = -2, -1, 0, 1$ and 2

Solution

1) $y = x - 2$, when $x = 0, 1, 2, 3, 4$ and 5

Step 1: Determination of table of value

X	0	1	2	3	4	5
y	$0 - 2 = -2$	$1 - 2 = -1$	$2 - 2 = 0$	$3 - 2 = 1$	$4 - 2 = 2$	$5 - 2 = 3$

Step 2: Plotting

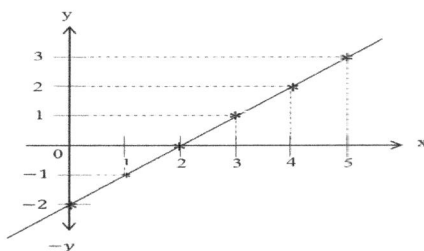

2) $y = 2x$, when $x = -2, -1, 0, 1$ and 2

Step 1: Determination of table of value

x	-2	-1	0	1	2
y	$2(-2) = -4$	$2(-1) = -2$	$2(0) = 0$	$2(1) = 2$	$2(2) = 4$

Step 2: Plotting

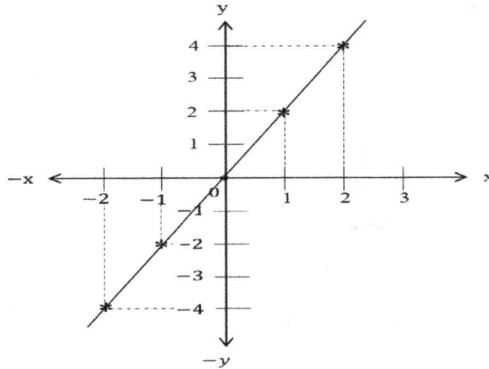

Example 4: Calculate the slope of the line below

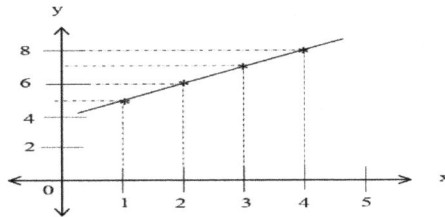

<u>Solution</u>

Step 1: Formula. $\text{Slope}(m) = \frac{y_2 - y_1}{x_2 - x_1}$

Step 2: Data substitution and solving

Left to right movement..........[Partial line]	Right to left movement..........[Complete line]
$\text{Slope}(m) = \frac{y_2 - y_1}{x_2 - x_1}$ Where; $y_2 = 7, y_1 = 6, x_2 = 3$ and $x_1 = 2$	$\text{Slope}(m) = \frac{y_2 - y_1}{x_2 - x_1}$ Where; $y_2 = 5, y_1 = 8, x_2 = 1$ and $x_1 = 4$ $\text{Slope}(m) = \frac{5-8}{1-4} = \frac{-3}{-3} = 1$

$\text{Slope}(m) = \frac{7-6}{3-2} = \frac{1}{1} = 1$	

Example 5: Using the graph below, calculate the slopes and derive the line equation

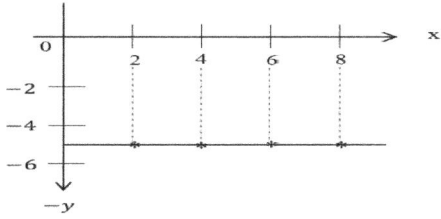

a) Calculation of slope

Step 1: Formula. $\text{Slope}(m) = \frac{y_2 - y_1}{x_2 - x_1}$

Step 2: Data substitution and solving

Complete line…[Right to left]	Partial Line…[Left to right]
$\text{Slope}(m) = \frac{y_2 - y_1}{x_2 - x_1}$, where; $y_2 = -5$, $y_1 = -5$, $x_2 = 2$ and $x_1 = 8$ $\text{Slope}(m) = \frac{-5-(-5)}{2-8} = \frac{0}{-6} = 0$	$\text{Slope}(m) = \frac{y_2 - y_1}{x_2 - x_1}$, where; $y_2 = -5$, $y_1 = -5$, $x_2 = 6$ and $x_1 = 2$ $\text{Slope}(m) = \frac{-5-(-5)}{6-4} = \frac{0}{2} = 0$

b) Derivation of line equation

Step 1: Formula. $y = mx + c$, where; $m = 0$ and $y = -5$

Step 2: Data substitution and solving

$y = mx + c$ ……………..…………...……… ………..……..[Substitute (m) and (c)]

$y = (0)x + (-5)$ ………………………………..……..………….…[Open brackets]

$y = 0 - 5$ …………………………… ………..……..……..……..[Subtract values]

103

$$y = -5$$

Example 6: At the start of the school year, a teacher had 240,000 sheets of copier paper to use. If 2,000 sheets of paper are used each day during a school year,

1) Write an equation describing the statement
2) How many sheets of paper will be left after 30 days of school?
3) How many days would it take to run out of papers

Solution

1) Write an equation describing the statement

Hint: The number of sheets of paper is the dependent variable (y) and the number of days is the independent variable(y). 420,000 sheets of paper is the constant and 2,000 sheet the slope. The term **'are used'** shows reduction and represents negative sign.

$$y = 420,000 - 2,000x$$

2) How many sheets of paper will be left after 30 days of school?

Step 1: Formula. $y = 420,000 - 2,000x$, where; $x = 30$

Step 2: Data substitution and solving

$y = 420,000 - 2,000(30)$ ....[Open bracket]

$y = 420,000 - 60,000$ ....[Solve]

$y = \mathbf{360,000}$ **Sheets of Paper**

3) How many days would it take to run out of paper

Hint: In order to be out of papers in means the dependent variable should be equal to zero$(y = 0)$.

Step 1: Formula. $y = 420,000 - 2,000x$, where; $y = 0$

Step 2: Data substitution and solving

$0 = 420,000 - 2,000x$ ....[Collect like terms]

$-2,000x = -420,000$ ....[Divide both sides by $(-2,000)$]

$x = \mathbf{210}$ **Days** ....[Justification. $210 \times 2,000 = 420,000$]

Example 7: A caterer charges 75 FCFA as fees to provide the equipment for a party and 7.50 FCFA per person for the food.

1) Write a function describing the statement
2) Sketch a graph illustrating the intercepts and slope

<u>Solution</u>

1) **Write a function describing the statement**...[Let catere cost be (y) and the number of people (x)]

$$y = 75 + 7.50x$$

2) **Sketch a graph illustrating the intercepts and slope**

Step 1: Determination of coordinate values

When.............................. [y = 0]	When ....[x = 0]
$y = 75 + 7.50x$ ....[Substitute y]	$y = 75 + 7.50x$ ....[Substitute y]
$0 = 75 + 7.50x$ ......[Collect like terms]	$y = 75 + 7.50(0)$ ....[Open bracket]
$7.50x = -75$ ...[Divide sides by (7.50)]	$y = \mathbf{75}$
$x = -10$	

Step 2: Plotting and calculation of gradient

Theoretical plot	Calculation
	$\text{Slope}(\triangle) = \frac{y_2 - y_1}{x_2 - x_1}$, where; $y_2 = 75$,
	$y_1 = 0$ $x_2 = 0$ and $x_1 = -10$
	$\text{Slope}(\Delta) = \frac{75-0}{0-(-10)}$ ....[Simplify]
	$\text{Slope}(\Delta) = \frac{75}{10}$ ....[Solve]
	$\text{Slope}(\Delta) = \mathbf{7.50}$

<u>**Test Your Understanding**</u>

Exercise 2: Calculate and plot the x and y intercept of the following equations

1) $3y - 2x = 6$..…..…..**[Answer: See Appendix]**

2) $x + 2y = -1$...…..… .**[Answer: See Appendix]**

Exercise 3: Determine the table of value of the following equations

1) $y = -2x - 1$, when $x = -2, -1, 0, 1$ and 2.......…...….......**[Answer: See Appendix]**

2) $y = -2$, when $x = -2, -1, 0, 1$ and 2...................…....….**[Answer: See Appendix]**

3) $y = x$, when $x = -2, -1, 0, 1$ and 2.........................…...**[Answer: See Appendix]**

Exercise 4: Plot the following equations

1) $y = -2x + 2$, when $x = -2, -1, 0, 1$ and 2..............…....….**[Answer: See Appendix]**

2) $y = x$, when $x = -4, -2, 0, 2$ and 4.............................**[Answer: See Appendix]**

3) $y = 4$, when $x = -2, -1, 0, 1$ and 2...............................[**Answer: See Appendix**]

Exercise 5: Without using graph, calculate the slope and intercept value of the following equations

1) $3x + 2y = 10$...[**Answer: $m = -\frac{3}{2}$ and $c = 5$**]

2) $-3y = -3x - 1$...[**Answer: $m = 1$ and $c = \frac{1}{3}$**]

3) $y = 2$...[**Answer: $m = 0$ and $c = 2$**]

Exercise 6: Write the equation of the following statements. Represent the dependent variable by (y), independent by (x) and constant by (c).

1) A worker is given 25,000 FCFA to start a job and 5,000 FCFA per day to complete the job. ....[**Answer: $y = 25,000 + 5,000x$**]

2) A student is having 1,000,000 FCFA as income and uses 2,000 FCFA per week

[**Answer: $y = 1,000,000 - 2,000x$**]

3) A business man with 200 litters of oil sells 10 litters daily. [**Answer: $y = 200 - 10x$**]

3.2 Quadratic Equation

A quadratic equation is a polynomial equation of second degree or power two. The general form of a quadratic equation is given below.

$ax^2 + bx + c = 0$ ....(1)

N/B: The elements (a) and (b) are coefficient of the variables (x^2) and (x) respectively and (c) is a constant. The value of (b) and (c) can be equal to zero but that of (a) should not be equal to zero. Therefore, a quadratic equation can take the following forms.

$ax^2 + bx = 0$ ....[When $c = 0$]

$ax^2 + c = 0$ ....[When $b = 0$]

$ax^2 = 0$ ....[When $b = 0$ and $c = 0$]

The solution of a quadratic equation is also called the roots of a quadratic equation and can be determined using the factorization method, formula method and graphical method.

3.2.1 Factorization method

The calculation of the roots of quadratic equation using the factorization method depends on the value of the first term (a).

1. Determination of Factorization Expression

a) When (a) is equal to 1

Here, we identify two numbers (p and q) such that their sum equals the coefficient of the second term $(p + q = b)$ and their product equals the third term $(p \times q = c)$. The factorization expression is given as seen below,

$$X^2 + bX + C = (X + p)(X + q) \text{..}(1)$$

N/B: The value (p and q) can be positive $(+p \text{ and } + q)$ and/or negative $[(-p \text{ and } - q), (+p \text{ and } - q) \text{ and } (-p \text{ and } + q)]$.

Example 1: Express the following quadratic equations in their factorization forms

1) $X^2 + 5X - 24$, with factors -3 and 8
2) $t^2 + 6t = 16$, with factors 2 and -8

Solution

1) $X^2 + 5X - 24$, with factors -3 and 8

Step 1: Formula. $X^2 + bX + C = (X + p)(X + q)$

Where; $p = -3$ and $q = 8$....[Position does no matter]

Step 2: Data substitution and solving

$X^2 + bX + C = (X + p)(X + q)$ ....[Change (p) to (-3) and (q) to (8)]

$X^2 + bX + C = (X + \{-3\})(X + 8)$ ....[Open inner bracket]

$X^2 + bX + C = (X - 3)(X + 8)$ ....[The expression can also be $(X + 8)(X - 3)$]

Step 3: Justification...[$(X - 3)(X + 8) = (X^2 + 5X - 24)$]

$(X - 3)(X + 8)$ ....[Open brackets]

$(X - 3)(X + 8) = X^2 + 8X - 3X - 24$ ....[Collect like terms]

$(X - 3)(X + 8) = X^2 + 5X - 24$

2) $t^2 + 6t = 16$, with factors 2 and -8

Step 1: Formula. $t^2 + bt + C = (t + p)(t + q)$, where; $p = 2$ and $q = -8$

Step 2: Data substitution and solving

$t^2 + bt + C = (t + p)(t + q)$ ....[Change (p) to (2) and (q) to (-8)]

$t^2 + bt + C = (t + 2)(t + \{-8\})$ ....[Open inner bracket]

$t^2 + bt + C = (t + 2)(t - 8)$ ....[The expression can also be $(t - 8)(t + 2)$]

b) When (a) is not equal to 1

Here, we identify two numbers (p and q) such that their sum equals the coefficient of the second term $(p + q = b)$ and their product equals the third term and first term $(p \times q = a \times c)$. To properly derive these numbers, the following procedures are recommended.

1) Multiply the first and third term $(a \times c)$. $(aX^2 + bX + c) = (X^2 + bX + a * c)$.
2) Identify two numbers (p and q) whose sum equals the coefficient of the second term $(p + q = b)$ and their product equals to the third term $(p \times q = c * a)$.
3) Determine the working factorization expression using the formula $[aX^2 + pX + qX + c]$.

N/B: The concept of highest common factor (HCF) and factorization by grouping are going to be applied in this section. The following factorization by group procedures is worth knowing.

$ax + ay + bx + by$ ....[Group like terms, group (a) elements and (b) elements]
$(ax + ay) + (bx + by)$ ....[Factorize common elements, factorize (a) and (b)]
$a(x + y) + b(x + y)$ ....[Final expression]
$a(x + y) + b(x + y) = (x + y)(a + b)$

Example 1: Find the highest common factor (HCF) of the expression $(7X^2 + 1X)(28X + 4)$

<u>**Solution**</u>

First Term...................$(7X^2 + 1X)$		Second term...........................$(28X + 4)$	
7	1	28	4
7 | 7	1 | 1	2 | 28	2 | 4
1	1	2 | 14	2 | 2
		7 | 7	1
$[7 \times 1 = 7^1 \times 1^1]$	$[1 = 1^1]$	1	$[2 \times 2 = 2^2]$
		$[2 \times 2 \times 7 = 2^2 \times 7^1]$	
HCF of $(7X^2 + 1X) = 1^1 = 1$. This means we are to factorize 1 from this term		HCF of $(28X + 4) = 2^2 = 4$. This means we are to factorize 4 from this term	

Example 2: Factorize the following expressions

1) $(10X - 8)$, with highest common factor (HCF) 2
2) $(-35X + 28)$, with highest common factor (HCF) 7

Solution

$(10X - 8X)$	$(-35X + 28)$
$(10X^2 - 8X)$ .....[Put $(2X)$ out of the bracket]	$(-35X + 28)$ ...[Put (7) out of the bracket]
$2X(10X^2 - 8X)$ [Divide bracket terms by $(2X)$]	$7(-35X + 28)$ [Divide bracket terms by (7)]
$2X\left(\frac{10X^2}{2X} - \frac{8X}{2X}\right)$ ....[Re-express equation]	$7\left(\frac{-35X}{7} + \frac{28}{7}\right)$ ....[Solve fractions]
$2X\left(\frac{10}{2} * X^{2-1} - \frac{8}{2} * X^{1-1}\right)$ ....[Solve fractions]	$-35X + 28 = 7(-5X + 4)$
$2X(5 * X - 4 * 1) = 2X(5X - 4)$	
N/B: X is added to 2 since both terms in the bracket do have X	**N/B:** X is not added to 7 since both terms in the bracket do not have X

Example 3: Express the following quadratic equations in their factorization forms

1) $6X^2 + 17X + 5$, with factors 2 and 15
2) $10t^2 - 43t + 28$, with factors -8 and -35

Solution

1) **$6X^2 + 17X + 5$, with factors 2 and 15.**[Remember, $(6X^2 + 17X + 5 = X^2 + 17X + 30)$]

Step 1: Formula. $aX^2 + bX + c = aX^2 + pX + qX + c$, where; $p = 2, q = 15, a = 6$ and $c = 5$

Step 2: Data substitution and solving

$aX^2 + bX + c = 6X^2 + 2X + 15X + 5$ ....[Perform common factor term grouping]

$aX^2 + bX + c = (6X^2 + 2X) + (15X + 5)$ ....[Factor HCF from each group]

$aX^2 + bX + c = 2X(3X + 1) + 5(3X + 1)$.[Solve. $\{a(x + y) + b(x + y) = (x + y)(a + b)\}$]

$aX^2 + bX + c = (3X + 1)(2X + 5)$ ....[$(6X^2 + 17X + 5) = (3X + 1)(2X + 5)$]

2) **$10t^2 - 43t + 28$, with factors -8 and -35**.....[$(10t^2 - 43t + 28 = t^2 - 43t + 280)$]

Step 1: Formula. $aX^2 + bX + c = aX^2 + pX + qX + c$

Where; $p = -8, q = -35, a = 10$ and $c = 28$

Step 2: Data substitution and solving

$aX^2 + bX + c = 10X^2 + (-8)X + (-35)X + 28$ ..[Perform common factor term grouping]

$aX^2 + bX + c = (10X^2 - 8X) + (-35X + 28)$ ....[Factor HCF from each group]

$aX^2 + bX + c = 2X(5X - 4) + 7(-5X + 4)$ ....[Make bracket values the same]

$aX^2 + bX + c = 2X(5X - 4) + (-7)(5X - 4)$ ....[Solve]

$aX^2 + bX + c = (5X - 4)(2X - 7)$ ....[$(10t^2 - 43t + 28) = (5X - 4)(2X - 7)$]

Test Your Understanding

Exercise 7: Give the factorization expression of the following quadratic equations

1) $x^2 + 2x - 8 = 0$, with factors -4 and 2...................**[Answer: $(x - 4)(x + 2) = 0$]**

2) $t^2 - 7t + 6 = 0$, with factors -1 and -6................**[Answer: $(t - 1)(t - 6) = 0$]**

3) $x^2 + 3x - 4 = 0$...**[Answer: $(x + 4)(x - 1) = 0$]**

4) $x^2 - 9 = 0$ ....**[Answer: $(x + 3)(x - 3) = 0$]**

Exercise 8: Give the factorization expression of the following quadratic equations

1) $2x^2 + 4x - 6 = 0$, with factors -2 and 6..............**[Answer: $(2x + 6)(x - 1) = 0$]**

2) $2x^2 + 7x + 3 = 0$...**[Answer: $(2x + 1)(x + 3) = 0$]**

3) $6x^2 + 5x - 6 = 0$.......................................**[Answer: $(2x + 3)(3x - 2) = 0$]**

4) $-2x^2 + 6x = 0$...**[Answer: $-2x(x - 3) = 0$]**

---

2. The Zero Product Rule

According to the zero product rule, if (a) and (b) are non-zero factors, therefore, (a) equal zero and (b) equal zero.

$$(a)(b) = 0 \rightarrow \begin{cases} (a) = 0 \\ (b) = 0 \end{cases}$$ ....(1)

Example 1: Determine the roots of the equation $(X - 7)(X + 2) = 0$

Solution

Step: Formula. $(a)(b) = 0 \rightarrow \begin{cases} (a) = 0 \\ (b) = 0 \end{cases}$, where; (a) = $(X - 7)$ and (b) = $(X + 2)$

Step 2: Data substitution and solving.......[Equate each to zero and solve for the variable (X)]

First Root.....................[X − 7]	Second Root.....................[X + 2]
X − 7 ....[Equate to zero (0)]	X + 2 ....[Equate to zero (0)]
X − 7 = 0 ....[Take (7) right]	X + 2 = 0 ....[Take (2) right]
Variable (X) = **7**	Variable (X) = **−2**
N/B: The two roots in ascending order are $(X_1 = -2)$ and $(X_2 = 7)$.	

Step 3: Justification.

Hint: The roots (-2 and 7) means the quadratic equation is zero when the variable is (-2) or (7). The given equation [$(X - 7)(X + 2) = X^2 - 5X - 14$]

When................................... $[X = -2]$	When....................................... $[X = 7]$
$X^2 - 5X - 14 = 0$ ....[Change (X) to (-2)]	$X^2 - 5X - 14 = 0$ ....[Change (X) to (7)]
$-2^2 - 5(-2) - 14 = 0$ ...[Work power]	$7^2 - 5(7) - 14 = 0$ ....[Work power]
$4 - 5(-2) - 14 = 0$ ....[Work bracket]	$49 - 5(7) - 14 = 0$ ....[Work bracket]
$4 + 10 - 14 = 0$ ......[Simplify left side]	$49 - 35 - 14 = 0$ ....[Simplify left side]
$0 = 0$	$0 = 0$

Example 2: Calculate the roots of the equation $(5t - 1)(2t + 5) = 0$

<u>Solution</u>

Step: Formula. $(a)(b) = 0 \rightarrow \begin{cases} (a) = 0 \\ (b) = 0 \end{cases}$, where; $(a) = (5t - 1)$ and $(b) = (2t + 5)$

Step 2: Data substitution and solving

First Root..............................[5t − 1]	Second Root..............................[2t + 5]
$5t - 1$ ....[Equate to zero (0)]	$2t + 5$ ....[Equate to zero (0)]
$5t - 1 = 0$ ....[Take (1) right]	$2t + 5 = 0$ ....[Take (5) right]
$5t = 1$ ....[Divide both sides by (5)]	$2t = -5$ ....[Divide both sides by (2)]
Variable (t) $= \frac{1}{5}$	Variable (t) $= -\frac{5}{2}$
N/B: The roots are $\left(t_1 = -\frac{5}{2}\right)$ and $\left(q_2 = \frac{1}{5}\right)$	

Example 3: Compute the roots of the equation $q(q + 7) = 0$

<u>Solution</u>

Step: Formula. $(a)(b) = 0 \rightarrow \begin{cases} (a) = 0 \\ (b) = 0 \end{cases}$, where; $(a) = q$ and $(b) = (q + 7)$

Step 2: Data substitution and solving

First Root..............................[q]	Second Root... ....[q + 7]
q ....[Equate to zero (0)]	q + 7 ....[Equate to zero (0)]
Variable (q) $= \mathbf{0}$	q + 7 = 0 ... .[Take (7) right]
	Variable (q) $= \mathbf{-7}$
N/B: The roots are $(q_1 = -7)$ and $(q_2 = 0)$	

3.2.2 Formula Method

The formula method treats the situation of $(a = 1)$ and $(a \neq 1)$ in the same manner. The general formula use to determine the roots of a quadratic equation using variable (X) is given below.

$$X = \frac{-b \pm \sqrt{b^2 - 4ac}}{2a}$$ ....(1)

N/B: When $(b \neq 0)$, the equation is called complete quadratic equation and when $(b = 0)$, it is called pure or incomplete quadratic equation. Remember that $(a \neq 0)$.

1. Discriminant

The value of discriminant (Δ) determines the nature of root to be obtained. A quadratic equation roots is generally of three types; real and unequal roots, real and equal root and complex root. Discriminant is calculated using the formula below.

$$\text{Discriminant}(\Delta) = b^2 - 4ac$$ ....(2)

The value of discriminant is not the same for all quadratic equations. The three possible results or roots of quadratic equations are given below.

Nature of Discriminant	Nature of Root
$\Delta > 0$ ..[Discriminant is greater than zero]	$X_1 \neq X_2$ ....[Real and distinct roots]
$\Delta = 0$ ....[Discriminant is equal to zero]	$X_1 = X_2$ ....[Real and equal roots]
$\Delta < 0$ ....[Discriminant is less than zero]	$X = a \pm bi$ ....[Complex root]

112

N/B: When discriminant is greater than zero, the results can be real, rational and unequal (perfect root) or real, irrational and unequal (imperfect roots). When discriminant is equal to zero, the result is real, rational and equal. When discriminant is less than zero, the result is non-real and we apply the concept of complex number to determine the possible roots.

Example 1: Determine the discriminant of the equation; $X^2 + 6X + 1 = 0$

Solution

Step 1: Formula. Discriminant(Δ) $= b^2 - 4ac$, where; $a = 1$, $b = 6$ and $c = 1$

Step 2: Data substitution and solving

Discriminant(Δ) $= 6^2 - 4(1)(1)$ ....[Work exponent]

Discriminant(Δ) $= 36 - 4(1)(1)$ ....[Open bracket]

Discriminant(Δ) $= 36 - 4$ ....[Solve equation]

Discriminant(Δ) $= \mathbf{32}$ ....[($\Delta > 0$), indication real and unequal roots]

Note: The value of the discriminant (32) is not a square number. This is because its square root value is not a whole number. Therefore, we will expect a real, irrational and unequal roots or result.

Example 2: Find the discriminant of the equation; $X^2 + 2X - 3 = 0$

Solution

Step 1: Formula. Discriminant(Δ) $= b^2 - 4ac$, where; $a = 1$, $b = 2$ and $c = -3$

Step 2: Data substitution and solving

Discriminant(Δ) $= 2^2 - 4(1)(-3)$ ....[Work exponent]

Discriminant(Δ) $= 4 - 4(1)(-3)$ ....[Open bracket]

Discriminant(Δ) $= 4 + 12$ ....[Solve equation]

Discriminant(Δ) $= \mathbf{16}$ ....[($\Delta > 0$), indication real and unequal roots]

Note: The value of the discriminant (16) is a square number. This is because its square root gives a whole number. Therefore, we will expect a real, rational and unequal roots or result.

Example 3: Find the discriminant of the equation $X^2 - 8X + 16 = 0$

Solution

Step 1: Formula. Discriminant(Δ) $= b^2 - 4ac$, where; $a = 1$, $b = -8$ and $c = 16$

Step 2: Data substitution and solving

Discriminant(Δ) $= -8^2 - 4(1)(16)$ ....[Work exponent]

Discriminant(Δ) $= 64 - 4(1)(16)$ ....[Open bracket]

Discriminant(Δ) = 64 − 64 ....[Solve equation]

Discriminant(Δ) = **0** ....[(Δ= 0), indication real and equal roots]

Example 4: Compute the discriminant of the equation $X^2 - 4X + 11 = 0$

Solution

Step 1: Formula. Discriminant(Δ) = $b^2 - 4ac$, where; a = 1, b = −4 and c = 11

Step 2: Data substitution and solving

Discriminant(Δ) = $-4^2 - 4(1)(11)$ ....[Work exponent]

Discriminant(Δ) = $16 - 4(1)(11)$ ....[Open bracket]

Discriminant(Δ) = $16 - 44$ ....[Solve equation]

Discriminant(Δ) = **−28** ....[(Δ< 0), indication complex roots]

Test Your Understanding

Exercise 9: Calculate the discriminant of the following quadratic equations

1) $2X^2 - 3X - 5 = 0$.......[**Answer: Δ= 49, indicating real and distinct roots (Δ> 0)**]

2) $-2t^2 - t + 4 = 0$ ....[**Answer: Δ= 33, indicating real and distinct roots (Δ> 0)**]

3) $2Y^2 + 4Y + 3 = 0$...............[**Answer: Δ= −8, indication complex root (Δ< 0)**]

---

2. **Sum and Product of Roots**

From the standard quadratic equation($aX^2 + bX + c = 0$), we can derive the sum of root and product of root of quadratic equation. This helps in justifying the calculated roots.

a) When using quadratic equation values

Sum of Roots = $-\dfrac{b}{a}$ ....(1)

Product of roots = $\dfrac{c}{a}$ ....(2)

b) When using quadratic roots value

Sum of Roots = $\alpha + \beta$ ....(1)

Product of roots = $\alpha\beta$ ....(2)

N/B: (α) represent the smaller root value and (β) represent the larger root value. The sum of roots and product of roots equation can be included in the standard quadratic equation as seen below.

$aX^2 + bX + c = 0$ ....[Divide both sides of equation by (a)]

$X^2 + \dfrac{b}{a}X + \dfrac{c}{a}$..[Substitute fractions]

$$X^2 + (\alpha + \beta)X + \alpha\beta = X^2 + (\text{Sum of Roots})X + (\text{Product of Roots}) \ldots\ldots\ldots\ldots(3)$$

Note: Comparing the quadratic equation and root values, we can obtain the following relationships; $\left(\text{Sum of Roots} = -\dfrac{b}{a} = \alpha + \beta\right)$ and $\left(\text{Product of Roots} = \dfrac{c}{a} = \alpha\beta\right)$.

Example 1: Determine the sum of roots of the equation; $4t^2 - 9t + 2 = 0$

Solution

Step 1: Formula. Sum of Roots$(\alpha + \beta) = -\dfrac{b}{a}$, where $b = -9$ and $a = 4$

Step 2: Data substitution and solving

Sum of Roots$(\alpha + \beta) = -\dfrac{b}{a}$ ....[Substitute data]

Sum of Roots$(\alpha + \beta) = -\dfrac{-9}{4} = -\left(-\dfrac{9}{4}\right)$ ....[Open bracket]

Sum of Roots$(\alpha + \beta) = \dfrac{9}{4}$

Example 2: Calculate the product of roots of the equation; $-2Y^2 - Y + 4 = 0$

Solution

Step 1: Formula. Product of roots$(\alpha\beta) = \dfrac{c}{a}$, where; $a = -2$ and $c = 4$

Step 2: Data substitution and solving

Product of roots$(\alpha\beta) = \dfrac{4}{-2}$ ....[Solve equation]

Product of roots$(\alpha\beta) = -2$

Example 3: Compute the sum of roots and product of roots of the equation; $X^2 + 2X - 3 = 0$

Solution

Sum of Roots	Product of Roots
Step 1: Formula. Sum of Roots$(\alpha + \beta) = -\dfrac{b}{a}$	Step 1: Formula. Product of roots$(\alpha\beta) = \dfrac{c}{a}$
Where; $b = 2$ and $a = 1$	Where; $a = 1$ and $c = -3$
Step 2: Data substitution and solving	Step 2: Data substitution and solving
Sum of Roots$(\alpha + \beta) = -\dfrac{b}{a}$	Product of roots$(\alpha\beta) = \dfrac{c}{a}$
Sum of Roots$(\alpha + \beta) = -\dfrac{2}{1}$	Product of roots$(\alpha\beta) = \dfrac{-3}{1}$
Sum of Roots$(\alpha + \beta) = -2$	Product of roots$(\alpha\beta) = -3$

3. Roots of a Quadratic Equation

After seeing how to calculate the discriminant and sum and product of roots, we are going to calculate the roots from a quadratic equation.

Example 1: Compute the roots of the quadratic equations $X^2 - 3X + 2 = 0$ and $-3t^2 + 5t + 12 = 0$.

Solution

1) Determine the roots of; $X^2 - 3X + 2 = 0$

Step 1: Formula. $X = \frac{-b \pm \sqrt{b^2 - 4ac}}{2a} = \frac{-b \pm \sqrt{\Delta}}{2a}$, where; $a = 1$, $b = -3$ and $c = 2$

Step 2: Determination of determinant

   1) Formula. Discriminant$(\Delta) = b^2 - 4ac$, where; $a = 1$, $b = -3$ and $c = 2$

   2) Data substitution and solving

      Discriminant$(\Delta) = (-3)^2 - 4(1)(2)$ ....[Work exponent]

      Discriminant$(\Delta) = 9 - 4(1)(2)$............................[Open brackets and solve]

      Discriminant$(\Delta) = 1$ ....[$(\Delta > 0)$, Real and distinct roots$(X_1 \neq X_2)$]

Step 3: Data substitution and solving

     $X = \frac{3 \pm \sqrt{1}}{2}$ ....[Work square root]

     $X = \frac{3 \pm 1}{2}$ ....[Determine of roots]

      $X_1 = \frac{3-1}{2} = 1$ ....[First root of quadratic equation]

      $X_2 = \frac{3+1}{2} = 2$ ....[Second root of quadratic equation]

2) Determine the roots of $-3t^2 + 5t + 12 = 0$

Step 1: Formula. $t = \frac{-b \pm \sqrt{b^2 - 4ac}}{2a}$, where; $a = -3$, $b = 5$ and $c = 12$

Step 2: Data substitution and solving

     $t = \frac{-(5) \pm \sqrt{(5)^2 - 4(-3)(12)}}{2(-3)}$ ....[Work exponent]

$$t = \frac{-(5)\pm\sqrt{25-4(-3)(12)}}{2(-3)} \quad \dots\dots\dots\dots\dots\dots\dots\dots\dots\dots\dots\dots \text{[Open brackets]}$$

$$t = \frac{-5\pm\sqrt{25+144}}{-6} = X = \frac{-5\pm\sqrt{169}}{-6} \quad \dots\dots\dots\dots\dots\dots\dots\dots \text{[Work square root]}$$

$$t = \frac{-5\pm13}{-6} \quad \dots\dots\dots\dots\dots\dots\dots\dots\dots\dots\dots\dots\dots \text{[Determine of roots]}$$

$$t_1 = \frac{-5+13}{-6} = -\frac{4}{3} \quad \dots\dots\dots\dots\dots\dots\dots\dots \text{[First root of quadratic equation]}$$

$$t_2 = \frac{-5-13}{-6} = 3 \quad \dots\dots\dots\dots\dots\dots\dots\dots \text{[Second root of quadratic equation]}$$

Example 2: Given the quadratic equation; $Q^2 = 2Q + 2$, answer the following questions

1) Determine the roots of the quadratic equation

2) Show that the sum and product of roots is same with the roots above

Solution

1) Determine the roots of the quadratic equation

Hint: We are required to transform the equation $[Q^2 = 2Q + 2]$ to the standard quadratic equation form of $[Q^2 - 2Q - 2 = 0]$ before solving the roots

Step 1: Formula. $Q = \frac{-b\pm\sqrt{b^2-4ac}}{2a}$, where; $a = 1$, $b = -2$ and $c = -2$

Step 2: Data substitution and solving

$$Q = \frac{-(-2)\pm\sqrt{(-2)^2-4(1)(-2)}}{2(1)} \quad \dots\dots\dots\dots\dots\dots\dots\dots \text{[Work exponent]}$$

$$Q = \frac{-(-2)\pm\sqrt{4-4(1)(-2)}}{2(1)} \quad \dots\dots\dots\dots\dots\dots\dots\dots \text{[Open brackets]}$$

$$Q = \frac{2\pm\sqrt{4+8}}{2} = Q = \frac{2\pm\sqrt{12}}{2} \quad \dots\dots\dots\dots\dots\dots\dots\dots \text{[Re-expressed equation]}$$

$$Q = \frac{2\pm\sqrt{12}}{2} = \frac{2\pm\sqrt{4\times3}}{2} = \frac{2\pm\sqrt{4}\times\sqrt{3}}{2} \quad \dots\dots\dots\dots\dots\dots \text{[Work square root]}$$

$$Q = \frac{2\pm 2\times\sqrt{3}}{2} = \frac{2\pm 2\sqrt{3}}{2} \quad \dots\dots\dots\dots\dots\dots\dots\dots \text{[Determine of roots]}$$

$$Q_1 = \frac{2-2\sqrt{3}}{2} = \frac{2}{2} - \frac{2\sqrt{3}}{2} = 1 - \sqrt{3} \quad \dots\dots\dots\dots\dots \text{[First root of quadratic equation]}$$

$$Q_2 = \frac{2+2\sqrt{3}}{2} = \frac{2}{2} + \frac{2\sqrt{3}}{2} = 1 + \sqrt{3} \quad \dots\dots\dots\dots\dots \text{[Second root of quadratic equation]}$$

2) Show that the sum and product of roots is same with the roots above

	Using roots values	Using equation values
Sum of Roots	Step 1: Formula. Sum of Roots = $\alpha + \beta$ Where; $\alpha = 1 - \sqrt{3}$ and $\beta = 1 + \sqrt{3}$ Step 2: Data substitution and solving Sum of Roots = $1 - \sqrt{3} + 1 + \sqrt{3}$	Step 1: Formula. Sum of Roots = $-\frac{b}{a}$ Where; $b = -2$ and $a = 1$

		Step 2: Data substitution and solving
	Sum of Roots $= 2$	Sum of Roots $= -\left(-\frac{2}{1}\right)$
		Sum of Roots $= \mathbf{2}$
Product of Roots	Step 1: Formula. Product of roots $= \alpha\beta$ Where; $\alpha = 1 - \sqrt{3}$ and $\beta = 1 + \sqrt{3}$ Step 2: Data substitution and solving Product of roots $= \left(1 - \sqrt{3}\right)\left(1 + \sqrt{3}\right)$ Product of roots $= 1 + \sqrt{3} - \sqrt{3} - \left(\sqrt{3}\right)^2$ Product of roots $= 1 - \left(\sqrt{3}\right)^2$ $= 1 - \left(3^{\frac{1}{2}}\right)^2$ Product of roots $= 1 - 3 = \mathbf{-2}$	Step 1: Formula. Product of roots $= \frac{c}{a}$ Where; $a = 1$ and $c = -2$ Step 2: Data substitution and solving Product of roots $= \frac{-2}{1}$ Product of roots $= \mathbf{-2}$

Example 3: Calculate the roots of the quadratic equations $X^2 - 12X + 36 = 0$ and $Q^2 - 8Q + 16 = 0$.

<u>**Solution**</u>

1) **Determine roots of $X^2 - 12X + 36 = 0$**

Step 1: Formula. $X = \frac{-b \pm \sqrt{b^2 - 4ac}}{2a}$, where; $a = 1$, $b = -12$ and $c = 36$

Step 2: Data substitution and solving

$X = \frac{-(-12) \pm \sqrt{(-12)^2 - 4(1)(36)}}{2(1)}$ ....[Work exponent]

$X = \frac{-(-12) \pm \sqrt{144 - 4(1)(36)}}{2(1)}$ ....[Open brackets]

$X = \frac{12 \pm \sqrt{144 - 144}}{2} = \frac{12 \pm \sqrt{0}}{2}$ ....[Work square root]

$X = \frac{12 \pm 0}{2}$ ....[Determine of roots]

$X_1 = \frac{12 - 0}{2} = \mathbf{6}$ ....[First root of quadratic equation]

$X_2 = \frac{12 + 0}{2} = \mathbf{6}$ ....[Second root of quadratic equation]

2) **Determine roots of $Q^2 - 8Q + 16 = 0$**

Step 1: Formula. $Q = \frac{-b \pm \sqrt{b^2 - 4ac}}{2a}$, where; $a = 1$, $b = -8$ and $c = 16$

Step 2: Data substitution and solving

$$Q = \frac{-(-8)\pm\sqrt{(-3)^2-4(1)(16)}}{2(1)} \quad \text{...................................[Work exponent]}$$

$$Q = \frac{-(-8)\pm\sqrt{64-4(1)(16)}}{2(1)} \quad \text{.................................[Open brackets]}$$

$$Q = \frac{8\pm\sqrt{64-64}}{2} = \frac{8\pm\sqrt{0}}{2} \quad \text{.............................[Work square root]}$$

$$Q = \frac{8\pm0}{2} \quad \text{..[Determine of roots]}$$

$$Q_1 = \frac{8-0}{2} = 4 \quad \text{...[First root of quadratic equation]}$$

$$Q_2 = \frac{8+0}{2} = 4 \quad \text{...[Second root of quadratic equation]}$$

Example 4: Compute the roots of the quadratic equations $X^2 + 2X + 5 = 0$ and $4t^2 - t^2 = 5$

<u>**Solution**</u>

1) **Determine roots of; $X^2 + 2X + 5 = 0$**

Step 1: Formula. $X = \frac{-b\pm\sqrt{b^2-4ac}}{2a}$, where; $a = 1$, $b = 2$ and $c = 5$

Step 2: Data substitution and solving

$$X = \frac{-(2)\pm\sqrt{(2)^2-4(1)(5)}}{2(1)} \quad \text{.......................................[Work exponent]}$$

$$X = \frac{-(2)\pm\sqrt{4-4(1)(5)}}{2(1)} \quad \text{.......................................[Open brackets]}$$

$$X = \frac{-2\pm\sqrt{4-20}}{2} = \frac{-2\pm\sqrt{-16}}{2} \quad \text{...........................[Re-expressed equation]}$$

$$X = \frac{-2\pm\sqrt{-16}}{2} = \frac{-2\pm\sqrt{16\times-1}}{2} = \frac{-2\pm\sqrt{16}\times\sqrt{-1}}{2} \quad \text{............[Work roots, remember, } \sqrt{-1} = i]$$

$$X = \frac{-2\pm4\times i}{2} = \frac{-2\pm4i}{2} \quad \text{..................................[Determine of roots]}$$

$$X_1 = \frac{-2-4i}{2} = \frac{-2}{2} - \frac{4i}{2} = -1 - 2i \quad \text{......................[First root of quadratic equation]}$$

$$X_2 = \frac{-2+4i}{2} = \frac{-2}{2} + \frac{4i}{2} = -1 + 2i \quad \text{....................[Second root of quadratic equation]}$$

2) **Determine roots of; $t^2 - 4t = -9$**

Hint: We are required to transform the equation $[t^2 - 4t = -9]$ to the standard quadratic equation form of $[t^2 - 4t + 9 = 0]$ before solving the roots.

Step 1: Formula. $t = \frac{-b\pm\sqrt{b^2-4ac}}{2a}$, where; $a = 1$, $b = -4$ and $c = 9$

Step 2: Data substitution and solving

$$t = \frac{-(-4)\pm\sqrt{(-4)^2-4(1)(9)}}{2(1)} \quad \text{..................................[Work exponent]}$$

$$t = \frac{-(-4)\pm\sqrt{16-4(1)(9)}}{2(1)} \quad \text{....................................[Open brackets]}$$

119

$$t = \frac{4\pm\sqrt{16-36}}{2} = \frac{4\pm\sqrt{-20}}{2} \quad \dots\dots\dots\dots\dots\dots\dots\dots\dots\dots\dots\text{[Re-expressed equation]}$$

$$t = \frac{4\pm\sqrt{-20}}{2} = \frac{4\pm\sqrt{4\times-5}}{2} = \frac{4\pm\sqrt{4}\times\sqrt{-5}}{2} \quad \dots\dots\dots[\ (\sqrt{-5}) \text{ is maintained as 5 is not a square}$$

number]

$$t = \frac{4\pm2\times\sqrt{-5}}{2} = \frac{4\pm2\sqrt{-5}}{2} \quad \dots\dots\dots\dots\dots\dots\dots\dots\dots\dots\text{[Determine of roots]}$$

$$t_1 = \frac{4-2\sqrt{-5}}{2} = \frac{4}{2} - \frac{2\sqrt{-5}}{2} = 2 - \sqrt{-5} \quad \dots\dots\dots\dots\dots\text{[First root of quadratic equation]}$$

$$t_2 = \frac{4+2\sqrt{-5}}{2} = \frac{4}{2} + \frac{2\sqrt{-5}}{2} = 2 + \sqrt{-5} \quad \dots\dots\dots\dots\dots\text{[Second root of quadratic equation]}$$

Test Your Understanding

Exercise 11: Determine the roots of the following quadratic equations

1) $5X^2 + 27X + 10 = 0$......................................[Answer: $X_1 = -\frac{14}{5}$ and $X_2 = \frac{9}{5}$]

2) $24t^2 - 46t = 18$...[Answer: $t_1 = -\frac{1}{3}$ and $t_2 = \frac{9}{4}$]

3) $3q^2 = 5q - 1$...[Answer: $q_1 = \frac{5-\sqrt{13}}{6}$ and $q_2 = \frac{5+\sqrt{13}}{6}$]

Exercise 12: Calculate the roots of the following quadratic equations

1) $t^2 = 4t - 4$..[Answer: $X_1 = 2$ and $X_2 = 2$]

2) $-2t - 9 = t^2 + 4t$...[Answer: $t_1 = -3$ and $t_2 = -3$]

3) $25q^2 = 40q - 16$..[Answer: $q_1 = \frac{4}{5}$ and $q_2 = \frac{4}{5}$]

Exercise1 3: Compute the roots of the following quadratic equations

1) $X^2 - 2X + 2 = 0$...[Answer: $X_1 = 1 - i$ and $X_2 = 1 + i$]

2) $t^2 - 4t + 5 = 0$.. ....[Answer: $t_1 = 2 - i$ and $t_2 = 2 + i$]

3.2.3 Graphical Method

To determine the roots of a quadratic equation graphically, we are required to change the quadratic equation to a quadratic function.

$$aX^2 + bX + c = 0 \quad \dots\dots\dots\dots\dots\dots\dots\dots\dots\dots\dots\dots\dots\dots\dots\text{[Quadratic Equation]}$$

$$y = f(X) = aX^2 + bX + c \quad \dots\dots\dots\dots\dots\dots\dots\dots\dots\dots\dots\dots\dots\text{[Quadratic function]}$$

In both equations, the first term is never zero($a \neq 0$) and the roots of a quadratic equation are any values that make the quadratic function equal to zero[$y = f(X) = 0$].

To determine these values, we are required to graph the quadratic function with the help of a table of value. The graph of a quadratic function is called parabola and is symmetry in nature. That is if the parabola is divided into two, the halves match each other. To graph the quadratic function and determine the roots, the following terms are worth knowing.

1) **Direction of Parabola Open**

The open of a parabola is determined by the sign of the first term coefficient ($\pm a$) and from the parabola open; we can determine the vertex point. A negative value shows downward open and maximum vertex point and a positive value shows upward open and minimum vertex point.

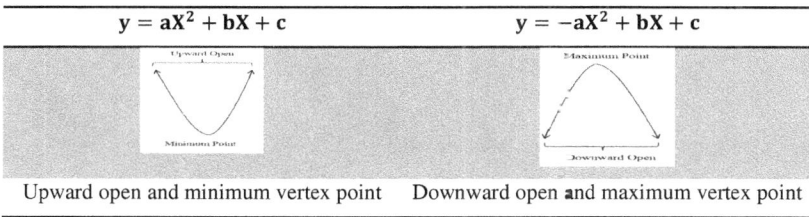

$y = aX^2 + bX + c$	$y = -aX^2 + bX + c$
Upward open and minimum vertex point	Downward open and maximum vertex point

2) **The Vertex Coordinates**

The vertex of a parabola is the maximum or minimum point of the parabola and is the only point on the parabola that is on the axis of symmetry. The axis of symmetry is a line that divides the parabola into two halve drawn from the vertex. The vertex coordinates is simply the value of (x) and (y) at the maximum or minimum point.

$$\text{Vertex} = (x, y) \dots\dots\dots\dots\dots\dots\dots\dots\dots\dots\dots\dots\dots\dots\dots\dots(1)$$

$$\text{Where; } x = -\frac{b}{2(a)} \dots\dots\dots\dots\dots\dots\dots\dots\text{[Equation of symmetry]}$$

$$y = aX^2 + Xb + c \dots\dots\dots\dots\dots\dots\dots\dots\text{[Quadratic function]}$$

The vertex coordinates determine the x-intercept of a parabola. The three possible vertex coordinates graphical display of a quadratic equation are given below.

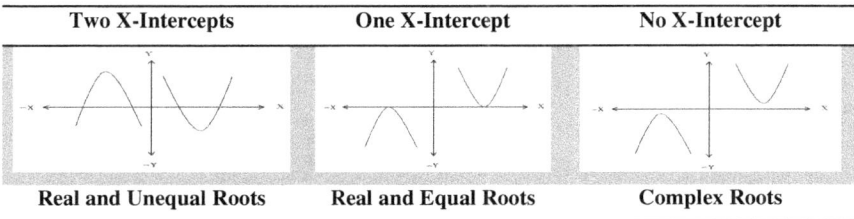

Two X-Intercepts	One X-Intercept	No X-Intercept
Real and Unequal Roots	**Real and Equal Roots**	**Complex Roots**

121

Example 1: Compute the vertex coordinates of the following quadratic equations

1) $X^2 - 2X - 3 = 0$
2) $-3X^2 - 6X + 4 = 0$

Solution

1) $X^2 - 2X - 3 = 0$

Step 1: Formula. Vertex Coordinates $= (x, y)$

Step 2: Determination of equation of symmetry (x)

1) Formula. Vertex Coordinates$(x) = -\frac{b}{2(a)}$, where; $b = -2$ and $a = 1$

2) Data substitution and solving

$\quad$ Vertex Coordinates$(x) = -\frac{(-2)}{2(1)}$ ....[Open bracket]

$\quad$ Vertex Coordinates$(x) = \frac{2}{2}$ ....[Solve fraction]

$\quad$ Vertex Coordinates$(x) = 1$ ....[x = 1]

Step 3: Determination of quadratic function value (y)[$X^2 - 2X - 3 = 0 \rightarrow y = X^2 - 2X - 3$]

1) Formula. $y = aX^2 + bX + c$, where; $x = 1$

2) Data substitution and solving

$\quad$ $y = 1^2 - 2(1) - 3$ ....[Work power and open bracket]

$\quad$ $y = 1 - 2 - 3$ ....[Solve equation]

$\quad$ Quadratic Function Value$(y) = -3$ ....[y = -3]

Step 4: Data substitution and solving ....[x = 1 and y = -3]

Algebraic Calculation	Graphical Illustration
Vertex Coordinates $= (x, y)$..[Substitute (x), (y)] $\quad$ Vertex Coordinates $= (1, -3)$ **N/B:** The theoretical graph of this coordinates is given beside.	

2) $-3X^2 - 6X + 4 = 0$

Step 1: Formula. Vertex Coordinates $= (x, y)$, where; $x = -\frac{b}{2(a)}$ and $y = aX^2 + bX + c$

Step 2: Data substitution and solving...[b = -6, a = -3]

$\left[x = -\dfrac{b}{2(a)}\right]$	$[y = aX^2 + bX + c]$	Graphical Illustration
$x = -\dfrac{(-6)}{2(-3)}$..[Open bracket] $x = \dfrac{6}{-6}$ ....[Solve] $x = -1$	$y = (-3)(-1)^2 +$ $(-6)(-1) + 4$ $y = (-3)(1) + (-6)(-1) + 4$ $y = -3 + 6 + 4$ $y = 7$	

Example 2: Calculate the vertex coordinates of the following quadratic equations

1) $X^2 + 2X + 1 = 0$
2) $Y + X^2 = 5$

Solution

1) $X^2 + 2X + 1 = 0$

Step 1: Formula. Vertex Coordinates $= (x, y)$, where; $x = -\dfrac{b}{2(a)}$ and $y = aX^2 + bX + c$

Step 2: Data substitution and solving..[b = 2 and a = 1]

$\left[x = -\dfrac{b}{2(a)}\right]$	$[y = aX^2 + bX + c]$	Graphical Illustration
$x = -\dfrac{2}{2(1)}$ ...[Open bracket] $x = -\dfrac{2}{2}$ ....[Solve] $x = -1$	$y = (1)(-1)^2 +$ $(2)(-1) + 1$ $y = (1)(1) + (2)(-1) + 1$ $y = 1 - 2 + 1$ $y = 0$	

2) $Y + X^2 = 5$..[$Y - X^2 = 5 \to Y = -X^2 + 5$]

Step 1: Formula. Vertex Coordinates $= (x, y)$, where; $x = -\dfrac{b}{2(a)}$ and $y = aX^2 + bX + c$

Step 2: Data substitution and solving...….....……....[b = 0 and a = −1]

$\left[x = -\frac{b}{2(a)}\right]$	[y = aX² + bX + c]	Graphical Illustration
$x = -\frac{0}{2(-1)}$ ...[Open bracket] $x = \frac{0}{2}$ ....[Solve] $x = \mathbf{0}$	$y = -(0)^2 + 5$ $y = -0 + 5$ $y = \mathbf{5}$	

Test Your Understanding

Exercise 14: Determine the vertex coordinates of the following quadratic equations

1) $X^2 + 6X - 9 = 0$ ....[Answer: x = −3 and y = −18]

2) $-X^2 - X + 6 = 0$ ....[Answer: x = −1 and y = 6]

3) $X^2 - 2X - 5$ ....[Answer: x = 1 and y = −6]

3) Roots of a Quadratic Equation

To determine the roots of a quadratic equation, we derive the table of value from which after plotting the values, we obtain a parabola.

Example 1: Compute the roots of the quadratic equation $2X^2 + 16 = 12X$

Solution

Step 1: Determination of vertex coordinates......[$2X^2 + 16 = 12X \rightarrow 2X^2 - 12X + 16 = 0$]

1) Formula. Vertex Coordinates = (x, y), where; $x = -\frac{b}{2(a)}$ and $y = aX^2 + bX + c$

2) Data substitution and solving...…...…......[b = −12 and a = 2]

$\left[x = -\frac{b}{2(a)}\right]$	[y = aX² + bX + c]
$x = -\frac{(-12)}{2(2)}$ ....[Open bracket] $x = \frac{12}{4}$ ....…......[Solve] $x = \mathbf{3}$	$y = 2(3)^2 - 12(3) + 16$ ....[Work power] $y = 2(9) - 12(3) + 16$ .....[Open bracket] $y = 18 - 36 + 16$ ....[Solve] $y = -2$
Vertex Coordinates = (3, −2)	

N/B: The vertex coordinates $(3, -2)$ shows that the vertex is minimum and the coordinate point is below the X-axis. The positive first term value (2) shows that the parabola will move upward and eventually cross the X-axis. We are required to propose values of X that will make the parabola to cross the X-axis.

Step 2: Derivation of table of value

Variable (x)	Working	Variable (y)	x	y
1	$2(1)^2 - 12(1) + 16 = 2 - 12 + 16$	6	1	6
2	$2(2)^2 - 12(2) + 16 = 8 - 24 + 16$	0	2	0
3	$2(3)^2 - 12(3) + 16 = 18 - 36 + 16$	-2	3	-2
4	$2(4)^2 - 12(4) + 16 = 32 - 48 + 16$	0	4	0
5	$2(5)^2 - 12(5) + 16 = 50 - 60 + 16$	6	5	6

Step 3: Plotting and determination of roots

Plotting	Explanation
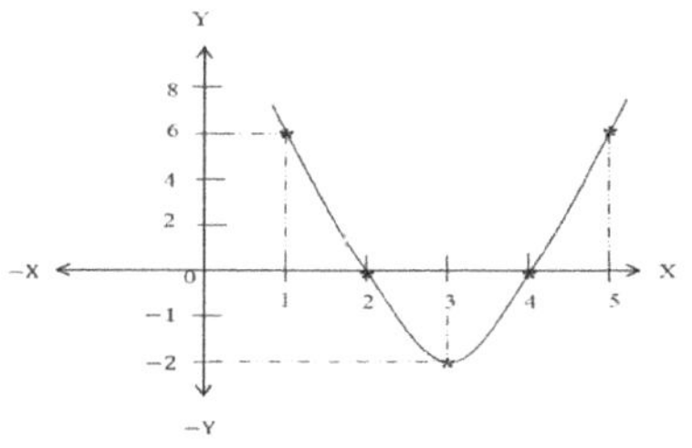	The X-axis is intercepted at two points, which is point 2 and point 4. The quadratic result is therefore **2** and **4**. The roots is justified using the formula method as seen below $$X = \frac{-(-12) \pm \sqrt{(-12)^2 - 4(2)(16)}}{2(2)}$$ $$X = \frac{12 \pm \sqrt{16}}{4} = X = \frac{12 \pm 4}{4}$$ $$X_1 = \frac{12 - 4}{4} = 2 \text{ and } X_2 = \frac{12 + 4}{4} = 4$$

Example 2: Determine the roots of the quadratic equation $X^2 - 10X + 25 = 0$

Solution

Step 1: Determination of vertex coordinates...$[y = X^2 - 10X + 25]$

1) Formula. Vertex Coordinates $= (x, y)$, where; $x = -\frac{b}{2(a)}$ and $y = aX^2 + bX + c$

2) Data substitution and solving...$[b = -10 \text{ and } a = 1]$

$\left[x = -\frac{b}{2(a)}\right]$	$[y = X^2 - 10X + 25]$
$x = -\frac{(-10)}{2(1)}$ ....[Open bracket] $x = \frac{10}{2}$ ....[Solve] $x = 5$	$y = 5^2 - 10(5) + 25$ ....[Work power] $y = 25 - 10(5) + 25$ ....[Solve] $y = 0$

N/B: The vertex coordinates $(5, 0)$ and the coordinate point is on the X-axis. The positive sign of the first term (1) indicates an upward open of the parabola (parabola is moving away

125

from the X-axis). Therefore, we can simply add two values before and after the X value (5) since more values will not change the result.

Step 2: Derivation of table of value

Variable (x)	Working	Variable (y)		x	y
3	$3^2 - 10(3) + 25 = 9 - 30 + 25$	4		3	4
4	$4^2 - 10(4) + 25 = 16 - 40 + 25$	1		4	1
5	$5^2 - 10(5) + 25 = 25 - 50 + 25$	0		5	0
6	$6^2 - 10(6) + 25 = 36 - 60 + 25$	1		6	1
7	$7^2 - 10(7) + 25 = 49 - 70 + 25$	4		7	4

Step 3: Plotting and determination of roots

Plotting	Explanation
	The parabola does not cut across but touches the X-axis at point 5. The result of the quadratic equation is therefore **5**. The value is justified using the factorization method below. $X^2 - 10X + 25 = 0 \rightarrow (X - 5)(X - 5) = 0$ $(X - 5) = 0 \rightarrow X_1 = 5$ $(X - 5) = 0 \rightarrow X_2 = 5$

Example 3: Solve the roots of the quadratic equation $-X^2 + 4X = 13$

<u>Solution</u>

Step 1: Determination of vertex coordinates..........$[-X^2 + 4X = 13 \rightarrow -X^2 + 4X - 13 = 0]$

1) Formula. Vertex Coordinates $= (x, y)$, where; $x = -\dfrac{b}{2(a)}$ and $y = aX^2 + bX + c$

2) Data substitution and solving...[b = 4 and a = −1]

$\left[x = -\dfrac{b}{2(a)}\right]$	$[y = -X^2 + 4X - 13]$
$x = -\dfrac{4}{2(-1)}$ ....[Open bracket] $x = \dfrac{4}{2}$ ....[Solve] $x = 2$	$y = -(2)^2 + 4(2) - 13$ ......[Work power] $y = -4 + 4(2) - 13$ ....[Open bracket] $y = -4 + 8 - 13$ ....[Solve] $y = -9$
Vertex Coordinates $= (2, -9)$	

N/B: The vertex coordinates $(2, -9)$ and the coordinate point is on the X-axis. The negative sign of the first term (-1) indicates a downward open of the parabola (parabola is moving away from the X-axis). Since there is no means of intersection, we add two values before and after the X value (2) to obtain a parabola shape.

126

Step 2: Derivation of table of value

Variable (x)	Working	Variable (y)	x	y
0	$-(0)^2 + 4(0) - 13 = -0 + 0 - 13$	13	0	13
1	$-(1)^2 + 4(1) - 13 = -1 + 4 - 13$	-10	1	-10
2	$-(2)^2 + 4(2) - 13 = -4 + 8 - 13$	-9	2	-9
3	$-(3)^2 + 4(3) - 13 = -9 + 12 - 13$	-10	3	-10
4	$-(4)^2 + 4(4) - 13 = -16 + 16 - 13$	-13	4	-13

Step 3: Plotting and determination of roots

Plotting	Explanation
	The parabola does not intersect the X-axis, indicating a complex root. $$X = \frac{-4 \pm \sqrt{4^2 - 4(-1)(-13)}}{2(-1)} = \frac{-4 \pm \sqrt{-36}}{-2} = \frac{-4 \pm \sqrt{36}\sqrt{-1}}{-2}$$ $$X = \frac{-4 \pm \sqrt{36}\sqrt{-1}}{-2} = \frac{-4 \pm 6i}{-2}$$ $$X_1 = 2 - 3i \text{ and } X_2 = 2 + 3i$$

Test Your Understanding

Exercise 15: Calculate the roots of the following quadratic equations graphically.

1) $3X^2 + 12X + 15$...............................…....................…..[Answer: See Appendix]

2) $X^2 + 6X + 8$ ....…..…........ [Answer: See Appendix]

3) $3X^2 = 3$...…....…[Answer: See Appendix]

3.3 Simultaneous Equation

Simultaneous equation is also called system of equations or an equation system. It is called simultaneous equation because the equations are solved at the same time. Simultaneous equation is defined as a set of two or more equations, each containing two or more variables whose values can at the same time satisfy all the equations in the set.

Simultaneous equation can be a combination of linear equations, linear equation and non-linear equation and/or non-linear equations. Simultaneous equation can be used using the substitution method, elimination method and graphical method.

127

3.3.1 Linear Simultaneous Equation

Linear simultaneous equation is made up of a set of linear equations. A linear equation is an equation of power one and when graphed , it gives a straight line and the general form of a linear equation is given as; $[y = mx + b]$.

1) Substitution Method
Here, we make a variable subject of formula and substitute the result and equation to solve for the value of a variable. To derive the value of the second variable, we substitute the value of the already calculated variable in any equation and solve for the unknown.

If we use equation (1) to make a variable subject of formula, the result will be substituted in equation (2). After determining the first variable value, the value is substituted in any equation (1 or 2) to find the value of the second variable.

Example 1: Making X subject of formula, solve the simultaneous equations below

$X + Y = 2$ ....(1)

$5X + Y = 28$ ....(2)

<u>Solution</u>

Step 1: Making variable (X) subject of formula

Equation (1)....................$[X + Y = 2]$	Equation (2)......................$[5X + Y = 28]$
$X + Y = 2$ ....[Take variable (Y) right] Variable $(X) = 2 - Y$	$5X + Y = 28$ ....[Take variable (Y) right] $5X = 28 - Y$ .....[Divide both sides by (5)] Variable $(X) = \dfrac{28-Y}{5}$

Step 2: Determine the value of non-subject making variable (Y)

Hint: Substitute $[X = 2 - Y]$ in equation (2) or substitute $\left[X = \dfrac{28-Y}{5}\right]$ in equation (1) to determine the value of variable (Y).

Using equation 1................$\left[X = \dfrac{28-Y}{5}\right]$	Using equation 2......................$[X = 2 - Y]$
$X + Y = 2$ ....[Change (X) to $\left(\dfrac{28-Y}{5}\right)$] $\left(\dfrac{28-Y}{5}\right) + Y = 2$ ....[Open bracket] $\dfrac{28-Y}{5} + Y = 2$ ......[Linearized equation]	$5X + Y = 28$ ....[Change (X) to $(2 - Y)$] $5(2 - Y) + Y = 28$ ....[Open bracket] $10 - 5Y + Y = 28$ ....[Collect like terms] $-4Y = 18$ ....[Divide both sides by (-4)]

$28 - Y + 5Y = 10$.[Collect like terms]	$\text{Variable}(Y) = -\frac{18}{4} = -\frac{9}{2}$
$4Y = -18$ …..[Divide both sides by (4)]	
Variable $(Y) = -\frac{9}{2}$	

Step 3: Determine the value of subject of formula variable (X)

$$\left[\text{Substitute } \left[-\frac{9}{2}\right] \text{ in any equations}\right]$$

Using equation 1	Using equation 2
$X + Y = 2$ …….[Change Y to $\left(-\frac{9}{2}\right)$]	$5X + Y = 28$ …………..[Change Y to $\left(-\frac{9}{2}\right)$]
$X + \left(-\frac{9}{2}\right) = 2$ ………[Open bracket]	$5X + \left(-\frac{9}{2}\right) = 28$ ………………[Open bracket]
$X - \frac{9}{2} = 2$ ……..[Linearized equation]	$5X - \frac{9}{2} = 28$ ……………..[Linearized equation]
$2X - 9 = 4$ ……..[Collect like terms]	$10X - 9 = 56$ ….. ……..…[Collect like terms]
$\text{Variable}(X) = \frac{13}{2}$	$\text{Variable}(X) = \frac{65}{10} = \frac{13}{2}$

Step 4: Justification……………………….....$\left[\text{Variable}(X) = \frac{13}{2} \text{ and Variable } (Y) = -\frac{9}{2}\right]$

Equation (1)….[X + Y = 2]	Equation (2)….[5X + Y = 28]
$X + Y = 2$.[Change $\left(X = \frac{13}{2}\right)$ and $\left(Y = -\frac{9}{2}\right)$]	$5X + Y = 28$.[Change $\left(X = \frac{13}{2}\right)$ and $\left(Y = -\frac{9}{2}\right)$]
$\frac{13}{2} + \left(-\frac{9}{2}\right) = 2$ …………[Open bracket]	$5\left(\frac{13}{2}\right) + \left(-\frac{9}{2}\right) = 28$ ………[Open bracket]
$\frac{13}{2} - \frac{9}{2} = 2$ ……..[Linearized equation]	$\frac{65}{2} - \frac{9}{2} = 28$ ….. …...[Linearized equation]
$2 = 2$	$28 = 28$

Example 2: Making Y subject of formula, solve the simultaneous equations below

$Y - X = 8$ ……………………………………….. .……………………………..(1)

$4X - 3Y = 26$ …………………………………………. ………...…………….(2)

<u>**Solution**</u>

Step 1: Making variable Y subject of formula

Equation (1)...................[$Y - X = 8$]	Equation (2)......................[$4X - 3Y = 26$]
$Y - X = 8$ ......[Take variable (X) right] Variable (Y) = 8 + X	$4X - 3Y = 26$ ....[Take (4X) right] $-3Y = 26 - 4X$.[Divide both sides by (-3)] Variable (Y) $= \frac{26-4X}{-3}$
N/B: Substitute $[Y = 8 + X]$ in equation (2)	**N/B**: Substitute $\left[Y = \frac{26-4X}{-3}\right]$ in equation (1)

Step 2: Determination of variable (X) value

Equation (1)..........................$\left[Y = \frac{26-4X}{-3}\right]$	Equation (2).........................[Y = 8 + X]
$Y - X = 8$ ....[Change (Y) to $\left(\frac{26-4X}{-3}\right)$] $\frac{26-4X}{-3} - X = 8$ ....[Linearized equation] $26 - 4X + 3X = -24$.[Collect like terms] $-X = -50$.[Multiply both sides by (-1)] Variable (X) = **50**	$4X - 3Y = 26$.....[Change (Y) to $(8 + X)$] $4X - 3(8 + X) = 26$ ......[Open bracket] $4X - 24 - 3X = 26$.[Collect like terms] Variable (X) = **50**

Step 3: Determine the value of variable (Y)..................[Substitute X = 50 in any equation]

Equation (1)..........................[X = 50]	Equation (2)........................... [X = 50]
$Y - X = 8$ ....[Change (X) to (50)] $Y - 50 = 8$ ....[Collect like terms] Variable (Y) = **58**	$4 - 3Y = 26$ ....[Change (X) to (50)] $4(50) - 3Y = 26$ ....[Open bracket] $200 - 3Y = 26$ ....[Collect like terms] $-3Y = -174$...[Divide both sides by (-1)] Variable (Y) = **58**

Step 4: Justification..[Both sides of equations must be the same]

Equation (1)...........[X = 50 and Y = 58]	Equation (2)............. [X = 50 and Y = 58]
$Y - X = 8$ ....[Substitute variables] $58 - 50 = 8$ ....[Simplify left side] **8 = 8**	$4X - 3Y = 26$ ....[Substitute variables] $4(50) - 3(58) = 26$ ....[Open bracket] $200 - 174 = 26$ ....[Simplify left sides] **26 = 26**

Example 3: Solve the simultaneous equations below using the substitution method.

$2X - 3Y = 9$ ....(1)

$2X + Y = 13$ ....(2)

Solution

Step 1: Making variables subject of formula

	Making X subject of formula	Making Y subject of formula
Equation 1	$2X - 3Y = 9$ ……...[Take 3Y to the right] $2X = 9 + 3Y$ ....[Divide both sides by (2)] $X = \frac{9+3Y}{2}$ ....[2X − 3Y = 9]	$2X - 3Y = 9$ ……...[Take 2X to the right] $-3Y = 9 - 2X$ ...[Multiply sides by (−1)] $3Y = 2X - 9$ ..[Divide both sides by (3)] $Y = \frac{2X-9}{3}$ ....[2X − 3Y = 9]
Equation 2	$2X + Y = 13$ ....[Take Y to the right] $2X = 13 - Y$ ...[Divide both sides by (2)] $X = \frac{13-Y}{2}$ ....[2X + Y = 13]	$2X + Y = 13$ ……...[Take 2X to the right] $Y = 13 - 2X$ ....[2X + Y = 13]

Step 2: Determination of variables values

Hint: Substitute $\left[X = \frac{9+3Y}{2}\right]$ in equation (2) and substitute $\left[X = \frac{13-Y}{2}\right]$ in equation (1) to find the value of variable (Y). Substitute $\left[Y = \frac{2X-9}{3}\right]$ in equation (2) and $[Y = 13 - 2X]$ in equation (1) to find the value of variable (X).

	Finding Value of Variable Y	Finding Value of Variable X
Equation 1	$2X - 3Y = 9$ ...[Change X to $\left(\frac{13-Y}{2}\right)$] $2\left(\frac{13-Y}{2}\right) - 3Y = 9$ …..[Open bracket] $\frac{26-2Y}{2} - 3Y = 9$..[Linearized equation] $26 - 2Y - 6Y = 18$ $-8Y = -8$ Variable (Y) = **1**	$2X - 3Y = 9$.[Change Y to (13 − 2X)] $2X - 3(13 - 2X) = 9$.[Open bracket] $2X - 39 + 6X = 9$ $8X = 48$ Variable (X) = **6**
Equation 2	$2X + Y = 13$ ...[Change X to $\left(\frac{9+3Y}{2}\right)$] $2\left(\frac{9+3Y}{2}\right) + Y = 13$ ....[Open bracket] $\frac{18+6Y}{2} + Y = 13$ [Linearized equation] $18 + 6Y + 2Y = 26$ Variable (Y) = **1**	$2X + Y = 13$.[Change Y to $\left(\frac{2X-9}{3}\right)$] $2X + \frac{2X-9}{3} = 13$ $6X + 2X - 9 = 39$ $8X = 48$ [Divide both sides by (8)] Variable (X) = **6**

Step 3: Justification ……………..[Left side of equation must be equal to right side of equation]

Using equation 1………..[X = 6 and Y = 1]	Using equation 2.. ………..[X = 6 and Y = 1]
2X − 3Y = 9 ……...[Substitute (X) and (Y)]	2X + Y = 13 ……….[Substitute (X) and (Y)]
2(6) − 3(1) = 9 …………..[Open bracket]	2(6) + 1 = 13 ……..………..[Open bracket]
12 − 3 = 9 ……….......[Simplify right side]	12 + 1 = 13 …………..[Simplify right side]
9 = 9 …………………[Values are correct]	13 = 13 ……………….[Values are correct]

Example 4: Solve using the simultaneous equations below

$$Y = X + 8 \text{ ……………………………………..…...…………………..(1)}$$
$$2X − 3Y = 16 \text{ …………………………………..…………….…...……..(2)}$$

<u>**Solution**</u>

Step 1: Re-expression equations in convenient expression

$$\begin{bmatrix} Y = X + 8 \\ 2X − 3Y = 16 \end{bmatrix} = \begin{matrix} Y − X = 8 \\ 2X − 3Y = 16 \end{matrix} = \begin{bmatrix} X − Y = −8 \\ 2X − 3Y = 16 \end{bmatrix}$$

Step 2: Making variable (X) subject of formula

First Expression …………..$\begin{bmatrix} Y − X = 8 \\ 2X − 3Y = 16 \end{bmatrix}$		Second Expression …………..$\begin{bmatrix} X − Y = −8 \\ 2X − 3Y = 16 \end{bmatrix}$	
Equation (1)	Equation (2)	Equation (1)	Equation (2)
Y − X = 8	2X − 3Y = 16	X − Y = −8	2X − 3Y = 16
−X = 8 − Y	2X = 16 + 3Y	X = −8 + Y	2X = 16 + 3Y
X = **Y − 8**	X = $\frac{16+3Y}{2}$	X = **Y − 8**	X = $\frac{16+3Y}{2}$
N/B: Substitute [X = Y − 8] in equation (2) and $\left[X = \frac{16+3Y}{2}\right]$ in equation (1)		**N/B:** Substitute [X = Y − 8] in equation (2) and $\left[X = \frac{16+3Y}{2}\right]$ in equation (1)	

Step 3: Determine the value of variable Y

First Expression		Second Expression	
Equation (1)	Equation (2)	Equation (1)	Equation (2)
Y − X = 8	2X − 3Y = 16	X − Y = −8	2X − 3Y = 16
Y − $\left(\frac{16+3Y}{2}\right)$ = 8	2(Y − 8) − 3Y = 16	$\frac{16+3Y}{2}$ − Y = −8	2(Y − 8) − 3Y = 16
2Y − 16 − 3Y = 16	2Y − 16 − 3Y = 16	16 + 3Y − 2Y = −16	2Y − 16 − 3Y = 16
−Y = 32	−Y = 32	Y = **−32**	−Y = 32
Y = **−32**	Y = **−32**		Y = **−32**

132

Step 4: Determination of value of variable X

First Expression		Second Expression	
Equation (1)	Equation (2)	Equation (1)	Equation (2)
$Y - X = 8$	$2X - 3Y = 16$	$X - Y = -8$	$2X - 3Y = 16$
$-32 - X = 8$	$2X - 3(-32) = 16$	$X - (-32) = -8$	$2X - 3(-32) = 16$
$-X = 40$	$2X + 96 = 16$	$X + 32 = -8$	$2X + 96 = 16$
$X = -40$	$2X = -80$	$X = -40$	$2X = -80$
	$X = -40$		$X = -40$

Step 5: Justification..…..….…........[X $= -40$ and Y $= -32$]

First Expression		Second Expression	
Equation (1)	Equation (2)	Equation (1)	Equation (2)
$Y - X = 8$	$2X - 3Y = 16$	$X - Y = -8$	$2X - 3Y = 16$
$-32 - (-40) = 8$	$2(-40) - 3(-32) = 16$	$-40 - (-32) = -8$	$2(-40) - 3(-32) = 16$
$-32 + 40 = 8$	$-80 + 96 = 16$	$-40 + 32 = -8$	$-80 + 96 = 16$
$8 = 8$	$16 = 16$	$-8 = -8$	$16 = 16$

Test Your Understanding
Exercise 16: Solve the following simultaneous equations using the substitution method.
1) $X + Y = 4$ and $3X + 5Y = 14$..................................…..…......[**Answer: X $= 3$ and Y $= 1$**]
2) $X + Y = 11$ and $4X + 3Y = 40$.. ....…..…......[**Answer: X $= 7$ and Y $= 4$**]
3) $Y = X + 8$ and $3X - 2Y = -21$.............................…..…......[**Answer: X $= -5$ and Y $= 3$**]

2) Elimination Method

Here, we identify a variable to remove from the equations and solve for the remaining variable. The value of the calculated variable is substituted in any equation to find the value of the eliminated (removed) variable.

To eliminate a variable, we must ensure that the variable is having same coefficient in both equations. After ensuring that, we add or subtract the equations using an appropriate mathematical sign ($+$ or $-$) such that the sum or difference of the eliminated variable must be equal to zero.

Example 1: Solve the simultaneous equations below using the elimination method.

$$3X + 2Y = 36 \dots\dots\dots\dots\dots\dots\dots\dots\dots\dots\dots\dots\dots\dots\dots\dots\dots\dots\dots\dots(1)$$

$$5X + 4Y = 64 \dots\dots\dots\dots\dots\dots\dots\dots\dots\dots\dots\dots\dots\dots\dots\dots\dots\dots\dots\dots(2)$$

Solution

Step 1: Pre-calculation

1) Equation Equivalence

Equalizing variable X coefficient	$\left.\begin{array}{l}3X + 2Y = 36\\5X + 4Y = 64\end{array}\right) = \left(\begin{array}{l}5[3X + 2Y = 36]\\3[5X + 4Y = 64]\end{array}\right. = \left(\begin{array}{l}\mathbf{15X + 10Y = 180}\\\mathbf{15X + 12Y = 192}\end{array}\right.$
Equalizing variable Y coefficient	$\left.\begin{array}{l}3X + 2Y = 36\\5X + 4Y = 64\end{array}\right) = \left(\begin{array}{l}4[3X + 2Y = 36]\\2[5X + 4Y = 64]\end{array}\right. = \left(\begin{array}{l}\mathbf{12X + 8Y = 144}\\\mathbf{10X + 8Y = 128}\end{array}\right.$

Step 2: Data substitution and solving

a) **Elimination** of X$\dots\dots\left.\begin{array}{l}[3X + 2Y = 36\\[5X + 4Y = 64\end{array}\right) = \left(\begin{array}{l}5[3X + 2Y = 36]\\3[5X + 4Y = 64]\end{array}\right. =$
$\left(\begin{array}{l}15X + 10Y = 180\\15X + 12Y = 192\end{array}\right]$

1) Determination of variable (Y) value

Hint: The mathematical sign of variable (X) in both equations are plus (+). To eliminate X, we should subtract the equation$[+X - (+X) = 0]$.

[Eq(2) − Eq(1)]	[Eq(1) − Eq(2)]
$15X - 15X + 12Y - 10Y = 192 - 180$	$15X - 15X + 10Y - 12Y = 180 - 192$
$2Y = 12$ ....[Divide both sides by (2)]	$-2Y = -12$ ....[Divide both sides by (−2)]
Variable (Y) = **6**	Variable (Y) = **6**

2) Determination of eliminated variable (X).........[Substitute value of (Y) in any equation]

Using equation (1)......[3X + 2Y = 36]	Using equation (2)...[5X + 4Y = 64]
$3X + 2Y = 36$ ....[Change (Y) to (6)]	$5X + 4Y = 64$ ....[Change (Y) to (6)]
$3X + 2(6) = 36$ ......[Open bracket]	$5X + 4(6) = 64$ ....[Open bracket]
$3X + 12 = 36$ ....[Collect like terms]	$5X + 24 = 64$ ......[Collect like terms]
$3X = 24$ ...[Divide both sides by (3)]	$5X = 40$ .....[Divide both sides by (5)]
Variable (X) = **8**	Variable (X) = **8**

b) **Elimination** of Y.......... $\begin{bmatrix} 3X + 2Y = 36 \\ 5X + 4Y = 64 \end{bmatrix} = \begin{pmatrix} 4[3X + 2Y = 36] \\ 2[5X + 4Y = 64] \end{pmatrix} =$

$\begin{pmatrix} 12X + 8Y = 144 \\ 10X + 8Y = 128 \end{pmatrix}$

1) Determination of variable (X) value

Hint: The mathematical sign of variable (Y) in both equations are plus(+). To eliminate X, we should subtract the equation $[+X - (+X) = 0]$.

[Eq(2) − Eq(1)]	[Eq(1) − Eq(2)]
$10X - 12X + 8Y - 8Y = 128 - 144$	$12X - 10X + 8Y - 8Y = 144 - 128$
$-2X = -16$ ...[Divide both sides by (−2)]	$2X = 16$ ....[Divide both sides by (2)]
Variable (X) = **8**	Variable (X) = **8**

2) Determination of eliminated variable (Y)......[Substitute value of (X) in any equation]

Using equation (1).......[12X + 8Y = 144]	Using equation (2)...[10X + 8Y = 128]
$12X + 8Y = 144$ .....[Change (X) to (8)]	$10X + 8Y = 128$ ...[Change (X) to (8)]
$12(8) + 8Y = 144$ ....[Open bracket]	$10(8) + 8Y = 128$ .....[Open bracket]
$96 + 8Y = 144$ ......[Collect like terms]	$80 + 8Y = 128$ ....[Collect like terms]
$8Y = 48$ ....[Divide both sides by (8)]	$8Y = 48$ .....[Divide both sides by (8)]
Variable (Y) = **6**	Variable (Y) = **6**

Step 3: Justification ....…..…...[Left side of equation must be equal to right side of equation]

Using equation 1....…......[X = 8 and Y = 6]	Using equation 2.. …........[X = 8 and Y = 6]
$3X + 2Y = 36$ .....[Substitute (X) and (Y)]	$5X + 4Y = 64$ ......[Substitute (X) and (Y)]
$3(8) + 2(6) = 36$ ....…......[Solve]	$5(8) + 4(6) = 64$ ....[Solve]
36 = 36 ....[Values are correct]	**64 = 64** ....[Values are correct]

Example 2: Solve the simultaneous equations below

$7X + Y = 25$ ....…..............(1)

$5X - Y = 11$... ....…..............(2)

<u>Solution</u>

a) Substitution Method

Step 1: Making variable (Y) subject of formula

Using equation 1	Using equation 2
$7X + Y = 25$ ....[Take 7X to the right] $Y = 25 - 7X$ ....[Equation 1]	$5X - Y = 11$ ....[Take 5S to the right] $-Y = 11 - 5X$ ....[Multiply equation by (-1)] $Y = 5X - 11$ ....[Equation 2]

Step 2: Determination of Non-subject of formula variable (X) value

Using equation (2)............[$Y = 25 - 7X$]	Using equation (1)..............[$Y = 5X - 11$]
$5X - Y = 11$ ....[Substitute Y] $5X - (25 - 7X) = 11$ ...[Open bracket] $5X - 25 + 7X = 11$ ....[Solve] Variable (X) = 3	$7X + Y = 25$ ....[Substitute Y] $7X + (5X - 11) = 25$ ....[Open bracket] $7X + 5X - 11 = 25$ ....[Solve] Variable (X) = 3

Step 3: Determination of subject of formula variable (Y) value

Using equation (1)......................[X = 3]	Using equation (2)......................[X = 3]
$7X + Y = 25$ ....[Substitute X] $7(3) + Y = 25$ ....[Open bracket] $21 + Y = 25$ ....[Collect like terms] Variable (Y) = 4	$5X - Y = 11$ ....[Substitute X] $5(3) - Y = 11$ ....[Open bracket] $15 - Y = 11$..............[Collect like terms] $-Y = -4$ ....[Multiply equation by (-1)] Variable (Y) = 4
N/B: $Y = 25 - 7X$ can be use as eq (1)	**N/B:** $Y = 5X - 11$ can be use as eq (2)

Step 4: Justification ....[Left side of equation must be equal to right side of equation]

Using equation 1...........[X = 3 and Y = 4]	Using equation 2.. ....[X = 3 and Y = 4]
$7X + Y = 25$ ......[Substitute (X) and (Y)] $7(3) + 4 = 25$ ....[Open bracket] $21 + 4 = 25$ ....[Simplify right side] **25 = 25** ....[Values are correct]	$5X - Y = 11$ ....[Substitute (X) and (Y)] $5(3) - 4 = 11$ ....[Open bracket] $15 - 4 = 11$ ....[Simplify right side] **11 = 11** ....[Values are correct]

b) Elimination method

Step 1: Pre-calculation

Eliminating (Y) Equation . $\begin{bmatrix} 7X + Y = 25 \\ 5X - Y = 11 \end{bmatrix}$.[No modification since coefficients are all (1)]

Eliminating (X) Equation.......... $\begin{bmatrix} 7X + Y = 25 \\ 5X - Y = 11 \end{bmatrix} = \begin{pmatrix} 5[7X + Y = 25] \\ 7[5X - Y = 11] \end{pmatrix} =$
$\begin{pmatrix} \mathbf{35X + 5Y = 125} \\ \mathbf{35X - 7Y = 77} \end{pmatrix}$

Step 2: Elimination of variable [Elimination Y means finding X and elimination X means finding Y]

Elimination of Y............ $\begin{bmatrix} 7X + Y = 25 \\ 5X - Y = 11 \end{bmatrix}$	Elimination of X................ $\begin{bmatrix} 35X + 5Y = 125 \\ 35X - 7Y = 77 \end{bmatrix}$
$(-Y) + (+Y) = 0 \ldots [\text{Eq}(2) + \text{Eq}(1)]$	$(+X) - (+X) = 0 \ldots [\text{Eq}(1) - \text{Eq}(2)]$
$5X + 7X - Y + Y = 11 + 25$	$35X - 35X + 5Y - (-7Y) = 125 - 77$
$12X = 36 \ldots [\text{Divide both sides by } (12)]$	$12Y = 48 \ldots\ldots\ldots [\text{Divide both sides by } (12)]$
Variable $(X) = \mathbf{3}$	Variable $(Y) = \mathbf{4}$

Step 3: Determination of eliminated variable value.................. ....[Use any equation]

	Finding variable Y [Variable $(X) = 3$]	Finding variable X [Variable $(Y) = 4$]
Equation (1) $7X + Y = 25$	$7X + Y = 25 \ldots\ldots [\text{Substitute }(X)]$ $7(3) + Y = 25 \ldots [\text{Open bracket}]$ $21 + Y = 25 \ldots [\text{Collect like terms}]$ Variable $(Y) = \mathbf{4}$	$7X + Y = 25 \ldots\ldots [\text{Substitute }(Y)]$ $7X + 4 = 25 \ldots [\text{Collect like terms}]$ $7X = 21 . \ [\text{Divide both sides by } (7)]$ Variable $(X) = \mathbf{3}$
Equation (2) $5X - Y = 11$	$5X - Y = 11 \ldots\ldots [\text{Substitute }(X)]$ $5(3) - Y = 11 \ldots\ldots [\text{Solve}]$ Variable $(Y) = \mathbf{4}$	$5X - Y = 11 \ldots\ldots [\text{Substitute }(Y)]$ $5X - 4 = 11 \ldots\ldots\ldots [\text{Solve}]$ Variable $(X) = \mathbf{3}$

Test Your Understanding

Exercise 17: Solve the following simultaneous equations below, using the elimination method

1) $3X + 2Y = 16$ and $4X + Y = 13$.............................[Answer: $X = 2$ and $Y = 5$]

2) $4X - 3Y = 5$ and $9X - 2Y = 16$.............................[Answer: $X = 2$ and $Y = 1$]

3) $2X + 3Y = 10$ and $-3X + 2Y = -41$....................[Answer: $X = 11$ and $Y = -4$]

4) $2X + \frac{1}{3}Y = 1$ and $3X + 5Y = 6$...............................[Answer: $X = \frac{1}{3}$ and $Y = 1$]

3) Graphical Method

To solve a linear simultaneous equation using the graphical method, we compute the coordinate pair of each equation and plot. The coordinate pair at the intersection point of the lines as the solution. The three possible plotting appearances using the graphical method are seen below.

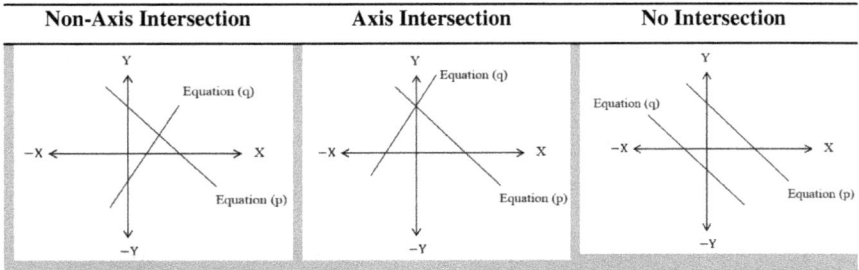

Non-Axis Intersection	Axis Intersection	No Intersection

Example 1: Solve the simultaneous equations below using the graphical method.

$$3Y + 2X = 12 \text{ ....(1)}$$

$$Y - X = -1 \text{ ....(2)}$$

Solution

Step 1: Determination of coordinates of variable (Y) and Variable (X)

	When [X = 0]	When [Y = 0]
Equation (1)	$3Y + 2X = 12$ .....[When X to 0]	$3Y + 2X = 12$ ..[When Y to 0]
	$3Y + 2(0) = 12$.[Open bracket]	$3(0) + 2X = 12$.[Open bracket]
	$3Y = 12$ ..[Divide sides by (3)]	$2X = 12$ ...[Divide sides by (2)]
	Variable (Y)Coordinate = **4**	Variable (X)Coordinate = **3**
N/B: Equation (1) pair coordinate is given as (**3, 4**)		

	When [X = 0]	When [Y = 0]
Equation (2)	$Y - X = -1$ ....[When X to 0]	$Y - X = -1$ .... [Change Y to 0]
	$Y - 0 = -1$ ....[Solve]	$0 - X = -1$ .... [Solve]
	Variable (Y)Coordinate = -1	Variable (X)Coordinate = 1
N/B: Equation (2) pair coordinate is given as (**1, −1**)		

138

Step 2: Plotting and determination of values of variables............[Variable X and variable Y]

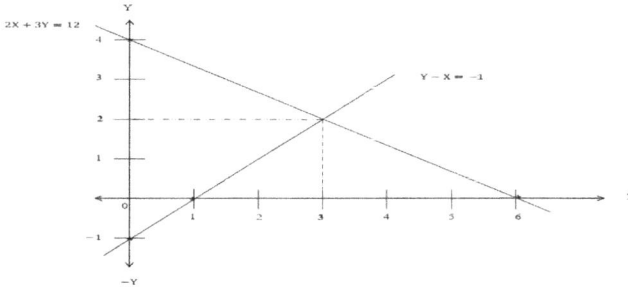

Note: Equation (1) and equation (2) intersect at the point where the variable (Y) is 2 and variable (X) is 3. Therefore, the roots of the equation are 2 and 3.

Example 2: Solve the simultaneous equations using the graphical method.

Y + 2X = 5 ....(1)

Y = 2X − 1 ....(2)

Solution

Step 1: Determination of coordinates of variable (Y) and Variable (X)

	When [X = 0]	When [Y = 0]
Equation (1)	Y + 2X = 5 ....[When X to 0] Y + 2(0) = 5 .....[Open bracket] Variable (Y)Coordinate = **5**	Y + 2X = 5 .....[When Y to 0] 0 + 2X = 5 ....[Solve] Variable (X)Coordinate $= \frac{5}{2} =$ **2. 5**
N/B: Equation (1) pair coordinate is given as **(2.5, 5)**		

	When [X = 0]	When [Y = 0]
Equation (2)	Y − 2X = −1 .[When X to 0] Y − 2(0) = −1 .[Open bracket] Variable (Y)Coordinate = **−1**	Y − 2X = −1 ... [Change Y to 0] 0 − 2X = −1 [Divide sides by (−2)] Variable (X)Coordinate $= \frac{1}{2}$
N/B: Equation (2) pair coordinate is given as **(0. 5, −1)**		

Step 2: Plotting and determination of values of variables...............[Variable X and variable Y]

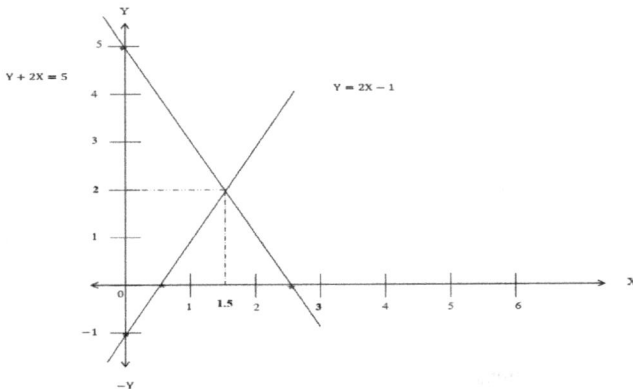

Note: Equation (1) and equation (2) intersect at the point where the variable (Y) is 2 and variable (X) is 1.5. Therefore, the roots of the equation are 1.5 and 2.

Example 3: Solve the simultaneous equations below using the graphical method.

$X - 2Y = -4$ ....(1)

$-3X + 6Y = 0$ ....(2)

Solution

Step 1: Determination of coordinates of variable (Y) and Variable (X)

	When [X = 0]	When [Y = 0]
Equation (1)	$X - 2Y = -4$ ....[When X to 0]	$X - 2Y = -4$ ....[When Y to 0]
	$0 - 2Y = -4$...[Divide sides by (−2)]	$X - 2(0) = -4$ ....[Open bracket]
	Variable (Y)Coordinate = **2**	Variable (X)Coordinate = **−4**
N/B: Equation (1) pair coordinate is given as $(-4, 2)$		

	When [X = 0]	When [Y = 0]
Equation (2)	$-3X + 6Y = 0$ .....[When X to 0]	$-3X + 6Y = 0$ ...... [Change Y to 0]
	$-3(0) + 6Y = 0$..........[Solve]	$-3X + 6(0) = 0$ ....[Solve]
	Variable (Y)Coordinate = **0**	Variable (X)Coordinate = **0**
N/B: Equation (2) pair coordinate is given as $(0, 0)$		

140

Step 2: Plotting and determination of values of variables............[Variable X and variable Y]

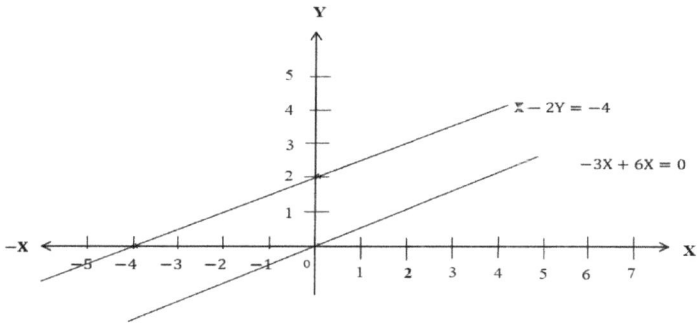

N/B: The lines of the two equations are parallel to each other. Since there is no meeting point, we conclude that there is no solution.

Example 4: Solve the simultaneous equations using the graphical method.

$$X + Y = 1 ....(1)$$
$$2X + 3Y = 3 ....(2)$$

Solution

Step 1: Determination of coordinates of variable (Y) and Variable (X)

	When [X = 0]	When [Y = 0]
Equation (1)	X + Y = 1 ....[When X to 0] 0 + Y = 1......................[Solve] Variable (Y)Coordinate = **1**	X + Y = 1 ....[When Y to 0] X + 0 = 1 ....[Solve] Variable (X)Coordinate = **1**
	N/B: Equation (1) pair coordinate is given as **(1, 1)**	

	When [X = 0]	When [Y = 0]
Equation (2)	2X + 3Y = 3 ....[When X to 0] 2(0) + 3Y = 3...............[Solve] Variable (Y)Coordinate = **1**	2X + 3Y = 3 ....[Change Y to 0] 2X + 3(0) = 3 ....[Solve] Variable (X)Coordinate = **1. 5**
	N/B: Equation (2) pair coordinate is given as **(1. 5, 1)**	

Step 2: Plotting and determination of values of variables.........[Variable X and variable Y]

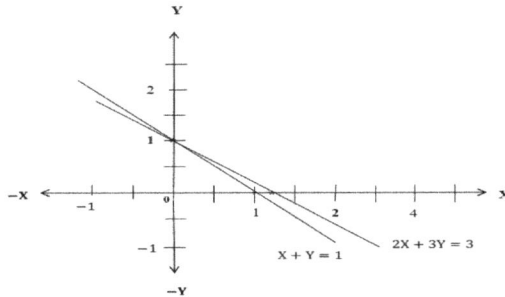

N/B: The equation lines meets at the Y-axis when Y is equal to 1. On the Y-axis, X is by default zero. Therefore, the result is Y = 1 and X = 0.

Test Your understanding

Exercise 18: Solve following simultaneous equations using the graphical method

1) $2Y - X = 4$ and $2Y + 3X = 12$....................................….....…..**[Answer: See Appendix]**

2) $X + Y = 5$ and $X - Y = 3$..….…..**[Answer: See Appendix]**

3.3.2 Non-Linear Simultaneous Equation

A non-linear simultaneous equation is made up of a linear and non-linear equation and/or non-linear equations. A non-linear equation is any equation that does not give a straight line when plotted and has an exponent greater than 1. A simplified form of a non-linear equation is given as; $[y = ax^2 + bz^2]$.

1) Substitution Method

The substitution method is mostly applied when a simultaneous equation is made up of a linear equation and a non-linear equation. Here, we use the linear equation and make a variable (X) the subject of formula. The resulted equation is then substituted in the non-linear equation to find the value of the other variable (Y). The value of (Y) is substituted in any equation to obtain the value of (X).

Example 1: Solve the simultaneous equations below using the substitution method.

$$X^2 + Y^2 = 20 \dots\dots\dots\dots\dots\dots\dots\dots\dots\dots\dots\dots\dots\dots\dots\dots\dots\dots\dots\dots\dots\dots\dots\dots\dots\dots(1)$$

$$X + Y = 6 \dots\dots\dots\dots\dots\dots\dots\dots\dots\dots\dots\dots\dots\dots\dots\dots\dots\dots\dots\dots\dots\dots\dots\dots\dots\dots\dots\dots\dots(2)$$

Solution

Step 1: Making (Y) subject of formula.............................. ....[Using equation (2)]

$\quad$ X + Y = 6 ....[Take (X) right]

$\quad$ Y = **6 − X**

Step 2: Determination of the working equation.........[Substitute (Y = 6 − X) in equation (1)]

$\quad$ $X^2 + Y^2 = 20$ ....[Change (Y) to (6 − X)]

$\quad$ $X^2 + (6 − X)^2 = 20$ ....[Work power]

$\quad$ $X^2 + [(6 − X)(6 − X)] = 20$ ....[Open bracket]

$\quad$ $X^2 + 36 − 12X + X^2 = 20$..[Collect like terms]

$\quad$ **$2X^2 − 12X + 16 = 0$** ....[Quadratic equation]

Step 3: Determination of value of variable (Y)........................[Using the formula method]

1) Formula. $\frac{-b \pm \sqrt{b^2 - 4ac}}{2a}$, where; a = 2, b = −12, c = 16 and

$\quad\quad$ $\Delta = b^2 − 4ac = −12^2 − 4(2)(16) = 144 − 128 = 16$.....[$\Delta > 0$]

2) Data substitution and solving

$\quad$ $\frac{-(-12) \pm \sqrt{16}}{2(2)}$ ....[Simplify denominator and work root]

$\quad$ $\frac{12 \pm 4}{4}$ ....[Find value of (X_1) and (X_2)]

$\quad\quad$ $X_1 = \frac{12-4}{4} = 2$ ....[First coordinate of (X)]

$\quad\quad$ $X_2 = \frac{12+4}{4} = 4$ ....[Second coordinate of (X)]

Step 4: Determination of the value of variable Y

Using equation 1	Using equation 2
When.............................. [X = 2]	When............................... [X = 2]
$(2)^2 + Y^2 = 20$ ....[Work exponent]	2 + Y = 6 ....[Take (2) right]
$4 + Y^2 = 20$ ....[Take (4) right]	Y = 6 − 2 ....[Simplify right side]
$Y^2 = 16$ ....[Square root both sides]	Variable(Y) = 4
Variable(Y) = **4**	
When.............................. [X = 4]	When............................... [X = 4]
$(4)^2 + Y^2 = 20$ ....[Work exponent]	4 + Y = 6 ....[Take (4) right]
$16 + Y^2 = 20$ ....[Solve]	Y = 6 − 4 ....[Simplify right side]
Variable(Y) = **2**	Variable(Y) = 2
N/B: The solutions of the simultaneous equation are (2, 4) and (4, 2)	

Step 5: Justification ....[Using equation 1]

When.........................[X = 2 and Y = 4]	When.........................[X = 4 and Y = 2]
$2^2 + 4^2 = 20$ ....[Work exponent]	$4^2 + 2^2 = 20$ ....[Work exponent]
$4 + 16 = 20$ ....[Simplify left side]	$16 + 4 = 20$ ....[Simplify left side]
$20 = 20$	$20 = 20$

Example 2: Solve the simultaneous equations below using the substitution method.

$$4X^2 + Y^2 = 13 \text{ ....(1)}$$

$$X^2 + Y^2 = 10 \text{ ....(2)}$$

Solution

Step 1: Making variable (X) subject of formula

Equation (1)...................[$4X^2 + Y^2 = 13$]	Equation (2)......................[$X^2 + Y^2 = 10$]
$4X^2 + Y^2 = 13$ ....[Take (Y^2) right] $4X^2 = 13 - Y^2$ ..[Divide both side by (4)] $X^2 = \frac{13-Y^2}{4}$ ......[Square root both sides] Variable $(X) = \sqrt{\frac{13-Y^2}{4}} = \left[\frac{13-Y^2}{4}\right]^{\frac{1}{2}}$	$X^2 + Y^2 = 10$ ....[Take (Y^2) right] $X^2 = 10 - Y^2$ ...[Square root both sides] Variable $(X) = \sqrt{10 - Y^2} = [10 - Y^2]^{\frac{1}{2}}$

Step 2: Determination of working equation

1) Using equation (1)..... [$4X^2 + Y^2 = 13$], where; Variable (X) $= \sqrt{10 - Y^2} = [10 - Y^2]^{\frac{1}{2}}$

$$4X^2 + Y^2 = 13 \text{ ....[Change X to } (10 - Y^2)^{\frac{1}{2}}]$$

$$4\left[(10 - Y^2)^{\frac{1}{2}}\right]^2 + Y^2 = 13 \text{ ....[Work Power. Remember}(a^n)^m = a^{nm}]$$

$$4(10 - Y^2) + Y^2 = 13 \text{ ....[Open bracket]}$$

$$40 - 4Y^2 + Y^2 = 13 \text{ ....[Collect like terms]}$$

$$-3Y^2 + 27 = 0 \text{ ....[Transform to standard from]}$$

$$3Y^2 + 0Y - 27 = 0 \text{ ....[Quadratic Equation]}$$

2) Using equation (2).......[$X^2 + Y^2 = 10$], where; Variable (X) $= \sqrt{\frac{13-Y^2}{4}} = \left[\frac{13-Y^2}{4}\right]^{\frac{1}{2}}$

$$X^2 + Y^2 = 10 \text{ ....[Change X to } \left(\frac{13-Y^2}{4}\right)^{\frac{1}{2}}]$$

$$\left[\left(\frac{13-Y^2}{4}\right)^{\frac{1}{2}}\right]^2 + Y^2 = 10 \text{ ....[Work Power. Remember}(a^n)^m = a^{nm}]$$

$\frac{13-Y^2}{4} + Y^2 = 10$ ....……..……….[Linearized equation]

$13 - Y^2 + 4Y^2 = 40$ ....…..………..[Collect like terms]

$3Y^2 - 27 = 0$ ....…...…………….......…..[Transform to standard from]

$3Y^2 + 0Y - 27 = 0$ ....…..…….……..[Quadratic Equation]

Step 3: Determination of value of variable (Y)...................….…..[Using the formula method]

3) Formula. $\frac{-b \pm \sqrt{b^2 - 4ac}}{2a}$, where; $a = 3$, $b = 0$, $c = -27$ and $\Delta = b^2 - 4ac = 0^2 -$
$4(3)(-27) = 324$

4) Data substitution and solving

$\frac{-0 \pm \sqrt{324}}{2(-3)}$ ....…. .[Simplify denominator and work root]

$\frac{-0 \pm 18}{-6}$ ....…....…….[Find value of (Y_1) and (Y_2)]

$Y_1 = \frac{-0 + 18}{-6} = -3$ .... ……………..[First coordinate of (Y)]

$Y_2 = \frac{-0 - 18}{-6} = 3$ .... …………. ..[Second coordinate of (Y)]

Step 4: Determination of value of variable (Y)

Equation 1	Equation 2
When............................. [Y = -3]	When... [Y = 3]
$4X^2 + Y^2 = 13$ ...[Change (Y) to(-3)]	$X^2 + Y^2 = 10$................[Change (Y) to(3)]
$4X^2 + (-3)^2 = 13$ ....[Work exponent]	$X^2 + (3)^2 = 10$..…........[Work exponent]
$4X^2 + 9 = 13$ …....[Collect like terms]	$X^2 + 9 = 10$ ....[Collect like terms]
$4X^2 = 4$ ....[Divide both sides by (4)]	$X^2 = 1$ ....[Square root both sides]
$X^2 = 1$ ....[Square root both sides]	$X = \sqrt{1}$ ....[Work square root]
$X = \sqrt{1}$ ....[Work square root]	Variable(X) = **1**
Variable(X) = **1**	

Step 4: Justification

Equation (1)................[$4X^2 + Y^2 = 13$]	Equation (2)..... ....[$X^2 + Y^2 = 10$]
Where; Y = 3 and X = 1	Where; Y = -3 and X = 1
$4(1)^2 + (3)^2 = 13$ ....[Work exponents]	$1^2 + (-3)^2 = 10$ ....[Work exponent]
$4 + 9 = 13$ ....[Simplify left side]	$1 + 9 = 10$ ....[Simplify left side]
13 = 13	**10 = 10**

Example 3: Solve the simultaneous equations below using the substitution method.

$$2X - Y = 7 \dots\dots\dots\dots\dots\dots\dots\dots\dots\dots\dots\dots\dots\dots\dots(1)$$

$$XY = 15 \dots\dots\dots\dots\dots\dots\dots\dots\dots\dots\dots\dots\dots\dots\dots\dots\dots(2)$$

<u>Solution</u>

Step 1: Making X subject of formula

Using equation (1)	Using equation (2)
$2X - Y = 7$ ....[Take (Y) right] $2X = 7 + Y$ .....[Divide both sides by (2)] Variable $(X) = \frac{7+Y}{2}$	$XY = 15$ ....[Divide both sides by (Y)] Variable $(X) = \frac{15}{Y}$

Step 2: Derivation of working equation

Using equation (1)....... $\left[$ Variable $(X) = \frac{15}{Y} \right]$	Using equation (2)....... $\left[$ Variable $(X) = \frac{7+Y}{2} \right]$
$2X - Y = 7$ ....[Change (X) to $\left(\frac{15}{Y}\right)$] $2\left(\frac{15}{Y}\right) - Y = 7$ ....[Linearized equation] $30 - Y^2 = 7Y$.[Standard from expression] $\mathbf{Y^2 + 7Y - 30 = 0}$	$XY = 15$ ....[Change (X) to $\left(\frac{7+Y}{2}\right)$] $\left(\frac{7+Y}{2}\right)Y = 15$ . ...[Linearized equation] $7Y + Y^2 = 30$.[Standard from expression] $\mathbf{Y^2 + 7Y - 30 = 0}$

Step 3: Determination of value of variable (Y)........................[Using the formula method]

1) Formula. $\frac{-b\pm\sqrt{b^2-4ac}}{2a}$, where; $a = 1$, $b = 7$, $c = -30$ and $\Delta = 169$.................[$\Delta > 0$]

2) Data substitution and solving

$$\frac{-7\pm\sqrt{169}}{2(1)} \dots\dots\dots\dots\dots\dots\dots\dots\dots\dots\dots[\text{Simplify denominator and work root}]$$

$$\frac{-7\pm13}{2} \dots\dots\dots\dots\dots\dots\dots\dots\dots\dots\dots\dots\dots[\text{Find value of } (Y_1) \text{ and } (Y_2)]$$

$$Y_1 = \frac{-7-13}{2} = -10 \dots\dots\dots\dots\dots\dots\dots\dots\dots[\text{First coordinate of } (Y)]$$

$$Y_2 = \frac{-7+13}{2} = \frac{3}{2} \dots\dots\dots\dots\dots\dots\dots\dots\dots[\text{Second coordinate of } (Y)]$$

Step 4: Determine the value of variable (X).................................[Using equation 2]

When ....[$Y = -10$]	When.................................$\left[Y = \frac{3}{2}\right]$
$XY = 15$ ....[Change (Y) to (-10)] $X(-10) = 15$.[Divide both sides by (-10)] Variable $(X) = -\frac{3}{2}$	$XY = 15$ ....[Change (Y) to $\left(\frac{3}{2}\right)$] $X\left(\frac{3}{2}\right) = 15$ ...[Divide both sides by $\left(\frac{3}{2}\right)$] Variable $(X) = \mathbf{10}$

Exercise 19: Solve the following simultaneous equations using the substitution method.

1) $2X + Y = 11$ and $2X^2 - Y^2 = 23$.......................[Answer: $(18, -25)$ and $(4, 3)$]

2) $X^2 + Y^2 = 17$ and $X + 4Y = 0$.............................[Answer: $(4, -1)$ and $(-4, 1)$]

2) Elimination Method

The elimination method is mostly used when the simultaneous equation is made up of non-linear equations. The solving procedure respects the same process as with linear situation.

Example 1: Eliminating (X), solve the following simultaneous equation

$3x^2 + 2y^2 = 35$ ....(1)

$4x^2 + 3y^2 = 48$ ....(2)

Solution

Step 1: Equivalence equation

$$\left[\begin{array}{l} 3x^2 + 2y^2 = 35 \\ 4x^2 + 3y^2 = 48 \end{array}\right) = \begin{array}{l} 4| \\ 3| \end{array}\left(\begin{array}{l} 3x^2 + 2y^2 = 35 \\ 4x^2 + 3y^2 = 48 \end{array}\right) = \left|\begin{array}{l} 12x^2 + 8y^2 = 140 \\ 12x^2 + 9y^2 = 144 \end{array}\right]$$

Step 2: Determine the value of variable (Y)

Eq(1) − Eq(2)	Eq(2) − Eq(1)
$-y^2 = -4$ ....[Divide sides by (-1)]	$y^2 = 4$ ....[Square root both sides]
$y^2 = 4$ ....[Square root both sides]	$y = \sqrt{4} = \pm 2$
$y = \sqrt{4} = \pm 2$	

Step 3: Determine the value of variable (X)....[Using equation (1). That is,($3x^2 + 2y^2 = 35$)]

When $y = 2$	When $y = -2$
$3x^2 + 2y^2 = 35$ ....[Substitute (y)]	$3x^2 + 2y^2 = 35$ ....[Substitute (y)]
$3x^2 + 2(2)^2 = 35$ ....[Work exponent]	$3x^2 + 2(-2)^2 = 35$ ......[Work exponent]
$3x^2 + 8 = 35$ ....[Collect like terms]	$3x^2 + 8 = 35$ ....[Collect like terms]
$3x^2 = 27$ ....[Divide sides by (3)]	$3x^2 = 27$ ....[Divide sides by (3)]
$x^2 = 9$ ....[Square root both sides]	$x^2 = 9$ ....[Square root both sides]
$x = \sqrt{9} = \pm 3$ .....[Solutions $(3,2)(-3,2)$]	$x = \sqrt{9} = \pm 3$ ... Solutions $(3, -2)(-3, -2)$]
N/B: The solutions of the simultaneous equation are; $(3,2)(-3,2)$ and	

$(3, -2)(-3, -2)$

Example 2: Eliminating (Y), solve the following simultaneous equation

$x^2 + y^2 - 13 = 0$ ....(1)

$x^2 - y^2 - 5 = 0$ ....(2)

Solution

$$\begin{bmatrix} x^2 + y^2 - 13 = 0 \\ x^2 - y^2 - 5 = 0 \end{bmatrix} = \begin{bmatrix} \mathbf{x^2 + y^2 = 13} \\ \mathbf{x^2 - y^2 = 5} \end{bmatrix}$$

Step 1: Determine the value of variable (Y)......................................[Eq(1) + Eq(2)]

$2x^2 = 18$ ....[Divide sides by (2)]

$x^2 = 9$ ....[Square root both sides]

$x = \sqrt{9} = \pm 3$ ....[(x = 3) and (x − 3)]

Step 2: Determine the value of variable (X).........[Using equation (2). That is,$(x^2 - y^2 = 5)$]

When y = 3	When y = −3
$x^2 - y^2 = 5$ ....[Substitute (y)]	$x^2 - y^2 = 5$ ....[Substitute (y)]
$x^2 - (3)^2 = 5$ ....[Work exponent]	$x^2 - (-3)^2 = 5$ ....[Work exponent]
$x^2 - 9 = 5$ ....[Collect like terms]	$x^2 - 9 = 5$ ....[Collect like terms]
$x^2 = 14$ ....,...[Square root both sides]	$x^2 = 14$ ....[Square root both sides]
Variable (x) = $\sqrt{14} \approx \pm 4$	Variable (x) = $\sqrt{14} \approx \pm 4$
N/B: The solutions of the simultaneous equation are; $(4,3)(-4,3)$ and $(4,-3)(-4,-3)$	

Test Your Understanding

Exercise 20: Solve the following simultaneous equation using the elimination method

1) $4x^2 - y^2 = 4$ and $4x^2 + y^2 = 4$...............................[**Answer: $(1,0)(-1,0)$**]
2) $3x^2 - 2y^2 = -5$ and $2x^2 - y^2 = -2$.[**Answer: $(1,2)(1,-2)$ and $(-1,2)(-1,-2)$**]

3) Graphical Method

Here, we apply the concept of parabola in graphing the non-linear equation. The three possible intersection expression are given below

148

Two Intersection	One Intersection	No Intersection
Two Solution	**One Solution**	**No Solution**

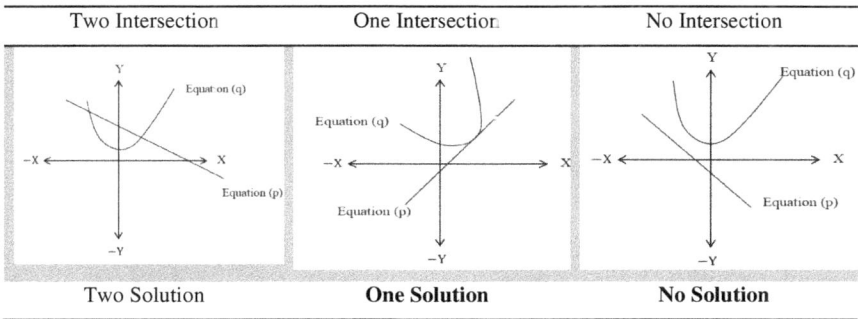

Example 1: Solve the simultaneous equation below using the graphical method

1) $Y = X^2 - 4$...(1)

2) $Y = -X - 2$...(2)

Solution

Step 1: Determine the coordinate pairs of linear equation...........................[$Y = -X - 2$]

When X = 0	When Y = 0	Illustration
$Y = -X - 2$.[Substitute X]	$Y = -X - 2$.[Substitute Y]	
$Y = -(0) - 2$ ....[Solve]	$0 = -X - 2$ ....[Solve]	
Variable (Y) = -2	Variable (X) = -2	

Step 2: Determination of vertex coordinate

Variable X	Variable Y	Illustration
$X = -\frac{b}{2(a)}$ Where; b = 0, a = 1 $X = -\frac{0}{2(1)}$ ....[Solve] Variable(X) = **0**	$Y = X^2 - 4$ ......[Substitute X] $Y = (0)^2 - 4$ ..[Work power] $Y = 0 - 4$ ....[Solve] Variable(Y) = -4	

Step 3: Determination of table of value and plotting

Table of value			Plotting [Linear and non-linear curve]
X	[Y = X² − 4]	Y	
3	$Y = (3)^2 - 4$	5	
2	$Y = (2)^2 - 4$	0	
1	$Y = (1)^2 - 4$	−3	
0	$\mathbf{Y = (0)^2 - 4}$	**−4**	
−1	$Y = (-1)^2 - 4$	−3	
−2	$Y = (-2)^2 - 4$	0	
−3	$Y = (-3)^2 - 4$	5	

N/B: The solutions of the simultenous equation are; $(-2, 0)$ and $(1, -3)$

Note: Using the substitution method (making Y subject of formula), the working equation is $[x^2 + x - 2 = 0]$ and solving the quadratic equation we obtain the roots$[(X = -2)$ and $(X = 1)]$. Solving for the values of variable (Y), [When $X = -2$, then $Y = 0$] and[When $X = 1$, then $Y = -3$]. The solution is therefore $(-2, 0)$ and $(1, -3)$.

Example 2: Solve the simultaneous equation using the graphical method

1) $Y = X^2 - 2X + 2$..(1)

2) $Y = 2X - 2$..(2)

Solution

Step 1: Determine the coordinate pairs of linear equation.......................[Y = 2X − 2]

When X = 0	When Y = 0	Illustration
$Y = 2X - 2$...[Substitute X]	$Y = 2X - 2$.[Substitute Y]	
$Y = 2(0) - 2$ ....[Solve]	$0 = 2X - 2$ ....[Solve]	
Variable (Y) = **−2**	Variable (X) = **1**	

150

Step 2: Determination of vertex coordinate

Variable X	Variable Y	Illustration
$X = -\dfrac{b}{2(a)}$ Where; $b = -2$, $a = 1$ $X = -\dfrac{-2}{2(1)}$ ....[Solve] Variable(X) $= 1$	$Y = X^2 - 2X + 2$.[Substitute X] $Y = (1)^2 - 2(1) + 2$ $Y = 1 - 2 + 2$ ....[Solve] Variable(Y) $= 1$	

Step 3: Determination of table of value and plotting

Table of value			Plotting [Linear and non-linear curve]
X	$[Y = X^2 - 2X + 2]$	Y	
3	$Y = (3)^2 - 2(3) + 2$	5	
2	$Y = (2)^2 - 2(2) + 2$	2	
1	$Y = (1)^2 - 2(1) + 2$	1	
0	$Y = (0)^2 - 2(0) + 2$	2	
−1	$Y = (-1)^2 - 2(-1) + 2$	5	

N/B: The solution of the simultenous equation is; $(2, 2)$

Note: Using the substitution method (making Y subject of formula), the working equation is $[x^2 - 4x + 4 = 0]$ with a discriminant of zero $[\Delta = 0]$ indicating unique solution. Solving the quadratic equation we obtain the root$[(X = 2)]$. Solving for the values of variable (Y), [When $X = 2$, then $Y = 2$]. The solution is$(2, 2)$.

Example 3: Solve the simultaneous equation below using the graphical method

1) $Y = X^2 + 1$...(1)

2) $Y = X - 1$...(2)

<u>Solution</u>

Step 1: Determine the coordinate pairs of linear equation.............................[Y = X − 1]

When X = 0	When Y = 0	Illustration
Y = X − 1 [Substitute X] Y = 0 − 1 ….…..[Solve] Variable (Y) = **−1**	Y = X − 1 [Substitute Y] 0 = X − 1 ……...[Solve] Variable (X) = **1**	

Step 2: Determination of vertex coordinate

Variable X	Variable Y	Illustration
$X = -\dfrac{b}{2(a)}$ Where; b = 0, a = 1 $X = -\dfrac{0}{2(1)}$ …..…..[Solve] Variable(X) = **0**	$Y = X^2 + 1$…...…..[Substitute X] $Y = (0)^2 + 1$…...[Work power] Y = 0 + 1 ……….…...…[Solve] Variable(Y) = **1**	

Step 3: Determination of table of value and plotting

Table of value			Plotting [Linear and non-linear curve]
X	$[Y = X^2 + 1]$	Y	
2	$Y = (2)^2 + 1$	5	
1	$Y = (1)^2 + 1$	2	
0	$Y = (0)^2 + 1$	**1**	
−1	$Y = (-1)^2 + 1$	2	
−2	$Y = (-2)^2 + 1$	5	
N/B: The is no solution since there is no intersection point betywen the curves			

Note: Using the substitution method (making Y subject of formula), the working equation is $[x^2 - x + 2 = 0]$ with a discriminant of zero $[\Delta = -7]$ indicating complex roots and graphically interpreted as no solution.

3.4 Inequality Equation

Inequality equations are equations with inequality sign separating the terms. The major inequality signs includes; less than($<$), greater than($>$), less than or equal to ($\leq$) and greater than or equal to($\geq$).

3.4.1 Graphical Illustration of Inequality Equation

When illustrating inequality equation graphically, we make use of bond circle when dealing with less than or equal to ($\leq$) and/or greater than or equal to ($\geq$) and non-bond circle when dealing with less than ($<$) and/or greater than ($>$) sign.

Example 1: Graphically represent ($X > -4$) and ($X \leq 4$)

Solution

$(X > -4)$	$(X \leq 4)$
N/B: Using non-bond circle and move from -4 right on the number line	**N/B**: Using bond circle and move from -4 left on the number line

Example 2: Graph $(-3 < X < 2)$ and $(-2 \leq X < 3)$

Solution

$(-3 < X < 2)$	$(-2 \leq X < 3)$

Example 3: Graphically represent $(0 \leq X \leq 5)$ and $(-4 \leq X < 0)$

153

$(0 \leq X \leq 5)$	$(-4 \leq X < 0)$

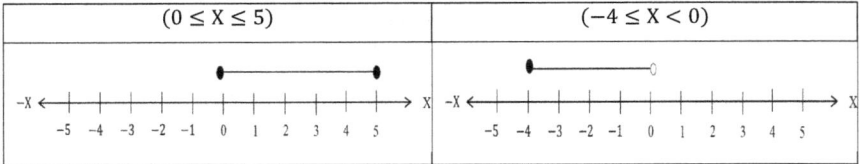

Example 4: Graphically represent the following on the same graph

1) $(X \leq -1)$ and $(X \geq 3)$

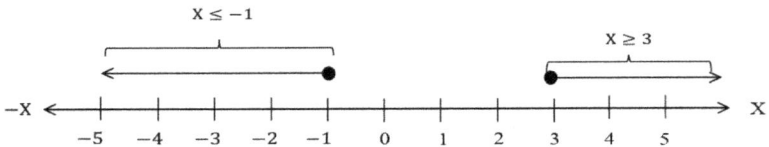

2) $(X > -3)$ and $(X \geq 0)$

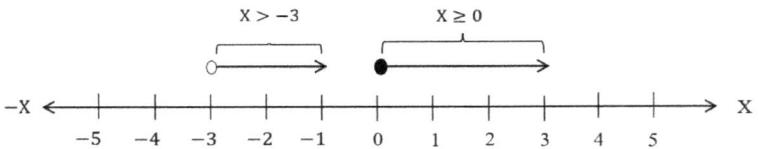

3) $(X \geq 2)$ and $(Y \leq 4)$

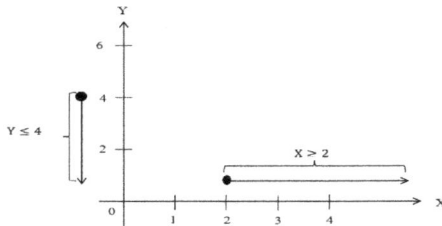

Solution

1) $(X \leq -1)$ and $(X \geq 3)$

2) $(X > -3)$ and $(X \geq 0)$

3) $(X \geq 2)$ and $(Y \leq 4)$

154

3.4.2 Inequality Equation with One variable

When solving inequality equations with one variable, the following principles are worth knowing. The addition and/or subtraction of a common value on both sides of the equation do not change the inequality sign.

	[$\leq$ and $\geq$]		[$<$ and $>$]	
Addition	$a + c \leq b + c = a + c \leq b + c$		$a + c < b + c = a + c < b + c$	
	$a + c \geq b + c = a + c \geq b + c$		$a + c > b + c = a + c > b + c$	
Subtraction	$a - c \leq b - c = a - c \leq b - c$		$a - c < b - c = a - c < b - c$	
	$a - c \geq b - c = a - c \geq b - c$		$a - c > b - c = a - c > b - c$	

When dealing with multiplication and division, the inequality sign changes when we multiply or divide both sides of equation by a negative value.

	[$\leq$ and $\geq$]		[$<$ and $>$]	
Multiplication	$a(c) \leq b(c) = ac \leq bc$	$a(c) \geq b(c) = ac \geq bc$	$a(c) < b(c) = ac < bc$	$a(c) > b(c) = ac > bc$
	$a(-c) \leq b(-c) = -ac \geq -bc$	$a(-c) \geq b(-c) = -ac \leq -bc$	$a(-c) < b(-c) = -ac > -bc$	$a(-c) > b(-c) = -ac < -bc$
Division	$\frac{a}{c} \leq \frac{b}{c} = \frac{a}{c} \leq \frac{b}{c}$	$\frac{a}{c} \geq \frac{b}{c} = \frac{a}{c} \geq \frac{b}{c}$	$\frac{a}{c} < \frac{b}{c} = \frac{a}{c} < \frac{b}{c}$	$\frac{a}{c} > \frac{b}{c} = \frac{a}{c} > \frac{b}{c}$
	$\frac{a}{-c} \leq \frac{b}{-c} = -\frac{a}{c} \geq -\frac{b}{c}$	$\frac{a}{-c} \geq \frac{b}{-c} = -\frac{a}{c} \leq -\frac{b}{c}$	$\frac{a}{-c} < \frac{b}{-c} = -\frac{a}{c} > -\frac{b}{c}$	$\frac{a}{-c} > \frac{b}{-c} = -\frac{a}{c} < -\frac{b}{c}$

N/B: When solving, we are recommended to respect the order of operation. This will help us avoid wrong mathematical manipulation.

Example 1: Solve $2(2X - 3) \geq X + 8$

Solution

$2(2X - 3) \geq X + 8$ ....[Open bracket]

$4X - 6 \geq X + 8$ ....[Add 6 on both sides]

$4X - 6 + 6 \geq X + 8 + 6 \Rightarrow 4X \geq X + 14$ ....[Subtract (X) on both sides]

$4X - X \geq X - X + 14 \Rightarrow 3X \geq 14$ ....[Divide both sides by (3)]

$\frac{3X}{3} \geq \frac{14}{3}$ ....[Solve equation]

$X \geq \frac{14}{3}$

Example 2: Compute $2(3 - Y) > 5(2 + Y)$

<u>**Solution**</u>

$2(3 - Y) > 5(2 + Y)$ ....[Open brackets]

$6 - 2Y > 10 + 5Y$ ....[Subtract 6 on both sides]

$6 - 6 - 2Y > 10 - 6 + 5Y \Rightarrow -2Y > 4 + 5Y$ ....[Subtract (5Y) on both sides]

$-2Y - 5Y > 4 + 5Y - 5Y \Rightarrow -7Y > 4$ ....[Divide both sides by (-7)]

$\frac{-7Y}{-7} > \frac{4}{-7}$ ..[Solve equation. Remember negative sign division changes the inequality sign]

$Y < -\frac{4}{7}$

Example 3: Solve the following inequalities

1) $-2(3 + K) < -44$

2) $\frac{X-9}{-4} \leq 2$

<u>**Solution**</u>

1) $-2(3 + K) < -44$

Method 1	Method 2
$-2(3 + K) < -44$ ....[Open bracket]	$-2(3 + K) < -44$ ...[Divide sides by (-2)]
$-6 - 2K < -44$ ......[Add (6) both sides]	$3 + K > 22$ ....[Subtract (3) both sides]
$-2K < -38$ ....[Divide sides by (-2)]	$K > 19$
$K > \frac{-38}{-2} > 19$	

2) $\frac{X-9}{-4} \leq 2$

Method 1	**Method 2**
$\frac{X-9}{-4} \leq 2$ ......[Multiply sides by (-4)]	$\frac{X-9}{-4} \leq 2 \Rightarrow -\frac{1}{4}(X - 9) \leq 2$.[Divide sides by $\left(-\frac{1}{4}\right)$]
$-4\left(\frac{X-9}{-4}\right) \leq 2(-4)$	$X - 9 \geq 2 \times -\frac{4}{1}X - 9 \geq -8$ ....[Add (8)]
$X - 9 \geq -8$ **X** $\geq$ **1**	**X** $\geq$ **1**

Example 4: Solve the following inequalities

1) $4 + 2(a + 5) < -2(-a - 4)$

2) $\frac{X}{5} \geq -\frac{6}{5}$

156

1) $4 + 2(a + 5) < -2(-a - 4)$

    $4 + 2(a + 5) < -2(-a - 4)$ ....[Open brackets]

    $2a + 14 < 2a + 8$ ....[Subtract 14 both sides]

    $2a < 2a - 6$ ....[Subtract (2a) both sides]

    $0 < -6 \equiv 0 > 6$ ....[No solution]

2) $\frac{X}{5} \geq -\frac{6}{5}$

Method 1	Method 2
$\frac{X}{5} \geq -\frac{6}{5}$ ....[Multiply both sides by 5]	$\frac{X}{5} \geq -\frac{6}{5} \equiv \left(\frac{1}{5}\right) X \geq -\frac{6}{5}$ ...[Divide sides by $\left(\frac{1}{5}\right)$]
$5 \left(\frac{X}{5}\right) \geq 5 \left(-\frac{6}{5}\right) \Rightarrow X \geq -\frac{30}{5}$ ...[Simplify]	$X \geq -\frac{6}{5} \times \frac{5}{1} \Rightarrow X \geq -\frac{30}{5}$ ....[Simplify]
$X \geq -6$	$X \geq -6$

Example 5: Solve the following inequalities

    1) $14 \leq 4X - 2 \leq 18$

    2) $-11 \leq 3Y - 2 \leq -5$

    3) $-7 < -\frac{3}{4}A - 1 \leq 11$

1) $14 \leq 4X - 2 \leq 18$

    $14 \leq 4X - 2 \leq 18$ ....[Add 2 on all sides]

    $14 + 2 \leq 4X - 2 + 2 \leq 18 + 2 \Rightarrow 16 \leq 4X \leq 20$ ....[Divide sides by (4)]

    $\frac{16}{4} \leq \frac{4X}{4} \leq \frac{20}{4}$ ....[Simplify fractions]

    $4 \leq X \leq 5$

2) $-11 \leq 3Y - 2 \leq -5$

    $-11 \leq 3Y - 2 \leq -5$ ....[Add (2) on all sides]

    $-11 + 2 \leq 3Y - 2 + 2 \leq -5 + 2 \Rightarrow -9 \leq 3Y \leq -3$ .....[Divide both sides by (3)]

    $\frac{-9}{3} \leq \frac{3Y}{3} \leq \frac{-3}{3}$ ....[Simplify fractions]

    $-3 \leq Y \leq -1$

3) $-7 < -\frac{3}{4}A - 1 \leq 11$

    $-7 < -\frac{3}{4}A - 1 \leq 11$ ....[Add (1) to all sides of equation]

$$-7 + 1 < -\frac{3}{4}A - 1 + 1 \leq 11 + 1 \Rightarrow -6 < -\frac{3}{4}A \leq 12 \text{ .[Divide all sides by } \left(-\frac{3}{4}\right)]$$

$$-6 \times -\frac{4}{3} > A \geq 12 \times -\frac{4}{3} \Rightarrow \frac{24}{3} > A \geq -\frac{48}{3} \text{ ....[Simplify fractions]}$$

$8 > A \geq -16$

Test Your Understanding

Exercise 23: Compute the following inequalities

1) $X + 10 \geq 20$ ....[**Answer: X $\geq$ 10**]

2) $-10 + Y \leq -4$...[**Answer: Y $\leq$ 6**]

3) $12 < -2A$...[**Answer: A $<$ −6**]

4) $-4Y \geq -16$ ....[**Answer: Y $\leq$ 4**]

5) $\frac{X}{-2} \leq -4$...[**Answer: X $\geq$ 8**]

Exercise 24: Solve the following inequalities

1) $-5(X + 2) \geq -3(X + 4)$...[**Answer: X $\leq$ −3**]

2) $3 - 5Y \leq 2(Y + 5)$ ....[**Answer: Y $\geq$ −1**]

3) $2A + 4 > 2(2 + A)$ ....[**Answer: No Solution**]

4) $2X > -8 + X$..[**Answer: X $>$ −8**]

5) $5 \geq \frac{D}{5} + 1$ ....[**Answer: D $\leq$ 20**]

Exercise 25: Solve the following inequalities

1) $2 \geq \frac{-2A+2}{-3} > -3$...[**Answer: $4 \geq A > -\frac{7}{2}$**]

2) $-21 \leq -\frac{2}{3}Y + 9 < 7$ ....[**Answer: 45 $\geq$ Y $>$ 3**]

3) $3 \leq 9 + X \leq 7$ ....[**Answer: −6 $\leq$ X $\leq$ −2**]

3.4.3 Inequality Equation with Two Variables

In inequality equation with two variables, we change the equality sign and calculate the equation coordinates by letting each variable to be zero in each equation.

Example 1: Graph the linear inequality equations below

1) $2X + 3Y \geq 6$

2) $4X - Y < 4$

Solution

Step 1: Determination of coordinates

	When Y= 0	When X = 0
Equation 1 $[2X + 3Y = 6]$	$2X + 3(0) = 6$ ....[Solve] $X = 3$ ....[X $\geq$ 3]	$2(0) + 3Y = 6$ ....[Solve] $Y = 3$ ....[X $\geq$ 2]
Equation 2 $[4X - Y = 4]$	$4X - 0 = 4$ ....[Solve] $X = 1$ ....[X < 1]	$4(0) - Y = 4$ ....[Solve] $Y = -4$ ....[Y < -4]

Step 2: Plotting and shading

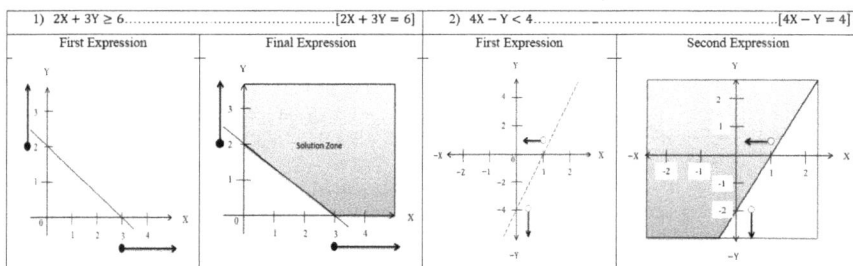

Example 2: Graph the linear inequality equations below

1) $4X \leq 8$

2) $Y \geq -3$

Solution

Step 1: Determination of coordinates

$4X \leq 8 \Rightarrow 4X = 8$ ....[X = 2 and by default Y = 0]

$Y \geq -3 \Rightarrow Y = -3$ ....[Y = -3 and by default X = 0]

Step 2: Plotting and shading

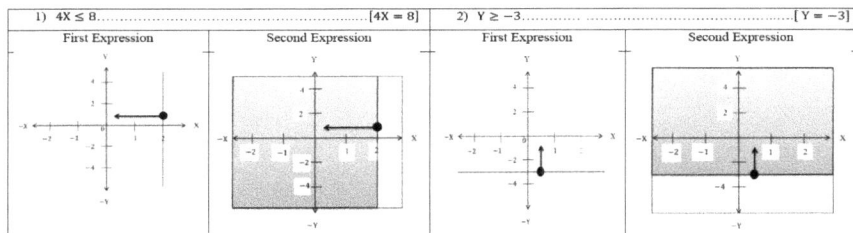

159

Example 3: Solve the simultaneous inequality equation below using the graphical method

1) $X \geq -3$ ....(1)

2) $Y \leq X + 2$ ....(2)

Solution

Step 1: Determination of coordinates

$X \geq -3 \Rightarrow X = -3$ ....[$X = -3$ and by default $Y = 0$]

$Y \leq X + 2 \Rightarrow Y = X + 2$ ....[$Y = 2$ and $X = -2$]

Step 2: Plotting and shading

Illustration	Explanation
	N/B: Intersection coordinate pair $(-3, -1)$ Equation 1.......[$X \geq -3 \Rightarrow X + 0Y = -3$] $-3 + 0(-1) = -3$ ....[Open bracket] $-3 + 0 = -3$ ....[Simplify left side] $\mathbf{-3 = -3}$ Equation 2.........[$Y = X + 2 \Rightarrow Y - X = 2$] $-1 - (-3) = 2$ ....[Open bracket] $-1 + 3 = 2$ ....[Simplify left side] $\mathbf{2 = 2}$

Example 4: Solve the following simultaneous inequality equation below.

1) $Y > \frac{1}{2}X + 1$

2) $Y \leq -\frac{1}{2}X + 3$

Solution

Step 1: Determination of coordinates

$Y > \frac{1}{2}X + 1 \Rightarrow Y - \frac{1}{2}X > 1 \Rightarrow Y - \frac{1}{2}X = 1$..................[$X = -2$ and by default $Y = 1$]

$Y \leq -\frac{1}{2}X + 3 \Rightarrow Y + \frac{1}{2}X \leq 3 \Rightarrow Y + \frac{1}{2}X = 3$ ....[$Y = 3$ and $X = 6$]

160

Step 2: Plotting and shading

Illustration	Explanation
	N/B: The intersection point coordinate pair is $(2, 2)$ Equation 1............$\left[Y > \frac{1}{2}X + 1 \Rightarrow Y - \frac{1}{2}X = 1 \right]$ $2 - \frac{1}{2}(2) = 1$ ....[Simplify left side] **$1 = 1$** Equation 2........$\left[Y \leq -\frac{1}{2}X + 3 \Rightarrow Y + \frac{1}{2}X = 3 \right]$ $2 + \frac{1}{2}(2) = 3$ ... .[Simplify left side] **$3 = 3$**

Example 5: Solve the following simultaneous inequality equation.

1) $Y + X \leq 3$

2) $Y - X \leq 3$

Solution

Step 1: Determination of coordinates

$Y + X \leq 3 \Rightarrow Y + X = 3$...[X = 3 and Y = 3]

$Y - X \leq 3 \Rightarrow Y - X = 3$ ....[Y = 3 and X = -3]

Step 2: Plotting and shading

Illustration	Explanation
	N/B: The intersection coordinate pair is $(0, 3)$ Equation 1.$\left[Y + X \leq 3 \Rightarrow Y + X = 3 \right]$ $3 + 0 = 3$ ....[Simplify left side] **$3 = 3$** Equation 2$\left[Y - X \leq 3 \Rightarrow Y - X = 3 \right]$ $3 - 0 = 3$ ....[Simplify left side] **$3 = 3$**
Test Your Understanding	
Exercise 26: Graph the linear inequality equations below 1) $Y \geq \frac{1}{2}X + 2$ ....[**See Appendix**] 2) $-2X > 6$ ....[**See Appendix**] 3) $4Y > 12$ ....[**See Appendix**]	

4) $3Y + 2X \leq 0$ ....[See Appendix]

Exercise 27: Solve the following simultaneous linear inequality equations below.

1) $2X + 4Y \leq -4$ and $2X + Y > 2$...............................[Answer: See Appendix]

2) $Y \geq 2X - 3$ and $Y < \frac{2}{3}X + 1$.......................................[Answer: See Appendix]

Exercise 28: Solve the following simultaneous linear inequality equations below.

1) $Y + X \leq 2$ and $-Y + X < 2$......................................[Answer: See Appendix]

2) $Y \geq 2 - X$ and $Y \leq -2 - X$....................................[Answer: See Appendix]

3.5 Absolute Value Equation

An absolute value equation is an equation in which the variable (a)is within an absolute value sign ($|a|$). The analysis of absolute value equation will be divided into simple absolute value and absolute value equation with inequality.

3.5.1   Simple Absolute Value Equation

Here, we talk of simple absolute value equation with equal to (=) sign. The following principles are worth knowing when solving simple absolute value equation.

When $|X| = 0$..[There is one solution]

When $|X| = -P$..[There is no solution]

When $|X| = +P \Rightarrow \begin{cases} X = -P \\ X = P \end{cases}$ ....[There are two solution]

When $|X| = |Y| \Rightarrow \begin{cases} X = -Y \\ X = Y \end{cases}$ ....[There are two solution]

Example 1: Solve $|Y - 7| = 3$

<u>**Solution**</u>

Step 1: Formula. $|X| = +P \Rightarrow \begin{cases} X = -P \\ X = P \end{cases}$, where; $X = Y - 7$ and $P = 3$

Step 2: Data substitution and solving

$X = -P$	$X = P$
$Y - 7 = -(3)$ ....[Open bracket]	$Y - 7 = 3$ ....[Add (7) both sides]
$Y - 7 = -3$ ....[Add (7) both sides]	$Y = 10$
$Y = 4$	

Example 2: Solve the absolute equation; $\frac{1}{4}|2X - 6| + 1 = 2$.

162

Solution

Step 1: Simplify equation

$\frac{1}{4}|2X - 6| + 1 = 2$ ....[Subtract (1) or both sides of the equation]

$\frac{1}{4}|2X - 6| = 1$ ....[Divide both sides by $\left(\frac{1}{4}\right)$]

$|2X - 6| = 4$ ....[Simplified form]

Step 2: Formula. $|X| = +P \Rightarrow \begin{cases} X = -P \\ X = P \end{cases}$, where; $X = 2X - 6$ and $P = 4$

Step 3: Data substitution and solving

$X = -P$	$X = P$
$2X - 6 = -(4)$ ....[Open bracket]	$2X - 6 = 4$ ....[Add (6) both sides]
$2X - 6 = -4$ ....[Add (6) both sides]	$2X = 10$ ....[Divide both sides by (2)]
$2X = 2$ ....[Divide both sides by (2)]	$X = 5$
$X = 1$	

Example 3: Compute the absolute value equation; $-3|X - 1| - 6 = 3$.

Solution

Step 1: Simplify equation

$-3|X - 1| - 6 = 3$ ....[Add (6) both sides]

$-3|X - 1| = 9$ ....[Divide both sides by (-3)]

$|X - 1| = -3$

N/B: The absolute value $|X - 1|$ is equal to negative value(-3), following the principle$(|X| = -P)$, we conclude that the equation is not having a solution.

Example 4: Calculate; $\left|12\left(X + \frac{1}{2}\right)\right| + 3 = 3$.

Solution

Step 1: Simplify equation

$\left|12\left(X + \frac{1}{2}\right)\right| + 3 = 3$ ....[Open bracket]

$|12X + 6| + 3 = 3$ ....[Subtract (3) on both sides]

$|12X + 6| = 0$ ....[Simplified form]

N/B: When $|X| = 0$, it shows the equation is having one solution. To solve the equation, we remove the absolute value sign, equate to zero and solve for the variable (unknown).

Step 2: Data substitution and solving

$12X + 6 = 0$ ....[Subtract (6) on both sides]

$12X = -6$ ....[Divide both sides by (12)]

$X = -\frac{1}{2}$

Example 5: Solve $|P + 10| = |3P - 2|$

Solution

Step 1: Formula. $|X| = |Y| \Rightarrow \begin{cases} X = -Y \\ X = Y \end{cases}$, where; $X = P + 10$ and $Y = 3P - 2$

Step 2: Data substitution and solving

$X = -Y$	$X = Y$
$P + 10 = -(3P - 2)$ ....[Open bracket]	$P + 10 = (3P - 2)$ ....[Open bracket]
$P + 10 = -3P + 2$ [Subtract 10 both sides]	$P + 10 = 3P - 2$...[Subtract 10 both sides]
$P = -3P - 8$ ....[Add (3P) both sides]	$P = 3P - 12$ ....[Subtract (3P) both sides]
$4P = -8$ ....[Divide both sides by (4)]	$-2P = -12$ ....[Divide both sides by (−2)]
$P = -2$	$P = 6$

Step 3: Justification

$P = -2$	$|-2 + 10| = |3(-2) - 2| \Rightarrow |8| = |8|$.[Only positive values are considered]
$P = 6$	$|6 + 10| = |3(6) - 2| \Rightarrow |16| = |16|$

Example 6: Solve $|5(X - 2)| = |3X|$

Solution

Step 1: Formula. $|X| = |Y| \Rightarrow \begin{cases} X = -Y \\ X = Y \end{cases}$, where; $X = 5(X - 2)$ and $Y = 3X$

Step 2: Data substitution and solving

$X = -Y$	$X = Y$
$5(X - 2) = -(3X)$..........[Open bracket]	$5(X - 2) = (3X)$..................[Open bracket]
$5X - 10 = -3X$ ......[Add (10) both sides]	$5X - 10 = 3X$ ....[Add (10) both sides]
$5X = -3X + 10$ ....[Add (3X) both sides]	$5X = 3X + 10$ ....[Subtract (3X) both sides]
$8X = 10$ ....[Divide sides by (8)]	$2X = 10$ ....[Divide sides by (2)]
$X = 5/4$	$X = 5$

Step 3: Justification

X = 5/4	$\left\|5\left(\frac{5}{4}-2\right)\right\| = \left\|3\frac{5}{4}\right\| \Rightarrow \left\|\frac{15}{4}\right\| = \left\|\frac{15}{4}\right\|$
X = 5	$\|5(5-2)\| = \|3(5)\| \Rightarrow \|15\| = \|15\|$

Example 7: Solve $|t - 1| = t^2 + 4t - 5$

Solution

Step 1: $|X| = +P \Rightarrow \begin{cases} X = -P \\ X = P \end{cases}$, where; $X = t - 1$ and $P = t^2 + 4t - 5$

Step 2: Data substitution and solving

$X = -P$	$X = P$
$t - 1 = -(t^2 + 4t - 5)$ ....[Open bracket]	$t - 1 = t^2 + 4t - 5$ ....[Collect like terms]
$t - 1 = -t^2 - 4t + 5$.[Collect like terms]	$t^2 + 3t - 4 = 0$ ......[Quadratic Equation]
$t^2 + 5t - 6 = 0$ ......[Quadratic Equation]	
Roots are $[-6$ and $1]$	**Roots are $[-4$ and $1]$**

Step 3: Determination of appropriate solution

$X = -P$	$t = -6$	$\|-6 - 1\| = -6^2 + 4(-6) - 5 \Rightarrow \|7\| = 7$
	$t = 1$	$\|1 - 1\| = 1^2 + 4(1) - 5 \Rightarrow \|0\| = 0$
$X = P$	$t = -4$	$\|-4 - 1\| = -4^2 + 4(-4) - 5 \Rightarrow \|5\| \neq -5$
	$t = 1$	$\|1 - 1\| = 1^2 + 4(1) - 5 \Rightarrow \|0\| = 0$

N/B: The appropriate solution is $X = -P$ since its roots when substituted in the original equation makes both sides equal.

Test Your Understanding

Exercise 29: Solve the following absolute value equations.

1) $|-5Y - 7| + 3 = 10$...**[Answer: $-\frac{14}{5}$ and 0]**

2) $2 - 7|X + 4| = -12$...**[Answer: -6 and -2]**

3) $|5X - 7| + 10 = 3$...**[Answer: No Solution]**

Exercise 30: Solve the following absolute value equations.

1) $|5Y + 8| = |2Y + 3|$...**[Answer: $-\frac{11}{7}$ and $-\frac{5}{3}$]**

2) $|3(X + 1)| = |7X|$...**[Answer: $-\frac{3}{10}$ and $\frac{3}{4}$]**

Exercise 31: Solve the following absolute value equations

1) $|X + 3| = X^2 - 4X - 3$...**[Answer: -1 and 6]**

2) $|t^2 + 9t + 14| = 0$...**[Answer: -2 and -7]**

3.5.2 Absolute Value Equation with Inequality

In absolute value equation with inequality, the equal to sign is changed to an inequality sign. When solving absolute value equation with inequality, the following principles are worth knowing.

When $\|X\| > 0$ ....[There is no solution]	
When $\|X\| \leq 0$ ....[There is one solution]	
When $\|X\| \geq 0$..........................[Solution is all real numbers ($\mathbb{R}$)]	
When $\|X\| > 0$.............[Solution is all real number ($\mathbb{R}$) except zero]	
When $\|X\| \leq -P$..[There is no solution]	
When $\|X\| \geq -P$.....................[Solution is all real numbers ($\mathbb{R}$)]	
When $\|X\| \leq +P$, then $- P \leq X \leq P$.....................[Two solutions]	
When $\|X\| \geq +P$, then $X \leq -P$ or $X \geq P$................[Two solutions]	

Example 1: Solve $\|2Y - 3\| \leq 5$

Solution

Step 1: Formula. $\|X\| \leq +P$, then $- P \leq X \leq P$, where; $X = 2Y - 3$ and $P = 5$

Step 2: Data substitution and solving

$-5 \leq 2Y - 3 \leq 5$ ....[Add 3 all sides of the equation]

$-5 + 3 \leq 2Y - 3 + 3 \leq 5 + 3$ ....[Simplify equation]

$-2 \leq 2Y \leq 8$ ....[Divide both sides by (2)]

$\frac{-2}{2} \leq \frac{2Y}{2} \leq \frac{8}{2}$ ....[Simplify fraction]

$\mathbf{-1 \leq Y \leq 4}$

Example 2: Solve $-5\|2X + 2\| - 3 \geq -3$

Solution

Step 1: Simplify equation

$-5\|2X + 2\| - 3 \geq -3$ ....[Add (3) on both sides]

$-5\|2X + 2\| \geq 0$.[Divide both sides by (-5). Remember negative change the inequality sign]

$\|2X + 2\| \leq 0$ ....[$\|X\| \leq 0$, indicating that the is one solution]

Step 2: Data substitution and solving

$2X + 2 = 0$ ....[Subtract (2) on both sides]

$2X = -2$ ....[Divide both sides by (2)]

$\mathbf{X = -1}$

Example 3: Compute $\|3(2Y - 1)\| \geq 15$

Solution

Step 1: Formula. $\|X\| \geq +P$, then $X \leq -P$ or $X \geq P$

Where; $X = 3(2Y - 1) = 6Y - 3$ and $P = 15$

166

Step 2: Data substitution and solving

$X \leq -P$	$X \geq P$
$6Y - 3 = -15$ .....[Add (3) on both sides]	$6Y - 3 = 15$ ... .[Add (3) on both sides]
$6Y = -12$ ....[Divide both side by (6)]	$6Y = 18$ ....[Divide both side by (6)]
$Y = -2$	$Y = 3$

Example 4: Solve the following absolute value equations with inequalities

1) $\left| \frac{1}{3}Y - \frac{2}{3} \right| \leq 1$

2) $3 \left| \frac{2}{2}X + 2 \right| + 6 > 15$

<u>Solution</u>

1) $\left| \frac{1}{3}Y - \frac{2}{3} \right| \leq 1$

Step 1: Formula. $|X| \leq +P$, then $-P \leq X \leq P$, where; $X = \frac{1}{3}Y - \frac{2}{3}$ and $P = 1$

Step 2: Data substitution and solving

$-1 \leq \frac{1}{3}Y - \frac{2}{3} \leq 1$ ....[Add $\left(\frac{2}{3}\right)$ on all sides of equation]

$-1 + \frac{2}{3} \leq \frac{1}{3}Y - \frac{2}{3} + \frac{2}{3} \leq 1 + \frac{2}{3}$ ....[Simplify equation]

$-\frac{1}{3} \leq \frac{1}{3}Y \leq \frac{5}{3}$ ....[Divide both sides by $\left(\frac{1}{3}\right)$]

$-\frac{3}{3} \leq Y \leq \frac{15}{3}$ ....[Simplify fractions]

$\mathbf{-1 \leq Y \leq 5}$

2) $3 \left| \frac{1}{2}Y + 2 \right| + 6 > 15$

Step 1: Simplify equation

$3 \left| \frac{1}{2}Y + 2 \right| + 6 > 15$ ....[Subtract (6) on both sides]

$3 \left| \frac{1}{2}Y + 2 \right| > 9$ ....[Divide both sides by (3)]

$\left| \frac{1}{2}Y + 2 \right| > 3$ ....[$|X| \geq +P$, then $X \leq -P$ or $X \geq P$. Two solutions]

Step 2: Formula. $|X| \geq +P$, then $X \leq -P$ or $X \geq P$, where; $X = \frac{1}{2}Y + 2$ and $P = 3$

$X \leq -P$	$X \geq P$
$\frac{1}{2}Y + 2 = -3$ ......[Subtract (2) both sides]	$\frac{1}{2}Y + 2 = 3$ ....[Subtract (2) both sides]
$\frac{1}{2}Y = -5$ ....[Divide both sides by $\left(\frac{1}{2}\right)$]	$\frac{1}{2}Y = 1$ ....[Divide both sides by $\left(\frac{1}{2}\right)$]
$Y = -5 \times \frac{2}{1} = \mathbf{-10}$	$Y = 1 \times \frac{2}{1} = \mathbf{2}$

Test Your Understanding

Exercise 32: Solve the following absolute value equations.

1) $|X + 5| \leq 3$..**[Answer: $-8 \leq X \leq -2$]**

2) $|Y - 2| < 6$ ....**[Answer: $-4 < Y < 8$]**

Exercise 33: Compute the following,

1) $|10 - X| \leq -4$...**[Answer: No solution]**

2) $|Y + 2| \geq -2$...**[Answer: All real numbers $(\mathbb{R})$]**

168

CHAPTER FOUR

MATRIX AND LINEAR EQUATION

4.1 Meaning of Matrix

A matrix (matrices) is a table of numbers arranged in rows and columns. Matrix is also seen as the combination of vectors. The size of a matrix is determined by the number of rows (m) and the number of columns (n) in the matrix. The order of matrix is given as[(rows x columns) or (nxm)]. The numbers in a matrix are called element and the different types of matrices can be classified in terms of shapes and elements.

1) Classification in Terms of Shape

In terms of shapes, matrix can be a square matrix (number of rows equal number of columns), non-square matrix (number of rows is not equal to number of columns), row matrix (only one row and many columns) and column matrix (one column and many rows).

Square Matrix	Non-Square Matrix		Row Matrix	Column Matrix
$A = \begin{bmatrix} a_{11} & a_{12} & a_{13} \\ a_{21} & a_{22} & a_{23} \\ a_{31} & a_{32} & a_{33} \end{bmatrix}$	$A = \begin{bmatrix} a_{11} & a_{12} \\ a_{21} & a_{22} \\ a_{31} & a_{23} \end{bmatrix}$	$A = \begin{bmatrix} a_{11} & a_{12} & a_{13} \\ a_{21} & a_{22} & a_{23} \end{bmatrix}$	$A = \begin{bmatrix} a_{11} & a_{12} & \cdots & a_{1n} \end{bmatrix}$	$A = \begin{bmatrix} a_{11} \\ a_{21} \\ \vdots \\ a_{m1} \end{bmatrix}$

N/B: When dealing with a square matrix, we make use of words such as diagonal elements and off-diagonal elements. Diagonal elements are elements in a square matrix having same rows and columns position $[a_{11}, a_{22}, a_{33}]$ and the sum of diagonal element is called the trace of a matrix.

2) Classification in Term of Elements

In terms of elements (numbers in a matrix), the different types of matrices are explained and illustrated below. In our illustration, we will mostly use (2x2) and (3x3) matrices.

1) Diagonal Matrix: A diagonal matrix is a matrix in which the diagonal elements are non-zero and the off-diagonal elements are zeros. A special type of diagonal matrix is identify matrix.

2) Identify and Null Matrix: An identify matrix is one in which the diagonal elements are 1 and the off-diagonal are zeros. A null matrix is one in which all the elements in the matrix are zero.

3) Symmetric and Non-Symmetric Matrix: Symmetric matrix is a matrix in which the opposite off-diagonal elements are the same. In a non-symmetric matrix, the opposite off-diagonal elements are not the same.

Diagonal Matrix		Identify (Unit) Matrix		Null Matrix	
$A = \begin{bmatrix} a_{11} & 0 \\ 0 & a_{22} \end{bmatrix}$	$A = \begin{bmatrix} a_{11} & 0 & 0 \\ 0 & a_{22} & 0 \\ 0 & 0 & a_{33} \end{bmatrix}$	$A = \begin{bmatrix} 1 & 0 \\ 0 & 1 \end{bmatrix}$	$A = \begin{bmatrix} 1 & 0 & 0 \\ 0 & 1 & 0 \\ 0 & 0 & 1 \end{bmatrix}$	$A = \begin{bmatrix} 0 & 0 \\ 0 & 0 \end{bmatrix}$	$A = \begin{bmatrix} 0 & 0 & 0 \\ 0 & 0 & 0 \\ 0 & 0 & 0 \end{bmatrix}$
(2x2)	(3x3)	(2x2)	(3x3)	(2x2)	(3x3)

Symmetric Matrix		Non-Symmetric Matrix	
$A = \begin{bmatrix} a_{11} & y \\ y & a_{22} \end{bmatrix}$	$A = \begin{bmatrix} a_{11} & x & y \\ x & a_{22} & z \\ y & z & a_{33} \end{bmatrix}$	$A = \begin{bmatrix} a_{11} & x \\ y & a_{22} \end{bmatrix}$	$A = \begin{bmatrix} a_{11} & y & p \\ x & a_{22} & k \\ z & q & a_{33} \end{bmatrix}$
(2x2)	(3x3)	(2x2)	(3x3)

4) Skew Symmetric Matrix: It is a symmetric matrix in which the diagonal elements are zero (0) and the opposite off-diagonal elements are the same in value but differ in mathematical signs ($\pm$)

2x2 Matrix		3x3 Matrix		
First Expression	Second Expression	First Expression	Second Expression	Third Expression
$A = \begin{bmatrix} 0 & -x \\ x & 0 \end{bmatrix}$	$A = \begin{bmatrix} 0 & x \\ -x & 0 \end{bmatrix}$	$A = \begin{bmatrix} 0 & x & y \\ -x & 0 & z \\ -y & -z & 0 \end{bmatrix}$	$A = \begin{bmatrix} 0 & -x & -y \\ x & 0 & -z \\ y & z & 0 \end{bmatrix}$	$A = \begin{bmatrix} 0 & \pm x & \pm y \\ \mp x & 0 & \pm z \\ \mp y & \mp z & 0 \end{bmatrix}$

Note: A given matrix rows and columns can be interchange and the resulted matrix is called the matrix transpose. The transpose of a square matrix (A) is denoted (A^T).

(2x2) Matrix		(3x3) Matrix	
Original Matrix	Transpose Matrix	Original Matrix	Transpose Matrix
$A = \begin{bmatrix} a & c \\ b & d \end{bmatrix}$	$A^T = \begin{bmatrix} a & b \\ c & d \end{bmatrix}$	$A = \begin{bmatrix} a & d & g \\ b & e & h \\ c & f & i \end{bmatrix}$	$A^T = \begin{bmatrix} a & d & g \\ d & e & h \\ g & f & i \end{bmatrix}$
N/B: Swap rows to columns or interchange the position of off-diagonal elements			

Example 1: Use the elements and their positions; $a_{11} = 2$, $a_{31} = 6$, $a_{23} = 16$, $a_{12} = 8$, $a_{22} = 10$, $a_{21} = 4$, $a_{33} = 18$, $a_{13} = 14$ and $a_{32} = 12$ to develop the following,

1) Row matrix using the second rows elements
2) A 3x3 matrix and bond the diagonal elements
3) A 2x3 matrix

Solution

1) Row matrix using the second rows elements

Hint: Element position first digit represents the rows and using the second row means will include only elements whose first position digit is 2. The second digit shows the columns and guides us in the chronological placement of row elements

Step 1: Identification of element

➢ Non-chronological identification: $a_{23} = 16$, $a_{22} = 10$ and $a_{21} = 4$

➢ Chronological identification: $a_{21} = 4$, $a_{22} = 10$ and $a_{23} = 16$

Step 2: Development of matrix...Let us call the matrix (M)]

$$M = [4 \quad 10 \quad 16]$$

2) A 3x3 matrix and bond the diagonal elements

Hint. The digit 3x3 means 3 rows and 3 columns matrix. Therefore, a 3x3 matrix can be seen as a combination of three rows vector or three columns vector

Column Vector Analysis				Row Vector Analysis			
1st vector	2nd vector	3rd vector	Matrix				Matrix ...[Call it matrix P]
(a_{n1})	(a_{n2})	(a_{n3})	Let's call it matrix W	1st vector	(a_{1m})	$[2 \quad 8 \quad 14]$	$P = \begin{bmatrix} 2 & 8 & 14 \\ 4 & 10 & 16 \\ 6 & 12 & 18 \end{bmatrix}$
$\begin{bmatrix} a_{11} = 2 \\ a_{21} = 4 \\ a_{31} = 6 \end{bmatrix}$	$\begin{bmatrix} a_{12} = 8 \\ a_{22} = 10 \\ a_{32} = 12 \end{bmatrix}$	$\begin{bmatrix} a_{13} = 14 \\ a_{23} = 16 \\ a_{33} = 18 \end{bmatrix}$	$W = \begin{bmatrix} 2 & 8 & 14 \\ 4 & 10 & 16 \\ 6 & 12 & 18 \end{bmatrix}$	2nd vector	(a_{2m})	$[4 \quad 10 \quad 16]$	
				3rd vector	(a_{3m})	$[6 \quad 12 \quad 18]$	

3) A 2x3 matrix

Hint: The matrix order 2x3 means 2 rows and 3 columns. Since the data given is having 3 rows, we can derive the following 2x3 matrices.

Matrices	First Matrix	Second Matrix	Third Matrix
Rows Utilized	1st and 2nd Rows	1st and 3rd Rows	2nd and 3rd Rows
Symbolic Expression	$A = \begin{bmatrix} a_{11} & a_{12} & a_{13} \\ a_{21} & a_{22} & a_{23} \end{bmatrix}$	$B = \begin{bmatrix} a_{11} & a_{12} & a_{13} \\ a_{31} & a_{32} & a_{33} \end{bmatrix}$	$C = \begin{bmatrix} a_{21} & a_{22} & a_{23} \\ a_{31} & a_{32} & a_{33} \end{bmatrix}$
Numerical Expression	$A = \begin{bmatrix} 2 & 8 & 14 \\ 4 & 10 & 16 \end{bmatrix}$	$B = \begin{bmatrix} 2 & 8 & 14 \\ 6 & 12 & 18 \end{bmatrix}$	$C = \begin{bmatrix} 4 & 10 & 16 \\ 6 & 12 & 18 \end{bmatrix}$

Example 2: Use the elements and their positions; $a_{11} = 1$, $a_{31} = 3$, $a_{23} = 5$, $a_{12} = 2$, $a_{22} = 4$, $a_{21} = 2$, $a_{33} = 6$, $a_{13} = 3$ and $a_{32} = 5$ to develop the following,

1) A 2x2 matrix using 1st and 3rd rows and 2nd and 3rd columns

2) A 3x2 matrix using 1st and 3rd columns

3) Transpose of a 3x3 matrix

Solution

1) A 2x2 matrix using 1st and 3rd rows and 2nd and 3rd columns

Hint: The matrix order 2x2 means the matrix should have 2 rows and 2 columns. Therefore the matrix should have four elements $[2x2 = 4]$.

Symbolic Expression	Symbolic and numerical Expression	Numerical Expression
$A = \begin{bmatrix} a_{12} & a_{13} \\ a_{22} & a_{23} \end{bmatrix}$	$A = \begin{bmatrix} a_{12} = 2 & a_{13} = 3 \\ a_{22} = 4 & a_{23} = 5 \end{bmatrix}$	$A = \begin{bmatrix} 2 & 3 \\ 4 & 5 \end{bmatrix}$

2) A 3x2 matrix using 1st and 3rd columns

Hint: The matrix order 3x2 means the matrix should have 3 rows and 2 columns. Therefore the matrix should have six elements $[3 \times 2 = 6]$.

Symbolic Expression	Symbolic and numerical Expression	Numerical Expression
$B = \begin{bmatrix} a_{11} & a_{13} \\ a_{21} & a_{23} \\ a_{31} & a_{33} \end{bmatrix}$	$B = \begin{bmatrix} a_{11} = 1 & a_{13} = 3 \\ a_{21} = 2 & a_{23} = 5 \\ a_{31} = 3 & a_{33} = 6 \end{bmatrix}$	$B = \begin{bmatrix} 1 & 3 \\ 2 & 5 \\ 3 & 6 \end{bmatrix}$

3) Transpose of a 3x3 matrix

Step 1: Derivation of original matrix

Symbolic Expression	Symbolic and numerical Expression	Numerical Expression
$C = \begin{bmatrix} a_{11} & a_{12} & a_{13} \\ a_{21} & a_{22} & a_{23} \\ a_{31} & a_{32} & a_{33} \end{bmatrix}$	$C = \begin{bmatrix} a_{11} = 1 & a_{12} = 2 & a_{13} = 3 \\ a_{21} = 2 & a_{22} = 4 & a_{23} = 5 \\ a_{31} = 3 & a_{32} = 5 & a_{33} = 6 \end{bmatrix}$	$C = \begin{bmatrix} 1 & 2 & 3 \\ 2 & 4 & 5 \\ 3 & 5 & 6 \end{bmatrix}$
N/B: Matrix C is a square matrix and is symmetric in nature. This is because the off-diagonal (bond) elements are equal.		

Step 2: Derivation of original matrix transpose

Original Matrix	First Row Swap	Second Row Swap	Third Row Swap	Transpose Matrix
$C = \begin{bmatrix} 1^* & 2^* & 3^* \\ 2^{**} & 4^{**} & 5^{**} \\ 3^{***} & 5^{***} & 6^{***} \end{bmatrix}$	$C = \begin{bmatrix} 1^* & 2^{**} & 3^{**} \\ 2^{**} & 4^{**} & 5^{**} \\ 3^{***} & 5^{**} & 6^{***} \end{bmatrix}$	$C = \begin{bmatrix} 1^* & 2^{**} & 3^{**} \\ 2^* & 4^{**} & 5^{**} \\ 3^* & 5^{**} & 6^{***} \end{bmatrix}$	$C = \begin{bmatrix} 1^* & 2^{**} & 3^{***} \\ 2^* & 4^{**} & 5^{***} \\ 3^* & 5^{**} & 6^{***} \end{bmatrix}$	$C^T = \begin{bmatrix} 1 & 2 & 3 \\ 2 & 4 & 5 \\ 3 & 5 & 6 \end{bmatrix}$
N/B: The original matrix and transpose matrix are the same. This indicate that the transpose of a symmetric matrix is equal to the matrix.				

Test Your Understanding

Exercise 1: Given that; $a_{11} = 2$, $a_{22} = 10$, $a_{13} = 6$, $a_{21} = -8$, $a_{23} = -12$ and $a_{12} = -4$, develop a 2x2 matrix (A) using,

Exercise 2: Given that; $a_{33} = -\frac{5}{2}$, $a_{21} = 0$, $a_{23} = -3$, $a_{11} = 5$, $a_{13} = -0.5$, $a_{12} = \frac{1}{2}$, $a_{32} = 0$, $a_{31} = 1$ and $a_{22} = -1$, develop the following,

1) A 3x2 matrix (A) using 2nd and 3rd columns.................. ....[Answer: See appendix]

2) 2x3 matrix (N) using 1st and 3rd rows......................... ....[Answer: See appendix]

3) 2x2 matrix (D) using 1st and 3rd rows and 1st and 2nd column[Answer: See appendix]

4) Transpose of the 3x3 matrix (P)...................................[Answer: See appendix]

4.2 Matrix Operation

Here, we are going to analyse the addition, subtraction, multiplication and division of matrix of the same sizes and different sizes.

1) Addition and Subtraction of Matrix

To add and/or subtract two matrices, the matrix should be of the same size (have the same rows and columns). Given two matrices $A = \begin{bmatrix} a & b \\ c & d \end{bmatrix}$ and $B = \begin{bmatrix} e & f \\ g & h \end{bmatrix}$, the addition and subtraction pattern is given as seen below.

$$A + B = \begin{bmatrix} a & b \\ c & d \end{bmatrix} + \begin{bmatrix} e & f \\ g & h \end{bmatrix} = \begin{bmatrix} (a+e) & (b+f) \\ (c+g) & (d+h) \end{bmatrix} \quad \ldots\ldots\ldots \quad \ldots\ldots\ldots\ldots\ldots\ldots\ldots\ldots(1)$$

$$A - B = \begin{bmatrix} a & b \\ c & d \end{bmatrix} + \begin{bmatrix} e & f \\ g & h \end{bmatrix} = \begin{bmatrix} (a-e) & (b-f) \\ (c-g) & (d-h) \end{bmatrix} \quad \ldots\ldots\ldots\ldots\ldots\ldots\ldots\ldots\ldots\ldots\ldots..\ldots\ldots\ldots(2)$$

N/B: From the above equation, we realized that we simply add or subtract elements in the corresponding position.

Example 1: Given that $A = \begin{bmatrix} 1 & 2 \\ 4 & 1 \end{bmatrix}$ and $B = \begin{bmatrix} 2 & 1 \\ 3 & 1 \end{bmatrix}$, calculate the A + B and A − B

Solution

$$A + B = \begin{bmatrix} 1 & 2 \\ 4 & 1 \end{bmatrix} + \begin{bmatrix} 2 & 1 \\ 3 & 1 \end{bmatrix} = \begin{bmatrix} (1+2) & (2+1) \\ (4+3) & (1+1) \end{bmatrix} \Rightarrow A + B = \begin{bmatrix} 3 & 3 \\ 7 & 2 \end{bmatrix}$$

$$A - B = \begin{bmatrix} 1 & 2 \\ 4 & 1 \end{bmatrix} - \begin{bmatrix} 2 & 1 \\ 3 & 1 \end{bmatrix} = \begin{bmatrix} (1-2) & (2-1) \\ (4-3) & (1-1) \end{bmatrix} \Rightarrow A - B = \begin{bmatrix} -1 & 1 \\ 1 & 0 \end{bmatrix}$$

Example 2: Given that $A = \begin{bmatrix} 3 & 1 \\ 2 & 0 \\ 1 & 4 \end{bmatrix}$ and $B = \begin{bmatrix} 4 & 1 \\ 3 & 2 \\ 0 & 1 \end{bmatrix}$, calculate the A + B and A − B

Solution

$$A + B = \begin{bmatrix} 3 & 1 \\ 2 & 0 \\ 1 & 4 \end{bmatrix} + \begin{bmatrix} 4 & 1 \\ 3 & 2 \\ 0 & 1 \end{bmatrix} = \begin{bmatrix} (3+4) & (1+1) \\ (2+3) & (0+2) \\ (1+0) & (4+1) \end{bmatrix} \Rightarrow A + B = \begin{bmatrix} 7 & 2 \\ 5 & 2 \\ 1 & 5 \end{bmatrix}$$

$$A - B = \begin{bmatrix} 3 & 1 \\ 2 & 0 \\ 1 & 4 \end{bmatrix} - \begin{bmatrix} 4 & 1 \\ 3 & 2 \\ 0 & 1 \end{bmatrix} = \begin{bmatrix} (3-4) & (1-1) \\ (2-3) & (0-2) \\ (1-0) & (4-1) \end{bmatrix} \Rightarrow A - B = \begin{bmatrix} -1 & 0 \\ -1 & -2 \\ 1 & 3 \end{bmatrix}$$

Test Your Understanding

Exercise 3: Given that $A = \begin{bmatrix} 3 & 2 \\ -2 & 5 \end{bmatrix}$ and $B = \begin{bmatrix} 2 & 4 \\ 4 & 0 \end{bmatrix}$, calculate the following

1) B + A...[Answer: See appendix]

2) B − A ....[Answer: See appendix]

Exercise 4: Given that $A = \begin{bmatrix} 3 & 4 & 0 \\ -2 & 0 & -4 \\ 1 & 1 & 3 \end{bmatrix}$ and $B = \begin{bmatrix} 3 & 1 & -6 \\ 1 & 0 & -3 \\ 2 & 3 & 4 \end{bmatrix}$, calculate the following

1) A + B...[Answer: See appendix]

2) B − A ....[Answer: See appendix]

---

2) Multiplication of Matrix

The multiplication of matrix can be done between a scalar (a number) and matrix, matrices of the same size and matrices of different sizes. Given two matrices with size (mxn) and (nxp), the product matrix must have the order (mxp).

$$\lambda \begin{bmatrix} a & b \\ c & d \end{bmatrix} = \begin{bmatrix} (\lambda \times a) & (\lambda \times b) \\ (\lambda \times c) & (\lambda \times d) \end{bmatrix} = \begin{bmatrix} a\lambda & b\lambda \\ c\lambda & d\lambda \end{bmatrix} \quad \dots\dots(1)$$

$$\begin{bmatrix} a & c \\ b & d \end{bmatrix}\begin{bmatrix} e \\ g \end{bmatrix} = \begin{bmatrix} (a \times e) + (c \times g) \\ (b \times e) + (d \times g) \end{bmatrix} = \begin{bmatrix} ae + cg \\ be + dg \end{bmatrix} \quad \dots\dots(2)$$

$$\begin{bmatrix} a & c \\ b & d \end{bmatrix}\begin{bmatrix} e & f \\ g & h \end{bmatrix} = \begin{bmatrix} (\{a \times e\} + \{b \times g\}) & (\{a \times f\} + \{b \times h\}) \\ (\{c \times e\} + \{d \times g\}) & (\{c \times f\} + \{d \times h\}) \end{bmatrix} = \begin{bmatrix} (ae + bg) & (af + bh) \\ (ce + dg) & (cf + dh) \end{bmatrix} \,..(3)$$

N/B: The above pattern is the same for other matrix sizes. It is worth knowing that matrix multiplication does not respect the commutative rule but follows the associative and distributive rules. The product of a non-identify matrix with an identity matrix is the non-identify matrix and the product of a non-null matrix and a null matrix is a null matrix.

$AB \neq BA$ ....[Commutative Rule]

$ABC = (AB)C = A(BC)$ ....[Associative Rule]

$A(B + C) = AB + AC$ ....[Distributive Rule]

$\lambda(AB) = (\lambda A)B = A(\lambda B)$ ....[Scalar(λ) Rule]

$OA = 0$ and $AO = 0$ ....[Null Matrix Rule]

$AI = A$ ....[Identity Matrix Rule]

$A^{-1}A = I$ ....[Inverse-Matrix Product Rule]

Note: To avoid complexity or wrong solving, it is recommended to maintain the proposed sequential order of matrices when determining matrix product.

Example 1: Given that $M = \begin{bmatrix} 5 & 2 \\ -4 & 4 \end{bmatrix}$, $\beta = -\frac{1}{10}$ and $\lambda = 2$, calculate βM and λM

Solution

1) $\beta M = -\frac{1}{10}\begin{bmatrix} 5 & 2 \\ -4 & 4 \end{bmatrix} = \begin{bmatrix} 5\left(-\frac{1}{10}\right) & 2\left(-\frac{1}{10}\right) \\ -4\left(-\frac{1}{10}\right) & 4\left(-\frac{1}{10}\right) \end{bmatrix} = \begin{bmatrix} -\frac{5}{10} & -\frac{2}{10} \\ \frac{4}{10} & -\frac{4}{10} \end{bmatrix} \Rightarrow \beta M \begin{bmatrix} -\frac{1}{2} & -\frac{1}{5} \\ \frac{2}{5} & -\frac{2}{5} \end{bmatrix}$

2) $\lambda M = 2\begin{bmatrix} 5 & 2 \\ -4 & 4 \end{bmatrix} = \begin{bmatrix} (2 \times 5) & (2 \times 2) \\ (2 \times -4) & (2 \times 4) \end{bmatrix} = \begin{bmatrix} 10 & 4 \\ -8 & 8 \end{bmatrix} \Rightarrow \lambda M = \begin{bmatrix} 5 & 2 \\ -4 & 4 \end{bmatrix}$

Example 2: Given that $Q = \begin{bmatrix} 3 & 1 & -5 \\ 2 & 0 & 4 \\ 1 & 6 & 3 \end{bmatrix}$ and $D = \begin{bmatrix} 3 \\ 2 \\ -2 \end{bmatrix}$, find the matrix QD

Solution

$QD = \begin{bmatrix} 3 & 1 & -5 \\ 2 & 0 & 4 \\ 1 & 6 & 3 \end{bmatrix}\begin{bmatrix} 3 \\ 2 \\ -2 \end{bmatrix} = \begin{bmatrix} [(3 \times 3) + (1 \times 2) + (-5 \times -2)] \\ [(2 \times 3) + (0 \times 2) + (4 \times -2)] \\ [(1 \times 3) + (6 \times 2) + (3 \times -2)] \end{bmatrix}$

$QD = \begin{bmatrix} [9 + 2 + 10] \\ [6 + 0 - 8] \\ [3 + 12 - 6] \end{bmatrix} = \begin{bmatrix} 21 \\ -2 \\ 9 \end{bmatrix}$

Example 3: Given that $P = \begin{bmatrix} 1 & 4 & -3 \\ 6 & 3 & 0 \end{bmatrix}$ and $D = \begin{bmatrix} 3 & -2 \\ 2 & -6 \\ 1 & 5 \end{bmatrix}$, calculate the matrix PD

$$PD = \begin{bmatrix} 1 & 4 & -3 \\ 6 & 3 & 0 \end{bmatrix} \begin{bmatrix} 3 & -2 \\ 2 & -6 \\ 1 & 5 \end{bmatrix} \dots\dots\dots\dots\dots\dots\text{[Perform row column multiplication]}$$

$$PD = \begin{bmatrix} \{(1\times3)+(4\times2)+(-3\times1)\} & \{(1\times-2)+(4\times-6)+(-3\times5)\} \\ \{(6\times3)+(3\times2)+(0\times1)\} & \{(6\times-2)+(3\times-6)+(0\times5)\} \end{bmatrix}$$

$$PD = \begin{bmatrix} \{3+8-3\} & \{-2-26-15\} \\ \{18+6+0\} & \{-12-18+0\} \end{bmatrix} \Rightarrow PD = \begin{bmatrix} 8 & -41 \\ 24 & -30 \end{bmatrix}$$

Example 4: Given that A = $\begin{bmatrix} 4 & -1 \\ 1 & 3 \end{bmatrix}$ and I = $\begin{bmatrix} 1 & 0 \\ 0 & 1 \end{bmatrix}$, show that AI = A

Solution

$$AI = \begin{bmatrix} 4 & -1 \\ 1 & 3 \end{bmatrix} \begin{bmatrix} 1 & 0 \\ 0 & 1 \end{bmatrix} = \begin{bmatrix} [(4\times1)+(-1\times0)] & [(4\times0)+(-1\times1)] \\ [(1\times1)+(3\times0)] & [(1\times0)+(3\times1)] \end{bmatrix}$$

$$AI = \begin{bmatrix} [4-0] & [0-1] \\ [1+0] & [0+3] \end{bmatrix} \Rightarrow AI = \begin{bmatrix} 4 & -1 \\ 1 & 3 \end{bmatrix}$$

Example 5: Given that; A = $\begin{bmatrix} 1 & 3 & 2 \\ -1 & 2 & 1 \\ 0 & 1 & 0 \end{bmatrix}$, B = $\begin{bmatrix} 4 & 1 & 2 \\ 1 & 0 & 1 \\ 3 & 1 & 5 \end{bmatrix}$, C = $\begin{bmatrix} \frac{1}{\sqrt{2}} & \frac{1}{\sqrt{2}} \\ \frac{1}{\sqrt{2}} & -\frac{1}{\sqrt{2}} \end{bmatrix}$ and D =

$\begin{bmatrix} \frac{3}{\sqrt{2}} & \frac{1}{\sqrt{2}} \\ \frac{3}{\sqrt{2}} & -\frac{1}{\sqrt{2}} \end{bmatrix}$, calculate AB and CD

Solution

1) **Calculation of AB**

$$AB = \begin{bmatrix} 1 & 3 & 2 \\ -1 & 2 & 1 \\ 0 & 1 & 0 \end{bmatrix} \begin{bmatrix} 4 & 1 & 2 \\ 1 & 0 & 1 \\ 3 & 1 & 5 \end{bmatrix} \dots\dots\dots\dots\dots\dots\text{[Perform row column multiplication]}$$

$$AB = \begin{bmatrix} \{(1\times4)+(3\times1)+(2\times3)\} & \{(1\times1)+(3\times0)+(2\times1)\} & \{(1\times2)+(3\times1)+(2\times5)\} \\ \{(-1\times4)+(2\times1)+(1\times3)\} & \{(-1\times1)+(2\times0)+(1\times1)\} & \{(-1\times2)+(2\times1)+(1\times5)\} \\ \{(0\times4)+(1\times1)+(0\times3)\} & \{(0\times1)+(1\times0)+(0\times1)\} & \{(0\times2)+(1\times1)+(0\times5)\} \end{bmatrix}$$

$$AB = \begin{bmatrix} \{4+3+6\} & \{1+0+2\} & \{2+3+10\} \\ \{-4+2+3\} & \{-1+0+1\} & \{-2+2+5\} \\ \{0+1+0\} & \{0+0+0\} & \{0+1+0\} \end{bmatrix} \Rightarrow AB = \begin{bmatrix} 13 & 3 & 15 \\ 1 & 0 & 5 \\ 1 & 0 & 1 \end{bmatrix}$$

2) **Calculation of CD**

$$CD = \begin{bmatrix} \frac{1}{\sqrt{2}} & \frac{1}{\sqrt{2}} \\ \frac{1}{\sqrt{2}} & -\frac{1}{\sqrt{2}} \end{bmatrix} \begin{bmatrix} \frac{3}{\sqrt{2}} & \frac{1}{\sqrt{2}} \\ \frac{3}{\sqrt{2}} & -\frac{1}{\sqrt{2}} \end{bmatrix} =$$

$$\begin{bmatrix} \{(\frac{1}{\sqrt{2}}\times\frac{3}{\sqrt{2}})+(\frac{1}{\sqrt{2}}\times\frac{3}{\sqrt{2}})\} & \{(\frac{1}{\sqrt{2}}\times\frac{1}{\sqrt{2}})+(\frac{1}{\sqrt{2}}\times-\frac{1}{\sqrt{2}})\} \\ \{(\frac{1}{\sqrt{2}}\times\frac{3}{\sqrt{2}})+(-\frac{1}{\sqrt{2}}\times\frac{3}{\sqrt{2}})\} & \{(\frac{1}{\sqrt{2}}\times\frac{1}{\sqrt{2}})+(-\frac{1}{\sqrt{2}}\times-\frac{1}{\sqrt{2}})\} \end{bmatrix}$$

$$CD = \begin{bmatrix} \left\{\frac{3}{\sqrt{2}\times\sqrt{2}} + \frac{3}{\sqrt{2}\times\sqrt{2}}\right\} & \left\{\frac{1}{\sqrt{2}\times\sqrt{2}} - \frac{1}{\sqrt{2}\times\sqrt{2}}\right\} \\ \left\{\frac{3}{\sqrt{2}\times\sqrt{2}} - \frac{3}{\sqrt{2}\times\sqrt{2}}\right\} & \left\{\frac{1}{\sqrt{2}\times\sqrt{2}} + \frac{1}{\sqrt{2}\times\sqrt{2}}\right\} \end{bmatrix} = \begin{bmatrix} \left\{\frac{3}{\sqrt{4}} + \frac{3}{\sqrt{4}}\right\} & \left\{\frac{1}{\sqrt{4}} - \frac{1}{\sqrt{4}}\right\} \\ \left\{\frac{3}{\sqrt{4}} - \frac{3}{\sqrt{4}}\right\} & \left\{\frac{1}{\sqrt{4}} + \frac{1}{\sqrt{4}}\right\} \end{bmatrix} = \begin{bmatrix} \left\{\frac{3}{2} + \frac{3}{2}\right\} & \left\{\frac{1}{2} - \frac{1}{2}\right\} \\ \left\{\frac{3}{2} - \frac{3}{2}\right\} & \left\{\frac{1}{2} + \frac{1}{2}\right\} \end{bmatrix}$$

$$CD = \begin{bmatrix} \left\{\frac{3}{2} + \frac{3}{2}\right\} & \left\{\frac{1}{2} - \frac{1}{2}\right\} \\ \left\{\frac{3}{2} - \frac{3}{2}\right\} & \left\{\frac{1}{2} + \frac{1}{2}\right\} \end{bmatrix} = \begin{bmatrix} \frac{6}{2} & 0 \\ 0 & 1 \end{bmatrix} \Rightarrow CD = \begin{bmatrix} 3 & 0 \\ 0 & 1 \end{bmatrix}$$

Example 6: Given that $A = \begin{bmatrix} 1 & -2 & 3 \\ 2 & 1 & 5 \end{bmatrix}$, $B = \begin{bmatrix} 4 \\ 3 \\ 2 \end{bmatrix}$ and $C = \begin{bmatrix} 3 \\ 1 \\ 2 \end{bmatrix}$, show that $A(B + C) = AB + AC$

Solution

$A(B + C)$	$AB + AC$
$\begin{bmatrix} 1 & -2 & 3 \\ 2 & 1 & 5 \end{bmatrix}\left(\begin{bmatrix} 4 \\ 3 \\ 2 \end{bmatrix} + \begin{bmatrix} 3 \\ 1 \\ 2 \end{bmatrix}\right) = \begin{bmatrix} 1 & -2 & 3 \\ 2 & 1 & 5 \end{bmatrix}\begin{pmatrix} 7 \\ 4 \\ 4 \end{pmatrix}$	$\begin{bmatrix} 1 & -2 & 3 \\ 2 & 1 & 5 \end{bmatrix}\begin{bmatrix} 4 \\ 3 \\ 2 \end{bmatrix} + \begin{bmatrix} 1 & -2 & 3 \\ 2 & 1 & 5 \end{bmatrix}\begin{bmatrix} 3 \\ 1 \\ 2 \end{bmatrix}$
$\begin{bmatrix} [(1 \times 7) + (-2 \times 4) + (3 \times 4)] \\ [(2 \times 7) + (1 \times 4) + (5 \times 4)] \end{bmatrix} = \begin{bmatrix} 11 \\ 38 \end{bmatrix}$	$\begin{bmatrix} (4 - 6 + 6) \\ (8 + 3 + 10) \end{bmatrix} + \begin{bmatrix} (3 - 2 + 6) \\ (6 + 1 + 10) \end{bmatrix} = \begin{bmatrix} 11 \\ 38 \end{bmatrix}$

Test Your Understanding

Exercise 5: If $A = \begin{bmatrix} 10 & -8 \\ 4 & -2 \end{bmatrix}$, $B = \begin{bmatrix} 2 \\ 1 \end{bmatrix}$, $C = \begin{bmatrix} 1 & 0 & -1 \\ 0 & 1 & -1 \\ -1 & -1 & 2 \end{bmatrix}$, $D = \begin{bmatrix} -\frac{1}{2} \\ -\frac{1}{2} \\ 1 \end{bmatrix}$, $E = \begin{bmatrix} 3 & 1 \\ 2 & 4 \\ 6 & 5 \\ 1 & 2 \end{bmatrix}$ and

$F = \begin{bmatrix} -3 & 1 \\ 4 & 2 \end{bmatrix}$, find;

1) AB..[Answer: See appendix]

2) DC..[Answer: See appendix]

3) EF..[Answer: See appendix]

Exercise 6: Given that $I = \begin{bmatrix} 1 & 0 \\ 0 & 1 \end{bmatrix}$, $A = \begin{bmatrix} 10 & -8 \\ 4 & -2 \end{bmatrix}$, $B = \begin{bmatrix} 1 & 2 \\ 4 & 1 \end{bmatrix}$ and $C = \begin{bmatrix} 2 & 1 \\ 3 & 1 \end{bmatrix}$, calculate the matrix

1) $\lambda I - A$..[Answer: See appendix]

2) $\alpha B + \beta C$..[Answer: See appendix]

4.3 Determinant of a matrix

Determinant is a scalar value calculated from a square matrix. The value of a determinant can be positive, negative and zero and the value determinant determine the existence or non-existence of matrix inverse. Matrix with determinant equal to zero have no inverse and the determinant of matrix (A) is denoted |A| or det(A). The following principles are worth knowing,

1) $\det(A) = |A^T|$...(1)
2) $|AB| = \det(A) \times \det(B)$ ....(2)
3) $\det(A_{2x2}) = \lambda_1 \times \lambda_2$ ....(3)

N/B: The symbol (λ) represents the Eigen value and for a 3x3 matrix, the determinant is $[\lambda_1 \times \lambda_2 \times \lambda_3]$.

4.3.1 Determinant of a 2x2 Matrix

Given a 2x2 matrix; $A = \begin{bmatrix} a_{11} & a_{12} \\ a_{21} & a_{22} \end{bmatrix} = \begin{bmatrix} a & b \\ c & d \end{bmatrix}$, the formula to compute the determinant and inverse of the matrix is given as seen below.

$$\det(A) = |A| = [a \times d] - [b \times c] \dots\dots\dots\dots\dots\dots\dots\dots\dots(1)$$

Example 1: Calculate the determinant of the matrix $A = \begin{bmatrix} 10 & -8 \\ 4 & -2 \end{bmatrix}$

<u>**Solution**</u>

Step 1: Formula. $\det(A) = [a \times d] - [b \times c]$, where; $a = 10, b = -8, c = 4$ and $d = -2$

Step 2: Application of formula

$|A| = [10 \times (-2)] - [4 \times (-8)]$ ....[Open inner brackets]

$|A| = [-20] - [-32]$ ....[Open main brackets]

$|A| = -20 + 32 = \mathbf{12}$ ....[Note that ($|A| \neq 0$), therefore, matrix is non-singular]

Example 2: Given that; $M = \begin{bmatrix} 5 & 2 \\ -1 & 3 \end{bmatrix}$, show that $\det(M) = \det(M^T)$.

<u>**Solution**</u>

Hint: The transpose of the given matrix is $M^T = \begin{bmatrix} 5 & -1 \\ 2 & 3 \end{bmatrix}$

det(M)	det(M^T)
$\det(M) = [(5 \times 3) - (2 \times -1)]$	$|M^T| = [(5 \times 3) - (-1 \times 2)]$
$\det(M) = [15 + 2]$ ....[Add values]	$|M^T| = [15 + 2]$ ....[Add values]
$\det(M) = \mathbf{17}$	$|M^T| = \mathbf{17}$
Note: The determinants are the same (17). We can conclude that $\det(M) = \det(M^T)$	

Example 3: Given that; $M = \begin{bmatrix} \lambda - 2 & -5 \\ 1 & \lambda + 4 \end{bmatrix}$, compute det(M).

Step 1: Formula. $|M| = $ [Product of diagonal elements] $-$ [Product of off diagonal elemts]

Step 2: Data substitution and solving

$|M| = [(\lambda - 2)(\lambda + 4)] - [-5 \times 1]$ ....[Simplify brackets]

$|M| = [\lambda^2 + 4\lambda - 2\lambda - 8] - [-5]$

$|M| = \lambda^2 + 2\lambda - 8 + 5$ ....[Solve]

$|M| = \lambda^2 + 2\lambda - 3$

Example 4: Given that $P = \begin{bmatrix} 2 & 3 \\ -1 & 1 \end{bmatrix}$ and $W = \begin{bmatrix} 4 & 2 \\ 5 & 4 \end{bmatrix}$, show that $\det(PW) = \det(P) \times \det(W)$

Solution

$\det(PW)$	$\det(P) + \det(W)$
$\|PW\| = \begin{bmatrix} 2 & 3 \\ -1 & 1 \end{bmatrix}\begin{bmatrix} 4 & 2 \\ 5 & 4 \end{bmatrix}$	$\det\begin{pmatrix} 2 & 3 \\ -1 & 1 \end{pmatrix} \times \det\begin{pmatrix} 4 & 2 \\ 5 & 4 \end{pmatrix}$
$\|PW\| = \begin{bmatrix} 23 & 16 \\ 1 & 2 \end{bmatrix} = (46) - (16)$	$[2+3] \times [16-10]$
$\|PW\| = 30$	$\det(P) \times \det(W) = 30$
Note: The values are the same (30), showing that the claim is true $[\det(PW) = \det(P) \times \det(W)]$	

Example 5: Given that $A = \begin{bmatrix} 2 & 5 \\ -1 & -4 \end{bmatrix}$, find the value of determinant using Eigen value (λ)

Solution

Hint: Since it is a two by two (2x2) matrix, we are going to get a maximum of two Eigen values.

Step 1: Formula. $\text{Det}(A) = \lambda_1 \times \lambda_2$

Step 2: Determination of (λ_1) and (λ_2)

1) Formula. $P(\lambda) = \det[\lambda I - A] = 0$, where; $A = \begin{bmatrix} 2 & 5 \\ -1 & -4 \end{bmatrix}$ and $I = \begin{bmatrix} 1 & 0 \\ 0 & 1 \end{bmatrix}$

2) Determine $\det[\lambda I - A]$

$\det\left[\lambda\begin{bmatrix} 1 & 0 \\ 0 & 1 \end{bmatrix} - \begin{bmatrix} 2 & 5 \\ -1 & -4 \end{bmatrix}\right] = 0$ ....[Perform scalar multiplication]

$\text{Det}\left[\begin{bmatrix} \lambda & 0 \\ 0 & \lambda \end{bmatrix} - \begin{bmatrix} 2 & 5 \\ -1 & -4 \end{bmatrix}\right] 0$ ....[Subtract matrices]

$\text{Det}\begin{bmatrix} \lambda - 2 & -5 \\ 1 & \lambda + 4 \end{bmatrix} = 0$ ....[Work determinant]

$[(\lambda - 2)(\lambda + 4)] - [(-5)(1)] = 0$ ....[Simplify equation]

$\lambda^2 + 2\lambda - 3 = 0$ ....[Quadratic equation]

N/B: To solve the quadratic equation as seen above, we will obtain $(\lambda_1 = -3)$ and $(\lambda_2 = 1)$.

Step 3: Data substitution and solving

$Det(A) = -3 \times 1$ ....[Solve equation]

$Det(A) = -3$ ....[Verify. $Det(A) = [(2 \times -4) - (5 \times -1)] = -8 + 5 = -3$]

<u>**Test Your Understanding**</u>

Exercise 7: What is the determinant of the following matrices

1) $G = \begin{bmatrix} -2 & 4 \\ -3 & 7 \end{bmatrix}$ ....[Answer: -2, having an inverse matrix]

2) $H = \begin{bmatrix} 1 & 0 \\ 0 & 1 \end{bmatrix}$ ....[Answer: 1, having an inverse matrix]

3) $K = \begin{bmatrix} 2 & 1 \\ 4 & 2 \end{bmatrix}$ ....[Answer: 0, does not have an inverse matrix]

4) $L = \begin{bmatrix} 1 & 1 \\ -i & i \end{bmatrix}$ ....[Answer: 2i, having an inverse matrix]

5) $D = \begin{bmatrix} \lambda - 3 & 1 \\ 1 & \lambda - 3 \end{bmatrix}$ ....[Answer: $\lambda^2 - 6\lambda + 8$]

Exercise 8: Determine the Eigen values of the following matrices

1) $A = \begin{bmatrix} 3 & -1 \\ -1 & 3 \end{bmatrix}$ ....[Answer: $\lambda_1 = 2$ and $\lambda_2 = 4$]

2) $M = \begin{bmatrix} 0 & -1 \\ 1 & 0 \end{bmatrix}$ ....[Answer: $\lambda_1 = -i$ and $\lambda_2 = i$]

Exercise 9: The Eigen values of a 2x2 matrix is given as; $\lambda_1 = 6$ and $\lambda_2 = 2$, what is the determinant of the matrix?.. [Answer: 12]

4.3.2 Determinant of 3x3 Matrix

Determinant of a 3x3 matrix can be derived using the row co-factor display method and/or the basket weave method. When determining the determinant of a 3x3 matrix, the following properties are worth knowing,

1) The determinant of a square matrix with all the element of a row or column zero is equal to zero.
2) The determinant of a square matrix with two identical rows or columns is zero.

The formula of 3x3 matrix (A) using the first row co-factor method is given below.

$det(A) = |A| = +a[C_{11}] - b[C_{12}] + c[C_{13}]$ ....(1)

Where the value of $[C_{11}]$, $[C_{12}]$ and $[C_{13}]$ is calculated from the given matrix as seen below

Cover 1st Row and 1st Column (C_{11})	Cover 1st Row and 2nd Column (C_{12})	Cover 1st Row and 3rd Column (C_{13})
$\begin{bmatrix} a & b & c \\ d & e & f \\ g & h & i \end{bmatrix} \Rightarrow C_{11} = \begin{bmatrix} e & f \\ h & i \end{bmatrix}$	$\begin{bmatrix} a & b & c \\ d & e & f \\ g & h & i \end{bmatrix} \Rightarrow C_{12} = \begin{bmatrix} d & f \\ g & i \end{bmatrix}$	$\begin{bmatrix} a & b & c \\ d & e & f \\ g & h & i \end{bmatrix} \Rightarrow C_{13} = \begin{bmatrix} d & e \\ g & h \end{bmatrix}$
$C_{11} = (e \times i) - (f \times h)$	$C_{12} = (d \times i) - (f \times g)$	$C_{13} = (d \times h) - (e \times g)$

N/B: The performing second co-factor and third co-factor yield the same value of determinant. The formula of determinant under the basket weave method is given as seen below.

$$\det(A) = \sum \text{of Negative Slope Lines} - \sum \text{of Positive Slope Lines} \dots\dots\dots\dots\dots\dots(2)$$

Using the basket weave method, different weave pattern can be applied to obtain the same determinant value. The different weave pattern includes; vertical expansion (only first and the third columns are expanded), horizontal expansion (only the first row and third row are expanded), vertical-horizontal expansion (only second column and second row are expanded) and two row expansion (first two columns are added after the third column).

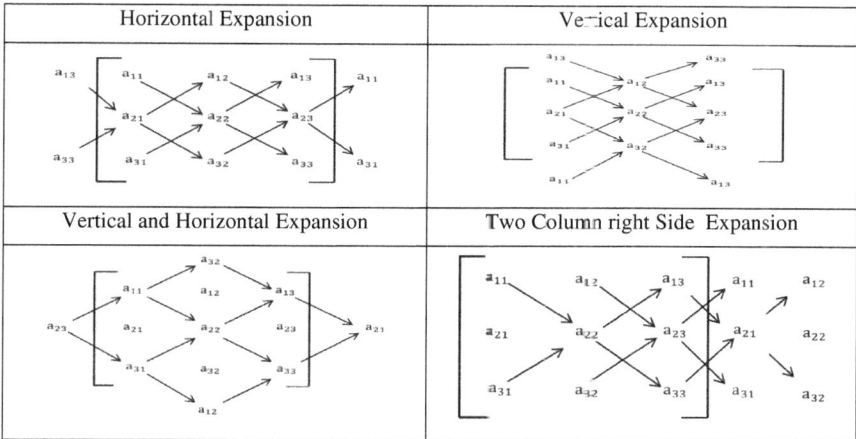

Horizontal Expansion	Vertical Expansion
Vertical and Horizontal Expansion	Two Column right Side Expansion

Note: Using the basket weave method, all the different weave patterns should give the same value and the basket weave method and row column co-factor display should give the same value of determinant. It is worth knowing that the determinant of a matrix is also the product of the matrix Eigen values.

Example 1: Given that $A = \begin{bmatrix} 5 & 6 & 2 \\ 0 & -1 & -8 \\ 1 & 0 & -2 \end{bmatrix}$, compute A| using the following,

1) First row and columns co-factor display

2) Second row and columns co-factor display

3) Third row and columns co-factor display

Solution

1) First row and columns co-factor display

Step 1: Formula. $|M| = +(a)[C_{11}] - (b)[C_{12}] + (c)[C_{13}]$

Step 2 Determine the value of $[C_{11}]$, $[C_{12}]$ and $[C_{13}]$

C_{11}	C_{12}	C_{13}
$A = \begin{bmatrix} 5 & 6 & 2 \\ 0 & -1 & -8 \\ 1 & 0 & -2 \end{bmatrix} = C_{11} = \begin{bmatrix} -1 & -8 \\ 0 & -2 \end{bmatrix}$	$A = \begin{bmatrix} 5 & 6 & 2 \\ 0 & -1 & -8 \\ 1 & 0 & -2 \end{bmatrix} = C_{12} = \begin{bmatrix} 0 & -8 \\ 1 & -2 \end{bmatrix}$	$A = \begin{bmatrix} 5 & 6 & 2 \\ 0 & -1 & -8 \\ 1 & 0 & -2 \end{bmatrix} = C_{13} = \begin{bmatrix} 0 & -1 \\ 1 & 0 \end{bmatrix}$
Where: a = 5	Where: b = 6	Where; c = 2

Step 3: Data substitution and solving

$|M| = +(5)\begin{bmatrix} -1 & -8 \\ 0 & -2 \end{bmatrix} - (6)\begin{bmatrix} 0 & -8 \\ 1 & -2 \end{bmatrix} + (2)\begin{bmatrix} 0 & -1 \\ 1 & 0 \end{bmatrix}$..[Work 2x2 matrices determinants]

$|M| = +(5)[(-1 \times -2) - (-8 \times 0)] - (6)[(0 \times -2) - (-8 \times 1)] + (2)[(0 \times 0) - (-1 \times 1)]$

$|M| = +(5)[2 - 0] - (6)[0 + 8] + (2)[0 + 1]$

$|M| = +(5)[2] - (6)[8] + (2)[1]$ ....[Open brackets]

$|M| = 10 - 48 + 2$ ....[Solve]

$|M| = -36$

2) Second row and columns co-factor display

Step 1: Formula. $|M| = -(a)[C_{21}] + (b)[C_{22}] - (c)[C_{23}]$

Step 2 Determine the value of $[C_{21}]$, $[C_{22}]$ and $[C_{23}]$

C_{21}	C_{22}	C_{23}
$A = \begin{bmatrix} 5 & 6 & 2 \\ 0 & -1 & -8 \\ 1 & 0 & -2 \end{bmatrix} = C_{21} = \begin{bmatrix} 6 & 2 \\ 0 & -2 \end{bmatrix}$	$A = \begin{bmatrix} 5 & 6 & 2 \\ 0 & -1 & -8 \\ 1 & 0 & -2 \end{bmatrix} = C_{22} = \begin{bmatrix} 5 & 2 \\ 1 & -2 \end{bmatrix}$	$A = \begin{bmatrix} 5 & 6 & 2 \\ 0 & -1 & -8 \\ 1 & 0 & -2 \end{bmatrix} = C_{23} = \begin{bmatrix} 5 & 6 \\ 1 & 0 \end{bmatrix}$
Where; a = 0	Where; b = -1	Where; c = -8

Step 3: Data substitution and solving

$|M| = -(0)\begin{bmatrix} 6 & 2 \\ 0 & -2 \end{bmatrix} + (-1)\begin{bmatrix} 5 & 2 \\ 1 & -2 \end{bmatrix} - (-8)\begin{bmatrix} 5 & 6 \\ 1 & 0 \end{bmatrix}$.[Work 2x2 matrices determinant]

$|M| = -(0)[(6 \times -2) - (2 \times 0)] + (-1)[(5 \times -2) - (2 \times 1)] - (-8)[(5 \times 0) - (6 \times 1)]$

$|M| = -(0)[-12 - 0] + (-1)[-10 - 2] - (-8)[0 - 6]$

$|M| = -(0)[-12] + (-1)[-12] - (-8)[-6]$ ....[Open brackets]

$|M| = 0 + 12 - 48$ ....[Solve]

$|M| = -36$

3) Third row and columns co-factor display

Step 1: Formula. $|M| = +(a)[C_{31}] - (b)[C_{32}] + (c)[C_{33}]$

Step 2 Determine the value of $[C_{31}]$, $[C_{32}]$ and $[C_{33}]$

C_{31}	C_{32}	C_{33}
$A = \begin{bmatrix} 5 & 6 & 2 \\ 0 & -1 & -8 \\ 1 & 0 & -2 \end{bmatrix} = C_{31} = \begin{bmatrix} 6 & 2 \\ -1 & -8 \end{bmatrix}$	$A = \begin{bmatrix} 5 & 6 & 2 \\ 0 & -1 & -8 \\ 1 & 0 & -2 \end{bmatrix} = C_{32} = \begin{bmatrix} 5 & 2 \\ 0 & -8 \end{bmatrix}$	$A = \begin{bmatrix} 5 & 6 & 2 \\ 0 & -1 & -8 \\ 1 & 0 & -2 \end{bmatrix} = C_{33} = \begin{bmatrix} 5 & 6 \\ 0 & -1 \end{bmatrix}$
Where; a = 1	Where; b = 0	Where; c = −2

Step 3: Data substitution and solving

$|M| = +(1) \begin{bmatrix} 6 & 2 \\ -1 & -8 \end{bmatrix} - (0) \begin{bmatrix} 5 & 2 \\ 0 & -8 \end{bmatrix} + (-2) \begin{bmatrix} 5 & 6 \\ 0 & -1 \end{bmatrix}$ ..[Work   2x2   matrices

determinant]

$|M| = +(1)[(6 \times -8) - (2 \times -1)] - (0)[(5 \times -8) - (2 \times 0)] + (-2)[(5 \times -1) - (6 \times 0)]$

$|M| = +(1)[-48 + 2] - (0)[-40 - 0] + (-2)[-5 - 0]$

$|M| = +(1)[-46] - (0)[-40] + (-2)[-5]$ ....[Open brackets]

$|M| = -46 - 0 + 10$ ....[Solve]

$|M| = -36$

Note: All the different rows column co-factor display gives the same value of determinant. This indicates that we can use any row column co-factor display in determining the determinant.

Example 2: Given that; $A = \begin{bmatrix} 2 & 3 & 4 \\ 5 & 6 & 7 \\ 8 & 9 & 1 \end{bmatrix}$, find det(A) using the basket weave method

Solution

Step 1: Formula. $\det(A) = \sum$ of Negative Slope Lines $- \sum$ of Positive Slope Lines

Step 2: Determination of line values

Vertical Expansion	$\begin{bmatrix} 7 \times 3 \times 8 = 168 \\ 1 \times 6 \times 2 = 12 \\ 4 \times 9 \times 5 = 180 \end{bmatrix}$ 168 + 12 + 180 = **360** $\begin{bmatrix} 5 \times 3 \times 1 = 15 \\ 8 \times 5 \times 4 = 192 \\ 2 \times 9 \times 7 = 126 \end{bmatrix}$ 15 + 192 + 126 = **333**
Horizontal Expansion	180 + 12 + 168 = **360** 15 + 192 + 126 = **333** (180) (12) (168) (15) (192) (126)

Vertical and horizontal expansion	
Two column right expansion	

Step 3: Data substitution and solving

$$\det(A) = 360 - 333 \dots\dots\dots\dots\dots\dots\dots\dots\dots\dots\dots\dots\dots\dots\dots\dots\text{[Sum values]}$$

$$\det(A) = 27$$

Example 3: Given that; $M = \begin{bmatrix} \lambda - 5 & -6 & -2 \\ 0 & \lambda + 1 & 8 \\ -1 & 0 & \lambda + 2 \end{bmatrix}$, compute $|M|$

Solution

a) **Row-column co-factor method**

Step 1: Formula. $|M| = +(a)[C_{21}] - (b)[C_{22}] + (c)[C_{23}]$

Step 2: Data substitution and solving

$$|M| = -(0)\begin{bmatrix} -6 & -2 \\ 0 & \lambda + 2 \end{bmatrix} + (\lambda + 1)\begin{bmatrix} \lambda - 5 & -2 \\ -1 & \lambda + 2 \end{bmatrix} - (8)\begin{bmatrix} \lambda - 5 & -6 \\ -1 & 0 \end{bmatrix}$$

$$|M| = +(\lambda + 1)[\{(\lambda - 5)(\lambda + 2)\} - \{-2 \times -1\}] - (8)[(0) - (-1 \times -6)]$$

$$|M| = +(\lambda + 1)[\{\lambda^2 - 3\lambda - 10\} - \{2\}] - (8)[-6]$$

$$|M| = +(\lambda + 1)[\lambda^2 - 3\lambda - 12] + 48$$

$$|M| = \lambda^3 - 3\lambda^2 - 12\lambda + \lambda^2 - 3\lambda - 12 + 48$$

$$|M| = \lambda^3 - 2\lambda^2 - 15\lambda + 36$$

b) **Basket weave method**

Step 1: Formula. $\det(A) = \sum$ of Negative Slope Lines $- \sum$ of Positive Slope Lines

Step 2: Determination of line values

$(8)(-6)(-1) = \mathbf{48}$

$(\lambda + 2)(\lambda + 1)(\lambda - 5) =$
$(\lambda^3 - 2\lambda^2 - 13\lambda - 10)$

$(-2)(0)(0) = \mathbf{0}$

$(0)(-6)(\lambda + 2) = \mathbf{0}$

$(-1)(\lambda + 1)(-2) = 2\lambda + 2$

$(\lambda - 5)(0)(8) = \mathbf{0}$

$$\begin{bmatrix} -1 & & \lambda + 2 \\ \lambda - 5 & -6 & -2 \\ 0 & \lambda + 1 & 8 \\ -1 & 0 & \lambda + 2 \\ \lambda - 5 & & -2 \end{bmatrix}$$

Step 3: Data substitution and solving

$$\det(A) = [48 + (\lambda^3 - 2\lambda^2 - 13\lambda - 10) + 0] - [0 + (2\lambda + 2) + 0]$$

$$\det(A) = [48 + \lambda^3 - 2\lambda^2 - 13\lambda - 10 + 0] - [0 + 2\lambda + 2 + 0]$$

$$\det(A) = [\lambda^3 - 2\lambda^2 - 13\lambda + 38] - [2\lambda + 2]$$

$$\det(A) = \lambda^3 - 2\lambda^2 - 13\lambda + 38 - 2\lambda - 2$$

$$\det(A) = \boldsymbol{\lambda^3 - 2\lambda^2 - 15\lambda + 36}$$

Test Your Understanding

Exercise 10: Compute the determinant of the following matrices,

1) $M = \begin{bmatrix} 2 & -1 & 0 \\ 0 & 3 & -2 \\ 1 & 0 & 1 \end{bmatrix}$ ....**[Answer: 8]**

2) $N = \begin{bmatrix} 5 & 0 & 6 \\ 1 & 0 & 4 \\ 10 & 0 & 7 \end{bmatrix}$ ....**[Answer: 0]**

3) $K = \begin{bmatrix} 1 & 2 & 3 \\ 1 & 2 & 3 \\ 4 & 5 & 6 \end{bmatrix}$ ....**[Answer: 0]**

4) $H = \begin{bmatrix} 1 & 3 & 4 \\ 2 & 0 & 1 \\ 3 & 1 & 0 \end{bmatrix}$ ....**[Answer: 16]**

Exercise 11: The Eigen values of a 3x3 matrix are; $\lambda_1 = 3$, $\lambda_2 = 3$ and $\lambda_3 = -4$, fine the determinant of the matrix...**[Answer: -36]**

4.4 Adjugate or Adjoint of a Matrix

The adjugate of a 2x2 matrix is derived by changing the position of the diagonal element and changing the sign of the off-diagonal elements. Using the matrix $A = \begin{bmatrix} a & c \\ b & d \end{bmatrix}$, the adjugate is given as;

$$\text{Adj}(A) = \begin{bmatrix} d & -c \\ -b & a \end{bmatrix} \quad \text{...(1)}$$

The adjugate of a 3x3 matrix can be determined using the row column co-factor method and basket weave method. The adjugate of a matrix (A) is the transpose of the co-factor matrix.

$$\text{Adj}(A) = [\text{Co} - \text{Factor Matrix}]^T \dots\dots\dots\dots\dots\dots\dots\dots\dots\dots\dots\dots\dots\dots\dots\dots\dots\dots\dots(2)$$

Using the row-column co-factor method, the co-factor matrix is a combination of minor matrices and the checkerboard signs (varies with rows). Using the matrix; $A = \begin{bmatrix} a_{11} & a_{12} & a_{13} \\ a_{21} & a_{22} & a_{23} \\ a_{31} & a_{32} & a_{33} \end{bmatrix} = \begin{bmatrix} a & b & c \\ d & e & f \\ g & h & i \end{bmatrix}$, the minor matrices and co-factor matrix is expressed below.

$$\text{Co} - \text{Factor} = \underbrace{\begin{bmatrix} \begin{bmatrix} e & f \\ h & i \end{bmatrix} & \begin{bmatrix} d & f \\ g & i \end{bmatrix} & \begin{bmatrix} d & e \\ g & h \end{bmatrix} \\ \begin{bmatrix} b & c \\ h & i \end{bmatrix} & \begin{bmatrix} a & c \\ g & i \end{bmatrix} & \begin{bmatrix} a & b \\ g & h \end{bmatrix} \\ \begin{bmatrix} b & c \\ e & f \end{bmatrix} & \begin{bmatrix} a & c \\ d & f \end{bmatrix} & \begin{bmatrix} a & b \\ d & e \end{bmatrix} \end{bmatrix}}_{\text{Minors Matrix}} = \underbrace{\begin{bmatrix} + & - & + \\ - & + & - \\ + & - & + \end{bmatrix}}_{\text{Checkerboard Sign}} =$$

$$\underbrace{\begin{bmatrix} +\begin{bmatrix} e & f \\ h & i \end{bmatrix} & -\begin{bmatrix} d & f \\ g & i \end{bmatrix} & +\begin{bmatrix} d & e \\ g & h \end{bmatrix} \\ -\begin{bmatrix} b & c \\ h & i \end{bmatrix} & +\begin{bmatrix} a & c \\ g & i \end{bmatrix} & -\begin{bmatrix} a & b \\ g & h \end{bmatrix} \\ +\begin{bmatrix} b & c \\ e & f \end{bmatrix} & -\begin{bmatrix} a & c \\ d & f \end{bmatrix} & +\begin{bmatrix} a & b \\ d & e \end{bmatrix} \end{bmatrix}}_{\text{Co-Factor Matrix}}$$

Using the basket weave method, the weave pattern of determining the co-factor matrix is illustrated below.

$$\text{Co} - \text{Factor} = \begin{bmatrix} a_{22} & a_{23} & a_{21} & a_{22} \\ a_{32} & a_{33} & a_{31} & a_{32} \\ a_{12} & a_{13} & a_{11} & a_{12} \\ a_{22} & a_{23} & a_{21} & a_{22} \end{bmatrix} = \begin{bmatrix} \begin{bmatrix} a_{22} & a_{23} \\ a_{32} & a_{33} \end{bmatrix} & \begin{bmatrix} a_{23} & a_{21} \\ a_{33} & a_{31} \end{bmatrix} & \begin{bmatrix} a_{21} & a_{22} \\ a_{31} & a_{32} \end{bmatrix} \\ \begin{bmatrix} a_{32} & a_{33} \\ a_{12} & a_{13} \end{bmatrix} & \begin{bmatrix} a_{33} & a_{31} \\ a_{13} & a_{11} \end{bmatrix} & \begin{bmatrix} a_{31} & a_{32} \\ a_{11} & a_{12} \end{bmatrix} \\ \begin{bmatrix} a_{12} & a_{13} \\ a_{22} & a_{23} \end{bmatrix} & \begin{bmatrix} a_{13} & a_{11} \\ a_{23} & a_{21} \end{bmatrix} & \begin{bmatrix} a_{11} & a_{12} \\ a_{21} & a_{22} \end{bmatrix} \end{bmatrix}$$

Example 1: Find the adjugate of the matrices; $A = \begin{bmatrix} 2 & 5 \\ 3 & 4 \end{bmatrix}$, $D = \begin{bmatrix} 3 & -3 \\ 2 & 1 \end{bmatrix}$ and $P = \begin{bmatrix} 5 & -2 \\ -1 & 3 \end{bmatrix}$

Solution

Step 1: Formula. $\text{Adj}(A) = \begin{bmatrix} d & -c \\ -b & a \end{bmatrix}$

Step 2: Data substitution and solving

$\text{Adj}(A) = \begin{bmatrix} 4 & -5 \\ -3 & 2 \end{bmatrix}$	$\text{Adj}(D) = \begin{bmatrix} 1 & 3 \\ -2 & 3 \end{bmatrix}$	$\text{Adj}(P) = \begin{bmatrix} 3 & 2 \\ 1 & 5 \end{bmatrix}$

Example 2: Given that; $P = \begin{bmatrix} 0 & 1 & 1 \\ 2 & 3 & -1 \\ -1 & 2 & 1 \end{bmatrix}$, fine the minor matrix using the co-factor method.

Solution

$$P = \begin{bmatrix} 0 & 1 & 1 \\ 2 & 3 & -1 \\ -1 & 2 & 1 \end{bmatrix} = \begin{bmatrix} \begin{bmatrix} 3 & -1 \\ 2 & 1 \end{bmatrix} & \begin{bmatrix} 2 & -1 \\ 1 & 1 \end{bmatrix} & \begin{bmatrix} 2 & 3 \\ -1 & 2 \end{bmatrix} \\ \begin{bmatrix} 1 & 1 \\ 2 & 1 \end{bmatrix} & \begin{bmatrix} 0 & 1 \\ -1 & 1 \end{bmatrix} & \begin{bmatrix} 0 & 1 \\ -1 & 2 \end{bmatrix} \\ \begin{bmatrix} 1 & 1 \\ 3 & -1 \end{bmatrix} & \begin{bmatrix} 0 & 1 \\ 2 & -1 \end{bmatrix} & \begin{bmatrix} 0 & 1 \\ 2 & 3 \end{bmatrix} \end{bmatrix}$$ [Work 2x2 matrices determinant]

$$P = \begin{bmatrix} [(3 \times 1) - (2 \times -1)] & [(2 \times 1) - (-1 \times -1)] & [(2 \times 2) - (3 \times -1)] \\ [(1 \times 1) - (1 \times 2)] & [(0 \times 1) - (1 \times -1)] & [(0 \times 2) - (1 \times -1)] \\ [(1 \times -1) - (1 \times 3)] & [(0 \times -1) - (1 \times 2)] & [(0 \times 3) - (1 \times 2)] \end{bmatrix}$$

$$P = \begin{bmatrix} [3+2] & [2-1] & [4+3] \\ [1-2] & [0+1] & [0+1] \\ [-1-3] & [0-2] & [0-2] \end{bmatrix} = \begin{bmatrix} 5 & 1 & 7 \\ -1 & 1 & 1 \\ -4 & -2 & -2 \end{bmatrix}$$

Example 3: Given that; $M = \begin{bmatrix} 1 & 1 & 1 \\ 1 & 2 & -3 \\ 2 & -1 & 3 \end{bmatrix}$, find the co-factor matrix using the row-column

co-factor method.

Solution

a) **First Method**

$$Co - Factor = \begin{bmatrix} 1 & 1 & 1 \\ 1 & 2 & -3 \\ 2 & -1 & 3 \end{bmatrix} = \begin{bmatrix} 2 & 3 & 1 & 2 \\ -1 & 3 & 2 & 1 \\ 1 & 1 & 1 & 1 \\ 2 & -3 & 1 & 2 \end{bmatrix}$$ ...[Calculate   minors

determinants]

$$Co - Factor =$$

$$\begin{bmatrix} [(2 \times 3) - (-3 \times -1)] & [(-3 \times 2) - (1 \times 3)] & [(1 \times -1) - (2 \times 2)] \\ [(-1 \times 1) - (3 \times 1)] & [(3 \times 1) - (2 \times 1)] & [(2 \times 1) - (-1 \times 1)] \\ [(1 \times -3) - (1 \times 2)] & [(1 \times 1) - (1 \times -3)] & [(1 \times 2) - (1 \times 1)] \end{bmatrix}$$

$$Co - Factor = \begin{bmatrix} [6-3] & [-6-3] & [-1-4] \\ [-1-3] & [3-2] & [2+1] \\ [-3-2] & [1+3] & [2-1] \end{bmatrix} = \begin{bmatrix} 3 & -9 & -5 \\ -4 & 1 & 3 \\ -5 & 4 & 1 \end{bmatrix}$$

b) **Second Method**

$$Co - Factor = \begin{bmatrix} 1 & 1 & 1 \\ 1 & 2 & -3 \\ 2 & -1 & 3 \end{bmatrix} = \begin{bmatrix} +\begin{bmatrix} 2 & -3 \\ -1 & 3 \end{bmatrix} & -\begin{bmatrix} 1 & -3 \\ 2 & 3 \end{bmatrix} & +\begin{bmatrix} 1 & 2 \\ 2 & -1 \end{bmatrix} \\ -\begin{bmatrix} 1 & 1 \\ -1 & 3 \end{bmatrix} & +\begin{bmatrix} 1 & 1 \\ 2 & 3 \end{bmatrix} & -\begin{bmatrix} 1 & 1 \\ 2 & -1 \end{bmatrix} \\ +\begin{bmatrix} 1 & 1 \\ 2 & -3 \end{bmatrix} & -\begin{bmatrix} 1 & 1 \\ 1 & -3 \end{bmatrix} & +\begin{bmatrix} 1 & 1 \\ 1 & 2 \end{bmatrix} \end{bmatrix}$$ ...[Solve minors]

$$Co - Factor = \begin{bmatrix} +[(2 \times 3) - (-3 \times -1)] & -[(1 \times 3) - (-3 \times 2)] & +[(1 \times -1) - (2 \times 2)] \\ -[(1 \times 3) - (1 \times -1)] & +[(1 \times 3) - (1 \times 2)] & -[(1 \times -1) - (1 \times 2)] \\ +[(1 \times -3) - (1 \times 2)] & -[(1 \times -3) - (1 \times 1)] & +[(1 \times 2) - (1 \times 1)] \end{bmatrix}$$

$$\text{Co} - \text{Factor} = \begin{bmatrix} +[6-3] & -[3+6] & +[-1-4] \\ -[3+1] & +[3-2] & -[-1-2] \\ +[-3-2] & -[-3-1] & +[2-1] \end{bmatrix} = \begin{bmatrix} +[3] & -[9] & +[-5] \\ -[4] & +[1] & -[-3] \\ +[-5] & -[-4] & +[1] \end{bmatrix}$$

$$\text{Co} - \text{Factor} = \begin{bmatrix} +[3] & -[9] & +[-5] \\ -[4] & +[1] & -[-3] \\ +[-5] & -[-4] & +[1] \end{bmatrix} = \begin{bmatrix} 3 & -9 & -5 \\ -4 & 1 & 3 \\ -5 & 4 & 1 \end{bmatrix}$$

Example 4: Given that; $Q = \begin{bmatrix} 1 & 3 & 7 \\ 4 & 2 & 3 \\ 1 & 2 & 1 \end{bmatrix}$, find the co-factor matrix using the basket weave

method.

Solution

$$Q = \begin{bmatrix} 1 & 3 & 7 \\ 4 & 2 & 3 \\ 1 & 2 & 1 \end{bmatrix} = \begin{bmatrix} 2 & 3 & 4 & 2 \\ 2 & 1 & 1 & 2 \\ 3 & 7 & 1 & 3 \\ 2 & 3 & 4 & 2 \end{bmatrix} = \begin{bmatrix} \begin{bmatrix} 2 & 3 \\ 2 & 1 \end{bmatrix} & \begin{bmatrix} 3 & 4 \\ 1 & 1 \end{bmatrix} & \begin{bmatrix} 4 & 2 \\ 1 & 2 \end{bmatrix} \\ \begin{bmatrix} 2 & 1 \\ 3 & 7 \end{bmatrix} & \begin{bmatrix} 1 & 1 \\ 7 & 1 \end{bmatrix} & \begin{bmatrix} 1 & 2 \\ 1 & 3 \end{bmatrix} \\ \begin{bmatrix} 3 & 7 \\ 2 & 3 \end{bmatrix} & \begin{bmatrix} 7 & 1 \\ 3 & 4 \end{bmatrix} & \begin{bmatrix} 1 & 3 \\ 4 & 2 \end{bmatrix} \end{bmatrix}$$

$$Q = \begin{bmatrix} [(2 \times 1) - (2 \times 3)] & [(3 \times 1) - (4 \times 1)] & [(4 \times 2) - (2 \times 1)] \\ [(2 \times 7) - (1 \times 3)] & [(1 \times 1) - (1 \times 7)] & [(1 \times 3) - (2 \times 1)] \\ [(3 \times 3) - (7 \times 2)] & [(7 \times 4) - (1 \times 3)] & [(1 \times 2) - (3 \times 4)] \end{bmatrix}$$

$$Q = \begin{bmatrix} [2-6] & [3-4] & [8-2] \\ [14-3] & [1-7] & [3-2] \\ [9-14] & [28-3] & [2-12] \end{bmatrix} = \begin{bmatrix} -4 & -1 & 6 \\ 11 & -6 & 1 \\ -5 & 25 & -10 \end{bmatrix}$$

Test Your Understanding

Exercise 12: Determine the adjugate of the following matrices.

1) $A = \begin{bmatrix} 10 & 6 \\ 7 & 5 \end{bmatrix}$..[Answer: See appendix]

2) $D = \begin{bmatrix} 1 & 4 \\ -3 & 1 \end{bmatrix}$..[Answer: See appendix]

3) $M = \begin{bmatrix} 2 & -1 \\ 0 & 5 \end{bmatrix}$..[Answer: See appendix]

Exercise 13: Given that; $N = \begin{bmatrix} 3 & 0 & 2 \\ 2 & 0 & -2 \\ 0 & 1 & 1 \end{bmatrix}$, find the minor matrix...[Answer: See appendix]

Exercise 14: Calculate the co-factor matrix of the following matrices.

1) $A = \begin{bmatrix} 2 & 0 & -4 \\ 5 & 0 & 1 \\ 4 & 0 & 6 \end{bmatrix}$..[Answer: See appendix]

2) $G = \begin{bmatrix} 3 & 5 & 6 \\ 1 & 2 & 4 \\ 3 & 5 & 6 \end{bmatrix}$..[Answer: See appendix]

4.5 Inverse of a Matrix

The inverse of a matrix is the product of the inverse of the determinant and the transpose of the co-factor matrix. The inverse of the matrix (A) denoted (A^{-1}) is calculated using the formula below.

$$A^{-1} = \frac{1}{|A|}[Co - Factor]^T = \frac{1}{|A|}Adj[A] \quad\quad\quad\quad\quad\quad\quad\quad (1)$$

Example 1: Given that; $A = \begin{bmatrix} 1 & -2 \\ 3 & -4 \end{bmatrix}$, show that $A^{-1}A = I$

Solution

Step 1: Determination of A^{-1}

1) Formula. $A^{-1} = \frac{1}{|A|}Adj[A]$, where; $Adj[A] = \begin{bmatrix} -4 & 2 \\ -3 & 1 \end{bmatrix}$ and Determinant$|A| = 2$

2) Data substitution and solving

$$A^{-1} = \frac{1}{2}\begin{bmatrix} -4 & 2 \\ -3 & 1 \end{bmatrix} = \begin{bmatrix} \left(-4 \times \frac{1}{2}\right) & \left(2 \times \frac{1}{2}\right) \\ \left(-3 \times \frac{1}{2}\right) & \left(1 \times \frac{1}{2}\right) \end{bmatrix} \Rightarrow A^{-1} = \begin{bmatrix} -\frac{4}{2} & \frac{2}{2} \\ -\frac{3}{2} & \frac{1}{2} \end{bmatrix}$$

Step 2: Data substitution and solving

$$A^{-1}A = \begin{bmatrix} -\frac{4}{2} & \frac{2}{2} \\ -\frac{3}{2} & \frac{1}{2} \end{bmatrix}\begin{bmatrix} 1 & -2 \\ 3 & -4 \end{bmatrix} = \begin{bmatrix} \left[\left(-\frac{4}{2}\times 1\right)+\left(\frac{2}{2}\times 3\right)\right] & \left[\left(-\frac{4}{2}\times -2\right)+\left(\frac{2}{2}\times -4\right)\right] \\ \left[\left(-\frac{3}{2}\times 1\right)+\left(\frac{1}{2}\times 3\right)\right] & \left[\left(-\frac{3}{2}\times -2\right)+\left(\frac{1}{2}\times -4\right)\right] \end{bmatrix}$$

$$A^{-1}A = \begin{bmatrix} \left[-\frac{4}{2}+\frac{6}{2}\right] & \left[\frac{8}{2}-\frac{8}{2}\right] \\ \left[-\frac{3}{2}+\frac{3}{2}\right] & \left[\frac{6}{2}-\frac{4}{2}\right] \end{bmatrix} = \begin{bmatrix} \frac{2}{2} & \frac{0}{2} \\ \frac{0}{2} & \frac{2}{2} \end{bmatrix} \Rightarrow A^{-1}A = \begin{bmatrix} 1 & 0 \\ 0 & 1 \end{bmatrix}$$

Example 2: Given that co-factor matrix is $M = \begin{bmatrix} 5 & -1 & 7 \\ 1 & 1 & -1 \\ -4 & 2 & -2 \end{bmatrix}$ and the determinant is 2,

determine the inverse of the matrix$[M^{-1}]$.

Step 1: Formula. $M^{-1} = \frac{1}{|M|} Adj[M]$, where; $Adj[M] = [m]^T = \begin{bmatrix} 5 & 1 & -4 \\ -1 & 1 & 2 \\ -7 & -1 & -2 \end{bmatrix}$

Step 2: Data substitution and solving

$$M^{-1} = \frac{1}{2} \begin{bmatrix} 5 & 1 & -4 \\ -1 & 1 & 2 \\ -7 & -1 & -2 \end{bmatrix} \ldots\ldots\ldots\ldots\ldots\ldots\ldots\ldots\ldots\ldots\ldots\ldots\text{[Multiply scalar and matrix]}$$

$$M^{-1} = \begin{bmatrix} \left(\frac{1}{2}\times 5\right) & \left(\frac{1}{2}\times 1\right) & \left(\frac{1}{2}\times -4\right) \\ \left(\frac{1}{2}\times -1\right) & \left(\frac{1}{2}\times 1\right) & \left(\frac{1}{2}\times 2\right) \\ \left(\frac{1}{2}\times 7\right) & \left(\frac{1}{2}\times -1\right) & \left(\frac{1}{2}\times -2\right) \end{bmatrix} = \begin{bmatrix} \frac{5}{2} & \frac{1}{2} & -\frac{4}{2} \\ -\frac{1}{2} & \frac{1}{2} & \frac{2}{2} \\ \frac{7}{2} & -\frac{1}{2} & -\frac{2}{2} \end{bmatrix}$$

Example 3: Given that the Eigen values of matrix $D = \begin{bmatrix} 1 & 1 & 0 \\ 0 & -2 & 1 \\ 0 & 0 & 3 \end{bmatrix}$ are $\lambda_1 = 1$, $\lambda_2 = -2$ and

$\lambda_3 = 3$, determine D^{-1}.

Step 1: Formula. $D^{-1} = \frac{1}{|D|} Adj[D]$, where; $|D| = 1 \times -2 \times 3 = -6$

$$Adj[D] = \begin{bmatrix} +\begin{bmatrix} -2 & 1 \\ 0 & 3 \end{bmatrix} & -\begin{bmatrix} 0 & 1 \\ 0 & 3 \end{bmatrix} & +\begin{bmatrix} 0 & -2 \\ 0 & 0 \end{bmatrix} \\ -\begin{bmatrix} 1 & 0 \\ 0 & 3 \end{bmatrix} & +\begin{bmatrix} 1 & 0 \\ 0 & 3 \end{bmatrix} & -\begin{bmatrix} 1 & 1 \\ 0 & 0 \end{bmatrix} \\ +\begin{bmatrix} 1 & 0 \\ -2 & 1 \end{bmatrix} & -\begin{bmatrix} 1 & 0 \\ 0 & 1 \end{bmatrix} & +\begin{bmatrix} 1 & 1 \\ 0 & -2 \end{bmatrix} \end{bmatrix}^T = \begin{bmatrix} -6 & 0 & 0 \\ -3 & 3 & 0 \\ 1 & -1 & -2 \end{bmatrix}^T = \begin{bmatrix} -6 & -3 & 1 \\ 0 & 3 & -1 \\ 0 & 0 & -2 \end{bmatrix}$$

Step 3: Data substitution and solving

$$D^{-1} = -\frac{1}{6} \begin{bmatrix} -6 & -3 & 1 \\ 0 & 3 & -1 \\ 0 & 0 & -2 \end{bmatrix} = \begin{bmatrix} -6\left(-\frac{1}{6}\right) & -3\left(-\frac{1}{6}\right) & 1\left(-\frac{1}{6}\right) \\ 0\left(-\frac{1}{6}\right) & 3\left(-\frac{1}{6}\right) & -1\left(-\frac{1}{6}\right) \\ 0\left(-\frac{1}{6}\right) & 0\left(-\frac{1}{6}\right) & -2\left(-\frac{1}{6}\right) \end{bmatrix} \ldots\ldots\text{[Open brackets]}$$

$$D^{-1} = \begin{bmatrix} \frac{6}{6} & \frac{3}{6} & -\frac{1}{6} \\ 0 & -\frac{3}{6} & \frac{1}{6} \\ 0 & 0 & \frac{2}{6} \end{bmatrix} = \begin{bmatrix} 1 & \frac{1}{2} & -\frac{1}{6} \\ 0 & -\frac{1}{2} & \frac{1}{6} \\ 0 & 0 & \frac{1}{3} \end{bmatrix}$$

Test Your Understanding

Exercise 16: The system of equation is given as;

$3X - Y + 6Z = 8$ ....(1)

$6Z = 3$ ....(2)

$2Z = 10$ ....(3)

The Eigen value equation is given as; $\lambda = (3 - \lambda)(-\lambda)(2 - \lambda)$. Find the inverse of the equation matrix (A)..............................[**Answer: No inverse. This is because** $|A| = 0$]

Exercise 17: The Eigen value equation of matrix (M) is given as; $\lambda = (\lambda - 4)(\lambda^2 + 4\lambda + 4)$.

Given that; $M = \begin{bmatrix} 1 & -3 & 3 \\ 3 & -5 & 3 \\ 6 & -6 & 4 \end{bmatrix}$, find M^{-1}. ....[**Answer: See appendix**]

4.6 Solving System of Equation using Matrix

System of equation such as simultaneous equation as seen above can also be solved using matrix. To solve system of equation using matrix, we are required to transform the system of equation to matrix form or matrix equation.

4.6.1 Matrix Equation

The matrix equation is made up of a coefficient matrix (A), variable vector (k) and constant vector (b). The matrix equation of a system of equations is the product of the coefficient matrix and variable vector which is equal to the constant vector.

$Ak = b$ ....[Matrix Equation]

The transformation process of a system of equations to matrix equation is given below.

Equation System	Matrix Equation
$\begin{aligned} a_{11}W + a_{12}X + a_{13}Y \cdots a_{1n}Z &= K_1 \\ a_{21}W + a_{22}X + a_{23}Y \cdots a_{2n}Z &= K_2 \\ a_{31}W + a_{32}X + a_{33}Y \cdots a_{3n}Z &= K_3 \\ \vdots \\ a_{m1}W + a_{m2}X + a_{m3}Y \cdots a_{mn}Z &= K_p \end{aligned}$	$\underbrace{\begin{bmatrix} a_{11} & a_{12} & a_{13} & \cdots & a_{1n} \\ a_{21} & a_{22} & a_{23} & \cdots & a_{2n} \\ a_{31} & a_{32} & a_{33} & \cdots & a_{3n} \\ \vdots & \vdots & \vdots & \vdots & \vdots \\ a_{m1} & a_{m2} & a_{m3} & \cdots & a_{mn} \end{bmatrix}}_{A} \underbrace{\begin{bmatrix} W \\ X \\ Y \\ \vdots \\ Z \end{bmatrix}}_{k} = \underbrace{\begin{bmatrix} K_1 \\ K_2 \\ K_3 \\ \vdots \\ K_p \end{bmatrix}}_{b}$

Example 1: Transform the equation system to its matrix equation.

First Equation	Second Equation
$5X + 4Y = 15$ ....(1) $2X - 3Y = -13$ ....(2)	$-12X + 8Y = -10$ ....(1) $5X - 20Y = 4$ ....(2) $X - Y = 15$ ....(3)

<u>**Solution**</u>

Step 1: Formula. Matrix Equation = Am = B

Step 2: Application of formula

First Equation	Second Equation
$\underbrace{\begin{bmatrix} 5 & 4 \\ 2 & -3 \end{bmatrix}}_{A} \underbrace{\begin{bmatrix} X \\ Y \end{bmatrix}}_{m} = \underbrace{\begin{bmatrix} 15 \\ -13 \end{bmatrix}}_{B}$	$\underbrace{\begin{bmatrix} -12 & 8 \\ 5 & -20 \\ 1 & -1 \end{bmatrix}}_{A} \underbrace{\begin{bmatrix} X \\ Y \end{bmatrix}}_{m} = \underbrace{\begin{bmatrix} -10 \\ 4 \end{bmatrix}}_{B}$

Example 2: Transform the equation system to its matrix equation

First Equation	Second Equation
$2X - 2Y + Z = 3$ ....(1) $3X + Y - Z = 7$ ....(2) $X - 3Y + 2Z = 0$ ....(2)	$3X + X - 2Z = 2$ ....…...............(1) $X - 2Y + Z = 3$ ....…...(2)

<u>**Solution**</u>

Step 1: Formula. Matrix Equation = Am = B

Step 2: Application of formula

First Equation	Second Equation
$\underbrace{\begin{bmatrix} 2 & -2 & 1 \\ 3 & 1 & -1 \\ 1 & -3 & 2 \end{bmatrix}}_{A} \underbrace{\begin{bmatrix} X \\ Y \\ Z \end{bmatrix}}_{m} = \underbrace{\begin{bmatrix} 3 \\ 7 \\ 0 \end{bmatrix}}_{B}$	$\underbrace{\begin{bmatrix} 3 & 1 & -2 \\ 1 & -2 & 1 \end{bmatrix}}_{A} \underbrace{\begin{bmatrix} X \\ Y \\ Z \end{bmatrix}}_{m} = \underbrace{\begin{bmatrix} 2 \\ 3 \end{bmatrix}}_{B}$

Test Your Understanding

Exercise 18: Write the equation form of the following equation systems

First System	Second System	Third System
$3X - 4Y = 9$ ....(1) $-2X + Y = -3$ ....(2)	$3Y = 2$ ....…...(1) $-5 - 3Y = 1$ ....(2) $6X = 6$ ....…...(3)	$\frac{1}{2}X - 3Y + Z = 2$ ....(1) $Y - 4Z = \frac{3}{5}$ ....…..(2)

[**Answer:** ....…......See Appendix]

Exercise 19: Write the equation form of the following equation systems

First System	Second System	Third System
$X + 2Y + Z = 9$ ....(1) $2X + Y + Z = 7$ ....(2)	$X + 3Y + Z = 1$ ....(1) $2X - Y + Z = 5$ ....…..(2)	$X - 2Y + 4Z = 44$ ....(1) $8Y - 14Z = -35$ ....…..(2)

| $X + Y + 3Z = 10$ ....(3) | $-2X + 2Y - Z = -8$ ... ...(3) | $3X - 9Y = 61$ ....(3) |

[Answer: ....See Appendix]

Exercise 20: Derive the equation systems from the following matrix equations

Equation 1	Equation 2	Equation 3
$\begin{bmatrix} 0 & -7 \\ 2 & 1 \end{bmatrix}\begin{bmatrix} X \\ Y \end{bmatrix} = \begin{bmatrix} 2 \\ 5 \end{bmatrix}$	$\begin{bmatrix} 2 & 1 & 1 \\ 1 & 3 & -1 \\ 1 & 1 & 1 \end{bmatrix}\begin{bmatrix} X \\ Y \\ Z \end{bmatrix} = \begin{bmatrix} 3 \\ 7 \\ 2 \end{bmatrix}$	$\begin{bmatrix} 2 & 3 & 1 \\ 1 & 2 & 3 \end{bmatrix}\begin{bmatrix} X \\ Y \\ Z \end{bmatrix} = \begin{bmatrix} 2 \\ 4 \end{bmatrix}$

[Answer: ....See Appendix]

4.6.2 Coefficient matrix and Augmented Matrix

A coefficient matrix as seen above is made up of the coefficients from the system of equation variables. The addition of the constant vector to the coefficient matrix changes the coefficient matrix to augmented matrix. The similarity between coefficient matrix and augmented matrix is that both ignore the variable vector.

Coefficient Matrix	Constant Matrix	Augmented Matrix
$\begin{bmatrix} a_{11} & a_{12} & a_{13} & \cdots & a_{1n} \\ a_{21} & a_{22} & a_{23} & \cdots & a_{2n} \\ a_{31} & a_{32} & a_{33} & \cdots & a_{3n} \\ \vdots & & \vdots & \vdots & \vdots \\ a_{m1} & a_{m2} & a_{m3} & \cdots & a_{mn} \end{bmatrix}$ A	$\begin{bmatrix} K_1 \\ K_2 \\ K_3 \\ \vdots \\ K_p \end{bmatrix}$ k	$\left[\begin{array}{ccccc|c} a_{11} & a_{12} & a_{13} & \cdots & a_{1n} & K_1 \\ a_{21} & a_{22} & a_{23} & \cdots & a_{2n} & K_2 \\ a_{31} & a_{32} & a_{33} & \cdots & a_{3n} & K_3 \\ \vdots & \vdots & & \vdots & \vdots & \vdots \\ a_{m1} & a_{m2} & a_{m3} & \cdots & a_{mn} & K_p \end{array}\right]$ A b
Only the coefficient of variables	Only results	Coefficients and results

Example 1: Construct an augmented matrix and coefficient matrix from the system of equation below.

$$2X + 4Y = 6 \ldots\ldots\ldots\ldots\ldots\ldots\ldots\ldots\ldots\ldots\ldots\ldots\ldots\ldots\ldots\ldots(1)$$

$$8X + 10Y = 12 \ldots\ldots\ldots\ldots\ldots\ldots\ldots\ldots\ldots\ldots\ldots\ldots\ldots(2)$$

Solution

Coefficient Matrix	Constant Matrix	Augmented Matrix
$\begin{bmatrix} 2 & 4 \\ 8 & 10 \end{bmatrix}$ A	$\begin{bmatrix} 6 \\ 12 \end{bmatrix}$ b	$\left[\begin{array}{cc|c} 2 & 4 & 6 \\ 8 & 10 & 12 \end{array}\right]$
Only the coefficient of variables	Only results	Coefficients and results

Example 2: Construct an augmented matrix and coefficient matrix from the system of equation below.

$2X - 4Y + Z = -4$ ....(1)

$4X - 8Y + 8Z = 5$ ....(2)

$-2X + 4Y - 3Z = 9$ ....(3)

<u>Solution</u>

Coefficient Matrix	Constant Matrix	Augmented Matrix
$\begin{bmatrix} 2 & -4 & 1 \\ 4 & -8 & 8 \\ -2 & 4 & -3 \end{bmatrix}$ A	$\begin{bmatrix} -4 \\ 5 \\ 9 \end{bmatrix}$ b	$\left[\begin{array}{ccc|c} 2 & -4 & 1 & -4 \\ 4 & -8 & 8 & 5 \\ -2 & 4 & -3 & 9 \end{array}\right]$
Only the coefficient of variables	Only results	Coefficients and results

---

Test Your Understanding

Exercise 21: Construct the augmented matrix and coefficient matrix of the following system of equations below.

Equation	Equation 2	Equation 3
$X + 2Y = 3$ ....(1) $3X - 4Y = 1$ ....(2)	$Y - 3Z = 5$ ....(1) $2X + 4Z = 9$ ....(2) $3X + Y + 7Z = -3$ ....(3)	$5X - 3Y = 8$ ....(1) $Z = -3$ ....(2)

[Answer: ....**See Appendix]**

4.6.3 Methods of Solving Linear Equations System using Matrix

The variables in a system of equation can be determined using the inverse method, the Cramer's method and Gaussian method (substitution and elimination). Our analysis will be limited to inverse matrix and Cramer's method.

$k = A^{-1}b$ ....[Inverse Matrix Method]

Value of Variable(n) $= \frac{det(\Delta_n)}{det(A)}$ ....[Cramer's Matrix Method]

Note: The symbol (Δ_n) is a derived matrix and changes with variables and (k) is a variable vector and should respect the placement of variables in the equation system.

a) Two variables situation [X and Y]..[Δ_X and Δ_Y]

Δ_X = [Constant Matrix and Y coefficient Vector] ....(1)

Δ_Y = [X coefficient Vector and Constant Matrix] ....(2)

b) Three variable Situation [X, Y and Z]...[Δ_X, Δ_Y and Δ_Z]

Δ_X = [Constant Matrix, Y Vector and Z vector] ....(1)

Δ_Y = [X Vector, Constant Matrix and Z Vector] ....(2)

Δ_X = [X Vetcor, Y Vector and Conctant Matrix] ....(3)

N/B: The swap pattern can be applied in situation of more than two or three variables in the equation system.

Example 1: Solve the simultaneous equation below using the inverse matrix method

$7X + Y = 25$ ....(1)

$5X - Y = 11$ ....(2)

$$\underline{\text{Solution}}$$

Step 1: Formula. $k = A^{-1}b$, ... $\left\{ \begin{matrix} 7X + Y = 25 \\ 5X - Y = 11 \end{matrix} \right\} = \begin{bmatrix} 7 & 1 \\ 5 & -1 \end{bmatrix} \begin{bmatrix} X \\ Y \end{bmatrix} = \begin{bmatrix} 25 \\ 11 \end{bmatrix} \right\}$.

Where; $A = \begin{bmatrix} 7 & 1 \\ 5 & -1 \end{bmatrix}$ and $b = \begin{bmatrix} 25 \\ 11 \end{bmatrix}$

Step 2: Determination of matrix inverse $[A^{-1}]$

1) Formula. $A^{-1} = \frac{1}{|A|} Adj[A]$

Where; $d = [7 \times (-1)] - [1 \times 5] = -12$ and $Adj[A] = \begin{bmatrix} -1 & -1 \\ -5 & 7 \end{bmatrix}$

2) Data substitution and solving

$$A^{-1} = -\frac{1}{12} \begin{bmatrix} -1 & -1 \\ -5 & 7 \end{bmatrix} = \begin{bmatrix} \{(-1) \times (-\frac{1}{12})\} & \{(-1) \times (-\frac{1}{12})\} \\ \{(-5) \times (-\frac{1}{12})\} & \{(7) \times (-\frac{1}{12})\} \end{bmatrix} = \begin{bmatrix} \frac{1}{12} & \frac{1}{12} \\ \frac{5}{12} & -\frac{7}{12} \end{bmatrix}$$

Step 3: Data substitution and solving

$$k = \begin{bmatrix} \frac{1}{12} & \frac{1}{12} \\ \frac{5}{12} & -\frac{7}{12} \end{bmatrix} \begin{bmatrix} 25 \\ 11 \end{bmatrix} = \begin{bmatrix} \{(25) \times (\frac{1}{12})\} + (11) \times (\frac{1}{12})\} \\ \{(25) \times (\frac{5}{12}) + (11) \times (-\frac{7}{12})\} \end{bmatrix}$$

$$k = \begin{pmatrix} X \\ Y \end{pmatrix} = \begin{bmatrix} \{\frac{25}{12} + \frac{11}{12}\} \\ \{\frac{125}{12} + -\frac{77}{12}\} \end{bmatrix} = \begin{pmatrix} X \\ Y \end{pmatrix} = \begin{bmatrix} 3 \\ 4 \end{bmatrix}$$ ....[**X = 3 and Y = 4**]

Example 2: Compute the value of X and Y using the Cramers method

$3X + 2Y = 36$ ....(1)

$5X + 4Y = 64$ ....(2)

Solution

Step 1: Formula. $X = \frac{|\Delta x|}{|A|}$ and $Y = \frac{|\Delta y|}{|A|}$

$$\underbrace{\left\{\begin{array}{l} 3X + 2Y = 36 \\ 5X + 4Y = 64 \end{array}\right.}_{\text{Equation System}} \Rightarrow \underbrace{X\binom{3}{5}}_{\text{X Vector}} + \underbrace{Y\binom{2}{4}}_{\text{Y Vector}} = \underbrace{\binom{36}{64}}_{\text{Constant Matrix}} \Rightarrow \overbrace{\underbrace{\begin{pmatrix} 3 & 2 \\ 5 & 4 \end{pmatrix}}_{A}\underbrace{\binom{X}{Y}}_{k} = \underbrace{\binom{36}{64}}_{b}}^{\text{Matrix Equation}}$$

Where; $A = \begin{pmatrix} 3 & 2 \\ 5 & 4 \end{pmatrix}$, $\Delta x = \begin{pmatrix} 36 & 2 \\ 64 & 4 \end{pmatrix} = \begin{pmatrix} 36 & 2 \\ 64 & 4 \end{pmatrix}$

$\underbrace{}_{\text{Constant Matrix}}\underbrace{}_{\text{Y Vector}}$

and $\Delta y = \begin{pmatrix} 3 & 36 \\ 5 & 64 \end{pmatrix} = \begin{pmatrix} 3 & 36 \\ 5 & 64 \end{pmatrix}$

$\underbrace{}_{\text{X Vector}}\underbrace{}_{\text{Constant Matrix}}$

Step 2: Data susbtitution and solving

$$X = \frac{|\Delta x|}{|A|} = \frac{\begin{pmatrix} 36 & 2 \\ 64 & 4 \end{pmatrix}}{\begin{pmatrix} 3 & 2 \\ 5 & 4 \end{pmatrix}} = \frac{[36\times4]-[2\times64]}{[3\times4]-[2\times5]} = \frac{144-128}{12-10} = \frac{16}{2} = 8 \dots\dots\dots\dots\dots\dots[\mathbf{X = 8}]$$

$$Y = \frac{|\Delta y|}{|A|} = \frac{\begin{pmatrix} 3 & 36 \\ 5 & 64 \end{pmatrix}}{\begin{pmatrix} 3 & 2 \\ 5 & 4 \end{pmatrix}} = \frac{[3\times64]-[36\times5]}{[3\times4]-[2\times5]} + \frac{192-180}{12-10} = \frac{12}{2} = 6 \dots\dots\dots\dots\dots\dots[\mathbf{Y = 6}]$$

Example 3: Solve the simultaneous equation below using the inverse matrix and Cramer method

$$3X - 2Y = 17 \dots\dots\dots\dots\dots\dots\dots\dots\dots\dots\dots\dots\dots\dots\dots\dots\dots(1)$$
$$4X + 5Y = -8 \dots\dots\dots\dots\dots\dots\dots\dots\dots\dots\dots\dots\dots\dots\dots\dots\dots(2)$$

Solution

a) Inverse Matrix Method

Step 1: Formula. $k = A^{-1}b$, $\dots\dots\dots\dots\dots$ $\left\{\begin{array}{l} 3X - 2Y = 17 \\ 4X + 5Y = -8 \end{array}\right\} = \begin{bmatrix} 3 & -2 \\ 4 & 5 \end{bmatrix}\begin{bmatrix} X \\ Y \end{bmatrix} = \begin{bmatrix} 17 \\ -8 \end{bmatrix}$.

Where; $A = \begin{bmatrix} 3 & -2 \\ 4 & 5 \end{bmatrix}$, $b = \begin{bmatrix} 17 \\ -8 \end{bmatrix}$, $|A| = 23$, $\text{Adj}[A] = \begin{bmatrix} 5 & 2 \\ -4 & 3 \end{bmatrix}$ and $A^{-1} = \begin{bmatrix} \frac{5}{12} & \frac{2}{23} \\ -\frac{4}{23} & \frac{3}{23} \end{bmatrix}$

Step 2: Data substitution and solving

$$k = \begin{bmatrix} \frac{1}{12} & \frac{1}{12} \\ \frac{5}{12} & -\frac{7}{12} \end{bmatrix}\begin{bmatrix} 17 \\ -8 \end{bmatrix} = \begin{bmatrix} \left(17\times\frac{5}{23}\right) + \left(-8\times\frac{2}{23}\right) \\ \left(17\times-\frac{4}{23}\right) + \left(-8\times\frac{3}{23}\right) \end{bmatrix}$$

$$k = \binom{X}{Y} = \begin{bmatrix} \left(\frac{85}{23} - \frac{16}{23}\right) \\ \left(-\frac{68}{23} - \frac{24}{23}\right) \end{bmatrix} = \begin{bmatrix} \left(\frac{69}{23}\right) \\ \left(-\frac{92}{23}\right) \end{bmatrix} = \begin{bmatrix} 3 \\ -4 \end{bmatrix} \dots\dots\dots\dots\dots[\mathbf{X = 3 \text{ and } Y = -4}]$$

196

b) Cramer Method

Step 1: Formula. $X = \dfrac{\det(\Delta_X)}{\det(A)}$ and $Y = \dfrac{\det(\Delta_Y)}{\det(A)}$

$$\begin{cases} 3X - 2Y = 17 \\ 4X + 5Y = -8 \end{cases} \Rightarrow X\begin{pmatrix} 3 \\ 4 \end{pmatrix} + Y\begin{pmatrix} -2 \\ 5 \end{pmatrix} = \begin{pmatrix} 17 \\ -8 \end{pmatrix} \Rightarrow \begin{pmatrix} 3 & -2 \\ 4 & 5 \end{pmatrix}\begin{pmatrix} X \\ Y \end{pmatrix} = \begin{pmatrix} 17 \\ -8 \end{pmatrix}$$

$$A = \begin{pmatrix} 3 & -2 \\ 4 & 5 \end{pmatrix}, \Delta_X = \begin{pmatrix} 17 & -2 \\ -8 & 5 \end{pmatrix}, \Delta_Y = \begin{pmatrix} 3 & 17 \\ 4 & -8 \end{pmatrix}$$

Step 2: Data susbtitution and solving

$$X = \frac{\det(\Delta_X)}{\det(A)} = \frac{\begin{pmatrix} 17 & -2 \\ -8 & 5 \end{pmatrix}}{\begin{pmatrix} 3 & -2 \\ 4 & 5 \end{pmatrix}} = \frac{[17 \times 5] - [-2 \times -8]}{[3 \times 5] - [-2 \times 4]} = \frac{85-16}{15+8} = \frac{69}{23} = 8 \dots\dots\dots\dots\dots\dots[\mathbf{X = 3}]$$

$$Y = \frac{\det(\Delta_Y)}{\det(A)} = \frac{\begin{pmatrix} 3 & 17 \\ 4 & -8 \end{pmatrix}}{\begin{pmatrix} 3 & -2 \\ 4 & 5 \end{pmatrix}} = \frac{[3 \times -8] - [17 \times 4]}{[3 \times 5] - [-2 \times 4]} + \frac{-24-68}{15+8} = \frac{-92}{23} = 6 \dots\dots\dots\dots\dots\dots[\mathbf{Y = -4}]$$

Example 4: Determine the value of X and Y from the equation system below using the Cramer method.

$$X_1 - 2X_2 + X_3 = 3 \dots\dots\dots\dots\dots\dots\dots\dots\dots\dots\dots\dots\dots\dots\dots\dots\dots\dots.(1)$$

$$2X_1 + X_2 - X_3 = 5 \dots\dots\dots\dots\dots\dots\dots\dots\dots\dots\dots\dots\dots\dots\dots\dots\dots\dots.(2)$$

$$3X_1 - X_2 + 2X_3 = 12 \dots\dots\dots\dots\dots\dots\dots\dots\dots\dots\dots\dots\dots\dots\dots.(3)$$

Solution

a) Using Cramer's Method

Step 1: Formula. $X_1 = \dfrac{|\Delta_{X1}|}{|A|}$, $X_2 = \dfrac{|\Delta_{X2}|}{|A|}$ and $X_3 = \dfrac{|\Delta_{X3}|}{|A|}$ ....Note:

Syetems of Equation

$$\left.\begin{matrix} X_1 - 2X_2 + X_3 = 3 \\ 2X_1 + X_2 - X_3 = 5 \\ 3X_1 - X_2 + 2X_3 = 12 \end{matrix}\right\}$$

Matrix Equation

$$X_1\underbrace{\begin{bmatrix} 1 \\ 2 \\ 3 \end{bmatrix}}_{X_1 \text{ Vector}} + X_2\underbrace{\begin{bmatrix} -2 \\ 1 \\ -1 \end{bmatrix}}_{X_2 \text{ Vector}} + X_3\underbrace{\begin{bmatrix} 1 \\ -1 \\ 2 \end{bmatrix}}_{X_3 \text{ Vector}} = \underbrace{\begin{bmatrix} 3 \\ 5 \\ 12 \end{bmatrix}}_{\text{Constant Matrix}} \Rightarrow \underbrace{\begin{bmatrix} 1 & -2 & 1 \\ 2 & 1 & -1 \\ 3 & -1 & 2 \end{bmatrix}}_{A} \underbrace{\begin{bmatrix} X_1 \\ X_2 \\ X_3 \end{bmatrix}}_{k} = \underbrace{\begin{bmatrix} 3 \\ 5 \\ 12 \end{bmatrix}}_{b}$$

$$\text{Where; } A = \begin{bmatrix} 1 & -2 & 1 \\ 2 & 1 & -1 \\ 3 & -1 & 2 \end{bmatrix}, \Delta_{X1} = \begin{bmatrix} 3 & -2 & 1 \\ 5 & 1 & -1 \\ 12 & -1 & 2 \end{bmatrix}, \Delta_{X2} = \begin{bmatrix} 1 & 3 & 1 \\ 2 & 5 & -1 \\ 3 & 12 & 2 \end{bmatrix}$$

$$\text{and } \Delta_{X3} = \begin{bmatrix} 1 & -2 & 3 \\ 2 & 1 & 5 \\ 3 & -1 & 12 \end{bmatrix}$$

Step 2: Determination of determinants of [A], [Δ_{X1}], [Δ_{X2}] and [Δ_{X3}]

$$|A| = \begin{bmatrix} 1 & -2 & 1 \\ 2 & 1 & -1 \\ 3 & -1 & 2 \end{bmatrix} = +1\begin{bmatrix} 1 & -1 \\ -1 & 2 \end{bmatrix} - (-2)\begin{bmatrix} 2 & -1 \\ 3 & 2 \end{bmatrix} + 1\begin{bmatrix} 2 & 1 \\ 3 & -1 \end{bmatrix} = 1 + 14 - 5 = \mathbf{10}$$

$$|\Delta_{X1}| = \begin{bmatrix} 3 & -2 & 1 \\ 5 & 1 & -1 \\ 12 & -1 & 2 \end{bmatrix} = +3\begin{bmatrix} 1 & -1 \\ -1 & 2 \end{bmatrix} - (-2)\begin{bmatrix} 5 & -1 \\ 12 & 2 \end{bmatrix} + 1\begin{bmatrix} 5 & 1 \\ 12 & -1 \end{bmatrix} = 3 + 44 - 17 = \mathbf{30}$$

$$|\Delta_{X2}| = \begin{bmatrix} 1 & 3 & 1 \\ 2 & 5 & -1 \\ 3 & 12 & 2 \end{bmatrix} = +1\begin{bmatrix} 5 & -1 \\ 12 & 2 \end{bmatrix} - (3)\begin{bmatrix} 2 & -1 \\ 3 & 2 \end{bmatrix} + 1\begin{bmatrix} 2 & 5 \\ 3 & 12 \end{bmatrix} = 22 - 21 + 9 = \mathbf{10}$$

$$|\Delta_{X3}| = \begin{bmatrix} 1 & -2 & 3 \\ 2 & 1 & 5 \\ 3 & -1 & 12 \end{bmatrix} = +1\begin{bmatrix} 1 & 5 \\ -1 & 12 \end{bmatrix} - (-2)\begin{bmatrix} 2 & 5 \\ 3 & 12 \end{bmatrix} + 3\begin{bmatrix} 2 & 1 \\ 3 & -1 \end{bmatrix} = 17 + 18 - 15 = $$
20

Step 3: Data susbtitution and solving

$$X_1 = \frac{|\Delta_{X1}|}{A} = \frac{30}{10} \dots\dots\dots\dots\dots\dots\dots\dots\dots\dots\dots\dots\dots\dots\dots [\mathbf{X_1 = 3}]$$

$$X_2 = \frac{|\Delta_{X2}|}{A} = \frac{10}{10} \dots\dots\dots\dots\dots\dots\dots\dots\dots\dots\dots\dots\dots\dots\dots [\mathbf{X_2 = 1}]$$

$$X_3 = \frac{|\Delta_{X3}|}{A} = \frac{20}{10} \dots\dots\dots\dots\dots\dots\dots\dots\dots\dots\dots\dots\dots\dots\dots [\mathbf{X_3 = 2}]$$

b) Inverse Matrix Method

Syetems of Equation   Matrix Equation

Step 1: Formula. $k = A^{-1}b$, ... $\left.\begin{array}{l} X_1 - 2X_2 + X_3 = 3 \\ 2X_1 + X_2 - X_3 = 5 \\ 3X_1 - X_2 + 2X_3 = 12 \end{array}\right\}$ $= \underbrace{\begin{bmatrix} 1 & -2 & 1 \\ 2 & 1 & -1 \\ 3 & -1 & 2 \end{bmatrix}}_{A} \underbrace{\begin{bmatrix} X_1 \\ X_2 \\ X_3 \end{bmatrix}}_{k} = \underbrace{\begin{bmatrix} 3 \\ 5 \\ 12 \end{bmatrix}}_{b}$.

Step 2: Determination of (A^{-1})

1) Formula. $A^{-1} = \frac{1}{|A|} \text{Adj}[A]$, where; $|A| = 10$......[Calculated in the Cramer's method]

$$\text{Adj}[A] = \begin{bmatrix} 1 & -1 & 2 & 1 \\ -1 & 2 & 3 & -1 \\ -2 & 1 & 1 & -2 \\ 1 & -1 & 2 & 1 \end{bmatrix}^T = \begin{bmatrix} \begin{bmatrix} 1 & -1 \\ -1 & 2 \end{bmatrix} & \begin{bmatrix} -1 & 2 \\ 2 & 3 \end{bmatrix} & \begin{bmatrix} 2 & 1 \\ 3 & -1 \end{bmatrix} \\ \begin{bmatrix} -1 & 2 \\ -2 & 1 \end{bmatrix} & \begin{bmatrix} 2 & 3 \\ 1 & 1 \end{bmatrix} & \begin{bmatrix} 3 & -1 \\ 1 & -2 \end{bmatrix} \\ \begin{bmatrix} -2 & 1 \\ 1 & -1 \end{bmatrix} & \begin{bmatrix} 1 & 1 \\ -1 & 2 \end{bmatrix} & \begin{bmatrix} 1 & -2 \\ 2 & 1 \end{bmatrix} \end{bmatrix}^T =$$

$$\begin{bmatrix} 1 & -7 & -5 \\ 3 & -1 & -5 \\ 1 & 3 & 5 \end{bmatrix}^T$$

2) Data substitution and solving

$$A^{-1} = \frac{1}{10}\begin{bmatrix} 1 & -7 & -5 \\ 3 & -1 & -5 \\ 1 & 3 & 5 \end{bmatrix}^T = \frac{1}{10}\begin{bmatrix} 1 & 3 & 1 \\ -7 & -1 & 3 \\ -5 & -5 & 5 \end{bmatrix}.\text{[Perform scalar matrix multiplication]}$$

$$A^{-1} = \begin{bmatrix} 1\left(\frac{1}{10}\right) & 3\left(\frac{1}{10}\right) & 1\left(\frac{1}{10}\right) \\ -7\left(\frac{1}{10}\right) & -1\left(\frac{1}{10}\right) & 3\left(\frac{1}{10}\right) \\ -5\left(\frac{1}{10}\right) & -5\left(\frac{1}{10}\right) & 5\left(\frac{1}{10}\right) \end{bmatrix} = \begin{bmatrix} \frac{1}{10} & \frac{3}{10} & \frac{1}{10} \\ -\frac{7}{10} & -\frac{1}{10} & \frac{3}{10} \\ -\frac{5}{10} & -\frac{5}{10} & \frac{5}{10} \end{bmatrix}$$

Step 3: Data substitution and solving

$$k = \begin{bmatrix} \frac{1}{10} & \frac{3}{10} & \frac{1}{10} \\ -\frac{7}{10} & -\frac{1}{10} & \frac{3}{10} \\ -\frac{5}{10} & -\frac{5}{10} & \frac{5}{10} \end{bmatrix} \begin{bmatrix} 3 \\ 5 \\ 12 \end{bmatrix} = \begin{bmatrix} \left(\frac{3}{10} + \frac{15}{10} + \frac{12}{10}\right) \\ \left(-\frac{21}{10} - \frac{5}{10} + \frac{36}{10}\right) \\ \left(-\frac{15}{10} - \frac{25}{10} + \frac{60}{10}\right) \end{bmatrix} \Rightarrow \begin{pmatrix} X_1 \\ X_2 \\ X_3 \end{pmatrix} = \begin{bmatrix} 3 \\ 1 \\ 2 \end{bmatrix} \quad ...[X_1 = 3,\ X_2 = 1\ \text{and}\ X_3 = 2]$$

Test Your Understanding

Exercise 22: Solve the simultaneous equation below using the Cramer method.

$2X - 3Y = 6$ ....(1)

$4X - 6Y = 10$ ....(2)

[Answer: No Solution since $|A| = 0$]

Exercise 23: Determine the value of X and Y in the simultaneous equation below using the inverse matrix method.

$2X - 3Y = 1$ ....(1)

$4X + 4Y = 2$ ....(2)

[Answer: $X = 1/2$ and $Y = 0$]

Exercise 24: Solve the value of X, Y and Z from the system of equation below.

$X + Z = 0$ ....(1)

$X - 3Y = 1$ ....(2)

$4Y - 3Z = 3$ ....(3)

[Answer: $X = 1, Y = 0$ and $Z = -1$]

CHAPTER FIVE

SURDS AND LOGARITHM

5.1 SURDS

A surd is defined as any number that cannot be simplified to remove a root (square root, cube root etc). It is also defined as the root of an integer that cannot be further simplified, making it to be called an irrational root. It is often assumed that a surd is the squared root of non-squared numbers. Generally, surd is expressed as seen below

$$\sqrt[a]{X} \dots\dots\dots\dots\dots\dots\dots\dots\dots\dots\dots\dots\dots\dots\dots\dots\dots\dots\dots\dots\dots\dots(1)$$

Where; The root $\left(\sqrt{}\right)$ is called radical

(a) is called the power

(X) is called radicand

5.1.1 Simplification of Surd

Simplification of surd is the reduction of a given surd to an alternative form or transformation of the root of a non-squared number into a surd. The alternative form is often the lowest surd expression of the original surd.

Example 1: Simplify the surd $\sqrt{18}$

Solution

Hint: The radicand (18) is not a squared number. Here, we are required to transform the non-squared number root into a surd.

Step 1: Outlining and identification

Factors of 18 are; 1, 2, 3, 6, 9 and 18. The highest square number in the factor is (9) and which if multiplied by (2) will give (18). Therefore, the new surds are surd 9 and surd 2.

Step 2: Segmentation and simplification

$$\sqrt{18} = \sqrt{9} \times \sqrt{2} \dots\dots\dots\dots\dots\dots\dots\dots\dots\dots\text{[Work root of the square number]}$$
$$3 \times \sqrt{2} \dots\dots\dots\dots\dots\dots\dots\dots\dots\dots\dots\dots\text{[Multiply values]}$$
$$\sqrt{18} = 3\sqrt{2}$$

Example 2: Simplify the surd $4\sqrt{27}$

Solution

Step 1: Outlining and identification

Factors of 27 are; 1, 3, 9 and 27. The highest square number in the factor is (9) and which if multiplied by (3) will give (27). Therefore, the new surds are surd 9 and surd 3.

Step 2: Segmentation and simplification

$4\sqrt{27} = 4\left(\sqrt{9} \times \sqrt{3}\right)$ ....[Work root of the square number]

$4\left(3 \times \sqrt{3}\right)$ ....[Simplify bracket]

$4\left(3\sqrt{3}\right)$ ....[Multiply values]

$4\sqrt{27} = 12\sqrt{3}$

Example 3: Simplify the surd $\sqrt{-25}$

Solution

Hint: The surds value (25) is a square number. Therefore there is no need to look for the factors since the value is itself the highest square number factor. In situation of negative value, the negative sign is having by default a coefficient 1.

Step 1: Outlining and identification

Factors of 25 are 1, 5 and 25. The highest square number in the factor is (25) and which if multiplied by (1) will give (25). Therefore, the new surds are surds 25, surds 1 and surds -1.

Step 2: Segmentation and simplification

$\sqrt{-25} = \sqrt{25} \times \sqrt{1} \times \sqrt{-1}$ ....[Work square roots of positive squared numbers]

$5 \times 1 \times \sqrt{-1}$ ....[Multiply values]

$5 \times \sqrt{-1}$ ....[Work square root of negative value. remember $\sqrt{-1} = i$]

$5 \times i$ ....[Multiply elements]

$\sqrt{-25} = 5i$

Example 4: Simplify $-\sqrt{12}$

Solution

Step 1: Outlining and identification

Factors of 12 are 1, 2, 3, 4, 6 and 12. The highest square number in the factor is (4) and which if multiplied by (3) will give (12). Therefore, the new surds are surds 4 and surds 3.

201

Step 2: Segmentation and simplification

$$-\sqrt{12} = -\left[\sqrt{4} \times \sqrt{3}\right] \quad\text{...................[Work the square root of the squared number]}$$

$$-\left[2 \times \sqrt{3}\right] \quad\text{...[Simplify bracket]}$$

$$-\left[2\sqrt{3}\right] \quad\text{..[Open bracket]}$$

$$-\sqrt{12} = -2\sqrt{3}$$

Example 5: Simplify the surd $-\sqrt{-200}$

<u>**Solution**</u>

Step 1: Outlining and identification

The highest factor of 200 that is a square number is 100 which if multiplied by 2 will give 200. Therefore, the surds are surds 100, surds 2 and surds -1 since 200 is having a negative sign. For simplification, the non-squared number should be positioned last.

Step 2: Segmentation and simplification

$$-\sqrt{-200} = -\left[\sqrt{100} \times \sqrt{-1} \times \sqrt{2}\right] \quad\text{...............[Work square roots of squared elements]}$$

$$-\left[10 \times i \times \sqrt{2}\right] \quad\text{...[Simplify bracket]}$$

$$-\left[10i\sqrt{2}\right] \quad\text{..[Open bracket]}$$

$$-\sqrt{-200} = -10i\sqrt{2}$$

<u>Test Your Understanding</u>

Exercise 1: Simplify the following surds

1) $\sqrt{40}$...**[Answer: $2\sqrt{10}$]**

2) $\frac{3}{5}\sqrt{75}$...**[Answer: $3\sqrt{3}$]**

3) $\sqrt{-16}$...**[Answer: $4i$]**

4) $-\sqrt{196}$...**[Answer: -14]**

5) $-\sqrt{-169}$...**[Answer: $-13i$]**

5.1.2 Addition Properties of Surds

The addition of surd is possible if and only if the radicands are the same. In situations where they are not the same, we are required to use the principle of surds simplification as seen above to establish the equality of radicands. The addition of surds property is given below.

$$a\sqrt{c} + b\sqrt{c} = (a + b)\sqrt{c} \quad\text{...(1)}$$

Example 1: Evaluate $5\sqrt{4} + 3\sqrt{4}$

<u>Solution</u>

Step 1: Formula. $a\sqrt{c} + b\sqrt{c} = (a + b)\sqrt{c}$, where; $a = 5$, $b = 3$ and $c = 4$

Step 2: Data substitution and solving

$5\sqrt{4} + 3\sqrt{4}$ ....[Add non squared values]

$(5 + 3)\sqrt{4}$ ....[Work square root of 4]

$(5 + 3)2$ ....[Open bracket and solve]

$5\sqrt{4} + 3\sqrt{4} = \mathbf{16}$

Example 2: Solve $3\sqrt{9} + \sqrt{9}$

<u>Solution</u>

Step 1: Formula. $a\sqrt{c} + b\sqrt{c} = (a + b)\sqrt{c}$, where; $a = 3$, $b = 1$ and $c = 9$

Step 2: Data substitution and solving

$3\sqrt{9} + 1\sqrt{9}$ ....[Add non squared values]

$(3 + 1)\sqrt{9}$ ....[Work square root of 9]

$(3 + 1)3$ ....[Open bracket and solve]

$3\sqrt{9} + \sqrt{9} = \mathbf{12}$

Example 3: Evaluate $2\sqrt{3} + 4\sqrt{12}$

<u>Solution</u>

Hint: The radicands are not the same or the values in the roots (radicals) are not the same, making it inappropriate to apply the addition property. We are required to equalize them before adding the surds. It is advisable to change the large radicand (12) to the small radicand (3).

Step 1: Changing radicand (12) to (3)...[Note: $\sqrt{12} = 2\sqrt{3}$]

Step 2: Formula. $a\sqrt{c} + b\sqrt{c} = (a + b)\sqrt{c}$, where; $a = 2$, $b = 4 \times 2 = 8$ and $c = 3$.

Step 3: Data substitution and solving

$2\sqrt{3} + 4\sqrt{12} = 2\sqrt{3} + 8\sqrt{3}$ ....[Apply the addition property]

$(2 + 8)\sqrt{3}$ ....[Simplify bracket]

$2\sqrt{3} + 4\sqrt{12} = \mathbf{10\sqrt{3}}$

Exercise 2: Evaluate the following surds

1) $11\sqrt{3} + 8\sqrt{3}$..[Answer: $19\sqrt{3}$]

2) $9\sqrt{2} + \sqrt{18}$...[Answer: $12\sqrt{2}$]

3) $\sqrt{300} + \sqrt{75}$...[Answer: $15\sqrt{3}$]

5.1.3 Subtraction Properties of Surds

The subtraction of surd is possible if and only if the radicands are the same. In situations where they are not the same, we are required to use the principle of surds simplification as seen above to establish the equality of radicands. The subtraction of surds property is given below.

$$a\sqrt{c} - b\sqrt{c} = (a - b)\sqrt{c} \dots\dots\dots\dots\dots\dots\dots\dots\dots\dots\dots\dots\dots\dots\dots\dots(1)$$

Example 1: Evaluate the following surds

1) $4\sqrt{25} - 2\sqrt{25}$

2) $7\sqrt{3} - 10\sqrt{3}$

Solution

1) $4\sqrt{25} - 2\sqrt{25}$

Step 1: Formula. $a\sqrt{c} - b\sqrt{c} = (a - b)\sqrt{c}$, where; a = 4, b = 2 and c = 25

Step 2: Data substitution and solving

$4\sqrt{25} - 2\sqrt{25}$ ....[Apply the subtraction property]

$(4 - 2)\sqrt{25}$ ....[Simplify bracket and work square root]

$(2)5$ ....[Open bracket]

$4\sqrt{25} - 2\sqrt{25} = \mathbf{10}$

2) $7\sqrt{3} - 10\sqrt{3}$

Step 1: Formula. $a\sqrt{c} - b\sqrt{c} = (a - b)\sqrt{c}$, where; a = 7, b = 10 and c = 3

Step 2: Data substitution and solving

$7\sqrt{3} - 10\sqrt{3}$ ....[Apply the subtraction property]

$(7 - 10)\sqrt{3}$ ....[Simplify bracket]

$(-3)\sqrt{3}$ ....[Open bracket]

$7\sqrt{3} - 10\sqrt{3} = \mathbf{-3\sqrt{3}}$

Example 2: Evaluate the following surds

1) $6\sqrt{3} - \sqrt{12}$

2) $\frac{3}{2}\sqrt{-4} - 3\sqrt{-4}$

Solution

1) $6\sqrt{3} - \sqrt{12}$

Step 1: Changing radicand (12) to (3)...[Note; $\sqrt{12} = 2\sqrt{3}$]

Step 2: Formula. $a\sqrt{c} - b\sqrt{c} = (a - b)\sqrt{c}$, where; a = 6, b = 2 and c = 3

Step 3: Data substitution and solving

$6\sqrt{3} - 2\sqrt{3}$ ....[Apply the subtraction property]

$(6 - 2)\sqrt{3}$ ....[Simplify bracket]

$(4)\sqrt{3}$ ....[Open bracket]

$6\sqrt{3} - \sqrt{12} = 4\sqrt{3}$

2) $\frac{3}{2}\sqrt{-4} - 3\sqrt{-4}$

Step 1: Formula. $a\sqrt{c} - b\sqrt{c} = (a - b)\sqrt{c}$, where; $a = \frac{3}{2}$, b = 3 and c = -4 = -1 × 4

Step 2: Data substitution and solving

$\frac{3}{2}\sqrt{-4} - 3\sqrt{-4}$ ....[Apply the subtraction property]

$\left(\frac{3}{2} - 3\right)\sqrt{-4}$ ....[Simplify bracket and radicand]

$\left(-\frac{3}{2}\right)\left[\sqrt{4} \times \sqrt{-1}\right]$ ....[Work roots]

$\left(-\frac{3}{2}\right)[2 \times i]$ ....[Open bracket]

$\frac{3}{2}\sqrt{-4} - 3\sqrt{-4} = -3i$

Example 3: Evaluate $\sqrt{150} - \sqrt{54}$

Solution

Hint: The radicands are not the same. Therefore, we are required to make them similar in order to apply the subtraction property.

$\sqrt{150} = \sqrt{25 \times 6} = \sqrt{25} \times \sqrt{6} = 5 \times \sqrt{6} = 5\sqrt{6}$...(1)

$\sqrt{54} = \sqrt{9 \times 6} = \sqrt{9} \times \sqrt{6} = 3 \times \sqrt{6} = 3\sqrt{6}$ ....(2)

Step 1: Formula. $a\sqrt{c} - b\sqrt{c} = (a - b)\sqrt{c}$, where; a = 5, b = 3 and c = 6

Step 2: Data substitution and solving

$5\sqrt{6} - 3\sqrt{6}$ ....[Apply the subtraction property]

$(5 - 3)\sqrt{6}$ ....[Simplify bracket]

$(2)\sqrt{6}$ ....[Open bracket]

$\sqrt{150} - \sqrt{54} = 2\sqrt{6}$

Test Your Understanding

Exercise 3: Evaluate the following surds

1) $\sqrt{343} - \sqrt{28}$..[Answer: $5\sqrt{7}$]

2) $2\sqrt{3} - 4\sqrt{3}$..[Answer: $-2\sqrt{3}$]

3) $2\sqrt{9} - \sqrt{9}$..[Answer: 3]

4) $\sqrt{18} - \frac{1}{3}\sqrt{50}$..[Answer: $\frac{4}{3}\sqrt{2}$]

5) $4\sqrt{-64} - 3\sqrt{-64}$..[Answer: $8i$]

5.1.4 The Product and Quotient Property of Surds

The product property of a surd talks about the multiplication of two or more surds and the quotient property of surd talks about the division of two or more surds. The general formulas used for the multiplication and division of surds are given below.

$$\sqrt{a \times b} = \sqrt{a} \times \sqrt{b} \dots\dots\dots\dots\dots\dots\dots\dots\dots\dots(1)$$

$$\sqrt{\frac{a}{b}} = \frac{\sqrt{a}}{\sqrt{b}} \dots\dots\dots\dots\dots\dots\dots\dots\dots\dots(2)$$

Example 1: Evaluate $\sqrt{32} \times \sqrt{2}$

Solution

Step 1: Formula. $\sqrt{a \times b} = \sqrt{a} \times \sqrt{b}$, where; a = 32 and b = 2

Step 2: Data substitution and solving

$\sqrt{32} \times \sqrt{2}$ ....[Put values in a single root]

$\sqrt{32 \times 2}$ ....[Multiply values]

$\sqrt{64}$ ....[Work root]

$\sqrt{32} \times \sqrt{2} = 8$

Example 2: Evaluate $\sqrt{\frac{100}{25}}$

Solution

Step 1: Formula. $\sqrt{\frac{a}{b}} = \frac{\sqrt{a}}{\sqrt{b}}$, where; a = 100 and b = 25

Step 2: Data substitution and solving

$$\sqrt{\frac{100}{25}} = \frac{\sqrt{100}}{\sqrt{25}} \text{ ....[Work square root since both radicands are squared numbers]}$$

$$\sqrt{\frac{100}{25}} = \frac{10}{5} = 2$$

Example 3: Evaluate $\sqrt{12} \times \sqrt{-3}$

<u>**Solution**</u>

Step 1: Formula. $\sqrt{a \times b} = \sqrt{a} \times \sqrt{b}$, where; a = 12 and b = −3

Step 2: Data substitution and solving

$$\sqrt{12} \times \sqrt{-3} \text{ ....[Put values in a single root]}$$

$$\sqrt{12 \times -3} \text{ ....[Multiply values]}$$

$$\sqrt{-36} \text{ ....[Apply simplification principle]}$$

$$\sqrt{36} \times \sqrt{-1} \text{ ....[Work squared root of radicands]}$$

$$\sqrt{12} \times \sqrt{-3} = 6i$$

Example 4: Evaluate $\sqrt{\frac{32}{144}}$

<u>**Solution**</u>

Step 1: Formula. $\sqrt{\frac{a}{b}} = \frac{\sqrt{a}}{\sqrt{b}}$,

Where; a = 32 (not a squared number) and b = 144 (Squared number)

Step 2: Data substitution and solving

$$\sqrt{\frac{32}{144}} = \frac{\sqrt{32}}{\sqrt{144}} \text{ ....[Apply simplification principles on numerator]}$$

$$\frac{\sqrt{16 \times 2}}{\sqrt{144}} = \frac{\sqrt{16} \times \sqrt{2}}{\sqrt{144}} \text{ ....[Work square root or squared number radicands]}$$

$$\frac{4 \times \sqrt{2}}{12} = \frac{4\sqrt{2}}{12} \text{ ....[Simplify fraction]}$$

$$\sqrt{\frac{32}{144}} = \frac{\sqrt{2}}{3}$$

Test Your Understanding

Exercise 4: Evaluate the following surds

1) $\sqrt{5} \times \sqrt{45}$.................................... ....[**Answer: 15**]

2) $\sqrt{2} \times \sqrt{-50}$...[**Answer: 10i**]

3) $\sqrt{\frac{400}{16}}$...[**Answer: 5**]

4) $\sqrt{\frac{225}{75}}$.................................... ....[**Answer: $\frac{3}{\sqrt{3}}$**]

5.1.5 Rationalization of Surds

Rationalization of surds is the process of removing surds in the denomination of an equation. The three main formulas worth knowing are given below.

$$\frac{b}{\sqrt{a}} = \frac{b}{\sqrt{a}} \times \frac{\sqrt{a}}{\sqrt{a}} \quad \text{..}(1)$$

$$\frac{c}{a+b\sqrt{n}} = \frac{c}{a+b\sqrt{n}} \times \frac{a-b\sqrt{n}}{a-b\sqrt{n}} \quad \text{...................................}(2)$$

$$\frac{c}{a-b\sqrt{n}} = \frac{c}{a-b\sqrt{n}} \times \frac{a+b\sqrt{n}}{a+b\sqrt{n}} \quad \text{...................................}(3)$$

Note: when dealing with rationalization of surds, we are required to apply the concept of difference of two squares. That is,

$$(a - b) \times (a + b) = a^2 - b^2 \quad \text{..}(4)$$

From equation (4), we realized that the application of direct multiplication on the denominator of equation (2) and equation (3) gives use the following.

$$\left(a + b\sqrt{n}\right) \times \left(a - b\sqrt{n}\right) = a^2 - \left(b\sqrt{n}\right)^2 = a^2 - (b)^2\left(\sqrt{n}\right)^2 \quad \text{......................}(5)$$

$$\left(a - b\sqrt{n}\right) \times \left(a + b\sqrt{n}\right) = a^2 - \left(b\sqrt{n}\right)^2 = a^2 - (b)^2\left(\sqrt{n}\right)^2 \quad \text{......................}(6)$$

$$\left(a + b\sqrt{n}\right) \times \left(a + b\sqrt{n}\right) = a^2 + 2ab\sqrt{n} + \left(b\sqrt{n}\right)^2 \quad \text{.................................}(7)$$

Example 1: Rationalized the surd $\frac{3}{\sqrt{5}}$

Solution

Step 1: Formula. $\frac{b}{\sqrt{a}} = \frac{b}{\sqrt{a}} \times \frac{\sqrt{a}}{\sqrt{a}}$, where; b = 3 and a = 5

Step 2: Data substitution and solving

$$\frac{3}{\sqrt{5}} = \frac{3}{\sqrt{5}} \times \frac{\sqrt{5}}{\sqrt{5}} \quad \text{...............................}\text{[Perform direct multiplication]}$$

$$\frac{3 \times \sqrt{5}}{\sqrt{5} \times \sqrt{5}} \quad \text{...............................}\text{[Simplify numerator and denominator]}$$

$$\frac{3\sqrt{5}}{\sqrt{5 \times 5}} = \frac{3\sqrt{5}}{\sqrt{25}} \quad \text{...............................}\text{[Work square root of denominator]}$$

$$\frac{3}{\sqrt{5}} = \frac{3\sqrt{5}}{5}$$

Example 2: Rationalized $\frac{3}{2+\sqrt{2}}$

Solution

Step 1: Formula. $\frac{c}{a+b\sqrt{n}} = \frac{c}{a+b\sqrt{n}} \times \frac{a-b\sqrt{n}}{a-b\sqrt{n}}$, where; c = 3, b = 1, a = 2 and n = 2

Step 2: Data substitution and solving

$$\frac{3}{2+\sqrt{2}} = \frac{3}{2+\sqrt{2}} \times \frac{2-\sqrt{2}}{2-\sqrt{2}} \dots\dots\dots\dots\dots\dots\dots\dots\dots\text{[Perform direct multiplication]}$$

$$\frac{3\times(2-\sqrt{2})}{(2+\sqrt{2})\times(2-\sqrt{2})} \dots\text{[Open numerator bracket and denominator is difference of two square]}$$

$$\frac{6-3\sqrt{2}}{(2)^2-(\sqrt{2})^2} \dots\dots\dots\dots\dots\dots\dots\dots\dots\dots\dots\dots\dots\text{[Work denominator root and exponent]}$$

$$\frac{6-3\sqrt{2}}{4-2} \dots\dots\dots\dots\dots\dots\dots\dots\dots\dots\dots\dots\dots\dots \text{....[Simplify denominator]}$$

$$\frac{3}{2+\sqrt{2}} = \frac{6-3\sqrt{2}}{2}$$

Example 3: Rationalized $\frac{2}{2\sqrt{3}-3}$

Solution

Step 1: Formula. $\frac{c}{a-b\sqrt{n}} = \frac{c}{a-b\sqrt{n}} \times \frac{a+b\sqrt{n}}{a+b\sqrt{n}}$, where; c = 2, a = 3, b = 2 and n = 3

Step 2: Data substitution and solving

$$\frac{2}{2\sqrt{3}-3} = \frac{2}{2\sqrt{3}-3} \times \frac{2\sqrt{3}+3}{2\sqrt{3}+3} \dots\dots\dots\dots\dots\dots\dots\dots\dots\text{[Perform direct multiplication]}$$

$$\frac{2\times(2\sqrt{3}+3)}{(2\sqrt{3}-3)\times(2\sqrt{3}+3)} \text{[Open numerator bracket and denominator is difference of two square]}$$

$$\frac{4\sqrt{3}+6}{(2\sqrt{3})^2-(3)^2} = \frac{4\sqrt{3}+6}{\left[(2)^2\times(\sqrt{3})^2\right]-(3)^2} \dots\dots\dots\dots\dots\text{[Work denominator root and exponent]}$$

$$\frac{4\sqrt{3}+6}{[4\times3]-9} = \frac{4\sqrt{3}+6}{12-9} \dots\dots\dots\dots\dots\dots\dots\dots\dots\dots\dots\dots\text{......[Simplify denominator]}$$

$$\frac{2}{2\sqrt{3}-3} = \frac{4\sqrt{3}+6}{3} = \frac{4}{3}\left(\sqrt{3}\right) + 2$$

Example 4: Rationalized the following surds

1) $5\sqrt{8} + \frac{6}{\sqrt{2}}$

2) $\frac{1}{\sqrt{2}+1} + \frac{1}{\sqrt{2}-1}$

Solution

1) $5\sqrt{8} + \frac{6}{\sqrt{2}}$

Step 1: Normalization of second element $\left[\frac{6}{\sqrt{2}}\right]$

1) Formula. $\frac{b}{\sqrt{a}} = \frac{b}{\sqrt{a}} \times \frac{\sqrt{a}}{\sqrt{a}}$, where; b = 6 and $\sqrt{a} = \sqrt{2}$

2) Data substitution and solving

$$\frac{6}{\sqrt{2}} = \frac{6}{\sqrt{2}} \times \frac{\sqrt{2}}{\sqrt{2}} \dots\dots\dots\dots\dots\dots\dots\dots\dots\dots\dots\text{[Perform direct multiplication]}$$

$$\frac{6\sqrt{2}}{(\sqrt{2})^2} = \frac{6\sqrt{2}}{2} \dots\dots\dots\dots\dots\dots\dots\dots\dots\dots\dots\dots \text{..........[Simplify equation]}$$

$$\frac{6}{\sqrt{2}} = 3\sqrt{2} \quad \dots\dots\dots\dots\dots\dots\dots\dots\dots\dots\dots\dots\dots\dots\text{[Simplified form]}$$

Step 2: Formula. $a\sqrt{c} + b\sqrt{c} = (a + b)\sqrt{c}$, where; $b\sqrt{c} = 3\sqrt{2}$ and $5\sqrt{8} = 10\sqrt{2}$

Step 3: Data substitution and solving

$$10\sqrt{2} + 3\sqrt{2} \quad \dots\dots\dots\dots\dots\dots\dots\dots\dots\dots\dots\dots\text{[Apply addition property]}$$

$$(10 + 3)\sqrt{2} \quad \dots\dots\dots\dots\dots\dots\dots\dots\dots\dots\dots\text{[Add values and open bracket]}$$

$$5\sqrt{8} + \frac{6}{\sqrt{2}} = 13\sqrt{2}$$

2) $\frac{1}{\sqrt{2}+1} + \frac{1}{\sqrt{2}-1}$

Hint: Since the denominators are not the same, we are going to apply the positive and negative denominator formula to reduce the equation before applying the addition property.

Step 1: Formula. $\left[\frac{c}{a+b\sqrt{n}} \times \frac{a-b\sqrt{n}}{a-b\sqrt{n}}\right] + \left[\frac{c}{a-b\sqrt{n}} \times \frac{a+b\sqrt{n}}{a+b\sqrt{n}}\right]$

Step 2: Data substitution and solving

$$\frac{1}{\sqrt{2}+1} + \frac{1}{\sqrt{2}-1} = \left[\frac{1}{\sqrt{2}+1} \times \frac{\sqrt{2}-1}{\sqrt{2}-1}\right] + \left[\frac{1}{\sqrt{2}-1} \times \frac{\sqrt{2}+1}{\sqrt{2}+1}\right] \quad \dots\dots\dots\text{[Perform direct multiplication]}$$

$$\left[\frac{\sqrt{2}-1}{(\sqrt{2}+1)\times(\sqrt{2}-1)}\right] + \left[\frac{\sqrt{2}+1}{(\sqrt{2}-1)\times(\sqrt{2}+1)}\right] \quad \dots\dots\dots\text{[Apply difference of two square principle]}$$

$$\left[\frac{\sqrt{2}-1}{(\sqrt{2})^2-(1)^2}\right] + \left[\frac{\sqrt{2}+1}{(\sqrt{2})^2-(1)^2}\right] = \left[\frac{\sqrt{2}-1}{2-1}\right] + \left[\frac{\sqrt{2}+1}{2-1}\right] \quad \dots\dots\dots\text{[Simplify denominator]}$$

$$\frac{\sqrt{2}-1}{1} + \frac{\sqrt{2}+1}{1} = \sqrt{2} - 1 + \sqrt{2} + 1 \quad \dots\dots\dots\dots\dots\dots\dots\text{[Simplify equation]}$$

$$\sqrt{2} + \sqrt{2} \quad \dots\dots\dots\dots\dots\dots\dots\dots\dots\dots\dots\dots\dots\text{[Apply addition property]}$$

$$(1 + 1)\sqrt{2} \quad \dots\dots\dots\dots\dots\dots\dots\dots\dots\dots\text{[Simplify bracket and multiply elements]}$$

$$\frac{1}{\sqrt{2}+1} + \frac{1}{\sqrt{2}-1} = 2\sqrt{2}$$

Example 5: Rationalized $\frac{\sqrt{3}+1}{\sqrt{3}-1}$

<u>**Solution**</u>

Step 1: Formula. $\frac{c}{a-b\sqrt{n}} = \frac{c}{a-b\sqrt{n}} \times \frac{a+b\sqrt{n}}{a+b\sqrt{n}}$, where; $c = \sqrt{3} + 1$, $b\sqrt{n} = \sqrt{3}$, and $a = 1$

Step 2: Data substitution and solving

$$\frac{\sqrt{3}+1}{\sqrt{3}-1} = \frac{\sqrt{3}+1}{\sqrt{3}-1} \times \frac{\sqrt{3}+1}{\sqrt{3}+1} \quad \dots\dots\dots\dots\dots\dots\dots\dots\text{[Perform direct multiplication]}$$

$$\frac{(\sqrt{3}+1)\times(\sqrt{3}+1)}{(\sqrt{3}-1)\times(\sqrt{3}+1)} \quad \dots\dots\dots\dots\dots\dots\dots\dots\dots\dots\dots\dots\dots\text{[Open brackets]}$$

$$\frac{(\sqrt{3})(\sqrt{3})+\sqrt{3}+\sqrt{3}+1}{(\sqrt{3})^2-(1)^1} = \frac{(\sqrt{3})^2+2\sqrt{3}+1}{3-1}$$ ....[Work exponent]

$$\frac{3+2\sqrt{3}+1}{3-1} = \frac{4+2\sqrt{3}}{2}$$ ....[Simplify equation]

$$\frac{\sqrt{3}+1}{\sqrt{3}-1} = 2 + \sqrt{3}$$

Example 6: Evaluate $\frac{\sqrt{48}+2\sqrt{27}}{\sqrt{12}}$

Solution

Step 1: Formula. $\frac{b}{\sqrt{a}} = \frac{b}{\sqrt{a}} \times \frac{\sqrt{a}}{\sqrt{a}}$, where; $b = \sqrt{12}$ and $a = \sqrt{48} - 2\sqrt{27}$......[Remember; $2 = \sqrt{4}$]

Step 2: Data substitution and solving

$$\frac{\sqrt{48}+\sqrt{4}\sqrt{27}}{\sqrt{12}} = \frac{\sqrt{48}+\sqrt{4}\sqrt{27}}{\sqrt{12}} \times \frac{\sqrt{12}}{\sqrt{12}}$$ ....[Perform direct multiplication]

$$\frac{\sqrt{12}(\sqrt{48}+\sqrt{4}\sqrt{27})}{\sqrt{12}\times\sqrt{12}} = \frac{(\sqrt{12\times48})+(\sqrt{12\times4\times27})}{\sqrt{12\times12}}$$ ....[Multiply values in squares]

$$\frac{(\sqrt{576})+(\sqrt{1,296})}{\sqrt{144}}$$ ....[Work roots since all our squared numbers]

$$\frac{24+36}{12}$$ ....[Solve]

$$\frac{\sqrt{48}+2\sqrt{27}}{\sqrt{12}} = 5$$

Test Your Understanding

Exercise 5: Rationalize the following surds

1) $(2 + \sqrt{3})(4 - \sqrt{12})$..[Answer: 2]

2) $\frac{3}{4-2\sqrt{6}}$.......................................[Answer: $-\frac{6+3\sqrt{6}}{4} = -\frac{3(2-\sqrt{6})}{4}$]

3) $\frac{1}{\sqrt{3}+\sqrt{7}}$...................................[Answer: $-\frac{\sqrt{3}-\sqrt{7}}{4} = \frac{\sqrt{7}-\sqrt{3}}{4}$]

4) $\frac{\sqrt{12}+2}{\sqrt{12}-2}$................................ [Answer: $\frac{4+\sqrt{12}}{2} = 2 + \sqrt{3}$]

5) $\frac{5+\sqrt{7}}{3-\sqrt{7}}$................................ ....[Answer: $11 + 4\sqrt{7}$]

6) $\sqrt{45} + \frac{20}{\sqrt{5}}$...[Answer: $7\sqrt{5}$]

7) $\frac{\sqrt{75}-\sqrt{27}}{\sqrt{3}}$................................ ....[Answer: 2]

5.2 LOGARITHM

Logarithm is defined as the exponent that indicates the power to which the base number is raised to produce a given number. Logarithm is used to find the value of unknown power or exponent. Generally, logarithm is made up of three components; the base (a), argument (Y) and constant or argument power (m). These components can take the form of a whole number, fraction and decimal.

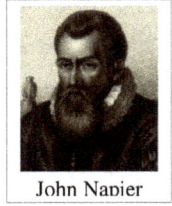

John Napier

$$m\, Log_a Y \dotfill (1)$$
$$Log_a Y^m \dotfill (2)$$

Equation (1) is the same as equation (2). The following properties of logarithm are worth knowing for easy understanding and calculations.

$$Log_a a = 1 \dotfill (3)$$
$$Log_a 1 = 0 \dotfill (4)$$
$$Log_a a^X = X \dotfill (5)$$
$$Log_a X^y = y\, Log_a X \dotfill (6)$$
$$a^{Log_a X} = X \dotfill (7)$$

Note: Due to simplicity, our analysis will be sub-divided into two, that is single logarithm analysis and compound logarithm analysis. The concept of single logarithm will help in compound logarithm analysis.

5.2.1 Single Logarithm Analysis

Here, we assume the existence of only one logarithm. In order to solve single logarithm problem, we are required to equate the logarithm equation to a variable called answer. Let's consider the logarithm equation ($Log_a Y$) and a variable (X), equating them will give us the equation below.

$$Log_a Y = X \dotfill (8)$$

From equation (8), we are demanded to transform the logarithm equation to an index equation. This index equation is often considered the working formula for the derivation of the value of logarithm.

$$Y = a^X \quad \dots\dots\dots\dots\dots\dots\dots\dots\dots\dots\dots\dots\dots\dots\dots(9)$$

From equation (9), we transform the argument (Y) into an index with base equal to the logarithm original base (a). Given that (Y) is transformed to the index (a^m), we obtain the equation below.

$$a^m = a^X \quad \dots\dots\dots\dots\dots\dots\dots\dots\dots\dots\dots\dots\dots\dots(10)$$

Finally, equate the powers and make the original log answer component (X) subject of the formula.

$$X = m \quad \dots\dots\dots\dots\dots\dots\dots\dots\dots\dots\dots\dots\dots\dots\dots(11)$$

Example 1: Find the value of $\text{Log}_2 32$

Solution

Step 1: Letting of Log equation to a variable. $\text{Log}_2 32 = X$

Step 2: Formula. $Y = a^X$, where; $Y = 32$ and $a = 2$

Step 3: Data substitution and solving

$32 = 2^X$ ....[Transform argument (32) to index with base (2)]

$2^5 = 2^X$ ....[Equate powers and make answer (X) subject of formula]

$X = 2$ ....[That is $\textbf{Log}_2 \textbf{ 32} = \textbf{2}$]

Example 2: Find the value of the following logarithm

1) $\frac{1}{2}\text{Log}_5 625$

2) $2\text{Log}_4 4$

Solution

1) $\frac{1}{2}\textbf{Log}_5 \textbf{625}$

Step 1: Letting of Log equation to a variable. $\frac{1}{2}\text{Log}_5 625 = X$...[Rem; $\frac{n}{m}\text{Log}_a X = \text{Log}_a X^{\frac{n}{m}}$]

Step 2: Formula. $Y = a^X$, where; $Y = 625^{\frac{1}{2}}$ and $a = 5$...............[Remember; $X^{\frac{n}{m}} = \left(\sqrt[m]{X}\right)^n$]

Step 3: Data substitution and solving

$\left(\sqrt[2]{625}\right)^1 = 5^X$ ....[Work root]

$(25)^1 = 5^X$ ....[Work left hand side of equation exponent]

$25 = 5^X$ .... [Transform argument (25) to index with base (5)]

$5^2 = 5^X$ ....[Equate powers and make answer (X) subject of formula]

$X = 2$ ....[That is $\frac{1}{2}\textbf{Log}_5 \textbf{625} = \textbf{2}$]

2) $2\text{Log}_4 4$

Step 1: Letting of Log equation to a variable. $2\text{Log}_4 4 = X$.........[Rem; $n\text{Log}_a X = \text{Log}_a X^n$]

Step 2: Formula. $Y = a^X$, where; $Y = 4^2$ and a = 4...................[Remember; $n^2 = n \times n$]

Step 3: Data substitution and solving

$4^2 = 4^X$ ....[Equate exponent]

$X = 2$ ....[That is $2\text{Log}_4 4 = 2$]

Remember: From equation (5), we saw that $(\text{Log}_a a^X = X)$, therefore, $(2\text{Log}_4 4 = \text{Log}_4 4^2 = 2)$. From equation (3), we saw that $(\text{Log}_a a = 1)$, therefore, $[2\text{Log}_4 4 = 2(\text{Log}_4 4) = 2(1) = 2]$.

Example 3: Evaluate the value of $\text{Log}_{\frac{1}{2}} 4$

<u>Solution</u>

Step 1: Letting of Log equation to a variable. $\text{Log}_{\frac{1}{2}} 4 = X$

Step 2: Formula. $Y = a^X$, where; $Y = 4$ and $a = \frac{1}{2}$

Step 3: Data substitution and solving

$4 = \left(\frac{1}{2}\right)^X$ ....[Transform right hand side to negative index]

$4 = 2^{-X}$ ....[Transform left hand side to index with base (2)]

$2^2 = 2^{-X}$ ....[Equate powers]

$-X = 2$ ....[Multiply both sides of equation by (-1)]

$X = -2$ ....[That is $\text{Log}_{\frac{1}{2}} 4 = X = -2$]

Example 4: Find the value of $\text{Log}_{0.1} 10$

<u>Solution</u>

Step 1: Letting of Log equation to a variable. $\text{Log}_{0.1} 10 = X$.

Remember: $0.1 = \frac{1}{10}$, therefore, $\text{Log}_{0.1} 10 = \text{Log}_{\frac{1}{10}} 10$

Step 2: Formula. $Y = a^X$, where; $Y = 10$ and $a = \frac{1}{10}$

Step 3: Data substitution and solving

$10 = \left(\frac{1}{10}\right)^X$ ....[Transform right hand side to negative index]

$10 = 10^{-X}$ ....[Transform left hand side to index, remember $(a = a^1)$]

$10^1 = 10^{-X}$ ....[Equate powers]

$-X = 1$ ....[Multiply both sides of equation by (-1)]

214

$X = -1$ ....[That is **$\text{Log}_{0.1} \, 10 = -1$**]

Example 5: Find the value of $\text{Log}_4 \left(\frac{1}{256}\right)$

<u>**Solution**</u>

Step 1: Letting of Log equation to a variable. $\text{Log}_4 \left(\frac{1}{256}\right) = X$.

Step 2: Formula. $Y = a^X$, where; $Y = \frac{1}{256}$ and $a = 4$

Step 3: Data substitution and solving

$\frac{1}{256} = 4^X$ ....[Transform left hand side denominator to index with base 4]

$\frac{1}{4^4} = 4^X$ ....[Transform left hand side to negative index. Remember $\left(\frac{1}{a^m} = a^{-m}\right)$]

$4^{-4} = 4^X$ ....[Equate powers]

$X = -4$ ....[That is **$\text{Log}_4 \left(\frac{1}{256}\right) = -4$**]

Example 6: Determine the value of $\text{Log}_5 \, 5^{\frac{2}{3}}$

<u>**Solution**</u>

Step 1: Letting of Log equation to a variable. $\text{Log}_5 \, 5^{\frac{2}{3}} = X$.

Step 2: Formula. $Y = a^X$, where; $Y = 5^{\frac{2}{3}}$ and $a = 5$

Step 3: Data substitution and solving

$5^{\frac{2}{3}} = 5^X$ ....[Equate powers]

$X = \frac{2}{3}$ ....[That is **$\text{Log}_5 \, 5^{\frac{2}{3}} = \frac{2}{3}$**]

<u>**Test Your Understanding**</u>

Exercise 6: Determine the value of the following

1) $\text{Log}_3 \, 81$...[**Answer: $X = 4$**]

2) $\frac{1}{3} \text{Log}_3 \, 27$...[**Answer: $X = 1$**]

3) $8\text{Log}_{10} \, 10$..[**Answer: $X = 8$**]

Exercise 7: Find the value of the following

1) $\text{Log}_{0.02} \, 2500$...[**Answer: $X = -2$**]

2) $\text{Log}_{\frac{1}{8}} \, 512$..[**Answer: $X = -3$**]

3) $\text{Log}_7 \left(\frac{1}{343}\right)$...[**Answer: $X = -3$**]

4) $\text{Log}_2 \, 2^{\frac{1}{2}}$ ....[**Answer: $X = \frac{1}{2}$**]

5.2.2 Compound Logarithm Analysis

Here, we are going to analyse situations where a given logarithm equation is made up of two or more logarithms. Our analysis will be limited to two logarithms equation combination. The two main properties applied here are the product (multiplication) property and quotient (division) property.

1) The Product Property

The product property is based on the assumption that the logarithm of a product is the sum of the logarithms. Given that the arguments of the logarithms are greater than zero, the product property is expressed as seen below.

$$Log_a mn = Log_a(m \times n) = Log_a m + Log_a n$$
....(1)

Example 1: Find the value of $Log_6 4 + Log_6 9$

Step 1: Simplification of equation
1) Formula. $Log_a m + Log_a n = Log_a(m \times n)$, where; $a = 6$, $m = 4$ and $n = 9$
2) Data substitution and solving

$Log_6 4 + Log_6 9$ ....[Maintain log base and multiply log arguments]

$Log_6(4 \times 9)$ ....[Simplify bracket]

$Log_6 36$ ....**[Simplified form]**

Step 2: Determination of equation value

1) Letting of Log equation to a variable. $Log_6 36 = X$
2) Formula. $Y = a^X$, where; $Y = 36$ and $a = 6$
3) Data substitution and solving

$36 = 6^X$ ....[Transform argument (36) to index with base (6)]

$6^2 = 6^X$ ....[Equate powers and make answer (X) subject of formula]

$X = 2$ ....[That is **$Log_6 4 + Log_6 9 = 2$**]

Example 2: Determine the value of $Log_2 5 + Log_2 \frac{16}{10}$

Solution

Step 1: Simplification of equation

1) Formula. $Log_a m + Log_a n = Log_a(m \times n)$, where; $a = 2$, $m = 5$ and $n = \frac{16}{10}$
2) Data substitution and solving

$$\text{Log}_2 5 + \text{Log}_2 \frac{16}{10} \quad ....\text{[Maintain log base and multiply log arguments]}$$

$$\text{Log}_2 \left(5 \times \frac{16}{10}\right) \quad ....\text{[Simplify bracket]}$$

$$\text{Log}_2 8 \quad ....\textbf{[Simplified form]}$$

Step 2: Determination of equation value

1) Letting of Log equation to a variable. $\text{Log}_2 8 = X$

2) Formula. $Y = a^X$, where; $Y = 8$ and $a = 2$

3) Data substitution and solving

$$8 = 2^X \quad ....\text{[Transform argument (9) to index with base (2)]}$$

$$2^3 = 2^X \quad ....\text{[Equate powers and make answer (X) subject of formula]}$$

$$X = 3 \quad ....\text{[That is } \textbf{Log}_2 \textbf{5} + \textbf{Log}_2 \frac{\textbf{16}}{\textbf{10}} = \textbf{3]}$$

Example 3: Find the value of $\text{Log}_3 15 + \text{Log}_3 0.6$

Solution

Step 1: Simplification of equation

3) Formula. $\text{Log}_a m + \text{Log}_a n = \text{Log}_a (m \times n)$, where; $a = 3$, $m = 15$ and $n = 0.6 = \frac{6}{10}$

4) Data substitution and solving

$$\text{Log}_3 15 + \text{Log}_3 \frac{6}{10} \quad ....\text{[Maintain log base and multiply log arguments]}$$

$$\text{Log}_3 \left(15 \times \frac{6}{10}\right) \quad ....\text{[Simplify bracket]}$$

$$\text{Log}_3 9 \quad ....\textbf{[Simplified form]}$$

Step 2: Determination of equation value

1) Letting of Log equation to a variable. $\text{Log}_3 9 = X$

2) Formula. $Y = a^X$, where; $Y = 9$ and $a = 3$

3) Data substitution and solving

$$9 = 3^X \quad ....\text{[Transform argument (9) to index with base (3)]}$$

$$3^2 = 3^X \quad ....\text{[Equate powers and make answer (X) subject of formula]}$$

$$X = 2 \quad ....\text{[That is } \textbf{Log}_3 \textbf{15} + \textbf{Log}_3 \textbf{0.6} = \textbf{2]}$$

Test Your Understanding

Exercise 8: Evaluate the following

1) $\text{Log}_4 2 + \text{Log}_4 32$...**[Answer: X = 3]**

2) $\text{Log}_8 32 + \text{Log}_8 2$...**[Answer: X = 2]**

3) $\text{Log} 50 + \text{Log} 20$...**[Answer: X = 3]**

2) The Quotient Property

The quotient property is based on the assumption that the logarithm of a quotient is the difference of their logarithms. Given that the arguments of the logarithms are greater than zero, the quotient property is expressed as seen below.

$$Log_a\frac{m}{n} = Log_a\left(\frac{m}{n}\right) = Log_a m - Log_a n \ \dots\dots\dots\dots\dots\dots\dots\dots\dots\dots\dots (2)$$

Example 1: Evaluate $Log_2 3 - Log_2 24$

Solution

Step 1: Simplification of logarithm equation

1) Formula. $Log_a m - Log_a n = Log_a\left(\frac{m}{n}\right)$, where; a = 2, m = 3 and n = 24

2) Data substitution and solving

$Log_2 3 - Log_2 24$ ....[Maintain logarithm base and divide arguments]

$Log_2\left(\frac{3}{24}\right)$ ....[Simplify bracket]

$Log_2\left(\frac{1}{8}\right)$ ....**[Simplified form]**

Step 2: Determination of equation value

1) Letting of Log equation to a variable. $Log_2\left(\frac{1}{8}\right) = X$

2) Formula. $Y = a^X$, where; $Y = \frac{1}{8}$ and a = 2

3) Data substitution and solving

$\frac{1}{8} = 2^X$ ....[Transform left hand side denomination to index with base (2)]

$\frac{1}{2^3} = 2^X$ ....[Transform left hand side of equation to negative index]

$2^{-3} = 2^X$ ....[Equate powers]

$X = -3$ ....[That is $Log_2 3 - Log_2 24 = -3$]

Example 2: Find the value of $3\ Log_5\ 5 - Log_5\ 5$

218

<u>**Solution**</u>

Step 1: Simplification of logarithm equation.......... [Rem; $3 \log_5 5 = \log_5 5^3 = \log_5 125$]

1) Formula. $\log_a m - \log_a n = \log_a \left(\frac{m}{n}\right)$, where; $a = 5$, $m = 125$ and $n = 5$

2) Data substitution and solving

$\log_5 125 - \log_5 5$ ....[Maintain base and divide arguments]

$\log_5 \left(\frac{125}{5}\right)$ ....[Simplify fraction]

$\log_5 25$ ....**[Simplified form]**

Step 2: Determination of equation value

1) Letting of Log equation to a variable. $\log_5 25 = X$

2) Formula. $Y = a^X$, where; $Y = 25$ and $a = 5$

3) Data substitution and solving

$25 = 5^X$ ....[Transform (25) in to index with base (5)]

$5^2 = 5^X$ ....[Equate powers]

$X = 2$ ....[That is **$3 \log_5 5 - \log_5 5 = 2$**]

Example 3: Find the value of $\frac{1}{2} \log_5 25 - \log_5 \sqrt{625}$

<u>**Solution**</u>

Step 1: Simplification of logarithm equation.............. [Remember; $\frac{1}{2} \log_5 25 = \log_5 \sqrt{25}$]

1) Formula. $\log_a m - \log_a n = \log_a \left(\frac{m}{n}\right)$, where; $a = 5$, $m = \sqrt{25}$ and $n = \sqrt{625}$

2) Data substitution and solving

$\log_5 \sqrt{25} - \log_5 \sqrt{625}$ ....[Work roots]

$\log_5 5 - \log_5 25$ ....[Maintain base and divide arguments]

$\log_5 \left(\frac{5}{25}\right)$ ....[Simplify fraction]

$\log_5 \left(\frac{1}{5}\right)$ ....**[Simplified form]**

Step 2: Determination of equation value

1) Letting of Log equation to a variable. $\log_5 \left(\frac{1}{5}\right) = X$

2) Formula. $Y = a^X$, where; $Y = \frac{1}{5}$ and $a = 5$

3) Data substitution and solving

$\frac{1}{5} = 5^X$ ....[Transform $\left(\frac{1}{5}\right)$ into negative index with base (5)]

$5^{-1} = 5^X$ ....[Equate powers]

$X = -1$ ....[That is $\frac{1}{2}$ **Log$_5$ 25** − **Log$_5$ $\sqrt{625}$** = −1]

Example 4: Solve Log$_2$ 0.5 − Log$_2$ $\left(\frac{1}{8}\right)$

Solution

Step 1: Simplification of logarithm equation................... [Remember; Log$_2$ 0.5 = Log$_2$ $\frac{1}{2}$]

1) Formula. Log$_a$m − Log$_a$n = Log$_a$ $\left(\frac{m}{n}\right)$, where; a = 2, m = $\frac{1}{2}$ and n = $\frac{1}{8}$

2) Data substitution and solving

Log$_2$ $\left(\frac{1}{2}\right)$ − Log$_2$ $\left(\frac{1}{8}\right)$ ....[Maintain base and divide arguments]

Log$_2$ $\left(\frac{1}{2} \times \frac{8}{1}\right)$ ....[Simplify bracket]

Log$_2$4 ....**[Simplified form]**

Step 2: Determination of equation value

1) Letting of Log equation to a variable. Log$_2$4 = X

2) Formula. Y = a^X, where; Y = 4 and a = 2

3) Data substitution and solving

$4 = 2^X$ ....[Transform (4) in to index with base (2)]

$2^2 = 2^X$ ....[Equate powers]

$X = 2$ ....[That is **Log$_2$ 0.5** − **Log$_2$ $\left(\frac{1}{8}\right)$** = 2]

Test Your Understanding

Exercise 10: Find the value of the following

1) Log$_3$ 162 − Log$_3$ 6...**[Answer: X = 3]**

2) Log$_6$ 5 − Log$_6$ 1080...**[Answer: X = −3]**

Exercise 11: Evaluate the following logarithm

1) $\frac{1}{3}$Log$_2$ 512 − Log$_2$ $\frac{1}{2}$...**[Answer: X = 4]**

2) Log$_2$ 40 − Log$_2$ 2.5..**[Answer: X = 4]**

3) Log$_4\sqrt{144}$ − Log$_4\sqrt{9}$...**[Answer: X = 1]**

4) $\frac{1}{3}$Log$_{\frac{1}{2}}$ 1,000 − Log$_{\frac{1}{2}}$ 10...**[Answer: X = 0]**

---

3) Combination of Product and Quotient Properties

Here, we are going to analyse situations where the product and quotient properties are included in the same logarithm equation.

220

Example 1: Find the value of $\text{Log}_2 48 + \text{Log}_2 3 - \text{Log}_2 9$

Solution

Step 1: Simplification of logarithm equation

a) Method 1

Technique 1	Technique 2
$(\text{Log}_2 48 + \text{Log}_2 3) - \text{Log}_2 9$ …..[Simplify] $\text{Log}_2(48 \times 3) - \text{Log}_2 9$ [Multiply value] $\text{Log}_2 144 - \text{Log}_2 9$ …....[Divide arguments] $\text{Log}_2 \left(\frac{144}{9}\right) = \text{Log}_2 16$ …[Simplified form]	$\frac{\text{Log}_2 48 + \text{Log}_2 3}{\text{Log}_2 9}$ … …..[Simplify numerator] $\frac{\text{Log}_2(48 \times 3)}{\text{Log}_2 9} = \frac{\text{Log}_2 144}{\text{Log}_2 9}$.[Divide arguments] $\text{Log}_2 \left(\frac{144}{9}\right) = \text{Log}_2 16$ ..[Simplified form]

b) Method 2

$\text{Log}_2 48 + \text{Log}_2 3 - \text{Log}_2 9$ …………….…………...…..[Factorization of (Log_2)]

$\text{Log}_2[(48 \times 3)/9]$ …………………………… ………..[Multiply inner bracket values]

$\text{Log}_2[144/9]$ ………………………… ……………...…….[Simplify fraction]

$\text{Log}_2 16$ …………………………… ……………...…..…[Simplified form]

Step 2: Determination of equation value

1) Letting of Log equation to a variable. $\text{Log}_2 16 = X$

2) Formula. $Y = a^X$, where; $Y = 16$ and $a = 2$

3) Data substitution and solving

$16 = 2^X$ …………………………….……..[Transform (16) to index with base (2)]

$2^4 = 2^X$ ……………….…………………….………..…….[Equate powers]

$X = 4$ ……………………………….[That is $\text{Log}_2 48 + \text{Log}_2 3 - \text{Log}_2 9 = 4$]

Example 2: Determine the value of $2\text{Log}_5 10 + 3\text{Log}_5 2 - \text{Log}_5 32$

Solution

Step 1: Simplification of logarithm equation

$2\text{Log}_5 10 + 3\text{Log}_5 2 - \text{Log}_5 32$ …..[Remove constant by making arguments index]

$\text{Log}_5 10^2 + \text{Log}_5 2^3 - \text{Log}_5 32^1$ ……………………………...…[Work exponents]

$\text{Log}_5 100 + \text{Log}_5 8 - \text{Log}_5 32$ …………….……..[Segment and simplify equation]

$Log_5(100 \times 8) - Log_5 32$ ....[Multiply bracket values]

$Log_5 800 - Log_5 32$ ....[Maintain logarithm base and divide arguments]

$Log_5\left(\frac{800}{32}\right) = Log_5 25$ ....[Simplified form]

Step 2: Determination of equation value

1) Letting of Log equation to a variable. $Log_5 25 = X$

2) Formula. $Y = a^X$, where; $Y = 25$ and $a = 5$

3) Data substitution and solving

$25 = 5^X$ ....[Transform (25) to index with base (5)]

$5^2 = 5^X$ ....[Equate powers]

$X = 2$ ....[That is $2Log_5 10 + 3Log_5 2 - Log_5 32 = 2$]

Example 3: Determine the value of $Log_2 36 + \frac{1}{2}Log_2 256 - 2Log_2 48$

Solution

Step 1: Simplification of logarithm equation

Remember; $\frac{1}{2}Log_2 256 = Log_2 \sqrt{256} = Log_2 16$ and $2Log_2 48 = Log_2 48^2 = Log_2 2,304$

$Log_2 36 + Log_2 16 - Log_2 2,304$ ....[Segment equation and apply properties]

$Log_2(36 \times 16) - Log_2 2,304$ ....[Simplify bracket and divide arguments]

$Log_2\left(\frac{576}{2,304}\right)$ ....[Simplify fraction]

$Log_2\left(\frac{1}{4}\right)$ ....[Simplified form]

Step 2: Determination of equation value

1) Letting of Log equation to a variable. $Log_2\left(\frac{1}{4}\right) = X$

2) Formula. $Y = a^X$, where; $Y = \frac{1}{4}$ and $a = 2$

3) Data substitution and solving

$\frac{1}{4} = 2^X$ ....[Transform denominator (4) to index with base (2)]

$\frac{1}{2^2} = 2^X$ ....[Transform $\left(\frac{1}{2^2}\right)$ to negative index]

$2^{-2} = 2^X$ ....[Equate powers]

$X = -2$ ....[That is $Log_2 36 + \frac{1}{2}Log_2 256 - 2Log_2 48 = -2$]

Example 4: Evaluate $4Log_x 2 - 3Log_x 4 + Log_x 6$

222

<u>Solution</u>

Hint: Since the base of the logarithm is unknown, we are going to limit our analysis to the reduced form of the logarithm equation.

$4\text{Log}_x 2 - 3\text{Log}_x 4 + \text{Log}_x 6$ ....[Remove constant by making arguments index]

$\text{Log}_x 2^4 - \text{Log}_x 4^3 + \text{Log}_x 6^1$ ....[Work exponents]

$\text{Log}_x 16 - \text{Log}_x 64 + \text{Log}_x 6$ ....[Segment and simplify equation]

$\text{Log}_x \left(\frac{16}{64}\right) + \text{Log}_x 6$ ....[Simplify bracket values]

$\text{Log}_x \left(\frac{1}{4}\right) + \text{Log}_x 6$ ....[Maintain logarithm base and multiply arguments]

$\text{Log}_x \left(\frac{1}{4} \times 6\right) = \text{Log}_x \left(\frac{6}{4}\right) = \text{Log}_x \left(\frac{3}{2}\right)$ ....[Simplified form]

Test Your Understanding

Exercise 12: Find the value of the following.

1) $\text{Log}_5 25 + \text{Log}_5 5 - \text{Log}_5 5$..[**Answer: X = 2**]

2) $\text{Log}_2 32 - \text{Log}_2 8 + \text{Log}_2 4$...[**Answer: X = 4**]

3) $\text{Log } 4 + \text{Log } 5 - \text{Log } 2$.. ....[**Answer: X= 1**]

Exercise 13: Determine the value of the following

1) $2\text{Log}_2 2 - \frac{1}{2}\text{Log}_2 64 + \text{Log}_2 \left(\frac{1}{2}\right)$..[**Answer: X − 2**]

2) $\frac{1}{3}\text{Log}_2 8 + \frac{1}{2}\text{Log}_2 36 - \frac{1}{2}\text{Log}_2 9$................................ ....[**Answer: X = 2**]

Note: In logarithm, we can have situations where there is more than one base, that is a given logarithm equation can have different bases. It is worth knowing that in order to solve a given logarithm problem, all bases must be the same. Thus we are required to change other bases to a desired base to make them the same. Let a, b and x be positive real numbers, such that $(a \neq 1)$ and $(b \neq 1)$, the formula to change logarithm base is given as seen below.

$\text{Log}_a x = \frac{\text{Log}_b x}{\text{Log}_b a}$ ....[Changing from base (a) to base (b)]

$\text{Log}_a x = \frac{\text{Log}_{10} x}{\text{Log}_{10} a} = \frac{\text{Log } x}{\text{Log } a}$ ....[Changing from base (a) to base (10)]

$\text{Log}_a x = \frac{\text{In } x}{\text{In } a}$ ....[Changing from base (a) to natural logarithm (In or e)]

CHAPTER SIX

ARITHMETIC PROGRESSION AND GEOMETRIC PROGRESSION

The analysis of arithmetic progression and geometric progression is based on the concept of sequence and series. A sequence is an arrangement or set of numbers or terms in a definite order according to some rules. The addition of all numbers or terms of a sequence gives what is called a series. A sequence can be finite or infinite. Finite sequence is a sequence that has a first term and a last time. Our analysis will be limited to the finite sequence.

6.1 Arithmetic Progression

Arithmetic progression is a sequence in which each term except the first term is obtained by adding a fixed number called common difference to the preceding term. Our analysis will be limited to finite arithmetic progression. Finite arithmetic progression is defined as an arithmetic progression that uses finite sequence.

6.1.1 Common Difference

Common difference is a fixed number or constant that is added to a preceding term or number to obtain the succeeding term of an arithmetic progression. The common difference is denoted (d) and can be a positive (leading to increasing sequence) or negative (decreasing sequence) value. Given two terms, the common difference is seen as the difference between the succeeding term and the preceding term.

$$\text{Common Difference}(d) = a_{n+1} - a_n \dots\dots\dots\dots\dots\dots\dots\dots\dots\dots\dots\dots\dots\dots\dots\dots\dots\dots\dots\dots\dots\dots(1)$$

$$\text{Where; } (a_{n+1}) \text{ represents the succeeding term or number}$$

$$(a_n) \text{ represents the preceding term or number}$$

From equation (1) above, the formula of succeeding terms can derived by making succeeding terms the subject of formula as seen below.

$$\text{Succeding Term } (a_{n+1}) = \text{Common Difference}(d) + \text{Preceding Term}(a_n) \dots\dots\dots(2)$$

Example 1: Given that the first term of an arithmetic progression is 3 and the second term is 10, compute the value of the common difference.

Solution

Step 1: Formula. Common Difference$(d) = a_{n+1} - a_n$, where; $a_{n+1} = 10$ and $a_n = 3$

Step 2: Data substitution and solving

> Common Difference(d) $= 10 - 3$ ....[Subtract values]
>
> Common Difference(d) $= \mathbf{7}$

Step 3: Interpretation: The common difference value (7) means a given succeeding term or number will be by default greater than its preceding value by 7.

Example 2: Find the common difference of an arithmetic progression using the following sequence;

1) $5, 15, 25, 35......$
2) $15, 11, 7, 3.......$
3) $\frac{3}{4}, \frac{5}{8}, \frac{1}{2}, \frac{3}{8}.........$
4) $-14, -10, -6, -2.........$

Solution

Hint: Here, we can separate the sequence in to segments from which we can chose any of the segments to determine the value of the common difference.

1) $\mathbf{5, 15, 25, 35......}$

Step 1: Formula. Common Difference(d) $= a_{n+1} - a_n$

Step 2: Data substitution and solving

Segment 1 [5, 15]	Segment 2 [15, 25]	Segment 3 [25, 35]
$d = 15 - 5 = \mathbf{10}$	$d = 25 - 15 = \mathbf{10}$	$d = 35 - 25 = \mathbf{10}$
Note: The common difference of all segments must be the same (10)		

2) $\mathbf{15, 11, 7, 3.......}$

Step 1: Formula. Common Difference(d) $= a_{n+1} - a_1$

Step 2: Data substitution and solving

Segment 1 [15, 11]	Segment 2 [11, 7]	Segment 3 [7, 3]
$d = 11 - 15 = \mathbf{-4}$	$d = 7 - 11 = \mathbf{-4}$	$d = 3 - 7 = \mathbf{-4}$
Note: The common difference of all segments must be the same (-4)		

3) $\frac{3}{4}, \frac{5}{8}, \frac{1}{2}, \frac{3}{8}.........$

Step 1: Formula. Common Difference(d) $= a_{n+1} - a_n$

Step 2: Data substitution and solving

Segment 1 $\left[\frac{3}{4}, \frac{5}{8}\right]$	Segment 2 $\left[\frac{5}{8}, \frac{1}{2}\right]$	Segment 3 $\left[\frac{1}{2}, \frac{3}{8}\right]$
$d = \frac{5}{8} - \frac{3}{4} = -\frac{1}{8}$	$d = \frac{1}{2} - \frac{5}{8} = -\frac{1}{8}$	$d = \frac{3}{8} - \frac{1}{2} = -\frac{1}{8}$
Note: The common difference of all segments must be the same $\left(-\frac{1}{8}\right)$		

4) $-14, -10, -6, -2\ldots\ldots\ldots$

Step 1: Formula. Common Difference$(d) = a_{n+1} - a_n$

Step 2: Data substitution and solving

Segment 1 $[-14, -10]$	Segment 2 $[-10, -6]$	Segment 3 $[-6, -2]$
$d = -10 - (-14) = 4$	$d = -6 - (-10) = 4$	$d = -2 - (-6) = 4$
Note: The common difference of all segments must be the same (4)		

Example 3: Given that the first term of an arithmetic progression is 5, determine the first five terms assuming that,

1) The common difference is 2
2) The common difference is -2

<u>**Solution**</u>

1) **The common difference is 2**

Step 1: Formula. Succeding Term $(a_{n+1}) = $ Common Difference$(d) +$ Preceding Term(a_n)

Step 2: Data substitution and solving

First Term	Second Term	Third Term	Fourth Term	Fifth Term
5 (Given)	$a_2 = d + a_1$	$a_3 = d + a_2$	$a_4 = d + a_3$	$a_5 = d + a_4$
	$a_2 = 2 + 5$	$a_3 = 2 + 7$	$a_4 = 2 + 9$	$a_5 = 2 + 11$
	$a_2 = 7$	$a_3 = 9$	$a_4 = 11$	$a_5 = 13$

2) **The common difference is -2**

Step 1: Formula. Succeding Term $(a_{n+1}) = $ Common Difference$(d) +$ Preceding Term(a_n)

Step 2: Data substitution and solving

First Term	Second Term	Third Term	Fourth Term	Fifth Term
5 (Given)	$a_2 = d + a_1$	$a_3 = d + a_2$	$a_4 = d + a_3$	$a_5 = d + a_4$
	$a_2 = -2 + 5$	$a_3 = -2 + 3$	$a_4 = -2 + 1$	$a_5 = -2 + (-1)$
	$a_2 = 3$	$a_3 = 1$	$a_4 = -1$	$a_5 = -3$

6.1.2 The nth Term of an Arithmetic Progression

After knowing how to determine the common difference, it is possible to calculate the value of a given term (n^{th} term) when the first term and common difference are known or certain. The formula for calculating individual terms of an arithmetic progression for the first four terms is given below.

The First Term $(a_1) = a_1 = a_1 + (1 - 1)d$..(1)

The Second Term $(a_2) = a_1 + d = a_1 + (2 - 1)d$.......................................(2)

The Third Term $(a_3) = a_1 + 2d = a_1 + (3 - 1)d$........ ....(3)

The Second Term $(a_4) = a_1 + 3d = a_1 + (4 - 1)d$.......................................(4)

Generally, from a given first term (a_1) of an arithmetic progression and the common difference (d), the value of any term or n^{th} term (a_n) can be derived using the general formula below.

The nth Term$(a_n) = a_1 + (n - 1)d$ ....(5)

Where; (a_n) represent the nth term of an arithmetic sequence

        (a_1) represent the first term of an arithmetic sequence

        (d) represent the common difference of an arithmetic sequence

        (n) represent the number of terms in an arithmetic sequence

227

Example 1: Find the 5th term of an arithmetic progression given that the first term is 5 and the common difference is 3.

Solution

Step 1: Formula. $a_n = a_1 + (n-1)d$, where; $a_1 = 5$, $n = 5$ and $d = 3$

Step 2: Data substitution and solve

$a_5 = 5 + (5-1)3$ ....[Simplify bracket]

$a_5 = 5 + (4)3$ ....[Open bracket]

$a_5 = 5 + 12$ ....[Add values]

The 5th Term$(a_5) = 17$

Step 3: Justification

First Term	Second Term	Third Term	Fourth Term	Fifth Term
(a_1)	(a_2)	(a_3)	(a_4)	(a_5)
$5 + 0 = 5$	$5 + 3 = 8$	$8 + 3 = 11$	$11 + 3 = 14$	$14 + 3 = 17$
Note: The value of fifth term (17) is equal to the calculated value above (17)				

Example 2: Find the 13th term using the sequence; 7, 17, 27,....... and the 20th term using the sequence; $1/6$, $1/4$, $1/3$ ···.

Solution

a) Determination of 13th term

Step 1: Formula. $a_n = a_1 + (n-1)d$

Where; $a_1 = 7$, $n = 13$ and $d = 17 - 7 = 10$ or $27 - 17 = 10$

Step 2: Data substitution and solving

$a_{13} = 7 + (13-1)10$ ....[Simplify bracket]

$a_{13} = 7 + (12)10$ ....[Open bracket]

$a_{13} = 7 + 120$ ....[Add values]

The 13th Term$(a_{13}) = 127$

b) Determination of 20th term

Step 1: Formula. $a_n = a_1 + (n-1)d$, where; $a_1 = \frac{1}{6}$, $n = 20$ and $d = \frac{1}{4} - \frac{1}{6} = \frac{1}{12}$ or $\frac{1}{3} - \frac{1}{4} = \frac{1}{12}$

Step 2: Data substitution and solving

$a_{20} = \frac{1}{6} + (20-1)\frac{1}{12}$ ....[Simplify bracket]

$a_{20} = \frac{1}{6} + (19)\frac{1}{12}$ ....[Open bracket]

$a_{20} = \frac{1}{6} + \frac{19}{12}$ ....[Solve equation using the concept of LCM (12)]

The 20th Term $(a_{20}) = \frac{21}{12}$

Example 3: Find the first term given that the common difference of an arithmetic progression is -4 and the 25th term is -79.

Solution

a) Method 1 [Using the general term formula]

Step 1: Formula. $a_n = a_1 + (n-1)d$, where; $a_n = -79$, $n = 25$ and $d = -4$

Step 2: Data substitution and solving

$-79 = a_1 + (25 - 1) - 4$ ....[Simplify bracket]

$-79 = a_1 + (24) - 4$ ....[Open bracket]

$-79 = a_1 - 96$ ....[Make (a_1) subject of formula]

First Term$(a_1) = -79 + 96$ ....[Solve equation]

First Term$(a_1) = \mathbf{17}$

b) Method 2 [Using specified formula]

Step 1: Formula. First Term$(a_1) = a_n - (n-1)d$, where; $a_n = -79$, $n = 25$ and $d = -4$

Step 2: Data substitution and solving

First Term$(a_1) = -79 - (25 - 1) - 4$ ....[Simplify bracket]

First Term$(a_1) = -79 - (24) - 4$ ....[Open bracket]

First Term$(a_1) = -79 + 96$ ....[Solve equation]

First Term$(a_1) = \mathbf{17}$

Example 4: Find the number of terms in an arithmetic progression given that; $a_1 = 5$, $d = 25$ and $a_n = 130$.

Solution

c) Method 1 [Using the general term formula]

Step 1: Formula. $a_n = a_1 + (n-1)d$, where; $a_1 = 5$, $d = 25$ and $a_n = 130$.

Step 2: Data substitution and solving

$130 = 5 + (n-1)25$ ....[Open bracket]

$130 = 5 + 25n - 25$ ....[Collect like terms together]

$25n = 130 - 5 + 25$ ....[Simplify right hand side of equation]

$25n = 150$ ....[Divide both sides of equation by (25)]

Number of Terms$(n) = \mathbf{6}$

d) Method 2 [Using specified formula]

Step 1: Formula. Number of Terms$(n) = \frac{a_n - (a_1)}{d} + 1$, where; $a_1 = 5, d = 25$ and $a_n = 130$.

Step 2: Data substitution and solving

Number of Terms$(n) = \frac{130-5}{25} + 1$ ....[Simplify numerator]

Number of Terms$(n) = \frac{125}{25} + 1$ ....[Solve fraction]

Number of Terms$(n) = 5 + 1$ ....[Add values]

Number of Terms$(n) = \mathbf{6}$

Example 5: Determine the common difference of an arithmetic progression given that the 9th equal -20 and the first term equal 4.

Solution

a) **Method 1 [Using the general term formula]**

Step 1: Formula. $a_n = a_1 + (n-1)d$, where; $a_1 = 4$, n =9 and $a_9 = -20$.

Step 2: Data substitution and solving

$-20 = 4 + (9 - 1)d$ ....[Simplify bracket]

$-20 = 4 + (8)d$ ....[Open bracket]

$-20 = 4 + 8d$ ....[Collect like terms together]

$8d = -20 - 4$ ....[Simplify right side of equation and divide both sides by (8)]

Common Difference$(d) = \mathbf{-3}$

b) **Method 2 [Using specified formula]**

Step 1: Formula. Common Difference$(d) = \frac{a_n - (a_1)}{n-1}$, where; $a_1 = 4$, n =9 and $a_n = a_9 = -20$.

Step 2: Data substitution and solving

Common Difference$(d) = \frac{-20-(4)}{9-1}$ ....[Simplify numerator and denominator]

Common Difference$(d) = \frac{-24}{8}$ ....[Solve fraction]

Common Difference$(d) = \mathbf{-3}$

Example 6: Consider that the 4th term of an arithmetic progression is -7 and the 7th term is 11. Determine the first term and common difference of the arithmetic progression.

Solution

Step 1: Derivation of equations

a) Determination of first equation [4th term equation]

   1) Formula. $a_n = a_1 + (n-1)d$, where; $a_n = -7$ and n = 4

230

2) Data substitution and solving

$$-7 = a_1 + (4 - 1)d \text{ ....[Simplify bracket]}$$
$$-7 = a_1 + (3)d \text{ ....[Open bracket]}$$
$$-7 = a_1 + 3d \text{ ....[Re-arranged equation]}$$
$$a_1 + 3d = -7 \text{ ....[Equation 1]}$$

b) Determination of second equation [7th term equation]

  1) Formula. $a_n = a_1 + (n - 1)d$, where; $a_n = 11$ and $n = 7$

  2) Data substitution and solving

$$11 = a_1 + (7 - 1)d \text{ ....[Simplify bracket]}$$
$$11 = a_1 + (6)d \text{ ....[Open bracket]}$$
$$11 = a_1 + 6d \text{ ....[Re-arranged equation]}$$
$$a_1 + 6d = 11 \text{ ....[Equation 2]}$$

Step 2: Determination of unknown [a_1 and d]

$$a_1 + 3d = -7 \text{ ....(1)}$$
$$a_1 + 6d = 11 \text{ ....(2)}$$

Note: Here, we apply the simultaneous equation elimination solving technique to obtain the numerical value of the unknowns. Subtract equation (1) from equation (2).

$$a_1 + 6d - (a_1 + 3d) = 11 - (-7) \text{ ....[Open equation brackets]}$$
$$a_1 + 6d - a_1 - 3d = 11 + 7 \text{ ....[Collect like terms together]}$$
$$3d = 18 \text{ ....[Divide both sides of equation by (3)]}$$
$$\text{Common Difference}(d) = \mathbf{6}$$

Note: The numerical value of the first term is being derived by substituting the numerical value of common difference in any equation or in any n^{th} term equations.

a) **Method 1 [Using simultaneous equation technique]**

Using Equation (1)	Using Equation (2)
$a_1 + 3d = -7$ ....[Substitute d value]	$a_1 + 6d = 11$ ....[Substitute d value]
$a_1 + 3(6) = -7$ ....[Open bracket]	$a_1 + 6(6) = 11$ ....[Open bracket]
$a_1 + 18 = -7$ ....[Collect like terms]	$a_1 + 36 = 11$ ....[Collect like terms]
First Term$(a_1) = -7 - 18$ ....[Solve]	First Term$(a_1) = 11 - 36$ ....[Solve]
First Term$(a_1) = \mathbf{-25}$	First Term$(a_1) = \mathbf{-25}$

b) **Method 2 [Using general term formula]**

Using 4th term information	Using 7th term information
Step 1: Formula. $a_n = a_1 + (n - 1)d$	Step 1: Formula. $a_n = a_1 + (n - 1)d$
Where; $a_4 = -7, n = 4$ and $d = 6$	Where; $a_7 = 11, n = 7$ and $d = 6$
Step 2: Data substitution and solving	Step 2: Data substitution and solving
$-7 = a_1 + (4 - 1)6$ ...[Simplify bracket]	$11 = a_1 + (7 - 1)6$ ....[Simplify bracket]
$-7 = a_1 + (3)6$ ....[Open bracket]	$11 = a_1 + (6)6$ ....[Open bracket]
First Term$(a_1) = -7 - 18$ ......[Solve]	First Term$(a_1) = 11 - 36$ ....[Solve]
First Term$(a_1) = -25$	First Term$(a_1) = -25$

Example 7: Find the 20th term of the arithmetic progression whose 3rd term is 7 and 8th term is 17

Solution

Step 1: Formula. $a_{20} = a_1 + (n - 1)d$, where; $n = 20$, $a_1 = $ Unknown and $d = $ Unknown

Step 2: Determination of unknowns

a) Determination of equations

Using 3rd term data

1) Formula. $a_n = a_1 + (n - 1)d$, where; $a_n = a_3 = 7$ and $n = 3$
2) Data substitution and solving

$7 = a_1 + (3 - 1)d$ ....[Simplify bracket]

$7 = a_1 + (2)d$ ....[Open bracket and make re-arranged equation]

$a_1 + 2d = 7$ ....[Equation 1]

Using 8th term data

1) Formula. $a_n = a_1 + (n - 1)d$, where; $a_n = a_8 = 17$ and $n = 8$
2) Data substitution and solving

$17 = a_1 + (8 - 1)d$ ....[Simplify bracket]

$17 = a_1 + (7)d$ ....[Open bracket and make re-arranged equation]

$a_1 + 7d = 17$ ....[Equation 2]

b) Determination of unknowns

$a_1 + 2d = 7$ ....(1)

$a_1 + 7d = 17$ ....(2)

Note: Solve the equations simultaneously, we derive the values; $a_1 = 3$ and $d = 2$

Step 3: Data substitution and solving [$a_1 = 3$ and $d = 2$]

20th Term$(a_{20}) = 3 + (20 - 1)2$ ....[Simplify bracket]

20th Term$(a_{20}) = 3 + (19)2$ ....…......…..[Solve]

20th Term$(a_{20}) = \mathbf{41}$

Example 8: The first three terms of an arithmetic progression are: 2, X and 18. Find X.

<u>**Solution**</u>

Step 1: Formula. $a_n = a_1 + (n-1)d$, where; $a_n = a_3 = 18$, $d = X - 2$, $n = 3$ and $a_1 = 2$

Step 2: Data substitution and solving

$18 = 2 + (3-1)(X-2)$ ....[Multiply brackets]

$18 = 2 + [3X - 6 - X + 2]$ ....…...........[Open bracket]

$18 = 2 + 3X - 6 - X + 2$ ....…....…........[Collect like terms together]

$2X = 20$ ....[Divide both sides by (2)]

Second Term$(X) = \mathbf{10}$ ....[The common difference is therefore equal to 8]

Example 9: Outline the sequence whose nth term is $4n + 5$. Prove mathematically that the sequence derived is in arithmetic progression.

<u>**Solution**</u>

Hint: Here, the nth term general formula $[a_n = a_1 + (n-1)d]$ will not be used since an nth term equation is given $[4n + 5]$. The sequences are derived by simply substituting variable (n) by the desired number of terms numerical values.

Step 1: Formula. $a_n = 4n + 5$

Step 2: Data substitution and solving [Assume; $n = 1, 2, 3, \ldots \ldots \ldots$]

$a_1 = 4(1) + 5 = \mathbf{9}$ ....…........[When n = 1]

$a_2 = 4(2) + 5 = \mathbf{13}$ ....[When n = 2]

$a_3 = 4(3) + 5 = \mathbf{17}$ ....[When n = 2]

[Note: The sequence are; 9, 13, 17,.......... ......n]

Exercise 8: The nth term of an arithmetic progression is 25. Assume the first term is 13 and the common difference is 4, determine the number of terms......................[Answer: n = 4]

Exercise 9: The third term of an arithmetic progression is 13 and the fourth term is 19. Determine the value of the first term and common difference....[Answer: $a_1 = 7$ and d = 3]

Exercise 10: The 9th term of an arithmetic progression is 30 and the 17th term is 50. Find the first three terms. ....[Answer: $a_1 = 10$, $a_2 = 25/2$ and $a_3 = 15$]

6.1.3 Sum of n Terms of an Arithmetic Progression

After determining the nth term of an arithmetic progression, we may also be required to determine the sum of a number of terms (that is sum of n terms). The general formula for calculating the sum of a given number of terms in an arithmetic progression denoted (S_n) also called arithmetic series is given as;

$$S_n = \frac{n}{2}[2a_1 + (n-1)d] \dots\dots\dots\dots\dots\dots\dots\dots\dots\dots\dots(1)$$

Example 1: Find the sum of an arithmetic progression using the following data; $a_1 = 3$, n = 16 and d = 8.

<u>Solution</u>

Step 1: Formula. $S_n = \frac{n}{2}[2a_1 + (n-1)d]$, where; $a_1 = 3$, n = 16 and d = 8

Step 2: Data substitution and solving

$$S_n = \frac{16}{2}[2(3) + (16-1)8] \dots\dots\dots\dots\dots\dots\dots\dots[\text{Simplify inner bracket}]$$

$$S_n = \frac{16}{2}[2(3) + (15)8] \dots\dots\dots\dots\dots[\text{Simplify fraction and open inner bracket}]$$

$$S_n = 8[6 + 120] \dots\dots\dots\dots\dots\dots\dots\dots\dots[\text{Add values and open bracket}]$$

$$S_n = 1,008$$

Example 2: Given that $S_n = 41,583$, $a_1 = 3$ and d = 3, determine the value of number of terms in the arithmetic progression.

<u>Solution</u>

Step 1: Formula. $S_n = \frac{n}{2}[2a_1 + (n-1)d]$, where; $S_n = 41,583$, $a_1 = 3$ and d = 3

Step 2: Data substitution and solving

$$41,583 = \frac{n}{2}[2(3) + (n-1)3] \dots\dots\dots\dots[\text{Multiply both sides of equation by (2)}]$$

$$83,166 = n[2(3) + (n-1)3] \dots\dots\dots\dots[\text{Open inner brackets and simplify bracket}]$$

$$83,166 = n[3 + 3n] \ldots\ldots\ldots\text{[Open equation bracket and equate equation to zero]}$$

$$83,166 - 3n - 3n^2 = 0 \ldots\ldots\text{[Multiply equation by } (-1) \text{ and re-arranged equation]}$$

$$3n^2 + 3n - 83,166 = 0 \ldots\ldots\ldots\ldots\ldots\ldots\ldots\ldots\ldots\ldots\ldots\text{[Quadratic equation]}$$

Note: Since we are having a quadratic equation, we can apply any standard technique to solve. Here, we are going to use the formula method. Know that $[a = 3, b = 3$ and $c = -83,166]$.

1) Formula. $n = \frac{-b \pm \sqrt{b^2 - 4ac}}{2a}$, where; $a = 3, b = 3$ and $c = -83,166$

2) Data substitution and solving

$$n = \frac{-3 \pm \sqrt{3^2 - 4(3)(-83,166)}}{2(3)} \ldots\ldots\ldots\ldots\ldots\ldots\text{[Work exponent and open brackets]}$$

$$n = \frac{-3 \pm \sqrt{9 + 997,992}}{6} \ldots\ldots\ldots\ldots\ldots\ldots\ldots\ldots\ldots\ldots\text{[Add values and work root]}$$

$$n = \frac{-3 \pm 999}{6} \ldots\ldots\ldots\ldots\ldots\ldots\ldots\ldots\ldots\ldots\ldots\ldots\ldots\ldots\ldots\text{[Solve equation]}$$

$$n_1 = \frac{-3 + 999}{6} = 166 \ldots\ldots\ldots\ldots\ldots\ldots\text{[Acceptable number of terms]}$$

$$n_2 = \frac{-3 - 999}{6} = -167 \ldots\ldots\ldots\ldots\ldots\ldots\text{[Unacceptable number of terms]}$$

Example 3: Considering that the sum of 16 terms in an arithmetic progression is 1,008 and the first term is 3, determine the value of the common difference.

Solution

a) Method 1 [Using sum of term formula]

Step 1: Formula. $S_n = \frac{n}{2}[2a_1 + (n - 1)d]$, where; $S_n = S_{16} = 1,008, a_1 = 3$ and $n = 16$

Step 2: Data substitution and solving

$$1,008 = \frac{16}{2}[2(3) + (16 - 1)d] \text{ .[Multiply both sides by (2) and simplify inner bracket]}$$

$$2,016 = 16[2(3) + (15)d] \ldots\ldots\ldots\ldots\ldots\ldots\ldots\ldots\ldots\ldots\text{[Open inner brackets]}$$

$$2,016 = 16[6 + 15d] \ldots\ldots\ldots\ldots\ldots\ldots\ldots\ldots\ldots\ldots\text{[Open equation bracket]}$$

$$2,016 = 96 + 240d \ldots\ldots\ldots\ldots\ldots\ldots\ldots\ldots\ldots\text{[Collect like terms together]}$$

$$240d = 2,016 - 96 \ldots\ldots\ldots\ldots\ldots\ldots\ldots\text{[Divide both sides of equation by (240)]}$$

Common Difference $(d) = 8$

b) **Method 2 [Using specified formula]**

Step 1: Formula. Common Difference $(d) = \frac{2S_n - 2a_1(n)}{n^2 - 1}$

Where; $S_n = S_{16} = 1,008, a_1 = 3$ and $n = 16$

Step 2: Data substitution and solving

Common Difference (d) $= \frac{2(1,008)-2(3)(16)}{16^2-1}$ ....[Open numerator brackets]

Common Difference (d) $= \frac{2,016-96}{16^2-1}$ ....[Work denominator exponent]

Common Difference (d) $= \frac{2,016-96}{256-1}$ ....[Solve equation]

Common Difference (d) $= 8$ ....**[Rounded value to no decimal place]**

Example 4: Find the value of the first term given that; $S_{31} = 1,116$ and $d = -1$

<u>Solution</u>

a) Method 1 [Using sum of terms formula]

Step 1: Formula. $S_n = \frac{n}{2}[2a_1 + (n-1)d]$, where; $S_n = S_{31} = 1,116$, $d = -1$ and $n = 31$

Step 2: Data substitution and solving

$1,116 = \frac{31}{2}[2a_1 + (31-1) - 1]$ ....[Multiply both sides of equation by (2)]

$2,232 = 31[2a_1 + (31-1) - 1]$ ....[Simplify inner bracket]

$2,232 = 31[2a_1 + (30) - 1]$ ....[Open inner bracket]

$2,232 = 31[2a_1 - 30]$ ....[Open equation bracket]

$2,232 = 62a_1 - 930$ ....[Collect like terms together]

$62a_1 = 2,232 + 930$ ....[Add values and divide both sides of equation by (62)]

First Term$(a_1) = 51$

b) **Method 2 [Using specified formula]**

Step 1: Formula. First Term$(a_1) = \frac{2S_n - (n^2-n)d}{2n}$, where; $S_n = S_{31} = 1,116$, $d = -1$ and $n = 31$

Step 2: Data substitution and solving

First Term$(a_1) = \frac{2(1,116)-(31^2-31)-1}{2(31)}$ ....[Work exponent]

First Term$(a_1) = \frac{2(1,116)-(961-31)-1}{2(31)}$ ....[Simplify inner bracket]

First Term$(a_1) = \frac{2(1,116)-(930)-1}{2(31)}$ ....[Open brackets]

First Term$(a_1) = \frac{2,232+930}{62}$ ....[Solve equation]

First Term$(a_1) = 51$

Example 5: The sum of 25 terms of an arithmetic progression is 2,825 and the sum of 50 terms is 11,275. Determine the first term and the common difference.

Solution

Step 1: Determination of equations

a) Using sum of 25 terms

1) Formula. $S_n = \frac{n}{2}[2a_1 + (n-1)d]$, where; $S_1 = S_{25} = 2,825$ and $n = 25$

2) Data substitution and solving

$$2,825 = \frac{25}{2}[2a_1 + (25-1)d] \ldots\ldots\ldots\text{[Simplify fraction and inner bracket]}$$

$$2,825 = 12.5[2a_1 + 24d] \ldots\ldots\ldots\ldots\ldots\ldots\ldots\text{[Open bracket]}$$

$$2,825 = 25a_1 + 300d \ldots\ldots\ldots\ldots\ldots\ldots\ldots\text{[Re-arranged equation]}$$

$$25a_1 + 300d = 2,825 \ldots\ldots\ldots\ldots\ldots\ldots\ldots\text{[Equation 1]}$$

b) Using sum of 50 terms

1) Formula. $S_n = \frac{n}{2}[2a_1 + (n-1)d]$, where; $S_n = S_{50} = 11,275$ and $n = 50$

2) Data substitution and solving

$$11,275 = \frac{50}{2}[2a_1 + (50-1)d] \ldots\ldots\ldots\text{[Simplify fraction and inner bracket]}$$

$$11,275 = 25[2a_1 + 49d] \ldots\ldots\ldots\ldots\ldots\ldots\ldots\text{[Open bracket]}$$

$$11,275 = 50a_1 + 1,225d \ldots\ldots\ldots\ldots\ldots\ldots\text{[Re-arranged equation]}$$

$$50a_1 + 1,225d = 11,275 \ldots\ldots\ldots\ldots\ldots\ldots\text{[Equation 2]}$$

Step 2: Determination of unknown [a_1 and d]

$$25a_1 + 300d = 2,825 \ldots\ldots\ldots\ldots\ldots\ldots\ldots\ldots\ldots\ldots(1)$$

$$50a_1 + 1,225d = 11,275 \ldots\ldots\ldots\ldots\ldots\ldots\ldots\ldots\ldots(2)$$

Note: Here, we solve the equations simultaneously. Assume we use the elimination method; we multiply equation (1) by (50) and equation 2 by (25). Doing this, we obtain the equations below.

$$1,250a_1 + 15,000d = 141,250 \ldots\ldots\ldots\ldots\ldots\ldots\ldots\ldots(3)$$

$$1,250a_1 + 30,625d = 281,875 \ldots\ldots\ldots\ldots\ldots\ldots\ldots\ldots(4)$$

To eliminate the first term (a_1), subtract equation (3) from equation (4), doing that we get,

$1,250a_1 + 30,625d - (1,250a_1 + 15,000d) = 281,875 - 141,250$...[Open bracket]

$1,250a_1 + 30,625d - 1,250a_1 - 15,000d = 281,875 - 141,250$..[Collect   like terms]

$15,625d = 140,625$ ....[Divide both sides of equation by (15,625)]

Common Difference(d) = **9**

Note: To determine the value of the first term (a_1), we can apply the simultaneous equation technique or the sum of terms technique.

Simultaneous technique [Use equation 1]	Sum of term technique [use 25 terms data]

$25a_1 + 300d = 2,825$ ......[Substitute d by 9]
 $25a_1 + 300(9) = 2,825$ ....[Open bracket]
 $25a_1 + 2,700 = 2,825$ ....[Collect like terms]
 $25a_1 = 125$ ....[Divide both sides by 25]
 First Term $(a_1) = 5$
	1) Formula. $S_n = \frac{n}{2}[2a_1 + (n-1)d]$
	     Where; d =9
	 2) Data substitution and solving
	     $2,825 = \frac{25}{2}[2a_1 + (25-1)9]$ ....[Simplify]
	     $2,825 = 25a_1 + 2,700$ ......[Solve for (a_1)]
	     First Term $(a_1) = 5$

Test Your Understanding

Exercise 11: Determine the sum of 50 terms using the following sequence; 5, 14, 23, 32, 41,..[**Answer: $S_{50} = 11,275$**]

Exercise 12: Given that $S_{35} = -1,785$ and $d = -4$, compute the value of a_1
[**Answer: $a_1 = 17$**]

Exercise 13: Consider that the sum of 16 terms is 624 and the first term is 9. Determine the value of the common difference...[**Answer: $d = 4$**]

Exercise 14: The sum of nth terms is 2,325. Given that the first term is 5 and the common difference is equal to the first term, determine the number of terms...........[**Answer: $n = 30$**]

Exercise 15: Given that $S_5 = 50$ and $S_{15} = -225$, fine a_1 and d
.[**Answer: $a_1 = 20$ and $d = -5$**]

Exercise 16: How many terms are required such that the sum of the terms will be greater than 275, given that the first term is 5 and the common difference is 5.............[**Answer: $n > 10$**]

6.1.4 Combination of nth Terms and Sum of n Terms

We can equally have situations where information related to n^{th} term and sum of n terms of an arithmetic progression are given for the determination of other components.

Example 1: Given that the 10th term of an arithmetic progression is 45 and the sum of the first 10 terms is 270. Find the value of the following

1) First term and common difference.
2) The sum of the first 15 terms

1) First term and common difference

Hint: The question is made up of two concepts, the nth term and sum of term. Applying each separately to look for the value of the first term will give us an equation with two unknown (a_1 and d). Combining the two concepts, we can apply the notion of simultaneous equation to solve for the first term.

Step 1: Linearization of equations

a) Linearization of nth term equation

1) Formula. $a_n = a_1 + (n - 1)d$, where; $a_n = 45$ and n = 10
2) Data substtution and solving

$$45 = a_1 + (10 - 1)d \dots\dots\dots\dots\dots\text{[Simplify bracket]}$$
$$45 = a_1 + (9)d \dots\dots\dots\dots\dots\text{[Open bracket]}$$
$$45 = a_1 + 9d \dots\dots\dots\dots\dots\text{[Re-arranged equation]}$$
$$a_1 + 9d = 45 \dots\dots\dots\dots\dots\text{[Equation 1]}$$

b) Linearization of sum of term equation

1) Formula. $S_n = \frac{n}{2}[2a_1 + (n - 1)d]$, where; $S_n = 270$ and n = 10
2) Data substitution and solving

$$270 = \frac{10}{2}[2a_1 + (10 - 1)d] \dots\dots\dots\text{[Simplify fraction and inner bracket]}$$
$$270 = 5[2a_1 + (9)d] \dots\dots\dots\dots\dots\text{[Open inner bracket]}$$
$$270 = 5[2a_1 + 9d] \dots\dots\dots\dots\dots\text{[Open equation bracket]}$$
$$270 = 10a_1 + 45d \dots\dots\dots\dots\dots\text{[Re-arranged equation]}$$
$$10a_1 + 45d = 270 \dots\dots\dots\dots\dots\text{[Equation 2]}$$

Step 2: Application of simultaneous equation

$$a_1 + 9d = 45 \dots\dots\dots\dots\dots(1)$$
$$10a_1 + 45d = 270 \dots\dots\dots\dots\dots(2)$$

Note: Multiply equation (10) by 10 and equation (2) by 1 to eliminate the first term (a_1) in the simultaneous equation or multiply equation (1) by 45 and equation (2) by 9 to eliminate the common difference (d) in the simultaneous equation.

Elimination of the first term (a_1)	Eliminate the common difference (d)
$10a_1 + 90d = 450$ ....(1)	$45a_1 + 405d = 2,025$ ....(1)
$10a_1 + 45d = 270$ ....(2)	$9a_1 + 405d = 2.430$ ....(2)

Subtract equation (1) from equation (2) and solve the equation to determine the value of common difference (d).

$$10a_1 + 45d - (10a_1 + 90d) = 270 - 450 \dots\dots\dots\dots\dots\text{[Open bracket]}$$

$$10a_1 + 45d - 10a_1 - 90d = 270 - 450 \dots\dots\dots\text{[Collect like terms together]}$$

$$-45d = -180 \dots\dots\dots\dots\dots\dots\text{[Divide both sides of equation by } (-45)]$$

Common Difference(d) = **4**

Note: Since the common difference value is known, we substitute its numerical value in equation (1) or equation (2) to derive the value of the first term. The first term value can still be derived by substituting the common difference value in the nth term equation or sum of term equation.

$$a_1 + 9d = 45 \dots\dots\dots\dots\dots\dots\dots\dots\dots\text{[Substitute d by 4]}$$

$$a_1 + 9(4) = 45 \dots\dots\dots\dots\dots\dots\dots\dots\dots\text{[Open bracket]}$$

$$a_1 + 36 = 45 \dots\dots\dots\dots\dots\dots\dots\dots\text{[Collect like terms together]}$$

First Term(a_1) = **9**

2) The sum of the first 15 terms [That is S_{15}]

Step 1: Formula. $S_{15} = \frac{n}{2}[2a_1 + (n - 1)d]$, where; $a_1 = 9$, $d = 4$ and $n = 15$

Step 2: Data substitution and solving

$$S_{15} = \frac{15}{2}[2(9) + (15 - 1)4] \dots\dots\dots\text{[Simplify fraction and open inner bracket]}$$

$$S_{15} = 7.5[18 + 56] \dots\dots\dots\dots\dots\dots\dots\text{[Simplify equation bracket]}$$

$$S_{15} = 7.5[74] \dots\dots\dots\dots\dots\dots\dots\dots\text{[Open equation bracket]}$$

Sum of the First 15 Terms (S_{15}) = **555**

Exercise 2: The sum of 5 terms in an arithmetic progression is 100 and the sum of 10 terms is 75. Determine the value of the 6th term.

Solution

Step 1: Formula. $a_n = a_1 + (n - 1)d$

Where; $a_n = a_6$, $n = 6$, $d = $ Unknown and $a_1 = $ Unknown

Step 2: Determination of Unknown

a) Determination of equations

Using the sum of 5 terms data

1) Formula. $S_n = \frac{n}{2}[2a_1 + (n - 1)d]$, where; $S_n = S_5 = 100$ and $n = 5$

2) Data substitution and solving

$$100 = \frac{5}{2}[2a_1 + (5 - 1)d] \text{ .[Simplify inner bracket and simplify fraction]}$$

$$100 = 2.5[2a_1 + 4d] \ldots\ldots\ldots [\text{Open bracket and re-arranged equation}]$$
$$5a_1 + 10d = 100 \ldots\ldots\ldots\ldots\ldots\ldots\ldots\ldots\ldots\ldots [\textbf{Equation 1}]$$

Using the sum of 10 terms data

1) Formula. $S_n = \frac{n}{2}[2a_1 + (n-1)d]$, where; $S_n = S_{10} = 75$ and $n = 10$

2) Data substitution and solving

$$75 = \frac{10}{2}[2a_1 + (10-1)d] \ldots [\text{Simplify inner bracket and simplify fraction}]$$

$$75 = 5[2a_1 + 9d] \ldots\ldots\ldots\ldots [\text{Open bracket and re-arranged equation}]$$

$$10a_1 + 45d = 75 \ldots\ldots\ldots\ldots\ldots\ldots\ldots\ldots\ldots [\textbf{Equation 2}]$$

b) Determination of common difference (d)

$$5a_1 + 10d = 100 \ldots\ldots\ldots\ldots\ldots \ldots\ldots\ldots \ldots\ldots\ldots\ldots\ldots\ldots.(1)$$

$$10a_1 + 45d = 75 \ldots\ldots\ldots\ldots \ldots\ldots\ldots \ldots\ldots\ldots\ldots\ldots\ldots.(2)$$

Note: Solving the equations simultaneously, we obtain the values, $\mathbf{a_1 = 30}$ and $\mathbf{d = -5}$

Step 3: Data substitution and solving

$$a_6 = 30 + (6-1) - 5 \ldots\ldots\ldots\ldots\ldots \ldots\ldots\ldots \text{Simplify and open bracket}]$$

$$a_6 = 30 - 25 \ldots\ldots\ldots\ldots\ldots\ldots\ldots\ldots\ldots\ldots\ldots\ldots\ldots.[\text{Solve equation}]$$

The nth Term $(a_6) = \mathbf{5}$

Example 3: The arithmetic progression has a first term 3 and common difference d. Given that the nth term is 93 and the sum of the first n terms is 768, find the value of n and d.

Solution

Hint: Solving the unknown separately using the nth term and sum of term concepts lends to equation with not only two unknown (n and d) but also a product of two variables (nd) making it difficult to solve. Therefore, we use the substitution method to determine the unknown since the both concepts have a common element $(n - 1)d$.

Step 1: Determination of the common element value

1) Formula. $a_n = a_1 + (n-1)d$, where; $a_n = 93$ and $a_1 = 3$

2) Data substitution and solving

$$93 = 3 + (n-1)d \ldots\ldots\ldots\ldots\ldots\ldots\ldots\ldots [\text{Collect like terms together}]$$

$$(n-1)d = 93 - 3 \ldots\ldots\ldots\ldots\ldots\ldots\ldots\ldots \ldots\ldots\ldots.[\text{Subtract values}]$$

$$(n-1)d = \mathbf{90}$$

Step 2: Determination of the unknown values

a) **Numerical value of number of nth terms (n)**

1) Formula. $S_n = \frac{n}{2}[2a_1 + (n-1)d]$, where; $S_n = 768$, $a_1 = 3$ and $(n-1)d = 90$

2) Data substitution and solving

$768 = \frac{n}{2}[2(3) + 90]$ ....[Open inner bracket]

$768 = \frac{n}{2}[6 + 90]$ ....[Simplify bracket and multiply both side by (2)]

$1,536 = n[96]$ ....[Divide both sides of equation by (96)]

Number of nth Terms(n) = **16**

b) Numerical value of the common difference (d)

1) Method 1 [Using sum of term concept]

Step 1: Formula. $S_n = \frac{n}{2}[2a_1 + (n-1)d]$, where; $S_n = 768$, $a_1 = 3$ and n = 16

Step 2: Data substitution and solving

$768 = \frac{16}{2}[2(3) + (16-1)d]$ ....[Simplify fraction and inner brackets]

$768 = 8[6 + 15d]$ ....[Open equation bracket]

$768 = 48 + 120d$ ....[Collect like terms together]

$120d = 768 - 48$ ....[Subtract values and divide both sides of equation by (120)]

Common Difference(d) = **6**

2) Method 2 [Using n^{th} term concept]

Step 1: Formula. $a_n = a_1 + (n-1)d$, where: $a_n = 93$, $a_1 = 3$ and n = 16

Step 2: Data substitution and solving

$93 = 3 + (16-1)d$ ....[Simplify bracket]

$93 = 3 + (15)d$ ....[Open bracket and collect like terms together]

$15d = 93 - 3$ ....[Subtract values and divide both sides of equation by (15)]

Common Difference(d) = **6**

Test Your Understanding

Exercise 17: The 2nd term of an arithmetic progression is 9 and the sum of 4 terms is 50. Determine the first term and common ratio.........................[**Answer: $a_1 = 2$ and d = 7**]

Exercise 18: The sum of four terms in an arithmetic progression is 44 and the sum of 10 terms is 170. Determine the 15th term of the arithmetic progression...............[**Answer: $a_{15} = 36$**]

Exercise 19: The 7th term of an arithmetic progression is -5 and the 12th term is -30.

Determine the sum of an arithmetic progression given that n is 10. ....[**Answer: $S_{10} = 25$**]

Exercise 20: The nth term of an arithmetic progression is 15 and the sum of n terms is 225. Given that the first term is 3, determine the following,

1) The 5th term….....................…............................. ….........[**Answer: $a_5 = 5$**]
2) The sum of 12 terms….....................…...........................[**Answer: $S_{25} = 63$**]

6. 2 Geometric Progression

Geometric sequence is a sequence of numbers where each term or number after the first is derived by multiplying the previous term by a fixed or constant non-zero number called the common ratio. Geometric progression is also called geometric sequence.

6.2.1 Common Ra

The common ratio is denoted (r) and is the quotient between the succeeding term $(a_{n+1})$and the preceding term (a_n). The formula for common ratio is given as seen below,

$$\text{Common Ratio (r)} = \frac{\text{Succeeding term } (a_{n+1})}{\text{Preceding term } (a_n)} \dots\dots\dots\dots\dots\dots\dots\dots\dots\dots\dots(1)$$

From equation (1), the value of other terms (succeeding terms) given the first term (preceding term) is calculated using the formula below.

$$\text{Succeeding term } (a_{n+1}) = \text{Common Ratio (r)} \times \text{Preceding term } (a_n) \dots\dots\dots\dots(2)$$

The common ratio can be positive, negative, whole number and/or fraction. The following properties are true about the common ratio of a geometric progression.

1) If the common ratio is positive (+), the terms derived will all be the same sign as the initial term or first term.
2) If the common ratio is negative (−), the terms will alternate or rotate between positive and negative.
3) If the common ratio is greater than $1(r > 1)$, the terms derived will be more than proportionally increasing towards positive infinity (when initial term is positive) or negative infinity (when initial term is negative).
4) If the common ratio is between negative one and positive one but not zero $[(-1 \leq r > 0)$ or $(0 < r \geq 1)]$, the terms derived will be more than proportionally decreasing towards zero.
5) If the common ratio is equal to one $(r = 1)$, the terms derived will be there same. This indicates a constant sequence. That is all values in the sequence are there same.

6) If the common ratio is equal to negative one ($r = -1$), the terms derived will be there same but with alternating signs.

Note: It is advisable to avoid putting common ratio in decimal. A sequence is in geometric progression if and only if the common ratios of all the segments are equal or the same.

Example 1: In a geometric progression, the first term is 5 and the second term is 15. Determine the value of the common ratio.

Solution

Step 1: Formula. Common Ratio $(r) = \frac{\text{Succeeding term } (a_{n+1})}{\text{Preceding term } (a_n)}$

Where; Succeeding term $(a_{n+1}) = 15$ and Preceding term $(a_n) = 5$

Step 2: Data substitution and solving

Common Ratio $(r) = \frac{15}{5}$ ....[Solve fraction]

Common Ratio $(r) = \mathbf{3}$

Step 3: Interpretation: The common ratio 3 means the second term is 3 times larger than the first term. That is the second term is derived by multiplying the first term by 3. That is $[5 \times 3 = 15]$.

Example 2: Determine the common ratio from the following sequence; 2, 16, 128, 1024..........

Solution

Hint: The given sequence can be segmented as follows; [2, 16], [16, 128] and [128, 1024]. Any of these segments can be used to determine the common ratio.

Step 1: Formula. Common Ratio $(r) = \frac{\text{Succeeding term } (a_{n+1})}{\text{Preceding term } (a_n)}$

Step 2: Data substitution and solving

Segment 1	Segment 2	Segment 3
$(a_{n+1}) = 16, (a_n) = 2$	$(a_{n+1}) = 128, (a_n) = 16$	$(a_{n+1}) = 1024, (a_n) = 128$
$r = \frac{16}{2} = 8$	$r = \frac{128}{16} = 8$	$r = \frac{1024}{128} = 8$

Example 3: Assume the first term of a geometric progression is 2 and the common ratio is 3, outline the first five terms.

Solution

Step 1: Formula. Succeeding term $(a_{n+1}) =$ Common Ratio $(r) \times$ Preceding term (a_n)

Step 2: Data substitution and solving

First term (a_1)	Second term (a_2)	Third term (a_3)	Fourth term (a_4)	Fifth term (a_5)
2 (Given)	$a_2 = r \times a_1$	$a_3 = r \times a_2$	$a_4 = r \times a_3$	$a_5 = r \times a_4$
2	$a_2 = 3 \times 2$	$a_3 = 3 \times 6$	$a_4 = 3 \times 18$	$a_5 = 3 \times 54$
	$a_2 = 6$	$a_3 = 18$	$a_4 = 54$	$a_5 = 162$
Note: The first five terms of the geometric progression are; 2, 6, 18, 54 and 162.				

Step 3: Complementary Explanation

1) The common ratio is positive (+2). The terms derived (+2, +6, +18, +54 and +162) are all having the sign (+) of the initial term (+2).

2) The common ratio is greater than one ($r > 1$). The terms derived shows an exponential growth towards positive infinity since the first term is positive (+2, +6, +18, +54, +162, ...+∞). The exponential increment is illustrated in the table below.

Values	2	6	18	54	162
Change	$2 - 0$	$6 - 2$	$18 - 6$	$54 - 18$	$162 - 54$
	2	**4**	**12**	**36**	**108**

Example 4: Assume the first term of a geometric progression is 3 and the common ratio is −2, outline the first five terms.

Solution

Step 1: Formula. Succeeding term (a_{n+1}) = Common Ratio (r) × Preceding term (a_n)

Step 2: Data substitution and solving

First term (a_1)	Second term (a_2)	Third term (a_3)	Fourth term (a_4)	Fifth term (a_5)
3 (Given)	$a_2 = r \times a_1$	$a_3 = r \times a_2$	$a_4 = r \times a_3$	$a_5 = r \times a_4$
3	$a_2 = -2 \times 3$	$a_3 = -2 \times -6$	$a_4 = -2 \times 12$	$a_5 = -2 \times 24$
	$a_2 = -6$	$a_3 = 12$	$a_4 = -24$	$a_5 = 48$
Note: The first five terms of the geometric progression are 3, −6, 12, −24 and 48				

Step 3: Complementary Explanation. The common ratio is negative(−2). The terms derived are rotate between positive and negative(−6, 12, −24 and 48).

Example 5: The first term of a geometric progression is 4 and the second term is 2. Determine the following;

1) The value of the common ratio

2) Outline the first five terms

Solution

1) The value of the common ratio

Step 1: Formula. Common Ratio $(r) = \dfrac{\text{Succeeding term } (a_{n+1})}{\text{Preceding term } (a_n)}$

Where; Succeeding term $(a_{n+1}) = 2$ and Preceding term $(a_n) = 4$

Step 2: Data substitution and solving

Common Ratio $(r) = \dfrac{2}{4}$ ....[Solve fraction]

Common Ratio $(r) = \dfrac{1}{2}$

2) Outline the first five terms

Step 1: Formula. Succeeding term (a_{n+1}) = Common Ratio $(r) \times$ Preceding term (a_n)

Step 2: Data substitution and solving

First term (a_1)	Second term (a_2)	Third term (a_3)	Fourth term (a_4)	Fifth term (a_5)
4 (Given)	$a_2 = r \times a_1$	$a_3 = r \times a_2$	$a_4 = r \times a_3$	$a_5 = r \times a_4$
4	$a_2 = \frac{1}{2} \times 4$ $a_2 = 2$	$a_3 = \frac{1}{2} \times 2$ $a_3 = 1$	$a_4 = \frac{1}{2} \times 1$ $a_4 = \frac{1}{2}$	$a_5 = \frac{1}{2} \times \frac{1}{2}$ $a_5 = \frac{1}{4}$

Step 3: Complementary Explanation: The common ratio is $\left(\frac{1}{2}\right)$ which is between $(-1$ and $1)$. The terms derived $\left(2, 1, \frac{1}{2} \text{ and } \frac{1}{4}\right)$ are driving towards zero.

Example 6: Given that the first term (a_1) of a geometric progression is 4. Determine the first five terms assuming that,

1) The common ratio is 1
2) The common ratio is -1

Solution

1) The common ratio is 1

Step 1: Formula. Succeeding term (a_{n+1}) = Common Ratio $(r) \times$ Preceding term (a_n)

Step 2: Data substitution and solving

First term (a_1)	Second term (a_2)	Third term (a_3)	Fourth term (a_4)	Fifth term (a_5)
4 (Given)	$a_2 = r \times a_1$	$a_3 = r \times a_2$	$a_4 = r \times a_3$	$a_5 = r \times a_4$
4	$a_2 = 1 \times 4$ $a_2 = 4$	$a_3 = 1 \times 4$ $a_3 = 4$	$a_4 = 1 \times 4$ $a_4 = 4$	$a_5 = 1 \times 4$ $a_5 = 4$

Step 3: Complementary Explanation. The common ratio is positive one (+1). The terms derived are there same (4, 4, 4 and 4) indication a constant sequence. Here the change between terms is equal to zero.

2) The common ratio is −1

Step 1: Formula. Succeeding term (a_{n+1}) = Common Ratio (r) × Preceding term (a_n)

Step 2: Data substitution and solving

First term (a_1)	Second term (a_2)	Third term (a_3)	Fourth term (a_4)	Fifth term (a_5)
4 (Given)	$a_2 = r \times a_1$	$a_3 = r \times a_2$	$a_4 = r \times a_3$	$a_5 = r \times a_4$
4	$a_2 = -1 \times 4$ $a_2 = -4$	$a_3 = -1 \times -4$ $a_3 = 4$	$a_4 = -1 \times 4$ $a_4 = -4$	$a_5 = -1 \times -4$ $a_5 = 4$

Step 3: Complementary Explanation. The common ratio is negative one (−1). The terms derived are constant alternating sequence (−4, 4, −4 and 4). That is a combination of positive and negative constant values.

Test Your Understanding

Exercise 21: Determine the common ratio of a geometric progression given that $a_n = 2$ and $a_{n+1} = -6$...**[Answer: r = −3]**

Exercise 22: Given that the first term of a geometric progression is $\frac{1}{2}$ and the second term is $\frac{1}{4}$, determine the common ratio..**[Answer: $r = \frac{1}{2}$]**

Exercise 23: Find the common ratio from the following sequence; $-\frac{1}{3}, 1, -3, 9$

[Answer: r = −3]

Exercise 24: Give that $a_1 = 6$ and r = 1/2, determine the first six terms of the geometric progression. ....**[Answer: 6, 3, 3/2 , 3/4 and 3/8]**

Exercise 25: Which of the following sequence is not in a geometric progression? Justify your answer mathematically

1) 5, 8, 11, 14,....
2) 4, 12, 36, 108. ....
3) 4, −2, 1, $-\frac{1}{2}, \frac{1}{4}$......
4) $\frac{1}{3}, -\frac{1}{9}, \frac{1}{27}, -\frac{1}{81}$ ......

[Answer: The sequence 5, 8, 11, 14.... Since it common ratio is not fixed. That is it common ratio varies from, $\frac{8}{5}$, to $\frac{11}{8}$, and $\frac{14}{11}$]

6.2.2 The nth Term of a Geometric Progression

After the determination of the common ratio, we are also required to know how to determine the terms of a geometric progression. The specific formula for determining the first four terms of a given geometric progression is given below.

The First Term $(a_1) = a_1 \times r^0 = a_1 r^0$...(1)

The Second Term $(a_2) = a_1 \times r = a_1 r$...(2)

The Third Term $(a_3) = a_1 \times r = a_1 r^2$...(3)

The fourth Term $(a_4) = a_1 \times r = a_1 r^3$...(4)

From the above equations, we realized that the exponent or power of the common ratio of a given term is lesser than the term by 1. Therefore, the nth term common ratio power will be $(n-1)$. The general formula for determining a given term when the first term and the common ratio $(r \neq 0)$ are given is seen below.

The nth Term$(a_n) = a_1 \times r^{n-1} = a_1 r^{n-1}$ ....(5)

Example 1: Find the 10th term of a geometric progression given that the common ratio is 2 and the first term is 2.

Solution

Step 1: Formula. $a_{10} = a_1 \times r^{n-1}$, where; $a_1 = 2, r = 2$ and $n = 10$

Step 2: Data substitution and solving

$a_{10} = 2 \times 2^{10-1}$ ....[Simplify exponent]

$a_{10} = 2 \times 2^9$ ....[Work exponent]

$a_{10} = 2 \times 512$ ....[Multiply values]

10th Term$(a_{10}) = \mathbf{1,024}$

Example 2: Find the 7th term of a geometric progression using the sequence; 4, -2, 1.............

Solution

Step 1: Formula. $a_n = a_1 \times r^{n-1}$, where; $a_1 = 4, r = -\frac{1}{2}$ and $n = 7$

Step 2: Data substitution and solving

$a_7 = 4 \times -\left[\frac{1}{2}\right]^{7-1}$ ....[Simplify exponent]

$a_7 = 4 \times -\left[\frac{1}{2}\right]^6$ ....[Re-arranged equation]

248

$a_7 = 4 \times - \left[\frac{1^6}{2^6}\right]$ ....[Work exponent]

$a_7 = 4 \times - \left[\frac{1}{64}\right]$ ....[Open bracket and multiply values]

$a_7 = -\frac{4}{64}$ ....[Simplify fraction to lowest form]

$7^{th}\ Term(a_n) = -\frac{1}{16}$

Example 3: Given that the 4th term of a geometric progression is 40 and the common ratio is 2. Compute the value of the first term.

<u>Solution</u>

a) **Method 1 [Using nth term formula]**

Step 1: Formula. $a_n = a_1 \times r^{n-1}$, where; $a_n = a_4 = 40, r = 2$ and $n = 4$

Step 2: Data substitution and solving

$40 = a_1 \times 2^{4-1}$ ....[Simplify exponent]

$40 = a_1 \times 2^3$ ....[Work exponent]

$40 = a_1 \times 8$ ....[Divide both sides of equation by (8)]

$1^{st}\ Term(a_1) = 5$

b) **Method 2 [Using specified formula]**

Step 1: Formula. $1^{st}\ Term(a_1) = \frac{a_n}{r^{n-1}}$, where; $a_n = a_4 = 40, r = 2$ and $n = 4$

Step 2: Data substitution and solving

$1^{st}\ Term(a_1) = \frac{40}{2^{4-1}}$ ....[Simplify denominator exponent]

$1^{st}\ Term(a_1) = \frac{40}{2^3}$ ....[Simplify denominator exponent]

$1^{st}\ Term(a_1) = \frac{40}{8}$ ....[Solve equation]

$1^{st}\ Term(a_1) = 5$

Example 4: The 6th term of a geometric progression is 972 and the first term is 4. Determine the value of the common ratio.

<u>Solution</u>

a) **Method 1 [Using n^{th} term formula]**

Step 1: Formula. $a_n = a_1 \times r^{n-1}$, where; $a_n = a_6 = 972, a_1 = 4$ and $n = 6$

Step 2: Data substitution and solving

$972 = 4 \times r^{6-1}$ ....[Simplify exponent]

$972 = 4 \times r^5$ ....[Divide both sides of equation by (4)]

$r^5 = 243$ ....[Multiply both sides of equation by exponent$(1/5)$]

$Common\ Ratio(r) = 243^{1/5}$ ....[Work exponent]

Common Ratio(r) = **3**

b) **Method 2 [Using specified formula]**

Step 1: Formula. Common Ratio(r) = $\left[\frac{a_n}{a_1}\right]^{\frac{1}{n-1}}$	Step 1: Formula. Common Ratio(r) = $\sqrt[n-1]{\frac{a_n}{a_1}}$
Where: $a_n = a_6 = 972$, $a_1 = 4$ and $n = 6$	
Step 2: Data substitution and solving	Step 2: Data substitution and solving
Common Ratio(r) = $\left[\frac{972}{4}\right]^{\frac{1}{6-1}}$ ....[Simplify]	Common Ratio(r) = $\sqrt[6-1]{\frac{972}{4}}$ ......[Simplify]
Common Ratio(r) = $\left[\frac{972}{4}\right]^{\frac{1}{5}}$	Common Ratio(r) = $\sqrt[5]{\frac{972}{4}}$
Common Ratio(r) = $[243]^{\frac{1}{5}}$.[Work exponent]	Common Ratio(r) = $\sqrt[5]{243}$.[Work 5th root]
Common Ratio(r) = **3**	Common Ratio(r) = **3**

Example 5: Given that the nth term of a geometric progression is 1,280 and the first term is 5. Considering that the common ratio is 2, determine the number of terms.

Solution

a) **Method 1 [Using n^{th} term formula]**

Step 1: Formula. $a_n = a_1 \times r^{n-1}$, where; $a_n = a_6 = 1,280$, $a_1 = 5$ and $r = 2$

Step 2: Data substitution and solving

$1,280 = 5 \times 2^{n-1}$ ....[Divide both sides of equation by (5)]

$256 = 2^{n-1}$ ....[Log both sides of equation]

$(n - 1)Log2 = Log\,256$ ....[Divide both sides of equation by Log 2]

$n - 1 = \frac{Log\,256}{Log\,2}$ ....[Work Log and Simplify fraction]

$n - 1 = 8$ ....[Take (−1) to the right side of the equation]

Number of Terms(n) = **9**

b) **Method 2 [Using specified formula]**

Step 1: Formula. Number of Terms(n) = $\frac{Log\left[\frac{a_n}{a_1}\right]}{Log\,r} + 1$

Where; $a_n = a_6 = 1,280$, $a_1 = 5$ and $r = 2$

Step 2: Data substitution and solving

$$\text{Number of Terms(n)} = \frac{Log\left[\frac{1,280}{5}\right]}{Log\ 2} + 1 \ \text{.........} \ \text{...........[Simplify numerator fraction]}$$

$$\text{Number of Terms(n)} = \frac{Log[256]}{Log\ 2} + 1 \ \text{..[Work log]}$$

$$\text{Number of Terms(n)} = \frac{2.408239965}{0.301029995} + 1 \ \text{...... ......[Simplify fraction and add value]}$$

$$\text{Number of Terms(n)} = 9$$

Example 6: The first three terms of a geometric progression are; 2, X, 18. Fine the value of X.

Solution

Step 1: Formula. $a_n = a_1 \times r^{n-1}$, where; $a_n = a_3 = 18$, $a_1 = 2$, $n = 2$ and $r = \frac{X}{2}$

Step 2: Data substitution and solving

$$18 = 2 \times \left(\frac{X}{2}\right)^{3-1} \ \text{...........[Simplify exponent and divide both sides of equation by (2)]}$$

$$9 = \left(\frac{X}{2}\right)^2 \ \text{...[Open bracket]}$$

$$9 = \frac{X^2}{2^2} \ \text{...[Work equation denominator exponent]}$$

$$9 = \frac{X^2}{4} \ \text{...[Multiply both sides of the equation by (4)]}$$

$$X^2 = 36 \ \text{............................[Square root both sides of equation and solve equation]}$$

$$\text{Second Term (X)} = 6 \ \text{...........................[The common ratio is therefore equal to 3]}$$

Example 7: The 5th term of the geometric progression is 405 and the 8th term is 10,935. Determine the value of the first term and the common ratio.

Solution

Step 1: Determination of equations

  a) Using the 5th term data

  1) Formula. $a_n = a_1 \times r^{n-1}$, where; $a_n = a_5 = 405$ and $n = 5$

  2) Data substitution and solving

  $$405 = a_1 \times r^{5-1} \ \text{......................... ....[Simplify equation exponent]}$$

  $$405 = a_1 \times r^4 \ \text{...[Re-arranged equation]}$$

  $$a_1 r^4 = 405 \ \text{... ....[Equation 1]}$$

  b) Using the 8th term data

  1) Formula. $a_n = a_1 \times r^{n-1}$, where; $a_n = a_5 = 10,935$ and $n = 8$

  2) Data substitution and solving

  $$10,935 = a_1 \times r^{8-1} \ \text{.....................................[Simplify equation exponent]}$$

  $$10,935 = a_1 \times r^7 \ \text{....................... ....[Re-arranged equation]}$$

251

$$a_1 r^7 = 10{,}935 \dots\dots\dots\dots\dots\dots\dots\dots\dots\dots\dots\dots[\text{Equation 2}]$$

Step 2: Determination of first term and common ratio

$$a_1 r^4 = 405 \dots\dots\dots\dots\dots\dots\dots\dots\dots\dots\dots\dots\dots\dots\dots(1)$$

$$a_1 r^7 = 10{,}935 \dots\dots\dots\dots\dots\dots\dots\dots\dots\dots\dots\dots\dots(2)$$

Note: To determine the values of (a_1) and (r) we divide equation (2) by equation (1). To perform this division, we make use of the concept of power.

$$\frac{a_1 r^7}{a_1 r^4} = \frac{10{,}935}{405} \dots\dots\dots\dots\dots[\text{Re-arranged the left hands side of the equation}]$$

$(a_1{}^{1-1})(r^{7-4}) = \frac{10{,}935}{405}$.[Simplify left hand side exponents and simplify fraction]

$$r^3 = 27 \dots\dots\dots\dots\dots\dots\dots\dots\dots\dots\dots[\text{Root 3 both sides of the equation}]$$

Common Ratio$(r) = \sqrt[3]{27}$ ....[Work root 3]

Common Ratio$(r) = \mathbf{3}$

Note: The value of first term (a_1) is derived by substituting the numerical value of common ratio (3) in any of the equation or formula of nth terms seen above.

Using the simultaneous technique [Equation 1]	Using the formula technique
1) Formula. $a_1 r^4 = 405$, where; $r = 3$	1) Formula. $a_n = a_1 \times r^{n-1}$, where; $r = 3$
2) Data substitution and solving	2) Data substitution and solving
$a_1(3^4) = 405$ ....[Work exponent]	$405 = a_1 \times 3^{5-1}$ ......[Work exponent]
$a_1(81) = 405$.[Divide both sides by 81]	$405 = a_1(81)$ ..[Divide both sides by 81]
First Term $(a_1) = \mathbf{5}$	First Term $(a_1) = \mathbf{5}$

Test Your Understanding

Exercise 26: Find the 4th term of a geometric progression using the series; 5, 15, 45, 135, 405...[**Answer: $a_4 = 135$**]

Exercise 27: Find the 5th term of a geometric progression given that $a_1 = 1$ and $r = 1/2$.

[**Answer: $a_5 = 1/16$**]

Exercise 28: The 5th term of a geometric progression is 2,500. Given that the common ratio is 5, determine the first term...[**Answer: $a_1 = 4$**]

Exercise 29: The 6th term of a geometric progression is 200,000. Assuming that the first term is 2, determine the common ratio..[**Answer: r = 10**]

Exercise 30: The nth term of a geometric progression is 98,415 and the common ratio is 3. Determine the number of terms given that the first term is 5....... ....[**Answer: n = 10**]

Exercise 31: The 4th term of a geometric progression is 3 and the 6th term is $\frac{3}{4}$. Determine the value of the 2nd term..[**Answer: $a_2 = 12$**]

6.2.3 The Sum of n Terms of Geometric Progression

The sum of n terms of a geometric progression is also known as the geometric series. Given that (a_1) is the first term, (n) is the number of terms and (r) the common ratio, the general formula for a geometric series is given as seen below.

$$S_n = \frac{a_1[1-r^n]}{1-r} \quad \text{...(1)}$$

$$S_n = \frac{a_1[r^n-1]}{r-1} \quad \text{...(2)}$$

Note: Generally, equation (1) is used when $(|r| < 1)$ or $(r \neq 1)$ and equation (2) is used when $(|r| > 1)$. For simplicity, equation (1) is widely used in the calculation of sum of n term of a geometric progression.

Example 1: Find the sum of 5 terms in a geometric progression given that the first term is 2 and the common ratio is 2.

<u>**Solution**</u>

Step 1: Formula. $S_n = \frac{a_1[r^n-1]}{r-1}$, where; $S_n = S_5$, $a_1 = 2$, $n = 5$ and $r = 2$

Step 2: Data substitution and solving

$$S_5 = \frac{2[2^5-1]}{2-1} \quad \text{...[Work numerator exponent]}$$

$$S_5 = \frac{2[32-1]}{2-1} \quad \text{...[Simplify numerator and denominator]}$$

$$S_5 = \frac{2[31]}{1} \quad \text{...[Open bracket and solve fraction]}$$

$$S_5 = 62$$

Example 2: The sum of four terms in a geometric progression is -40 and the common ratio is -3. Find the value of the first term.

Solution

a) Method 1 [**Using the n sum of terms formula**]

Step 1: Formula. $S_n = \frac{a_1[r^n - 1]}{r-1}$, where; $S_n = S_4 = -40$, n = 4 and r = −3

Step 2: Data substitution and solving

$-40 = \frac{a_1[-3^4 - 1]}{-3-1}$ ....[Work numerator exponent]

$-40 = \frac{a_1[81-1]}{-3-1}$ ....[Simplify numerator and denominator]

$-40 = \frac{a_1[80]}{-4}$ ....[Multiply both sides of equation by (−4)]

$a_1[80] = 160$ ....[Divide both sides of equation by 80]

First Term $(a_1) = $ **2**

b) **Method 2 [Using a specified formula]**

Step 1: Formula. First Term $(a_1) = \frac{S_n(r-1)}{r^n - 1}$, where; $S_n = S_4 = -40$, n = 4 and r = −3

Step 2: Data substitution and solving

First Term $(a_1) = \frac{-40(-3-1)}{-3^4-1}$ ....[Work denominator exponent]

First Term $(a_1) = \frac{-40(-3-1)}{81-1}$ ....[Simplify numerator and denominator]

First Term $(a_1) = \frac{-40(-4)}{80}$ ....[Open numerator bracket]

First Term $(a_1) = \frac{160}{80}$ ....[Solve fraction]

First Term $(a_1) = $ **2**

Example 3: The sum of n terms in a geometric progression is 3,069. Find the number of terms given that the first term is 3 and the common ratio is 2.

Solution

a) **Method 1[Using the n sum of terms formula]**

Step 1: Formula. $S_n = \frac{a_1[r^n - 1]}{r-1}$, where; $S_n = 3,069$, $a_1 = 3$ and r = 2

Step 2: Data substitution and solving

$3,069 = \frac{3[2^n - 1]}{2-1}$ ....[Simplify denominator]

$3,069 = \frac{3[2^n - 1]}{1}$ ....[Multiply both sides of equation by (1)]

$3,069 = 3[2^n - 1]$ ....[Divide both sides of equation by (3)]

$2^n - 1 = 1,023$ ....[Collect like terms together]

$2^n = 1,024$ ....[Log both sides of the equation]

n Log 2 = Log 1,024 ....[Divide both sides of equation by Log 2]

Number of Terms (n) $= \frac{\text{Log } 1,024}{\text{Log } 2}$ ....[Work log and solve fraction]

Number of Terms (n) $= 10$

b) **Method 2 [Using a specified formula]**

Step 1: Formula. Number of Terms (n) $= \frac{\text{Log}\left[\frac{S_n(r-1)}{a_1}+1\right]}{\text{Log } r}$, where; $S_n = 3,069$, $a_1 = 3$ and $r = 2$

Step 2: Data substitution and solving

Number of Terms (n) $= \frac{\text{Log}\left[\frac{3,069(2-1)}{3}+1\right]}{\text{Log } 2}$ ....[Simplify numerator fraction]

Number of Terms (n) $= \frac{\text{Log}[1,023+1]}{\text{Log } 2}$ ....[Simplify numerator bracket]

Number of Terms (n) $= \frac{\text{Log } 1,024}{\text{Log } 2}$ ....[Work log]

Number of Terms (n) $= \frac{3.010299957}{0.301029995}$ ....[Solve fraction]

Number of Terms (n) $= 10$

Test Your Understanding
Exercise 32: Find the sum of 7 terms of a geometric progression using the sequence; 2, 10, 50,...[Answer: $S_7 = 39,062$]
Exercise 33: Find S_5, given that $a_1 = 1/2$ and $r = 1/2$.................[Answer: $S_5 = 31/32$]
Exercise 34: The sum of 5 terms in a geometric progression is 12 and the common ratio is 3. Determine the value of the first term...[Answer: $a_1 = 1$]
Exercise 35: The sum of n terms is $363/243$ and first term is 1. Given that the common ratio is 1/3, determine the number of terms...[Answer: $n = 5$]
Exercise 36: The sum of 3 terms in a geometric progression is 124 and the sum of 6 terms is 15,624. Determine the value of the sum of 5 terms, given that the common ratio is 5...[Answer: $S_n = 3,124$]

6.2.4 Combination of nth term and sum of n terms of a geometric progression

We may have situations in which data from the nth term of a geometric progression and the sum of n terms are given to determine other components.

Example 1: How many terms in a geometric progression 4, 8, 16, 32,..... are needed so that the sum exceeds 35?

Solution

Step 1: Formula. $S_n = \frac{a_1[r^n-1]}{r-1}$, where; $a_1 = 4$, $r = 2$ and $S_n > 35$

Step 2: Data substitution and solving

$\frac{4(2^n-1)}{2-1} > 35$ ....[Simplify numerator and open numerator bracket]

$\frac{8^n-4}{1} > 35$ ....[Multiply both sides of equation by (1)]

$8^n - 4 > 35$ ....[Add (4) on both sides of the equation]

$8^n > 39$ ....…........[Log both sides of equation]

$n\, Log\, 8 > Log\, 39$ ....…...[Make (n) subject of the formula]

Number of Terms (n) $> \frac{Log\, 39}{Log\, 8}$ ....…..........[Work log and simplify fraction]

Number of Terms (n) > 2 ....…..........**[Rounded value to no decimal places]**

Step 3: Justification

Hint: The value (n > 2) means it is the sum of 3 or more terms that will be greater than 35. To justify this, we solve for sum of terms using 1 term and 2 terms (values less or equal to 2) and verify if their sum is equal or greater than 35. We maintain the value of the first term (4) and common ratio (2).

Using 1 term (n = 1)	Using 2 terms (n = 2)
$S_n = \frac{4[2^1-1]}{2-1}$ ....[Work exponent]	$S_n = \frac{4[2^2-1]}{2-1}$ ....[Work exponent]
$S_n = \frac{4[2-1]}{2-1}$ ....[Simplify]	$S_n = \frac{4[4-1]}{2-1}$ ....…......[Simplify]
$S_n = \frac{4}{1}$ ....…...[Solve]	$S_n = \frac{12}{1}$ ....…..............[Solve]
$S_n = 4$	$S_n = 12$
Note: Their values (4 and 12) are all less than 35. This indicates that only number of terms greater than 2 can generate a value greater than 35.	

Example 2: Given that the third term of a geometric progression is 32 and the fifth term is 512, find the value of first six terms.

Solution

Step 1: Formula. $S_n = \frac{a_1[r^n-1]}{r-1}$, where; $S_n = S_6$, n = 6 a_1 = Unknown and r = Unknown

Step 2: Determination of unknown

Hint: Use the term of geometric progression concept to determine individual term equation from which the first term and common ratio value will be derived.

1) Derivation of equations

Using third term data	Using fifth term data
1) Formula. $a_n = a_1 \times r^{n-1}$	1) Formula. $a_n = a_1 \times r^{n-1}$
Where; $a_n = a_3 = 32$ and $n = 3$	Where; $a_n = a_5 = 512$ and $n = 5$
2) Data substitution and solving	2) Data substitution and solving
$32 = a_1 \times r^{3-1}$ ....[Simplify exponent]	$512 = a_1 \times r^{5-1}$ ....[Simplify exponent]
$32 = a_1 \times r^2$ ....[Multiply values]	$512 = a_1 \times r^4$ ....[Multiply values]
$32 = a_1 r^2$ ....:[Re-arranged equation]	$512 = a_1 r^4$ ....:[Re-arranged equation]
$a_1 r^2 = 32$ ....**[Equation 1]**	$a_1 r^4 = 512$ ....**[Equation 2]**

2) Solving of equations to derive unknowns

$$a_1 r^2 = 32 \dots\dots\dots\dots\dots\dots\dots\dots\dots\dots\dots\dots\dots\dots\dots\dots\dots(1)$$

$$a_1 r^4 = 512 \dots\dots\dots\dots\dots\dots\dots\dots\dots\dots\dots\dots\dots\dots\dots\dots(2)$$

Note: Divide equation (2) by equation (1), will help eliminate one unknown, giving the possibility to determine the value of the other unknown.

$\frac{a_1 r^4}{a_1 r^2} = \frac{512}{32}$ ....[Re-arranged left hand side of equation]

$(a_1{}^{1-1})(r^{4-2}) = \frac{512}{32}$ ....[Simplify left hand side exponent]

$(a_1{}^0)(r^2) = \frac{512}{32}$ ....[Simplify left hand side of equation. Remember $(a^0 = 1)$]

$r^2 = \frac{512}{32}$ ....[Simplify fraction and square root both sides of equation]

Common Ratio (r) $= \sqrt{16}$ ....[Work square root]

Common Ratio (r) $= $ **4**

Note: To obtain the value of the first term (a_1), we simply substitute the numerical value of the common ratio $(r = 4)$ in any of the equations above

Using equation 1	Using equation 2
$a_1 r^2 = 32$ ....[Substitute r with 4]	$a_1 r^4 = 512$ ....[Substitute r with 4]
$a_1(4)^2 = 32$ ....[Work exponent]	$a_1(4)^4 = 512$ ....[Work exponent]
$a_1 16 = 32$ [Divide both sides by 16]	$a_1 256 = 512$ [Divide both sides by 256]
First Term$(a_1) = $ **2**	First Term$(a_1) = $ **2**

Step 3: Data substitution and solving $[n = 6 \; a_1 = 2 \text{ and } r = 4]$

Sum of First Six Terms $(S_6) = \frac{2[4^6 - 1]}{4-1}$ ....[Work exponent and simplify equation]

Sum of First Six Terms $(S_6) = \frac{2[4,095]}{3}$ ....[Open bracket and solve fraction]

Sum of First Six Terms $(S_6) = 2,730$

PART TWO

STATISTICS

CHAPTER SEVEN

THE CONCEPT OF STATISTICS

7.1 Meaning of Statistics

Statistics is defined as a scientific and systematic method and procedure used to facilitate the collection, classification, organization, summarization, presentation, transformation, analysis, and interpretation of data for the purpose of making better decisions and derivation of valid conclusions or generalization. Generally statistics is used for three purposes or functions,

a) Condensation Function: statistics is used to reduce complexity or huge data to a simple and understanding level.

b) Comparison Function: Statistics is used to measure and exposed the similarities and differences of elements under study.

c) Forecasting Function: Statistics through estimation and prediction is used to tell with high level of certainty the trend of event or element under study.

7.2 Important Concept of Statistics

7.2.1 Statistical Data

Data is a set of unprocessed observations, values, elements or objects under consideration. It is also seen as a representation of facts, concepts or instructions in a formalized manner suitable for communication, interpretation, or processing by humans or machine. It is worth knowing that data becomes information only after it is been processed.

a) **Types of Data:** The three major types of data are; time series data, cross sectional data and panel data. It is worth knowing that for simplicity, data can be sub-divided based on specific criteria such as; measurability (quantitative and qualitative data), Collection (Primary and secondary data) and values limit (discrete and continuous data).

1) **Primary Data:** These are data collected directly from the source or population/sample. It is considered as first hand collected data or data that has not been used before. For example data collected through questionnaire, interview and discussion.

2) **Secondary Data:** These are data collected from an already used data. It is considered as seconded handed data or data collected from already existing data base. For example data collected from data register/ documents and web sides.

3) **Qualitative (Categorical) Data:** These are data that are none-numerical in nature and capture feelings, believes, perception and opinion. Qualitative data can be sub-divided into; nominal data and ordinal data. For example data on gender, occupation, religion, political party etc.

4) **Quantitative Data:** These are data that are numerical or measurable in nature. Major examples of quantitative data include Gross National Product (GNP) data, Gross Domestic Product (GDP) data, Number of Birth and death data, Net export data etc.

Note: The idea of discrete and continuous data will be seen below. Time series data are data collected over time in a fixed interval period. The quantity of cocoa production of Cameroon from 2000 to 2018 is an example of time series data. Cross sectional data are data collected over an area in a given time period. The quantity of cocoa production in Meme division is an example of cross sectional data.

b) **Quantity of data**: The quantity of data depends on the data source. Data can be collected from two major sources; population or sample.

1) Population. A population or universe is defined as the entire group of observations or items from which data is to be collected.

2) Sample is defined as a fraction of the population from which data is to be collected and result generalized to the general population. The process of drawing a required sample from a population is called sampling and the list of the entire population from which the sample is drawn is called sample frame.

Note: Generally, data from the population are of higher quality than data from a sample. The quality of data from a sample depends on the sampling technique, researcher intellectual ability, and financial and human resources availability. Data from the population is of higher quality but is also costly and time consuming.

c) **Data Collection**: Data collection is defined as a scientific and systematic gathering of facts from a variety but sources of interest. Data can be collected through; questionnaire, interview, discussion, observation, documents, records/tapes, and internet.

d) **Data Organization and Summarization:** Data organization and summarization reduces data complexity and ease data understanding and interpretation and can be performed using tabular, mathematical and graphical techniques.

7.2.2 Statistical Analysis

Statistical analysis is defined as the processing of data with the help of statistical software or package to obtain information, draw conclusion and make decision. The two major method of analyzing data include; descriptive statistical method and inferential statistical method and to an extend probability method.

a) **Descriptive Analysis:** It involves the utilization of numerical (tabulation) and graphical (graphs and chart) means to organize, summarize and present data to reveal data pattern and other important characteristic of interest in a meaningful manner as observed. That is the construction of graphs, charts, and tables with data collected. Major mathematical concepts applied in descriptive statistics include; measures of central tendency, measures of variation, and percentiles.

b) **Inferential (Inductive) Analysis:** It is the process of using sample data to make estimates and test hypotheses about the characteristics of the population. Inferential statistics uses sample data to make estimates, decisions and conclusion, predictions and generalizations about the population. Inferential analysis examines the correction (correction analysis) and cause-effect (regression analysis) between elements of study which are not ignored in descriptive statistics.

Note: Descriptive and inferential statistics are interrelated as one is required to use descriptive statistics to organize and summarize data obtained before applying inferential statistics.

7.3 Methods of Data Representation

It is statistically necessary and significant to summarize analysis and present collected data into visual form for easy understanding and interpretation. Data presentation is defined as the visual organization, classification and summarization of collected data. Statistical data can be presented using two main techniques; that is tables and graphs.

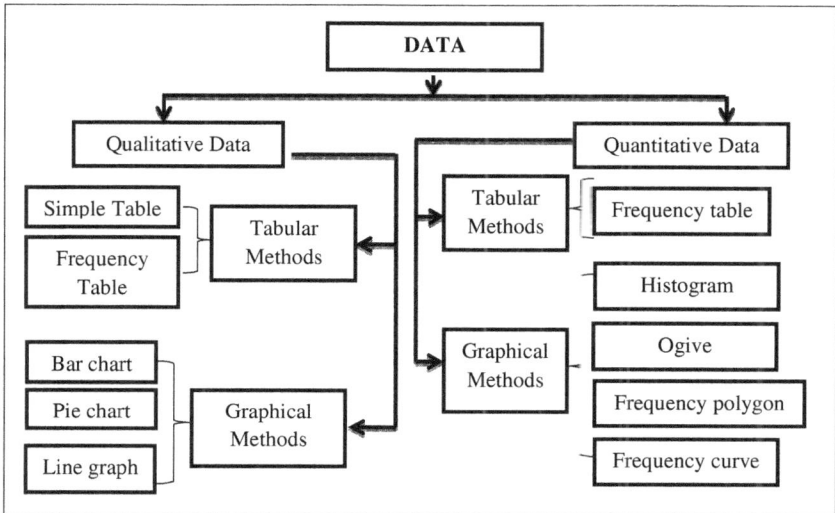

7.3.1 Tabular Methods of Data Representation

Tabulation is the process of summarizing and arranging collected data into columns and rows for easy understanding and interpretation and for quick identification of desired information. Tabulation reduces data complexity, facilitate comparison of related facts or issues and calculation of various statistical measures. Major types of tabulation include; simple table, cross tabulation and frequency table, but our focus will be limited to frequency table (frequency distribution table).

A frequency table or frequency distribution table is a table that shows individual or group (classes or intervals) of collected data with their respective count or frequency.

7.3.1.1 Important Concepts of Frequency Table

1) Number of Class intervals

It is the number of segments a data is sub- divided. Generally, the number of class interval should be between 5 and 20, class interval should be continuous (intermediate intervals without observation must be included), class interval must be exhaustive (class intervals must be enough to accommodate the entire data) and class interval should have equal width or size. Class intervals can take three major forms;

a) **Exclusive Class Interval:** It is a class interval in which the upper limit of the preceding class is also the lower limit of the succeeding class.

263

b) **Inclusive Class Interval:** It is a class interval in which the preceding class upper limit is not considered as the lower limit of the succeeding class.

c) **Open-ended class Interval:** It is a class interval in which the lower limit of the first class and the upper limit of the last class are unknown.

Exclusive class interval			Inclusive class interval			Open – ended class interval		
a − b	b − c	c − d	a − b	c − d	e − f	Below − b	b − c	c − above
Exist over-lap			Exist no over-lap			Unknown class limits (some)		

Note: In a given problem, the class interval can be given as well as can't be given. When the class interval is not given, we are required to determine the number of required class interval to include all the data before obtaining the class interval.

Number of class interval(K) = 1 + 3.322 [Log n] ………………………………....………(1)

Number of class inetval = $2^K \geq n$ …………………………………………………....…………(2)

It is worth knowing that (n) is the data set size and the value obtained must be rounded to the nearest whole number. Equation (2) is good to apply when "n" is small and is complex and time consuming when "n" is large as the try-and-error concept is used in determining the value of K. Therefore, equation (1) is advisable to use than equation (2).

Example 1: Determine the number of class interval from the given data distributions below.

12	14	19	18	15	15	18	17	20	27
22	23	22	21	33	28	14	18	16	13

<u>Solution</u>

Step 1: Formula and data. K = 1 + 3.322 [Log n] or $2^K \geq n$, where; n = 20

Step 2: Data substitution and solving.

Calculating the value of K using equation 1 {K = 1 + 3.322 [Log n]}

K = 1 + 3.322 [Log n] …………………………..[Substitute data into the formula]

K = 1 + 3.322 [Log 20] ……………………..…………………….[Log value (20)]

K = 1 + 3.322 [1.301029996] ………………..[Expand brackets by multiplication]

K = 1 + 4.322021646 ……………………………[Simplify equation by addition]

K = 5 …………………………………………..……………………..[Rounded value]

Calculating the value of K using equation 2 {$2^K \geq n$. That is $2^K > n$ or $2^K = n$ }

Step 1: Assumption. Let K ranges from 0, 1, 2, 3, 4, 5, 6....................the value of n (20)
Step 2: Substitute K range values into the formula and compare to n value

$2^0 = 1$ ....[1 < 20]	$2^3 = 8$ ....[8 < 20]
$2^1 = 2$ ....[2 < 20]	$2^4 = 16$ ....[16 < 20]
$2^2 = 4$ ....[4 < 20]	$2^5 = 32$ ....[32 > 20]

Step 3: Determine the value of K. **K = 5 since 32 is greater than 20.**

2) Class width or class Size

It is the number of element or members included in a given class. Mathematically, it is the difference between the upper limit and lower limit of a given class. When dealing with exclusive class interval, the real members includes the lower limit and the last number before the upper limit value. When dealing with inclusive class interval, the real members of the class includes both lower and upper limits of the class.

	Exclusive Class Interval		Inclusive Class Interval	
Class interval	$a - d$	$d - g$	$a - e$	$e - h$
Real class members	a, b and c	d, e and f	a, b, c and d	e, f, g and h

Note: When required to determine the class width or size, we are first demanded to determine the number of class interval. Class size is calculated using the formula below.

$$\text{Class Size} = \frac{\text{Highest Data Value} - \text{Lowest Data Value}}{\text{Number of Class Interval}} \text{..(3)}$$

Example 1: Compute the class size of the following data distribution if represented in a grouped frequency distribution table, assuming the number of class interval is 5.

14	14	13	15	11	15	10	20	4	13
2	4	0	20	10	13	8	19	16	7

Solution

Step 1: Formula and data. $\text{Class Size} = \frac{\text{Highest Data Value} - \text{Lowest Data Value}}{\text{Number of Class Interval}}$

Where; Highest Data Value = 20, Lowest Data Value = 0, Number of class interval = 5

Step 2: Data substitution and solving.

$$\text{Class Size} = \frac{\text{Highest Data Value} - \text{Lowest Data Value}}{\text{Number of Class Interval}} \text{.....[Substitute data into the formula]}$$

$$\text{Class Size} = \frac{20 - 0}{5} \text{....................[Simplify numerator and divide by denominator]}$$

Class Size = **4**

Step 3: Interpretation. The value 4 means a class interval should contain not less or more than 4 data values. That is the difference between the lower and upper limit must be equal to 4.

3) Class limits

After determining the number of class interval and class width, the next task is to identify the elements of a given class. Class limit is defined as the lowest and highest element of a given class interval. The lowest element or value is known as the lower class limit and the highest element or value is called upper class limits. The formula applied to determine class limits (lower and upper class limits) depends on the form of class interval in question.

a) **Exclusive Class Interval Situation:** Here, the first lower limit is by default the lowest data value and succeeding lower limit(s) is by default the upper limit value of their preceding class. The upper limit of a given class (say class "n") is calculated using the formula below.

$$\text{Upper Limit}_n = \text{Lower Limit}_n + (\text{Class size} + 1) \dots\dots\dots\dots\dots\dots\dots\dots(1)$$

b) **Inclusive Class Interval Situation:** Here, the first lower limit is the smallest data value and other lower and upper limits are calculated using the formulas below.

$$\text{Lower Limit}_n = \text{Upper Limit}_{n-1} + 1 \dots\dots\dots\dots\dots\dots\dots\dots\dots\dots(2)$$
$$\text{Upper Limits}_n = \text{Lower Limit}_n + \text{Class size} \dots\dots\dots\dots\dots\dots\dots\dots(3)$$

Example 1: Determine the value of lower and upper limits of a grouped frequency distribution table, assuming the number of class interval to be 5 and class size to be 4.

0	1	1	2	3	5	7	8	8	9
10	12	15	15	17	19	19	20	20	20

Solution

Step 1: Draft Work. Draw an empty class interval frame respecting the value of the number of class. Since the given number of class interval is 5, it means our frame will be made up of five rows.

Step 2: Calculation of lower and upper limits values.

Classes	Continuous grouped frequency situation		Continuous grouped frequency situation	
	Remember: The first lower limit by default is the smallest value of data in a distribution			
	Lower Limits $[LL_n = UL_{n-1}]$	Upper Limits $[UL = LL + (CS + 1)]$	Lower Limits $[LL_n = UL_{n-1} + 1]$	Upper Limits $[UL = LL + CS]$
1	**0** [Smallest value]	$0 + (4 + 1) = 5$	**0** [Smallest value]	$0 + 4 = 4$
2	$UL_{2-1} = UL_1 = 5$	$5 + (4 + 1) = 10$	$4 + 1 = 5$	$5 + 4 = 9$
3	$UL_{3-1} = UL_2 = 10$	$10 + (4 + 1) = 15$	$9 + 1 = 10$	$10 + 4 = 14$
4	$UL_{4-1} = UL_3 = 15$	$15 + (4 + 1) = 20$	$14 + 1 = 15$	$15 + 4 = 19$
5	$UL_{5-1} = UL_4 = 20$	$20 + (4 + 1) = 25$	$19 + 1 = 20$	$20 + 4 = 24$
	Note: LL = Lower limit, UL = Upper limit, CS = = Class size			

Note: It is worth knowing that the above table can be expressed horizontally. To do this, we apply signs such as "[" and "]", where **"[a"** means "a" is included, **"a["** means "a" is excluded, **"a]"** means "a" is included. The above table can be expressed horizontally as seen below.

Situation of grouped frequency distribution table that is continuous					
Class Interval	$[0 - 5[$	$[5 - 10[$	$[10 - 15[$	$[15 - 20[$	$[20 - 25[$
True Members	0 to 4	5 to 9	10 to 14	15 to 19	20 to 24
Situation of grouped frequency distribution table that is non-continuous					
Class Interval	$[0 - 4]$	$[5 - 9]$	$[10 - 14]$	$[15 - 19]$	$[20 - 24]$
True Members	0 to 4	5 to 9	10 to 14	15 to 19	20 to 24

4) Midpoints or Class Mark

It is the sum of lower and upper limit values divided by 2. The formula used for the calculation of class midpoints or class mark denoted [x] is given as;

$$\text{Midpoints}[x_i] = \frac{\text{Lower Limit} + \text{Upper Limit}}{2} \quad \ldots\ldots\ldots\ldots\ldots\ldots\ldots\ldots\ldots\ldots\ldots\ldots\ldots(1)$$

Note: The difference between two consecutive midpoints must be equal to the difference between upper limit and lower limit or class size.

Example 1: Use the class interval of a grouped continuous frequency distribution table below and determine the midpoints of each class.

267

Class interval	$[0-5[$	$[5-10[$	$[10-15[$	$[15-20[$

Step 1: Formula. Midpoints$[x_i] = \frac{\text{Lower Limit+Upper Limit}}{2}$

Step 2: Data substitution and solving.

Class Interval	$[0-5[$	$[5-10[$	$[10-15[$	$[15-20[$
Midpoints	$\frac{0+5}{2} = 2.5$	$\frac{5+10}{2} = 7.5$	$\frac{10+15}{2} = 12.5$	$\frac{15+20}{2} = 17.5$

5) Class boundary

The idea of class boundary is applied to change an inclusive class interval table (non-continuous table) to an exclusive class interval table (continuous table). It helps in eliminating any overlapping class limit. The formula for calculating class boundary is given as;

$$\text{Lower Class Boundary} = \text{Lower Limit} - \frac{d}{2} = \text{Lower Limit} - 0.5 \ldots\ldots\ldots\ldots\ldots(1)$$

$$\text{Upper Class Boundary} = \text{Upper Limit} + \frac{d}{2} = \text{Upper Limit} + 0.5 \ldots\ldots\ldots\ldots\ldots(2)$$

Where: **"d"** is the difference between two consecutive class limits

Example 1: Compute the class boundary of the following classes using the frequency table below

Class Interval	$[1-9]$	$[10-19]$	$[20-29]$	$[30-39]$

Solution

Step 1: Formula and data. Lower Class Boundary = Lower Limit − 0.5

Upper Class Boundary = Upper Limit + 0.5

Step 2: Data substitution and solving.

Class	LCL = Lower Limit − 0.5		UCL = Upper Limit + 0.5	
	Lower Limits	Lower Class Boundary	Upper Limits	Upper Class Boundary
1	1	$1 - 0.5 = 0.5$	9	$9 + 0.5 = 9.5$
2	10	$10 - 0.5 = 9.5$	19	$19 + 0.5 = 19.5$
3	20	$20 - 0.5 = 19.5$	29	$29 + 0.5 = 29.5$
4	30	$30 - 0.5 = 29.5$	39	$39 + 0.5 = 39.5$

Step 3: Replacing class interval with class boundary.

Class Interval	$[1-9]$	$[10-19]$	$[20-29]$	$[30-39]$
Class Boundary	$[0.5-9.5]$	$[9.5-19.5]$	$[19.5-29.5]$	$[29.5-39.5]$

6) Frequency

Frequency is defined in ungrouped data situation as the number of time a particular element or number occurs. In a grouped data situation, frequency is defined as the number of values or members in a specified class interval of a distribution.

Example 1: Use the following numbers and determine their respective frequencies

1	5	3	6	8	2
8	6	1	4	5	4
7	5	7	2	5	2

<u>Solution</u>

Step 1: Verification. Our distribution is not in order. Thus, we are required to put it in order by re-arranging the values in increasing or decreasing order before any further working.

1	1	2	2	2	3
4	4	5	5	5	5
6	6	7	7	8	8

Step 2: Determination of frequency. From the arranged table, we simply count the number of time a given number appears (tally) and note it as its frequency.

Numbers $[x_i]$	1	2	3	4	5	6	7	8	Total [n]
Tally	II	III	I	II	IIII	II	II	II	
Frequency $[f_i]$	2	3	1	2	4	2	2	2	18

Example 2: Use the distribution below and answer the following questions

8	9	0	23	4	8	12	2	15	3
4	20	1	18	0	5	9	14	8	3
16	24	2	17	8	2	0	6	21	7

Compute the frequency using the suggested class interval,

a) $[0-5[, [5-10[, [10-15[, [15-20[,$ and $[20-25[$.

b) $[0-5], [6-11], [12-17], [18-23]$ and $[24-29]$.

Solution

Step 1: Verification. Our distribution is not in order; therefore, we are required to re-arrange them to make it orderly.

0	0	0	1	2	2	2	3	3	4
4	5	6	7	8	8	8	8	9	9
12	14	15	16	17	18	20	21	23	24

Step 2: Determination of frequency. Here, we count the number of numbers or data that falls in class intervals or between lower and upper limit of each class.

Question a frequency determination			Question b frequency determination		
Class	Real members	Frequency	Class	Real members	Frequency
$[0-5[$	$0-4$	11	$[0-5]$	$0-5$	12
$[5-10[$	$5-9$	9	$[6-11]$	$6-11$	8
$[10-15[$	$10-14$	2	$[12-17]$	$12-17$	5
$[15-20[$	$15-19$	4	$[18-23]$	$18-23$	4
$[20-25[$	$20-25$	4	$[24-29]$	$24-29$	1
		$\sum f_i = 30$			$\sum f_i = 30$

Note: Frequency is sub-divided in to many types. The major types of frequency includes; relative frequency, cumulative frequency.

a) **Relative frequency:** It measures the portion of a given frequency in the total frequency. Mathematically, it is the frequency of a given class divided by the total frequency. The sum of relative frequency is equal to one (1) and when expressed in percentage it is equal to a hundred (100). The formula for relative frequency and percentage relative frequency is given as;

$$\text{Relative frequency} = \frac{\text{Individual frequency}(f_i)}{\text{Sum of frequency}(\sum f_i)} \quad \text{.......................................(1)}$$

$$\text{Percentage relative frequency} = \frac{\text{Individual frequency}(f_i)}{\text{Sum of frequency}(\sum f_i)} (100) \quad \text{....................(2)}$$

Example 1: Use the frequency distribution table below to derive a relative frequency and percentage relative frequency distribution tables.

Marks	12	14	16	18	20
Frequency	10	5	2	14	9

Solution

Step 1: Formula and data. Relative frequency $= \frac{\text{Individual frequency}(f_i)}{\text{Sum of frequency} \sum f_i}$ ....(1)

$$\text{Percentage relative frequency} = \frac{\text{Individual frequency}(f_i)}{\text{Sum of frequency}(\sum f_i)} (100) \ ......(2)$$

Where; $\sum f_i = 40$. That is $[10 + 5 + 2 + 14 + 9 = 40]$

Step 2: Data substitution and tabular derivation of relative frequency

Original Table		Relative frequency	Percentage relative frequency
Marks	Frequency		
12	10	$10/40 = 0.25$	$10/40\ (100) = 0.25(100) = 25\%$
14	5	$5/40 = 0.125$	$5/40\ (100) = 0.125(100) = 12.5\%$
16	2	$2/40 = 0.05$	$2/40\ (100) = 0.05(100) = 5\%$
18	14	$14/40 = 0.35$	$14/40\ (100) = 0.35(100) = 35\%$
20	9	$9/40 = 0.225$	$9/40\ (100) = 0.225(100) = 22.5\%$
	$\sum = 40$	$\sum = 1$	$\sum = 100\%$

b) **Cumulative frequency (CF):** Cumulative frequency is derived either by adding frequency from top to bottom (cumulative frequency greater than) or from bottom to top (cumulative frequency less than). When a given cumulative frequency value is divided by the sum of cumulative frequency, it is called relative cumulative frequency (RCF). When the relative cumulative frequency is multiply by 100, it is known as percentage relative cumulative frequency (PRCF). Their respective formulas are seen below.

$CF_n = f_n + CF_{n-1}$.................................... ....(1)

$CF_n = CF_n + f_{n-1}$.................................... ....(2)

$RCF = \frac{\text{Given cumulative frequency}}{\text{sum of cumulative frequency}}$ ....(3)

$PRCF = \frac{\text{Given cumulative frequency}}{\text{sum of cumulative frequency}} (100)$ ... .(4)

Where: CF_{n-1} denotes preceding cumulative frequency and f_{n-1} denotes preceding frequency. f_n denotes current frequency and CF_n denotes current cumulative frequency. Equation (1) is used for cumulative frequency greater than and equation (2) for cumulative frequency less than.

Example1: From the frequency distribution table below, determine the cumulative frequency greater than and cumulative frequency less than.

Class	[5 − 10[	[10 − 15[	[15 − 20[	[20 − 25[
Frequency	24	10	15	6

Solution

Step 1: Formula and data

$$CF_n = f_n + CF_{n-1}..(1)$$

$$CF_n = CF_n + f_{n-1}..(2)$$

Step 2: Tabular derivation of cumulative frequency greater and less than.

Cumulative frequency greater than				Cumulative frequency less than			
Class	f_i	$f_n + CF_{n-1}$	CF	Class	f_i	$CF_n + f_{n-1}$	CF
[5 − 10[	24	24 + 0 = 24	24	[5 − 10[	24	31 + 24 = 55	55
[10 − 15[	10	10 + 24 = 34	34	[10 − 15[	10	21 + 10 = 31	31
[15 − 20[	15	15 + 34 = 49	49	[15 − 20[	15	6 + 15 = 21	21
[20 − 25[	6	6 + 49 = 55	55	[20 − 25[	6	0 + 6 = 6	6
Sum	55				55		

Example 2: Compute the cumulative frequency, relative cumulative frequency and percentage relative cumulative frequency from the frequency distribution table below.

Class Interval	[0 − 10[	[10 − 20[	[20 − 30[	[30 − 40[	[40 − 50[
Frequency	12	11	10	4	13

Step 1: Formula and data.

$$CF_n = f_n + CF_{n-1} \quad , \quad RCF = \frac{\text{Given cumulative frequency}}{\text{sum of cumulative frequency}}, \quad PRCF =$$

$$\frac{\text{Given cumulative frequency}}{\text{sum of cumulative frequency}}(100)$$

Where; sum of cumulative frequency $= 50$ and $f_n =$ Curent frequency

Step 2: Data substitution and tabula derivation of frequencies.

Original table		CF	RCF	PRCF
Class	f_i	$f_n + CF_{n-1}$	$\frac{\text{Given cumulative frequency}}{\text{sum of cumulative frequency}}$	$\frac{\text{Given cumulative frequency}}{\text{sum of cumulative frequency}}(100)$
$0-10[$	12	$12 + 0 = 12$	$^{12}/_{155} = 0.077419354$	$\frac{12}{155}(100) = 7.741935484\%$
$10-20[$	11	$11 + 12 = 23$	$^{23}/_{155} = 0.148387096$	$\frac{23}{155}(100) = 14.83870968\%$
$20-30[$	10	$10 + 23 = 33$	$^{33}/_{155} = 0.212903225$	$\frac{33}{155}(100) = 21.29032258\%$
$30-40[$	4	$4 + 33 = 37$	$^{37}/_{155} = 0.238709677$	$\frac{37}{155}(100) = 23.87096774\%$
$40-50[$	13	$13 + 37 = 50$	$^{50}/_{155} = 0.322580645$	$\frac{50}{155}(100) = 32.25806452\%$
		$\Sigma = 155$	$\Sigma = 1$	$\Sigma = 100\%$

7.3.1.2 Construction of frequency table

Here, we are going to use the above concepts to solve practica. problem. We are going to construct both grouped and ungrouped frequency distribution tables.

Example 1: Draw an ungroup frequency distribution table using the data below representing scores,

0	2	1	0	0	2	1	0	2	3
2	2	3	3	2	3	1	4	3	2
2	1	2	0	3	2	2	1	0	3
3	2	2	2	1	0	2	1	0	1

Step 1: Re-arrangement of data.

0	0	0	0	0	0	0	0	1	1
1	1	1	1	1	1	2	2	2	2
2	2	2	2	2	2	2	2	2	2
2	3	3	3	3	3	3	3	3	4
Actual observations: 0, 1, 2, 3, and 4									

Step 2: Derivation of frequencies [Horizontal illustration]

Scores	0	1	2	3	4	Total
Frequency	8	8	15	8	1	40

Example 2: Draw an ungroup distribution table with frequency using the data below showing marks of students.

1	3	10	7	5
4	3	2	3	1
2	8	7	8	7
2	5	4	4	9

Solution

Step 1: Re-arrangement of data.

Re-arrangement in increasing order					Re-arrangement in decreasing order				
1	1	2	2	2	10	9	8	8	7
3	3	3	4	4	7	7	5	5	4
4	5	5	7	7	4	4	3	3	3
7	8	8	9	10	2	2	2	1	1

Step 2: Derivation of frequencies [Vertical illustration]

Increasing Order		Decreasing Order	
Marks	Frequency	Marks	Frequency
1	2	10	1
2	3	9	1
3	3	8	2
4	3	7	3
5	2	5	2
7	3	4	3
8	2	3	3
9	1	2	3
10	1	1	2
	N = 20		**N = 20**

Example 3: Construct a group frequency distribution table from the following data presented below.

1	1	1	2	2	2	2
3	3	3	3	3	4	4
4	4	4	4	5	5	5
6	6	6	7	7	8	10
11	12	15	18	19	19	20

Solution

Step 1: Calculation of class interval number.

1) Formula and data. $K = 1 + 3.322$ (Log n). Where $n = 35$

2) Data substitution and solving.

$\quad K = 1 + 3.322$ (Log 35)………….....………………......[Work log 35]

$\quad K = 1 + 3.322(1.544068044)$ …...[Multiply 3.322 by the value of log 35]

$\quad K = 1 + 5.129394043$ ………....[Simplify by adding values and round up]

$\quad \textbf{K = 6}$ …………….....………….......[Mean our table will have 6 rows]

Step 2: Calculation of size or width of class interval.

1) Formula and data. Class Size $= \dfrac{Range}{Number\ of\ class\ interval}$ or $S = \dfrac{Highest\ value - lowest\ value}{K}$

$\quad$ Where; Highest value = 20, owest value = 1 and K = 6

2) Data substitution and solving.

$\quad$ Class Size $= \dfrac{Highest\ value - lowest\ value}{K}$ …....[Substitute data into the formula]

$\quad$ Class Size $= \dfrac{20-1}{6}$ …………….........[Simplify numerator and divide by 6]

$\quad$ Class Size = 3 ……....[Meaning the difference between class limits is 3]

Step 3: Derivation of class limits

Continuous frequency distribution table				None-continuous frequency distribution table			
Note: The first lower limit is by default the lowest value of distribution.							
Lower limits are the upper limits of the pre-upper limits. Upper limits is given as **Upper limit = lower limit + (1 + class size)**				The lower limits are the next higher value of the upper limits. Upper limits is given as **Upper limit = lower limit + class size**			
LL	UL	Class	Frequency	LL	UL	Class	Frequency
1	1 + (1 + 3) = 5	1-5	18	1	1 + 3 = 4	1-4	18
5	5 + (1 + 3) = 9	5-9	9	5	5 + 3 = 8	5-8	9
9	9 + (1 + 3) = 15	9-15	3	9	9 + 3 = 12	9-12	3

15	13 + (1 + 3) = 17	15-17	1	13	13 + 3 = 16	13-16	1
17	17 + (1 + 3) = 21	17-21	4	17	17 + 3 = 20	17-20	4
21	21 + (1 + 3) = 25	21-25	0	21	21 + 3 = 24	21-24	0
			N = 35				**N = 35**

Example 4: Use the distribution of data below and construct a group frequency distribution table. Assume a class interval of 5.

11	18	12	16	0	3	15	11	4	2
18	7	9	12	2	1	1	0	5	9
13	6	1	9	3	15	19	20	18	19
2	1	7	8	0	9	3	2	4	0

Solution

Step 1: Calculation of size or width of class interval.

1) Formula and data. Class Size $= \dfrac{\text{Range}}{\text{Number of class interval}}$ or $S = \dfrac{\text{Highest value} - \text{lowest value}}{K}$

   Where; Highest value $= 20$, owest value $= 0$ and $K = 5$

2) Data substitution and solving.

   Class Size $= \dfrac{\text{Highest value} - \text{lowest value}}{K}$ …………..[Substitute data into the formula]

   Class Size $= \dfrac{20 - 0}{5}$……………………......…[Simplify numerator and divide by 5]

   Class size $= \mathbf{4}$

3) Interpretation: The value 4 means that the difference between the upper and lower limits is 4

Step 2: Re-arrangement of distribution in increasing or decreasing order.

Note: Re-arrangement help one to easily identify class interval members and hence frequencies. It makes it easier for the identification of omission, double counting and repetitions of data.

Re-arrangement in increasing order										Re-arrangement in decreasing order									
0	0	0	0	1	1	1	1	2	2	**20**	19	19	18	18	18	16	15	15	13
2	2	3	3	3	4	4	5	6	7	12	12	11	11	9	9	9	9	8	7
7	8	9	9	9	9	11	11	12	12	7	6	5	4	4	3	3	3	2	2
13	15	15	16	18	18	18	19	19	**20**	2	2	1	1	1	1	0	0	0	**0**

Step 3: Derivation of class limits. Where LL represents lower limit and UL represents upper limit

Continuous frequency distribution table				None-continuous frequency distribution table			
Remember: The first lower limit is by default the lowest value of distribution.							
LL	UL	Class	Frequency	LL	UL	Class	Frequency
0	$0 + (1 + 4) = 5$	0 - 5	17	0	$0 + 4 = 4$	0 - 4	17
5	$5 + (1 + 4) = 10$	5 - 10	9	5	$5 + 4 = 9$	5 - 9	9
10	$10 + (1 + 4) = 15$	10 - 15	5	10	$10 + 4 = 14$	10 - 14	5
15	$15 + (1 + 4) = 20$	15 - 20	8	15	$15 + 4 = 19$	15 - 19	8
20	$20 + (1 + 4) = 25$	20 - 25	1	20	$20 + 4 = 24$	20 - 24	1
			N = 40				**N = 40**

Example 5: Use the data below and construct a group frequency distribution table. Assume the class interval of; $[10 - 20[, [20 - 30[, [30 - 40[, [40 - 50[, [50 - 60[$ and $[60 - 70[$.

69	30	42	12	40
20	55	67	19	15
15	68	13	55	34
44	19	59	27	66

Solution

Step 1: Re-arrangement of data.

Re-arrangement in increasing order					Re-arrangement in decreasing order				
12	13	15	15	19	**69**	68	67	66	59
19	20	27	30	34	55	55	44	42	40
40	42	44	55	55	34	30	27	20	19
59	66	67	68	**69**	19	15	15	13	**12**

Step 2: Derivation of frequencies. Remember the sign

Class Interval	Frequency	True members of each class intervals
$[10 - 20[$	6	$[10 - 20[\rightarrow 10,11,12,13,14,15,16,17,18,$ and 19.
$[20 - 30[$	2	$[20 - 30[\rightarrow 20,21,22,23,24,25,26,27,28$ and 29.
$[30 - 40[$	2	$[30 - 40[\rightarrow 30,31,32,33,34,35,36,37,38,$ and 39.
$[40 - 50[$	3	$[40 - 50[\rightarrow 40,41,42,43,44,45,46,47,48$ and 49.
$[50 - 60[$	3	$[50 - 60[\rightarrow 50,51,52,53,54,55,56,57,58,$ and 59.
$[60 - 70[$	4	$[60 - 70[\rightarrow 60,61,62,63,64,65,66,67,68$ and 69.
	N = 20	

7.3.1.3 Frequency Distribution Table Correction

A normal or statistical frequency distribution table is a frequency table with complete and equal class interval, as well as completes frequency values. When all class intervals are not given or are unequal and when all frequencies are not unknown, the frequency distribution table is considered abnormal or non-statistical.

1) Unequal Class Sizes Correction

When the class size of a frequency distribution is not equal all through, it means one or two class sizes are less or more than the dominant (correct) class sizes. The following steps are suggested to correct such unequal situations.

a) Identify the dominant and abnormal class size interval. This is done by calculating individual class interval size.

b) Divide the abnormal class size by the dominant class size to obtain the value 'Z'. The value of 'Z' gives the number of class interval derived from an abnormal class interval. A 'Z' value of 2 means the abnormal class interval will become two normal class intervals.

c) Divide the given abnormal class interval frequency by the value of Z to obtain the corrected class interval frequency.

Example 1: Identify the abnormal class size and correct this abnormal situation if any exist.

Marks	$[0-10[$	$[10-20[$	$[20-50[$	$[50-60[$	$[60-70[$
Frequency	3	9	15	10	13

Solution

Step 1: Identify the dominant and abnormal class size interval. [Upper limits − Lower limits]

Marks	$[0-10[$	$[10-20[$	$[20-50[$	$[50-60[$	$[60-70[$
Class sizes	$10-0=$ **10**	$20-10=$ **10**	$50-20=$ **30**	$60-50=$ **10**	$70-60=$ **10**

Note: From class sizes calculation, we realized that the class size 10 is the dominant class size and the class size 30 is the abnormal. Therefore, the class interval $[20-50[$ is having an unequal class size from the rest of the class intervals and need to be corrected.

Step 2: Derivation of 'Z' value.

a) Formula and data. $Z = \frac{\text{Abnormal class size(ACS)}}{\text{Normal class size(NCS)}}$, where; $ACS = 30$ and $NCS = 10$

b) Data substitution and solving.

$$Z = \frac{\text{Abnormal class size(ACS)}}{\text{Normal class size(NCS)}} \quad \ldots\ldots\ldots\ldots\ldots\ldots\text{[Substitute data into the formula]}$$

$$Z = \frac{30}{10} \quad \ldots\ldots\ldots\ldots\ldots\ldots\ldots\ldots\ldots\ldots\ldots\ldots\ldots\ldots\ldots\ldots\ldots\text{[Simplify fraction by division]}$$

$$Z = \mathbf{3}$$

c) Interpretation: The value 3 means the abnormal class interval will be expanded into three normal class intervals. With normal class size is 10, expanding the abnormal class interval ($[20-50[$) to get a normal class interval gives; ($[20-30[$, $[30-40[$ and $[40-50[$).

Step 3: Derivation of real frequency. That is frequency of the adjusted abnormal class interval

a) Formula and data. Real Frequency $= \frac{\text{Frequency when class iretrval is abormal}}{\text{Value of Z}}$.

Where; Frequency when class inetrval is abormal $= 15$ and Value of $Z = 3$

b) Data substitution and solving.

$$\text{Real Frequency} = \frac{\text{Frequency when class inetrval is abormal}}{\text{Value of Z}} \quad \ldots\ldots\ldots\ldots\text{[Substitute data]}$$

$$\text{Real Frequency} = \frac{15}{3} \quad \ldots\ldots\ldots\ldots\ldots\ldots\ldots\ldots\ldots\ldots\ldots\ldots\text{[Simplify fraction by division]}$$

$$\text{Real Frequency} = \mathbf{5}$$

c) Interpretation: The value 5 means the new frequency of the expanded class interval, that is $[20-30[$, $[30-40[$ and $[40-50[$ is 5. Therefore, our new frequency distribution table is now given as,

Marks	$[0-10[$	$[10-20[$	$[20-30[$	$[30-40[$	$[40-50[$	$[50-60[$	$[60-70[$
Frequency	3	9	5	5	5	10	13

2) Incomplete Frequency Correction

We can equally have a situation in which the frequencies of a given class or classes are unknown. In such a situation, we are required to correct this problem before any further solving.

Example 1: Determine the value of the unknown frequency

Marks	$[0-10[$	$[10-20[$	$[20-30[$	$[30-40[$	$[40-50[$	Total
Frequency	5	8	15	10	f_5	50

Step 1: Identify the unknown frequency and calculate the sum of frequencies $[\sum f_i]$.

Note: The unknown frequency is (f_5) and the sum of frequencies is ($\sum f_i = 5 + 8 + 15 + 10 + f_5$.), which is also expressed as ($\sum f_i = N = 38 + f_5$).

Step 2: Determination of unknown frequency (f_5). Remember $[N = \sum f_i]$

    1) Formula. Total frequency(N) $= f_1 + f_2 + f_3 + f_4 + f_5$.

        Where: $f_1 = 5, f_2 = 8, f_3 = 15, f_4 = 10, f_5 = x$ and $N = 50$.

    2) Data substitution and solving.

        $50 = 5 + 8 + 15 + 10 + x$ ....[Simplify right hand side of equation]

        $50 = 38 + x$ ....[Make x subject of formula by collecting like terms]

        $x = 50 - 38$ ....[Solve equation]

        $x = 12$ ....**[The value of $f_5 = 12$]**

Step 3: Verification. Here we sum the frequencies and verify if it is equal to the given sum of frequency value $[N = \sum f_i = 50]$. The sum of frequencies $(5 + 8 + 15 + 10 + 12 = 50)$, Therefore conclude that the value of the unknown frequency $[f_5 = 12]$ is correct.

3) Incomplete Class Interval Correction

We can also be faced with a situation of incomplete class interval. To correct an incomplete class interval, we are required to know the midpoint values and frequencies. The midpoint values helps in the determination of class interval values.

Example 1: Determine the values of lower and upper limits of class intervals and fill the table below. Assume the frequency distribution table is a group and continuous frequency table.

Marks	[5 – 10[	[– –[	[– –[	[– –[	[– –[
Midpoints	7.5	12.5	17.5	22.5	27.5
Frequency	25	8	18	5	20

Note: Since midpoint formula is given as; $\left(\text{Midpoints} = \frac{\text{Lower limits} + \text{Upper limits}}{2}\right)$, we can calculate the class limits be making each of them the subject of formula.

    Upper limits $= 2(\text{Midpoints}) - \text{Lower limits}$ ....(1)

    Lower limits $= 2(\text{Midpoints}) - \text{Upper limits}$ ....(2)

Since our table is continuous, the lower limit of a given class is by default the upper limits it preceding class. Therefore, we are required only to calculate the value of upper limits.

Step 1: Formula. Upper limits = 2(Midpoints) − Lower limits

Step 2: Data substitution and solving

	Lower class limit	Upper class limit (Upper limits = 2(Midpoints) − Lower limits)
First Class	5	**10**
Second Class	10	2(12.5) − 10 = 25 − 10 = **15**
Third Class	15	2(17.5) − 15 = 35 − 15 = **20**
Fourth Class	20	2(22.5) − 20 = 45 − 20 = **25**
Firth Class	25	2(27.5) − 25 = 55 − 25 = **30**

Step 2: Filling of calculated class limits

Marks	[5 − 10[	[10 − 15[	[15 − 20[	[20 − 25[	[25 − 30[
Midpoints	7.5	12.5	17.5	22.5	27.5
Frequency	25	8	18	5	20

Test Your Understanding

Exercise 1: Using a number of class interval of 6, construct a grouped frequency distribution table using the following observations.

65	91	85	76	8	87	79	93
82	75	100	70	88	78	83	59
87	69	54	74	89	83	80	89
96	98	46	70	90	96	88	72

Exercise 2: Use the given marks values below and answer the following questions

5	8	12	15	8	19	5	15
19	24	5	30	8	19	30	30
5	15	19	8	19	42	15	12

1) Construct a discrete frequency distribution table with frequency
2) Construct a grouped continuous frequency distribution table using a class interval

number of 5 and a class size if 5.

3) Calculate the following; relative frequency, cumulative frequency, relative cumulative and percentage relative frequency.

Exercise 3: Correct the following frequency distribution tables

Class interval	$[10 - 20[$	$[20 - 30[$	$[30 - 40[$	$[40 - 70[$	Total
Frequency	1	12	f_3	75	110

arks	$[5 - 10[$	$[- -[$	$[- -[$	$[- - [$	$[- -[$
Midpoints	7.5	12.5	17.5	22.5	27.5
No of students	10	25	12	20	14

[Answers:...See appendix 1]

7.3.2 Graphical Methods of Data Representation

A graph is a visual form of display of statistical data and can takes the form of charts, plots and diagrams. Our study will be limited to; bar chart, pie chart, histogram, frequency polygon, Frequency curve, ogive and line diagram. Other graphical means includes; rectangles, square, cube, cylinders, spheres, pictograms, cartograms and Pareto chart. Generally, graphs are considered more attractive than tables and must be neatly drawn, have accurate and appropriate intervals, size and a name.

1) Bar Chart

Bar charts are group of separated vertical or horizontal rectangles representing categorical data and whose height indicates the frequency of data. The width of the bars must be the same and the height varies as it is influenced by frequency. Bars are separated from each other with equal space and the separation emphasizes the distinctness of categories.

a) **Simple Bar Chart:** It is used to represent qualitative data with single attribute. Here, each bar contains only information of a given or one qualitative data. That is a given bar is not sub-divided and is the simplest bar chart. To draw a simple bar chart, we respected following suggested steps.

1) **Examination:** Identify the highest and lowest frequency value and determine a given interval. The number of categories determines the number of bars to make up the bar chart.

282

2) **Drafting:** Draw an empty graph with both horizontal and vertical axis and decide on the bar chart style (vertical or horizontal bar chart). Divide the frequency axis into equal interval.

3) **Segmentation:** Divide the category axis into the number of bar stipulated in step (1) while respecting the concept of equal space between bars.

4) **Drawing:** Draw a vertical or horizontal line for each category to the height of its corresponding frequency.

Example 1: Draw a simple bar chart using the information below

Years	2000	2001	2002	2003
Production	45	40	20	55

<u>Solution</u>

Step 1: Examination. The highest and lowest frequencies are 55 and 10 respectively. Therefore we can select of 10. The number of categories (years) is four, indicating that our bar chart will be made up of four (4) bars.

Step 2: Drawing of chart

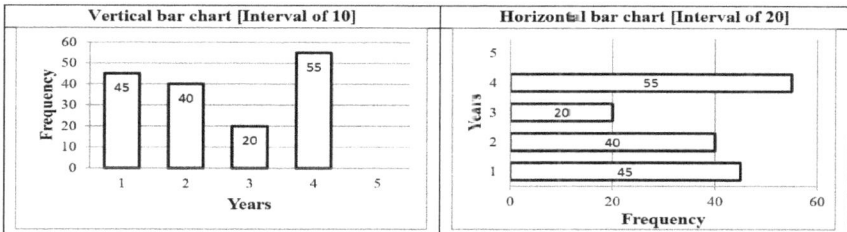

b) **Compound Bar Chart:** It shows several sub-grouped attributes of categories. Here each bar is sub-divided proportionally to the sub-group attributes in such a way that the sum of all sub-grouped attributes must be equal to the general category. The following suggested steps are used to draw a compound bar chart.

1) **Examination:** Identify the number of category to determine the number of bars to be included in the bar chart. Identify each category and their respective sub-grouped as well as the sum of each sub-groups. The sum value obtained will be used in the first plotting.

2) **Drafting and segmentation:** Draw an empty graph with both horizontal and vertical axis and decide on the bar chart style (vertical or horizontal). Divide

283

frequency axis into interval while the values of categories sub-groups. Divide category axis twice its stipulated segments to ensure equal separation.

3) **Frist drawing:** Draw the general category bar of each category using the sum of sub-group within a given category as the frequency of the general bar.

4) **Second drawing:** Sub-divide the general category bar with the appropriate frequency of the sub-grouped and shade or color differently to distinguish them.

Example 1: Draw a compound bar chart from the statistics of age group of three towns in Cameroon as seen below.

Towns	Children	Adult	Old
Bamenda	15	40	30
Buea	20	20	50
Douala	30	15	10

Solution

Step 1: Examination. The table has three categories (Bamenda. Buea and Douala), indicating that our chart will be made up of three bars. The three sub-groups are children, adult and old indicating that each bar will be sub-divided into three sub-sections.

Step 2: Frist drawing. Sum the values categories sub-group and use the sum to draw a simple bar chart. Bamenda frequency is 85, Buea frequency is 90 and Douala frequency is 55.

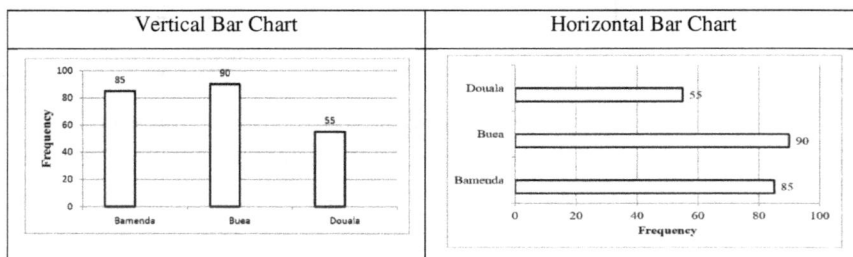

Vertical Bar Chart	Horizontal Bar Chart

Step 3: Second drawing. Divide each general category bar chart by their respective sub-group categories. Divide Bamenda frequency bar of height 85 into three sub-sections of 15, 40 and 30. Divide Buea frequency bar of height 90 into three sections of 20, 20 and 50. Divide Douala frequency bar of height 55 into three sub-sections of 30, 15 and 10.

Vertical compound bar chart	Horizontal bar chart

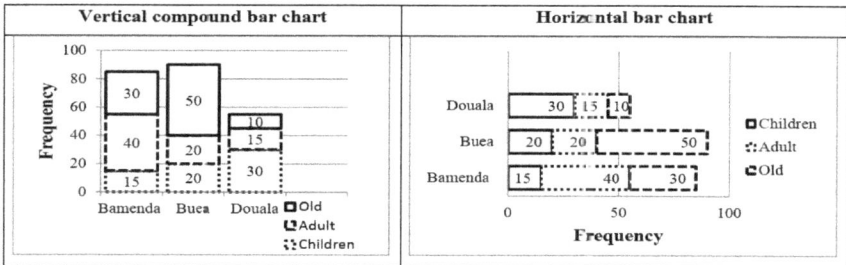

c) **Cluster or stacked bar chart**: This is a type of bar chart where several categorical bars are drawn on separate sub-groups.

1) **Examination.** Identify the number of general categories sub-group to determine the number of clustered bars to obtain and the number of general categories to determine the number of bars to exist in each category cluster.

2) **Drafting and segmentation.** Draw an empty graph with both horizontal and vertical axis. Divide the category axis twice the stipulated number and the frequency axis using a pre-determine interval depending on the desired graph size.

3) **Grouping.** Collect similar sub-groups from each category and place the name of the sub-group on the segmented category axis. That is, the general category is not placed on the category axis.

4) **Second segmentation.** Equally divide the already divided category axis under the named sub-group in to the number of collected similar sub-group.

5) **Drawing.** Under each sub-group, draw connected bars with respect to each general sub-group frequency value.

Example 1:   Draw a cluster bar chart from three families expenditure statistics presented below.

Expenditures	Family A	Family B	Family C
Food expenditure	50	40	60
Education expenditure	20	20	50
Health expenditure	40	15	10

Solution

Step 1: Examination. Our table shows three general categories sub-group (food, education and health) indicating that our graph is to be made up of three groups of cluster bars. The table also shows three general categories (family A, family B and family C) indicating that each cluster is to be made up of three bars.

Step 2: **Drafting and segmentation.** Draw an empty graph with both horizontal and vertical axis. Divide the category axis in to six and name three division food, education and health. The frequency axis is divided with an interval of 10.

Step 3: **Grouping.** Food equals 50 (family A), 40 (family B) and 60 (family C). Education equals 20 (family A), 20 (family B) and 50 (family C). Health equals 40(family A), 15 (family B) and 10 (Family C).

Step 4: **Second segmentation and drawing.** Divide each sub-grouped named in the category axis into three equal halves each representing family A, family B and family C. Under each sub-group, draw bars relating the grouping in (step 3) above.

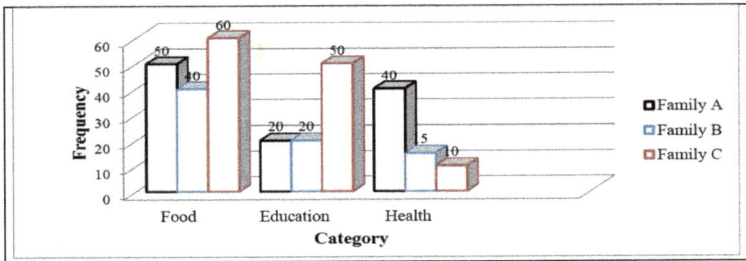

2) Pie Chart

Pie charts also known as circular diagram is defined as a circle that is sub-divided into segments which are proportional to the values of data. It is worth knowing that pie charts are use to represent mostly categorical data and can represent only one sub-group at a time. That is, it is statistically inappropriate to have a compound or multiple pie chart as the case with bar chart. However, two or more pie charts may be constructed side by side for comparison or to study the change over time. Pie chart can be expressed in terms of degree or percentage degree as seen below.

$$\text{Individual Item degree(IID)} = \frac{\text{Value of individual item}}{\text{Total value of items}}(360) \dots\dots\dots\dots\dots\dots\dots(1)$$
$$\text{Individual item percentage degree (IIPD)} = \frac{\text{Individual item degree}}{360}(100) \dots\dots\dots\dots(2)$$

The following suggested steps below are worth considering when drawing a pie chart;

1) Formula. State the formula of calculating individual item degree from the general degree.

2) Determine the degree of individual item

3) Draw an empty circle and partition or segment the circle according to the individual degrees.

Note: To partition the circle with ease, it is advisable to start with the degree that is equal to an angel [90°], or sum of angels [(180°), (270°) and (360°)], half of angel [45°]

Example 1: Construct a pie chart in degree and percentage degree showing the distribution of population using the statistics of towns and their population in the table below.

Countries	Yaoundé	Buea	Bamenda	Douala	Kumba
Production	34	20	25	40	15

Solution

Step 1: Formula and data.

Individual Item degree(IID) = $\frac{\text{Value of individual item}}{\text{Total value of items}}$ (360),

where; Total value of items = 134

Individual item percentage degree (IIPD) = $\frac{\text{Individual item degree}}{360}$ (100)

Step 2: Determine the degree and percentage degree of individual item

Calculation of item degree	Calculation of item percentage degree
Formula. IID = $\frac{\text{Value of individual item}}{\text{Total value of items}}$ (360)	Formula. IIPD= $\frac{\text{Individual item degree}}{360}$ (100)
1) Yaoundé = 34/134 (360) = **91°**	1) Yaoundé = 91/360 (100) = **25%**
2) Buea = 20/134 (360) = **54°**	2) Buea = 54/360 (100) = **15%**
3) Bamenda = 25/134 (360) = **67°**	3) Bamenda = 67/360 (100) = **19%**
4) Douala = 40/134 (360) = **107°**	4) Douala = 107/360 (100) = **30%**
5) Kumba = 15/134 (360) = **40°**	5) Kumba = 40/360 (100) = **11%**
Σ = **360°**	Σ = **100%**
Note: The sum of degree is not equal to 360 because we are dealing with rounded values	

Step 3: Drawing of pie chart in degree and percentage degree

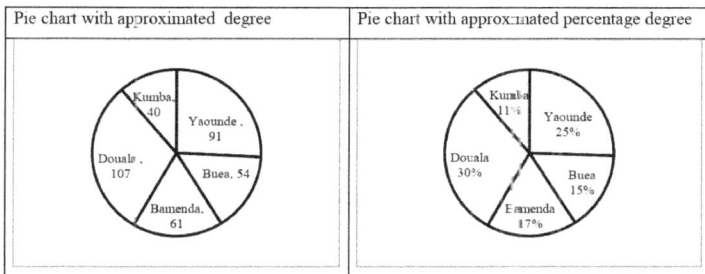

Pie chart with approximated degree	Pie chart with approximated percentage degree

287

Example 2: Construct a pie chart showing the distribution of cocoa production using the statistics of countries and their production in the table below.

Countries	Cameroon	Nigeria	Ghana	Brazil	Japan
Production quantity	60	40	40	80	20

Solution

Step 1: Formula. Individual Item degree $= \frac{\text{Value of individual item(VII)}}{\text{Total value of items(TVI)}}(360)$, where; TVI $=$ 240

Step 2: Determine the degree of individual item

1) Cameroon degree $= \frac{60}{240}(360) = 90°$.[Meaning Cameroon covers $90°$ out of $360°$]

2) Nigeria degree $= \frac{40}{240}(360) = 60°$..........[Meaning Nigeria covers $60°$ out of $360°$]

3) Ghana degree $= \frac{40}{240}(360) = 60°$............[Meaning Ghana covers $60°$ out of $360°$]

4) Brazil degree $= \frac{80}{240}(360) = 120°$.........[Meaning Brazil covers $120°$ out of $360°$]

5) Japan degree $= \frac{20}{240}(360) = 30°$..............[Meaning Japan covers $30°$ out of $360°$]

Step 3: Drawing of pie chart. Remember to calculate pie chart in percentage we use the formula

Pie chart in percentage $= \frac{\text{Individual item degree}}{360}(100)$ ....(1)

Note: During drawing, use different shading styles or colors in the circle to differentiate items

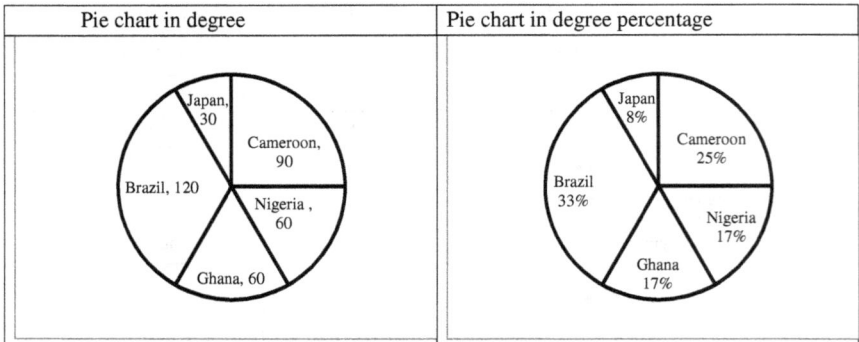

Pie chart in degree	Pie chart in degree percentage

3) Line graph

Line graph or Stick graph is a method in which lines or sticks are used to represent data and the line length is proportional to the frequency of category. The technique is similar to simple bar chart but differ in that lines are used instead of bars.

Example 1: Use the information in the table below and draw a stick graph.

Countries	Cameroon	Senegal	Mali	Togo
Number of death	200	150	300	75

Solution

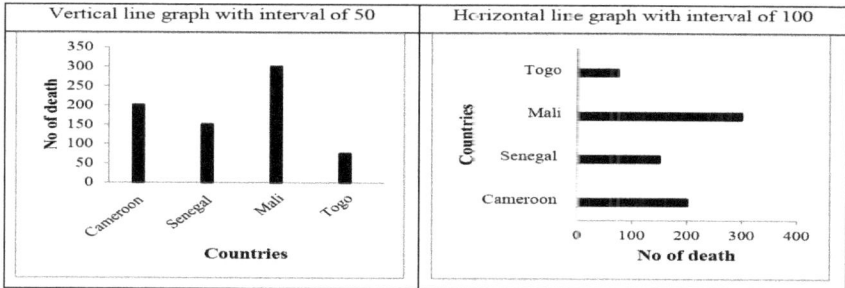

Vertical line graph with interval of 50 | Horizontal line graph with interval of 100

4) Histogram

Histogram is a block of joint vertical or horizontal rectangles representing grouped but continuous frequency data. The length represents the frequency and the width represents the class limits. The following suggested steps are worth considering when drawing a histogram;

1) **Examination:** Verify to make sure the data is grouped and continuous in nature and note the lower limit of the first class as it will help in segmenting the class limits axis.

2) **Naming and segmentation:** The y-axis is the frequency axis and x axis the class limits axis. The frequency axis should be segmented respecting the idea of interval. The class limits axis should respect the following pattern when segmenting;

   a) When the lower limit of the first class is zero (0), we consider only the upper limits of classes as the class limit intervals. This is because the default zero on the center of the graph also known as the original point is considered the value of the first lower class lower limit.

   b) When the lower limit of the first class is not zero (greater than 0), we consider the lower limit of only the first class and upper limits of all the classes (including the first class) as the class limit intervals

3) **Plotting.** When plotting, the following should be seen;

   a) When the first lower limit is zero (0), the first histogram bar must be connected to the frequency axis.

289

b) When the lower limit of the first class is greater than zero, the first histogram bar must not have any connection with the frequency axis.

Example 1: Draw a histogram from the grouped and continuous frequency distribution table below.

Marks	[0 − 10[	[10 − 20[	[20 − 30[	[30 − 40[	[40 − 50[	[50 − 60[
Frequency	10	4	20	15	30	10

Solution

Step 1: Examination. Our frequency distribution table is continuous in nature and the lower limit of the first class is zero (0).

Step 2: Naming and segmentation. Since the lower limit of the first class is zero (0), therefore, the intervals for the class limit will be; 10, 20, 30, 40, 50, and 60. That is only the upper limits of the classes.

Step 3: Plotting. Since the first lower limit equals zero (0), the resulting histogram must be connected with the frequency axis.

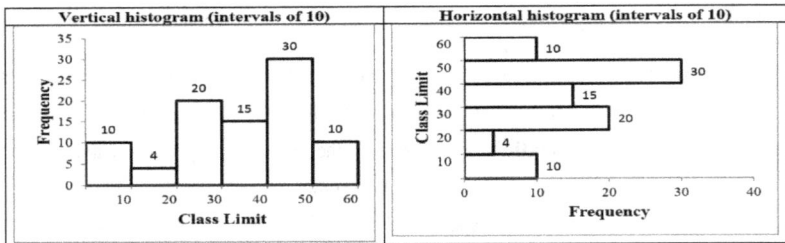

Example 2: Use the frequency distribution table below and draw a histogram.

Scores	[5 − 10[	[10 − 15[	[15 − 20[	[20 − 25[	[25 − 30[
Frequency	20	5	30	10	15

Solution

Step 1: Examination. Our frequency distribution table is continuous in nature and the lower limit of the first class is greater than zero (5).

Step 2: Naming and segmentation. Since the lower limit of the first class is greater than zero (5), therefore, the intervals for the class limit will be; 5, 10, 15, 20, 25 and 30. That is the lower limit of the first class (5) and the upper limits of the classes (10, 15, 20, 25 and 30).

290

Step 3: Plotting. Since the first lower limit equals is greater than zero (5), the resulting histogram must not be connected with the frequency axis.

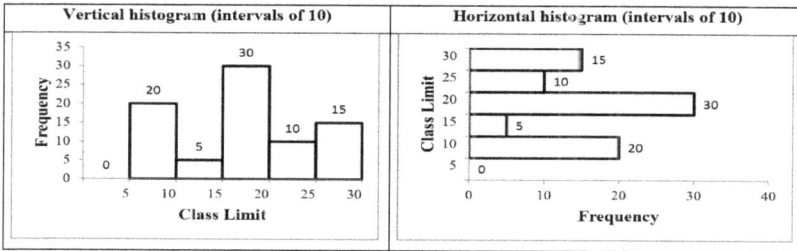

| Vertical histogram (intervals of 10) | Horizontal histogram (intervals of 10) |

5) Frequency Polygon

Frequency polygon is simply a set of upper extreme center histogram points joined up by straight lines. When drawing a frequency polygon, the first and the last parts of the polygon are to be brought to the horizontal axis at a distance equal to half of the class width. The frequency polygon is drawn using two methods; the class limits technique and the midpoints technique.

1) **Class Limits Technique:** Here, frequencies and the upper limits of a frequency distribution table are used to draw a histogram. Suggested steps used for the drawing of frequency polygon are given as follows;

   a) Examination: Verify to ensure that the frequency distribution is continuous in nature.

   b) Histogram drawing: Draw a histogram using the suggested steps used in drawing histogram seen above.

   c) Marking and linking: Identify and mark the top center or middle of each histogram bars. Connect these points with a line (frequency polygon).

2) **Midpoints Technique:** The midpoint of classes and frequencies are used. Suggested steps used for the drawing of frequency polygon are given as follows;

   a) Examination: Verify to ensure that the frequency distribution is continuous in nature.

   b) Midpoints derivation. Derive the midpoint of each class. Remember midpoint equal lower limit plus upper limit all divided by 2.

   c) Plotting: Plot the values of midpoints against their corresponding frequencies. Mark their points of intersection and connect these points with a line (frequency polygon).

Example 1: Construct a frequency polygon using the frequency distribution table below. [Using the class limit technique]

Marks	[0 – 20[	[20 – 40[	[40 – 60[	[60 – 80[	[80 – 100[	[100 – 120[
No of students	15	20	35	30	40	5

Solution

Note: Since the frequency distribution table is continuous in nature, we draw the histogram and frequency polygon directly [Draw a histogram, marks the top of each histogram bar and connect the mark points with a free hand].

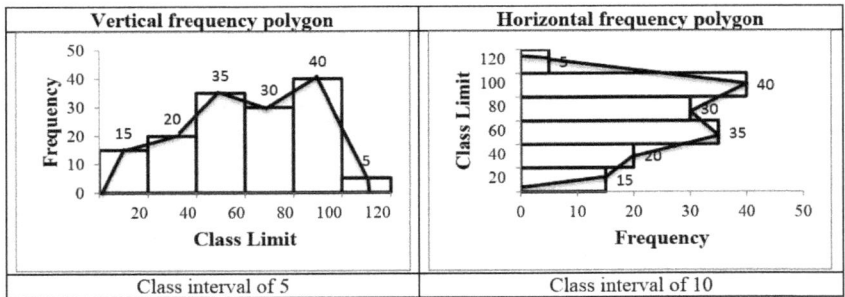

Vertical frequency polygon	Horizontal frequency polygon
Class interval of 5	Class interval of 10

Example 2: Draw a frequency polygon from the given frequency distribution table below. [Use the midpoint technique].

Marks	[0 – 15[	[15 – 30[	[30 – 45[	[45 – 60[	[60 – 75[
Frequency	3	40	20	36	10

Solution

Step 1: Examination. The given frequency distribution table is continuous in nature. Hence we move to the next step.

Step 2: Midpoints derivation and plotting.

Midpoints derivation			Drawing of frequency polygon
Original table		**Working table**	
Class	$[f_i]$	Midpoint values	
$[0 - 15[$	3	$\frac{0+15}{2} = 7.5$	
$[15 - 30[$	40	$\frac{15+30}{2} = 22.5$	
$[30 - 45[$	20	$\frac{30+45}{2} = 37.5$	
$[45 - 60[$	36	$\frac{45+60}{2} = 52.5$	
$[60 - 75[$	10	$\frac{60+75}{2} = 67.5$	

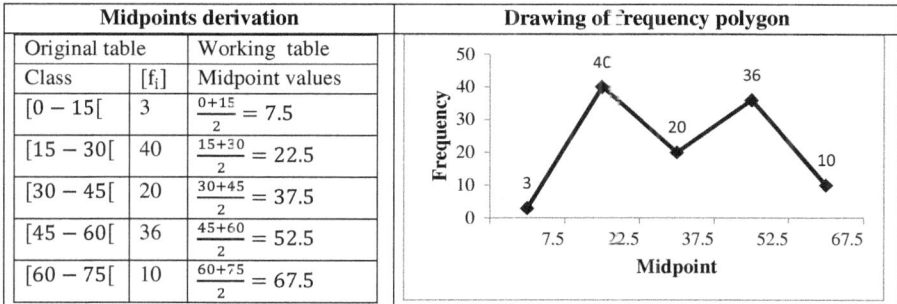

6) Cumulative frequency curve or Ogive

An ogive is simply a graph that represents the cumulative frequencies of a frequency distribution. The cumulative frequency (greater than cumulative frequency or lesser than cumulative frequency) and the class limit or class boundaries are used in drawing an ogive. The following are suggested steps to follow in drawing the cumulative frequency curve or ogive

1) **Examination:** Verify to ensure that the frequency distribution table is continuous in nature.
2) **Derivation of cumulative frequency:** Here, we derive cumulative frequency greater than and cumulative frequency lesser than.
3) **Drafting and segmentation:** Consider the y-axis the frequency axis and x- axis the class limit axis (consider only the upper class limits).
4) **Plotting:** Plot the cumulative frequency greater than values and their corresponding frequency to derive the greater than ogive curve. Plot the cumulative frequency lesser than values and their corresponding frequency to derive the lesser than ogive.

Example 1: Use the frequency distribution table below and draws an ogive.

Marks	20 - 30	30 - 40	40 - 50	50 - 60	60 - 70	70 - 80	80 - 90	90 - 100
Frequency	4	6	13	25	32	19	8	3

Solution

Step 1: Examination. Our frequency distribution table is continuous in nature.

Step 2: Derivation of cumulative frequency.

Marks	20 - 30	30 - 40	40 - 50	50 - 60	60 - 70	70 - 80	80 - 90	90 - 100
Frequency	4	6	13	25	32	19	8	3
[CF >]	4	10	23	48	80	99	107	110
[CF <]	110	106	100	87	62	30	11	3

Where; [CF >] is cumulative frequency greater than and [CF <] is cumulative frequency lesser than.

Step 3: Plotting. Use an interval of 10 for both frequency and class limit.

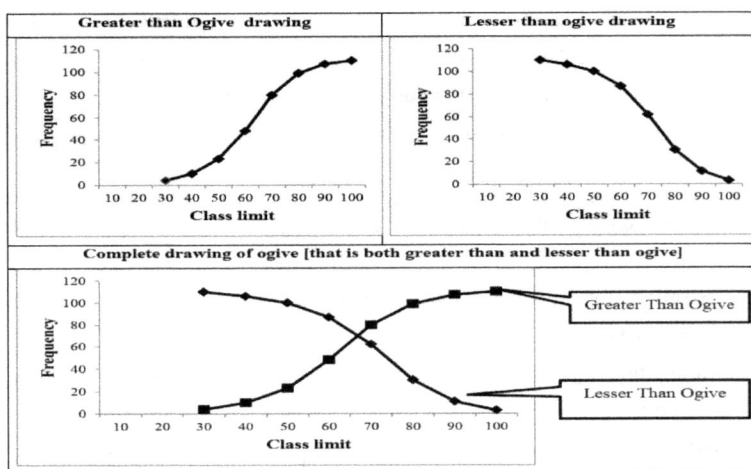

Test Your Understanding

Exercise 4: The table below shows hypothetical information about the main killing sicknesses in the less developed countries.

Countries	Cancer	HIV	Malaria
Cameroon	20	20	40
Nigeria	15	30	15
Ghana	25	40	55

1) Draw a simple bar chart, pie chart and stick graph showing countries sicknesses
2) Draw a compound bar chart and a grouped bar chart of the information

Exercise 5: From the given frequency distribution table below, answer the following questions

Class interval	$[0 - 10[$	$[10 - 20[$	$[20 - 30[$	$[30 - 40[$	$[40 - 60[$	$[60 - 70[$
Frequency	12	5	10	15	10	20

Work required

1) Is the frequency distribution statistical? Justify your answer
2) Draw the following; Histogram, Frequency polygon, Frequency Curve and Ogives

Exercise 6: Use the frequency distribution table below and answer the question given below

Class interval	$[5 - 14[$	$[15 - 24[$	$[25 - 34[$	$[35 - 44[$	$[45 - 54[$	Total
Frequency	2	10	4	8	f_5	30

Draw the following graphs;

1) Histogram,   2) Frequency polygon,   3) Frequency Curve,   4) Ogives

[Answers: ……………………………………………………………………….See appendix 1]

CHAPTER EIGHT

MEASUREMENT OF CENTRAL TENDENCY

8.1 Meaning of Central Tendency

Central tendency is defined as a score that indicate where the center of a distribution tends to be located. It is a single value that attempts to describe a set of data by identifying the central position within the data set. A measure of central tendency is also called measure of central location.

8.2 Measurement of Central Tendency

Major measurements of central tendency to be studied in this chapter include; mean or average, median and mode and some of their respective sub-types as illustrated below.

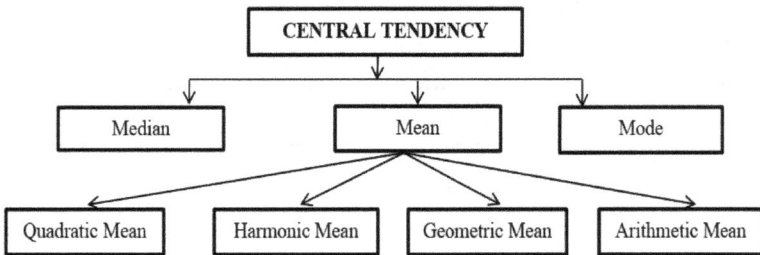

8.2.1 Mean

1) Arithmetic mean

Arithmetic mean or mean is defined as the sum of the observations divided by the number of observations. The formula used for calculating arithmetic mean depends on the nature of data. The following properties about arithmetic mean are worth knowing.

a) The sum of values $[\sum x_i]$ or the sum of the product of midpoints and frequency $[\sum f_i x_i]$ must equal the product of the number of observations and value of arithmetic $[n(\overline{X})]$ or product of sum of frequency and value of arithmetic mean $[N(\overline{X})]$. That is $[\sum x_i = n(\overline{X})]$ and/or $[\sum f_i x_i = N(\overline{X})]$.

b) The sum of two separate mean from separate data sets is given as; $\left[\overline{X}_{yz} = \frac{n_y\overline{X}_y + n_z\overline{X}_z}{n_y + n_y}\right]$

for discrete data without frequency or $\left[\overline{X}_{yz} = \frac{N_y\overline{X}_y + N_z\overline{X}_z}{N_y + N_y}\right]$ for discrete data with frequency and grouped data.

The formulas used in calculating the value of arithmetic mean for discrete data without frequency, discrete data with frequency and group data is giving as seen below.

$$\text{Arithmetic Mean}(\overline{X}) = \frac{1}{n}[\sum x_i] \equiv \overline{X} = \frac{\sum x_i}{n} \equiv \overline{X} = \frac{x_1 + x_2 + \cdots + x_z}{n} \quad \text{.......................(1)}$$

Where: $\overline{X}$ = Arithmetic Mean[AM], n = Number of Observation or items

$x_i = x_1 + x_2 + \cdots + x_z$ = Individual observations or items

$$\text{Arithmetic Mean}(\overline{X}) = \frac{\sum[f_i x_i]}{N} \equiv \overline{X} = \frac{\sum[f_i x_i]}{\sum f_i} \equiv \overline{X} = \frac{f_1(x_1) + f_2(x_2)\cdots f_z(x_z)}{f_1 + f_1 \cdots f_z} \quad \text{.......................(2)}$$

Where: $\overline{X}$ = Arithmetic Mean[AM], $N = \sum f_i = f_1 + f_1 \cdots f_z$ = Sum of frequencies

$x_i = x_1 + x_2 + \cdots x_z$ = Midpoints of observation class or intervals

$f_i x_i$ = Product of corresponding frequencies and their midpoints

$\sum[f_i x_i]$ = Sum of product of corresponding frequencies and their midpoints

Remember: The value of midpoint $[x]$ is calculated using the formula, $x = \frac{L_2 + L_1}{2}$

Note: Equation (1) is used for the calculation of mean for discrete data without frequency and equation (2) is used for discrete data with frequency as well as for group data.

Example 1: Calculate the mean score of the following scores, 2, 4, 6, 8 and 10.

Solution

Step 1: Formula. $\overline{X} = \frac{x_1 + x_2 + \cdots + x_z}{n}$, where; $n = 5, x_1 = 2, x_2 = 4, x_3 = 6, x_4 = 8, x_5 = 10$

Step 2: Data substitution and solving.

$\text{Arithmetic Mean}(\overline{X}) = \frac{x_1 + x_2 + \cdots + x_z}{n}$ ....[Substitute data into the formula]

$\text{Arithmetic Mean}(\overline{X}) = \frac{2 + 4 + 6 + 8 + 10}{5}$ ....[Sum numerator values]

$\text{Arithmetic Mean}(\overline{X}) = \frac{30}{5}$ ....[Simplify fraction by division]

$\text{Arithmetic Mean}(\overline{X}) = \mathbf{6}$

Example 2: Compute the value of mean from the given set of data; 1, 3,5,5,3, 4, 10 and 7.

Solution

Step 1: Formula and data. $\text{Arithmetic Mean}(\overline{X}) = \frac{\sum x_i}{n}$, where; $n = 8$ and $\sum x_i = ?$

Step 2: Derivation of unknown. That is; $\sum x_i$

a) Formula and data. $\sum x_i = x_1 + x_2 + x_3 + \cdots\cdots x_z$,

        Where; $x_1 = 1$, $x_2 = 3$, $x_3 = 5$, $x_4 = 5$, $x_5 = 3$, $x_6 = 4$, $x_7 = 10$, $x_8 = 7$

b) Data substitution and solving.

        $\sum x_i = 1 + 3 + 5 + 5 + 3 + 4 + 10 + 7$ ……………………..……….[Sum values]

        $\sum x_i = \mathbf{38}$

Step 2: Data substitution and solving.

        Arithmetic Mean$(\overline{X}) = \frac{38}{8}$ ……………………..……....[Simplify fraction by division]

        Arithmetic Mean$(\overline{X}) = \mathbf{4.75}$

Example 3: Compute the arithmetic mean (AM) using the information in the table below

Marks [x_i]	0	1	2	3	4	5
Frequency [f_i]	8	10	12	3	5	2

Solution

Step 1: Formula. Arithmetic Mean$(\overline{X}) = \frac{\sum[f_i x_i]}{N}$, where $N = 40$ and $\sum[f_i x_i] =?$

Step 2: Tabula derivation of the unknown. That is, $\sum[f_i x_i]$

Original Table		Working Table
[x_i]	[f_i]	$f_i x_i$
$x_1 = 0$	$f_1 = 8$	$f_1 x_1 = 8(0) = \mathbf{0}$
$x_2 = 1$	$f_2 = 10$	$f_2 x_2 = 10(1) = \mathbf{10}$
$x_3 = 2$	$f_3 = 12$	$f_3 x_3 = 12(2) = \mathbf{24}$
$x_4 = 3$	$f_4 = 3$	$f_4 x_4 = 3(3) = \mathbf{9}$
$x_5 = 4$	$f_5 = 5$	$f_5 x_5 = 5(4) = \mathbf{20}$
$x_6 = 5$	$f_6 = 2$	$f_6 x_6 = 2(5) = \mathbf{10}$
	$N = \mathbf{40}$	$\sum[f_i x_i] = \mathbf{73}$

Step 3: Data substitution and solving of arithmetic mean

        Arithmetic Mean$(\overline{X}) = \frac{73}{40}$ …………………..……….[Simplify fraction by division]

        Arithmetic Mean$(\overline{X}) = \mathbf{1.825}$

Example 4: Use the table below and compute the arithmetic mean of student's examination scores.

Marks	$[0-10[$	$[10-20[$	$[20-30[$	$[30-40[$	$[40-50[$	$[50-60[$

No of students	27	10	7	5	4	2

Solution

Step 1: Formula. Arithmetic Mean$(\overline{X}) = \frac{\Sigma[f_i x_i]}{N}$, where N = 55 and $\Sigma[f_i x_i] =?$

Step 2: Tabula derivation of the unknown. That is, $\Sigma[f_i x_i]$ and solving of AM

Original Table		Working Table	
Marks	$[f_i]$	$x_i =$ (Upper Limt + Lower Limit)/2	$f_i x_i$
$0 - 10$	$f_1 = 27$	$x_1 = (10 + 0)/2 = 10/2 = 5$	$f_1 x_1 = 27(5) = 135$
$10 - 20$	$f_2 = 10$	$x_2 = (20 + 10)/2 = 30/2 = 15$	$f_2 x_2 = 10(15) = 150$
$20 - 30$	$f_3 = 7$	$x_3 = (30 + 20)/2 = 50/2 = 25$	$f_3 x_3 = 7(25) = 175$
$30 - 40$	$f_4 = 5$	$x_4 = (40 + 30)/2 = 70/2 = 35$	$f_4 x_4 = 5(35) = 175$
$40 - 50$	$f_5 = 4$	$x_5 = 50 + 40/2 = 90/2 = 45$	$f_5 x_5 = 4(45) = 180$
$50 - 60$	$f_6 = 2$	$x_6 = (60 + 50)/2 = 110/2 = 55$	$f_6 x_6 = 2(55) = 110$
	N = 55		$\Sigma[f_i x_i] = 925$

Step 3: Data substitution and solving

Arithmetic Mean$(\overline{X}) = \frac{\Sigma[f_i x_i]}{N}$ ....[Substitute data into the formula]

Arithmetic Mean$(\overline{X}) = \frac{925}{55}$.................................[Simplify fraction by division]

Arithmetic Mean$(\overline{X}) = \mathbf{16.8182}$ ....[In four decimal places]

Example 4: The number of days that workers were missing from work due to sickness in a year was recorded as tabulated below. Determine the arithmetic mean from the table.

No of days missing	$[1 - 5]$	$[6 - 10]$	$[11 - 15]$	$[16 - 20]$	$[21 - 25]$
Frequency	12	11	10	4	3

Solution

Step 1: State formula. $\overline{X} = \frac{\Sigma[f_i x_i]}{N}$, where N = 40 and $\Sigma[f_i x_i] =?$

Step 2: Tabula derivation of the unknown. That is, $\Sigma[f_i x_i]$

Original Table		Working Table	
No of days	$[f_i]$	$x_i =$ (Upper Limt + Lower Limit)/2	$f_i x_i$
$1 - 5$	$f_1 = 12$	$x_1 = (5 + 1)/2 = 6/2 = 3$	$f_1 x_1 = 12(3) = 36$
$6 - 10$	$f_2 = 11$	$x_2 = (10 + 6)/2 = 16/2 = 8$	$f_2 x_2 = 11(8) = 88$

11 − 15	$f_3 = 10$	$x_3 = (15 + 11)/2 = 26/2 = 13$	$f_3x_3 = 10(13) = 130$
16 − 20	$f_4 = 4$	$x_4 = (20 + 16)/2 = 36/2 = 18$	$f_4x_4 = 4(18) = 72$
21 − 25	$f_5 = 3$	$x_5 = (25 + 21)/2 = 46/2 = 23$	$f_5x_5 = 3(23) = 69$
	N = 40		$\sum[f_ix_i] = \textbf{395}$

Step 3: Data substitution and solving.

$$\text{Arithmetic Mean}(\overline{X}) = \frac{\Sigma[f_ix_i]}{N} \quad \ldots\ldots\ldots\ldots\ldots\ldots\ldots\text{[Substitute data into the formula]}$$

$$\text{Arithmetic Mean}(\overline{X}) = \frac{395}{40} \ldots\ldots\ldots\ldots\ldots\ldots\ldots\ldots\text{[Simplify fraction by division]}$$

$$\text{Arithmetic Mean}(\overline{X}) = \mathbf{9.875}$$

Test Your Understanding

Exercise 1: Calculate the value of arithmetic mean from the given frequency distribution table below

Marks [x_i]	10	11	12	13	14	15
Frequency [f_i]	80	10	12	30	50	20

[Answer: Arithmetic mean = 12.099]

Exercise 2: From the table below, determine the arithmetic mean of marks.

Marks	$[10-10[$	$[20-30[$	$[30-40[$	$[40-50[$	$[50-60[$	$[60-70[$
No of students	2	10	7	5	4	2

[Answer: Arithmetic mean = 26.818]

Exercise 3: Compute the value of arithmetic mean from the frequency distribution table below.

Class Interval	$[0-9[$	$[10-19[$	$[20-29[$	$[30-39[$	$[40-49[$	$[50-59[$
Frequency	2	10	17	5	14	20

[Answer: Arithmetic mean = 36.118]

---

2) **Geometric mean**

The geometric mean of an observation [X_i or x_i] is the n^{the} root of the product of that observation. The formula used for the calculation of geometric mean is expressed in root form and log form. The following properties of geometric mean are worth knowing.

a) The log of geometric mean [$\log(GM)$] for a set of observations equals the arithmetic mean of the log values of the observation $\left[\frac{\Sigma\log(X_i)}{n}\right]$. That is $\left[\log(GM) = \frac{\Sigma\log(X_i)}{n}\right]$ and $\left[\log(GM) = \frac{\Sigma[f_i(\log x_i)]}{N}\right]$.

b) The product of nth observation [$X_1 \times X_2 \times \cdots X_z$] equals the nth power of their geometric mean $\left[(GM)^{n^{th}}\right]$. That is $\left\{[X_1 \times X_2 \times \cdots X_z] = (GM)^{n^{th}}\right\}$ or $\left\{[x_1{}^{f_1} \times x_1{}^{f_1} \times \cdots x_z{}^{f_z}] = (GM)^{N^{th}}\right\}$.

The formulas used for the calculation of geometric mean in both root and log form for; discrete data without frequency, discrete data with frequency and group data is given as;

$$\text{Geometric Mean(GM)} = \text{Antilog}\left[\frac{\sum(\log X_i)}{n}\right] \equiv GM = \text{Antilog}\left[\frac{\log X_1 + \log X_2 + \cdots \log X_z}{n}\right]$$

$$\text{Geometric Mean(GM)} = \sqrt[n^{th}]{\sum(X_i)} \equiv GM = \sqrt[n^{th}]{X_1 \times X_2 \times \cdots X_z}$$

$$....(1)$$

$$\text{Geometric Mean(GM)} = \left(\sum(X_i)\right)^{\frac{1}{n^{th}}} \equiv GM = (X_1 \times X_2 \times \cdots X_z)^{\frac{1}{n^{th}}}$$

Where: $\log X_i$ = Log of individual values, and n = number of observations

$$\text{Geometric Mean(GM)} = \text{Antilog}\left[\frac{\sum\{f_i(\log x_i)\}}{N}\right] \equiv GM = \text{Antilog}\left[\frac{\sum\{f_i(\log x_i)\}}{\sum f_i}\right]$$

$$\text{Geometric Mean(GM)} = \sqrt[N^{th}]{\sum \text{Product of} x_z{}^{f_z}} \equiv GM = \sqrt[N^{th}]{x_1{}^{f_1} \times x_2{}^{f_2} \times \cdots x_z{}^{f_z}}$$

$$.......(2)$$

$$\text{Geometric Mean(GM)} = \left(\sum \text{Product of} x_z{}^{f_z}\right)^{\frac{1}{N^{th}}} \equiv GM = \left(x_1{}^{f_1} \times x_2{}^{f_2} \times \cdots x_z{}^{f_z}\right)^{\frac{1}{N^{th}}}$$

Where: $f_i(\log x_i)$ = Product of frequencies and their respective midpoints

$\sum\{f_i(\log x_i)\}$ = Sum of product of frequencies and their respective midpoints

$\sum\{f_i(\log x_i)\} = f_1(\log x_1) + f_2(\log x_2) + \cdots f_z(\log x_z)$

Note: To antilog a value, we press the value → pressing 2ndf → press log. To n^{th} root a value $\left[\sqrt[n^{th}]{\text{Value}}\right]$, we press the value → pressing 2ndf → press y^x which shows a second function of $\sqrt[x]{y}$.

Example 1: Fine the geometric mean (GM) of the following scores, 45, 60, 48, 100 and 65

<u>**Solution**</u>

Step 1: Formula. Geometric Mean(GM) = Antilog$\left[\frac{\sum(\log x_i)}{n}\right]$, where; n = 5 and $\sum(\log x_i)$

Step 2: Tabula derivation of unknown. That is, $\sum(\log x_i)$

Original Table	Working Table
[x_i]	**$\log x_i$**
$x_1 = 45$	$\log x_1 = \log 45 = 1.653212514$
$x_2 = 60$	$\log x_2 = \log 60 = 1.77815125$
$x_3 = 48$	$\log x_3 = \log 48 = 1.681241237$
$x_4 = 100$	$\log x_4 = \log 100 = 2$
$x_5 = 65$	$\log x_5 = \log 65 = 1.812913357$
	$\sum(\log x_i) = 8.925518358$

302

Step 3: Data substitution and solving

Geometric Mean(GM) = Antilog $\left[\frac{\Sigma(\log x_i)}{n}\right]$ ....[Substitute data into the formula]

Geometric Mean(GM) = Antilog $\left[\frac{8.9255183\overline{5}8}{5}\right]$ ....[Simplify fraction by division]

Geometric Mean(GM) = Antilog[1.78510$\overline{3}$672] ....[Antilog value]

Geometric Mean(GM) = **60.9682** ....[In four decimal places]

Note: This problem can also be solved using alternative formulas illustrated above as seen below

Calculation of geometric mean using alternative formula 1

Step 1: Formula. Geometric Mean(GM) = $\sqrt[n^{th}]{X_1 \times X_2 \times \cdots X_z}$

Where; $x_1 = 45$, $x_2 = 60$, $x_3 = 48$, $x_4 = 100$, $x_5 = 65$

Step 2: Data substitution and solving

Geometric Mean(GM) = $\sqrt[n^{th}]{X_1 \times X_2 \times \cdots X_z}$ ....[Substitute data into the formula]

Geometric Mean(GM) = $\sqrt[5]{45 \times 60 \times 48 \times 100 \times 65}$. ....[Simply root values]

Geometric Mean(GM) = $\sqrt[5]{842,400,000}$ ....[Work fifth Root]

Geometric Mean(GM) = **60.9682** ....[In four decimal places]

Calculation of geometric mean using alternative formula 2

Step 1: Formula. Geometric Mean(GM) = $(X_1 \times X_2 \times \cdots X_z)^{\frac{1}{n^{th}}}$

Where; $x_1 = 45$, $x_2 = 60$, $x_3 = 48$, $x_4 = 100$, $x_5 = 65$

Step 2: Data substitution and solving

Geometric Mean(GM) = $(X_1 \times X_2 \times \cdots X_z)^{\frac{1}{n^{th}}}$ ....[Substitute data into the formula]

Geometric Mean(GM) = $(45 \times 60 \times 48 \times 100 \times 65)^{\frac{1}{5}}$.....[Work bracket and exponent]

Geometric Mean(GM) = $(842,400,000)^{0.2}$ ....[Solve equation]

Geometric Mean(GM) = **60.9682** ....[In four decimal places]

Note: When calculating the value of geometric mean using all three formulas, it is worth knowing that the value derived when using the log formula is slightly different from the value obtained when using root formula. This is because log reduces numerical value.

Example 2: Compute the value of geometric mean of the following observations, 4, 60, 8, 10 and 50

303

Solution

Step 1: Formula. Geometric Mean(GM) = Antilog$\left[\frac{\Sigma(\log x_i)}{n}\right]$, where; n = 5 and $\Sigma(\log x_i)$ =?

Step 2: Tabula derivation of unknown. That is, $\Sigma(\log x_i)$

Original Table	Working Table	Calculation of geometric mean
[x_i]	$\log x_i$	Step 1: Formula. GM = Antilog$\left[\frac{\Sigma(\log x_i)}{n}\right]$
$x_1 = 4$	log4 = 0.602059991	Where; $\Sigma(\log x_i) = 5.982271232$ and n = 5
$x_2 = 60$	log60 = 1.77815125	
$x_3 = 8$	log8 = 0.903089987	Step 2: Data substitution and solving
$x_4 = 10$	log10 = 1	GM = Antilog$\left[\frac{5.982271232}{5}\right]$ .....[Simplify fraction]
$x_5 = 50$	log50 = 1.698970004	GM = Antilog[1.196454246] .....[Antilog value]
	$\Sigma(\log x_i) = 5.982271232$	**GM = 15.72 ....[In two decimal places]**

Example 3: Fine the geometric mean of the following distribution presented below

Marks [x_i]	2	4	8	16
No of students [f_i]	2	3	3	2

Solution

Step 1: State formula. GM = Antilog$\left[\frac{\Sigma\{f_i(\log x_i)\}}{N}\right]$, where; N = 10 and $\Sigma\{f_i(\log x_i)\}$ =?

Step 2: Tabula derivation of the unknown. That is, $\Sigma\{f_i(\log x_i)\}$

Original Table		Working Table	
[x_i]	[f_i]	$\log x_i$	$f_i(\log x_i)$
$x_1 = 2$	$f_1 = 2$	$\log x_1 = \log 2 = 0.301029995$	2(0.301029995) = 0.602059991
$x_2 = 4$	$f_2 = 3$	$\log x_2 = \log 4 = 0.602059991$	3(0.602059991) = 1.806179974
$x_3 = 8$	$f_3 = 3$	$\log x_3 = \log 8 = 0.903089987$	3(0.903089987) = 2.709269961
$x_4 = 16$	$f_4 = 2$	$\log x_4 = \log 16 = 1.204119983$	2(1.204119983) = 2.408239965
	N = 10		$\Sigma\{f_i(\log x_i)\} = 7.525749892$

Step 3: Data substitution and solving

Geometric Mean(GM) = Antilog$\left[\frac{\Sigma\{f_i(\log x_i)\}}{N}\right]$ .....[Substitute data into the formula]

Geometric Mean(GM) = Antilog$\left[\frac{7.525749892}{10}\right]$ ....[Simplify fraction]

304

Geometric Mean(GM) = Antilog[0.752574989] ... .[Antilog value]

Geometric Mean(GM) = **5.6569** ....[In four decimal places]

N/B: Using the other formulas, the same value of geometric mean is obtained as seen below

Method 1: GM = $\sqrt[N^{th}]{x_1^{f_1} \times x_2^{f_2} \times \cdots x_z^{f_z}}$	Method 2: GM = $\left(x_1^{f_1} \times x_2^{f_2} \times \cdots x_z^{f_z}\right)^{\frac{1}{N^{th}}}$
GM = $\sqrt[10]{2^2 \times 4^3 \times 8^3 \times 16^2}$	GM = $(2^2 \times 4^3 \times 8^3 \times 16^2)^{\frac{1}{10}}$
GM = $\sqrt[10]{4 \times 64 \times 512 \times 256}$	GM = $(4 \times 64 \times 512 \times 256)^{0.1}$
GM = $\sqrt[10]{33,554,432}$	GM = $(33,554,432)^{0.1}$
Geometric Mean(GM) = **5.6569**	Geometric Mean(GM) = **5.6569**

Example 4: Compute the geometric mean (GM) of the following marks scored by students in general mathematics.

Marks	$[0-10[$	$[10-20[$	$[20-30[$	$[30-40[$
No of students	5	8	3	4

Solution

Step 1: State formula. GM = Antilog$\left[\frac{\sum\{f_i(\log x_i)\}}{N}\right]$, where; N = 20 and $\sum\{f_i(\log x_i)\}$ =?

Step 2: Tabula derivation of unknown. That is, $\sum\{f_i(\log x_i)\}$

Original Table			Working Table		
Marks	$[f_1]$	$x_i = \frac{L_1+L_2}{2}$	$\log x_i$	$f_i(\log x_i)$	
0-10	$f_1 = 5$	$x_1 = \frac{0+10}{2} = 5$	$\log 5 = 0.698970004$	3.494850022	
10-20	$f_2 = 8$	$x_2 = \frac{10+20}{2} = 15$	$\log 15 = 1.176091259$	9.408730072	
20-30	$f_3 = 3$	$x_3 = \frac{20+30}{2} = 25$	$\log 25 = 1.397940009$	4.193820026	
30-40	$f_4 = 4$	$x_4 = \frac{30+40}{2} = 35$	$\log 35 = 1.544068044$	6.176272177	
	N=20			$\sum$ = **23.2736723**	

Step 3: Data substitution and solving

Geometric Mean(GM) = Antilog$\left[\frac{\sum\{f_i(\log x_i)\}}{N}\right]$ ....[Substitute data into the formula]

Geometric Mean(GM) = Antilog$\left[\frac{23.2736723}{20}\right]$ ....[Simplify fraction]

Geometric Mean(GM) = Antilog[1.163683615] ....[Antilog value]

Geometric Mean(GM) $= \mathbf{14.5775}$ ....[In four decimal places]

Test Your Understanding

Exercise 4: Fine the geometric mean of the following distribution presented below

Marks [x_i]	12	14	18	26
No of students [f_i]	20	30	30	20

[Answer: Geometric mean = 16.567]

Exercise 5: Compute the geometric mean marks of students using the frequency distribution table below

Marks	$[20-40[$	$[40-60[$	$[60-80[$	$[80-100[$	$[100-120[$
Frequency	10	5	18	8	14

[Answer: Geometric mean = 67.725]

Exercise 6: Determine the value of geometric mean using the frequency distribution table below

Marks	$[5-15[$	$[16-26[$	$[27-37[$	$[38-48[$	$[49-59[$
Frequency	10	5	6	8	5

[Answer: Geometric mean = 24.734]

3) Harmonic mean

Harmonic mean of a set of observations $[X_1, X_2 \dots X_z]$ is defined as the reciprocal or inverse $\left[X_1^{-1}, X_2^{-1} .. X_z^{-1} \text{ or } \frac{1}{X_1}, \frac{1}{X_2}, \dots \frac{1}{X_z}\right]$ of the arithmetic average of the reciprocal of the given values $\left[\frac{\Sigma \frac{1}{X_i}}{n}\right]$. It is worth knowing that combined harmonic mean is calculated using the formula;

$$\text{Harmonic Mean }_{zy} = \frac{n_z + n_y}{\frac{n_z}{HM_z} + \frac{n_y}{HM_y}} \dots\dots\dots\dots\dots\dots\dots\dots\dots\dots\dots\dots\dots\dots\dots(1)$$

Where; n_z or N_z is the number of Z observation or sum of frequency of Z observation

n_y or N_y is the number of Y observation or sum of frequency of Y observation

HM_z and HM_y is the harmonic of observation Z and Y respectively

The formula used for calculating harmonic mean when dealing with; discrete data without frequency, discrete data with frequency and group data is giving as seen below.

$$\text{Harmonic Mean(HM)} = \frac{1}{\frac{1}{n}\left[\Sigma\left(\frac{1}{x_i}\right)\right]} \equiv HM = \frac{1}{\frac{\frac{1}{x_1} + \frac{1}{x_2} + \dots \frac{1}{x_z}}{n}} \dots\dots\dots\dots\dots\dots\dots\dots\dots(2)$$

$$\text{Harmonic Mean(HM)} = \frac{n}{\Sigma\left(\frac{1}{x_i}\right)} \equiv HM = \frac{n}{\frac{1}{x_1}+\frac{1}{x_2}\cdots\frac{1}{x_z}} \quad \ldots\ldots \ldots\ldots\ldots\ldots\ldots\ldots\ldots\ldots\ldots(3)$$

$$\text{Harmonic Mean(HM)} = \frac{1}{\frac{1}{N}\left[\Sigma\left(\frac{f_i}{x_i}\right)\right]} \equiv HM = \frac{1}{\frac{\frac{f_1}{x_1}+\frac{f_2}{x_2}+\cdots+\frac{f_z}{x_z}}{N}} \quad \ldots\ldots\ldots\ldots\ldots\ldots\ldots\ldots\ldots(4)$$

$$\text{Harmonic Mean(HM)} = \frac{N}{\Sigma\left[\frac{f_i}{x_i}\right]} \equiv HM = \frac{\Sigma f_i}{\frac{f_1}{x_1}+\frac{f_2}{x_2}+\cdots+\frac{f_z}{x_z}} \quad \ldots\ldots\ldots\ldots\ldots\ldots\ldots\ldots\ldots\ldots(5)$$

Example 1: Fine the harmonic mean of the following scores. 5,10,17,24 and 30

Solution

Step 1: State formula. $HM = \frac{1}{\frac{1}{n}\left[\Sigma\left(\frac{1}{x_i}\right)\right]}$, where; $n = 5$ and $\Sigma\left(\frac{1}{x_i}\right) = ?$

Step 2: Tabula derivation of the unknown. That is, $\Sigma\left(\frac{1}{x_i}\right)$ and calculation of harmonic mean

Original Table	Working Table	Calculation of Harmonic Mean
x_i	$\frac{1}{x_i}$	Step 1: Formula. $HM = \frac{1}{\frac{1}{n}\left[\Sigma\left(\frac{1}{x_i}\right)\right]}$
$x_1 = 5$	$\frac{1}{x_1} = \frac{1}{5} = 0.2$	Where; $\Sigma\left(\frac{1}{x_i}\right) = 0.433823529$ and $n = 5$
$x_2 = 10$	$\frac{1}{x_2} = \frac{1}{10} = 0.1$	Step 2: Data substitution and solving
$x_3 = 17$	$\frac{1}{x_3} = \frac{1}{17} = 0.058823529$	$HM = \frac{1}{\frac{1}{5}[0.433823529]}$.[Simplify denominator fraction]
$x_4 = 24$	$\frac{1}{x_4} = \frac{1}{24} = 0.041666666$	$HM = \frac{1}{0.2[0.433823529]}$.[Expand denominator bracket]
$x_5 = 30$	$\frac{1}{x_5} = \frac{1}{30} = 0.033333333$	$HM = \frac{1}{0.086764705}$ ....[Simplify equation fraction]
	$\Sigma\left(\frac{1}{x_i}\right) = \mathbf{0.433823529}$	$HM = \mathbf{11.52542373}$.......[Value of harmonic mean]

Example 2: Determine the value of harmonic mean of the following data. 14,8,25 and 30

Solution

Step 1: State formula. $\text{Harmonic Mean(HM)} = \frac{1}{\frac{1}{n}\left[\Sigma\left(\frac{1}{x_i}\right)\right]}$ OR $HM = \frac{n}{\Sigma\left(\frac{1}{x_i}\right)}$, where; $n = 4$

Step 2: Tabula derivation of the unknown. That is, $\Sigma\left(\frac{1}{x_i}\right)$ and calculation of Harmonic Mean(HM)

Original table	Working table	Calculation of Harmonic Mean(HM)	
x_i	$\frac{1}{x_i}$	$HM = \frac{1}{\frac{1}{n}\left[\Sigma\left(\frac{1}{x_i}\right)\right]}$	$HM = \frac{n}{\Sigma\left(\frac{1}{x_i}\right)}$
$x_1 = 14$	$\frac{1}{14} = 0.071428571$	$HM = \frac{1}{\frac{1}{4}[0.269761904]}$	$HM = \frac{4}{0.269761904}$
$x_2 = 8$	$\frac{1}{8} = 0.125$		$HM = \mathbf{14.8278906}$

$x_3 = 25$	$\frac{1}{25} = 0.04$	$HM = \frac{1}{0.25[0.269761904]}$	
$x_4 = 30$	$\frac{1}{24} = 0.033333333$	$HM = \frac{1}{0.067440476}$	
		$HM = \mathbf{14.8278906}$	
	$\Sigma = 0.269761904$		

Example 3: The table below presents the marks of students during the first team business mathematics in G.T.H.S Jakiri. From the table, determine the harmonic mean marks of these students.

Marks of students (x_i)	11	12	13	14	15
No of students (f_i)	3	7	8	5	2

Solution

Step 1: Stating of formula. Harmonic Mean(HM) $= \frac{1}{\frac{1}{N}\left[\Sigma\left(\frac{f_i}{x_i}\right)\right]}$, where: N = 25 and $\Sigma\left[\frac{f_i}{x_i}\right] = ?$

Step 2: Tabular derivation of unknown. That is $\left\{\Sigma\left(\frac{f_i}{x_i}\right)\right\}$ and calculation of

Harmonic Mean(HM)

Original Table		Working table	Calculation of Harmonic Mean
x_i	f_i	$\frac{f_i}{x_i}$	Step 1: Formula. Harmonic Mean(HM) $= \frac{1}{\frac{1}{N}\left[\Sigma\left(\frac{f_i}{x_i}\right)\right]}$
11	3	$^3/_{11} = 0.272727272$	Step 2: Data substitution and solving
12	7	$^7/_{12} = 0.583333333$	Where; $\Sigma\left(\frac{f_i}{x_i}\right) = 1.961921412$ and N = 25
13	8	$^8/_{13} = 0.615384615$	$HM = \frac{1}{\frac{1}{25}[1.961921412]}$ ....[Simplify denominator]
14	5	$^5/_{14} = 0.357142857$	$HM = \frac{1}{0.04[1.961921412]}$ .....[Expand denominator]
15	2	$^2/_{15} = 0.133333333$	$HM = \frac{1}{0.078476856}$ ....[Simplify fraction]
	$\Sigma = 25$	$\Sigma = \mathbf{1.961921412}$	$HM = \mathbf{12.7426103}$.[Value of harmonic mean]

Example 4: Use the table below and determine the value of harmonic mean (HM)

Classes	130-140	140-150	150-160	160-170	170-180	180-190
Frequency	3	12	23	14	6	4

Solution

Step 1: State formula. $HM = \dfrac{1}{\frac{1}{N}\left[\Sigma\left(\frac{f_i}{x_i}\right)\right]}$ OR $HM = \dfrac{N}{\Sigma\left[\frac{f_i}{x_i}\right]}$, where: $N = \Sigma f_i = 62$ and $\Sigma\left[\frac{f_i}{x_i}\right] = ?$

Step 2: Derivation of unknown. That is $\left[\Sigma\left(\frac{f_i}{x_i}\right)\right]$. It is worth knowing that since midpoints [X] are not given directly as seen in example 1 above, we are required to use the formula $\left[x = \frac{L_2+L_1}{2}\right]$ in deriving the values of respective midpoints.

Original Table		Working Table	
Classes	Frequency [f_i]	$x_i = \dfrac{L_2+L_1}{2}$	$\dfrac{f_i}{x_i}$
130 - 140	$f_1 = 3$	$x_1 = \dfrac{130+140}{2} = \dfrac{270}{2} = 135$	$\dfrac{f_1}{x_1} = \dfrac{3}{135} = 0.022222222$
140 - 150	$f_2 = 12$	$x_2 = \dfrac{140+150}{2} = \dfrac{290}{2} = 145$	$\dfrac{f_2}{x_2} = \dfrac{12}{145} = 0.08275862$
150 - 160	$f_3 = 23$	$x_3 = \dfrac{150+160}{2} = \dfrac{310}{2} = 155$	$\dfrac{f_3}{x_3} = \dfrac{23}{155} = 0.148387096$
160 - 170	$f_4 = 14$	$x_4 = \dfrac{160+170}{2} = \dfrac{330}{2} = 165$	$\dfrac{f_4}{x_4} = \dfrac{14}{165} = 0.084848484$
170 - 180	$f_5 = 6$	$x_5 = \dfrac{170+180}{2} = \dfrac{350}{2} = 175$	$\dfrac{f_5}{x_5} = \dfrac{6}{175} = 0.034285714$
180 - 190	$f_6 = 4$	$x_6 = \dfrac{180+190}{2} = \dfrac{370}{2} = 185$	$\dfrac{f_6}{x_6} = \dfrac{4}{185} = 0.021621621$
	$\Sigma f_i = N = 62$		$\Sigma\left(\dfrac{f_i}{x_i}\right) = 0.394123757$

Step 3: Data substitution and solving

Calculation of harmonic mean using formula 1

Harmonic Mean (HM) $= \dfrac{1}{\frac{1}{N}\left[\Sigma\left(\frac{f_i}{x_i}\right)\right]}$ ....[Substitute data into the formula]

Harmonic Mean (HM) $= \dfrac{1}{\frac{1}{62}[0.394123757]}$ ...[Simplify fraction of denominator]

Harmonic Mean (HM) $= \dfrac{1}{0.016129032[0.394123757]}$

.................................[Solve]

Harmonic Mean (HM) $= 157.31$ ....[In two decimal places]

Calculation of harmonic mean using formula 2

Harmonic Mean (HM) $= \dfrac{N}{\Sigma\left[\frac{f_i}{x_i}\right]}$ ....[Substitute data into the formula]

Harmonic Mean (HM) $= \dfrac{62}{0.394123757}$ ....[Simplify equation fraction]

Harmonic Mean (HM) $= 157.31$ ....[In two decimal places]

4) Quadratic mean

Quadratic mean of an observation is defined as the square root of the arithmetic mean of the square of observations. The formula for calculating quadratic mean for both discrete data without frequency and discrete with data frequency and/or grouped data is given as;

$$\text{Quadratic mean} = \sqrt{\frac{\sum x_i^2}{n}} \equiv \sqrt{\frac{x_1^2 + x_2^2 + \cdots x_z^2}{n}} \quad \ldots \ldots \text{[Discrete data without frequency]}$$

$$\text{Quadratic mean} = \sqrt{\frac{\sum [f_i(x^2{}_i)]}{N}} \equiv \sqrt{\frac{f_1 x_1^2 + f_2 x_2^2 + \cdots f_z x_z^2}{N}} . \text{[Discrete with frequency and grouped]}$$

Where: n represent the number of observation or items,

N represent sum of observations or items frequencies,

X_i^2 = represent square of actual value (discrete data without frequency) and square of midpoint value (discrete data with frequency and group data),

$f_i(x^2{}_i)$ represent product of observation frequencies and square of midpoint values.

310

Exercise 1: Compute the quadratic mean of the following observations; 2, 4, 6, 8 and 10.

Solution

Step 1: Formula and data. Quadratic mean $\left(\overline{X}_Q\right) = \sqrt{\frac{X_1{}^2 + X_2{}^2 + \cdots X_z{}^2}{n}}$, where; $n = 5$

Step 2: Data substitution and solving.

Quadratic mean $\left(\overline{X}_Q\right) = \sqrt{\frac{X_1{}^2 + X_2{}^2 + \cdots X_z{}^2}{n}}$ ....[Substitute data into the formula]

Quadratic mean $\left(\overline{X}_Q\right) = \sqrt{\frac{2^2 + 4^2 + 6^2 + 8^2 + 10^2}{5}}$ .....[Work out the exponents or power]

Quadratic mean $\left(\overline{X}_Q\right) = \sqrt{\frac{4 + 16 + 36 + 64 + 100}{5}}$ ..[Sum numerator and divide fraction]

Quadratic mean $\left(\overline{X}_Q\right) = \sqrt{44}$ ....[Work root 44]

Quadratic mean $\left(\overline{X}_Q\right) = \mathbf{6.63}$ ....**[In two decimal places]**

Example 2: From the given frequency distribution table below determine the quadratic mean

Marks of students	5	10	15	20
Frequency	3	5	2	8

Solution

Step 1: Formula and data. Quadratic mean$\left(\overline{X}_Q\right) = \sqrt{\frac{\sum[f_i(x^2{}_i)]}{N}} \equiv \sqrt{\frac{f_{=1}{}^2 + f_2 x_2{}^2 + \cdots f_z x_z{}^2}{N}}$, where $N = 18$

Step 2: Tabula determination of unknown $[\sum[f_i(x^2{}_i)]]$ and calculation of quadratic mean

Original table		Working table	
$[X_i]$	$[f_i]$	$X_i{}^2$	$f_i(x^2{}_i)$
5	3	$5^2 = 25$	$3(25) = 75$
10	5	$10^2 = 100$	$5(100) = 500$
15	2	$15^2 = 225$	$2(225) = 450$
20	8	$20^2 = 400$	$8(400) = 3,200$
	$N = 18$		$\sum = 4,225$

Step 3: Data substitution and solving

Quadratic mean $\left(\overline{X}_Q\right) = \sqrt{\frac{\sum[f_i(x^2{}_i)]}{N}}$ ....[Substitute data into the formula]

Quadratic mean $\left(\overline{X}_Q\right) = \sqrt{\frac{4,225}{18}}$ ....[Simplify square fraction]

Quadratic mean $(\overline{X}_Q) = \sqrt{234.7222222}$ ....[Work root]

Quadratic mean $(\overline{X}_Q) = \textbf{15.32}$................................**[In two decimal places]**

Example 3: Determine the quadratic mean from the given frequency distribution below

Class interval	$[0-10[$	$[10-20[$	$[20-30[$	$[30-40[$	$[40-50[$
Frequency	12	5	5	10	8

Solution

Step 1: Formula and data. Quadratic mean $(\overline{X}_Q) = \sqrt{\frac{\sum[f_i(x^2_i)]}{N}}$, where; N = 40 and $\sum[f_i(x^2_i)] = ?$

Step 2: Tabular determination of unknown

Original Table		Working Table		
Class Interval	f_i	x_i [Midpoint]	x^2_i	$f_i(x^2_i)$
$[0-10[$	12	$\frac{10+0}{2} = \frac{10}{2} = 5$	$5^2 = 25$	$12(25) = 300$
$[10-20[$	5	$\frac{10+20}{2} = \frac{30}{2} = 15$	$15^2 = 225$	$5(225) = 1,125$
$[20-30[$	5	$\frac{20+30}{2} = \frac{50}{2} = 25$	$25^2 = 625$	$5(625) = 3,125$
$[30-40[$	10	$\frac{30+40}{2} = \frac{70}{2} = 35$	$35^2 = 1,225$	$10(1,225) = 12,250$
$[40-50[$	8	$\frac{40+50}{2} = \frac{90}{2} = 45$	$45^2 = 2,025$	$8(2,025) = 16,200$
	$\sum = \textbf{40}$			$\sum = \textbf{33,000}$

Step 3: Data substitution and solving.

Quadratic mean $(\overline{X}_Q) = \sqrt{\frac{33,000}{40}}$ ....[Simplify equation by dividing fraction]

Quadratic mean $(\overline{X}_Q) = \sqrt{825}$ ....[Work root]

Quadratic mean $(\overline{X}_Q) = \textbf{28.72}$ ....**[In two decimal places]**

Test Your Understanding

Exercise 10: Compute the value of quadratic mean from the following frequency distribution tables.

a) Frequency distribution table 1

Marks of students	100	200	300	400
Frequency	30	50	20	80

[Answer: Quadratic mean = 306.41]

b) Frequency distribution table 2

Class interval	$[0-10[$	$[10-20[$	$[20-30[$	$[30-40[$	$[40-50[$
Frequency	6	5	4	10	2

[Answer: Quadratic mean = 27.27]

c) Frequency distribution table 3

Class interval	$[0-5[$	$[6-11[$	$[12-17[$	$[18-23[$	$[24-29[$
Frequency	2	15	5	10	4

[Answer: Quadratic mean = 15.95]

Note: The empirical relation between arithmetic mean, geometric mean and harmonic mean shows that arithmetic mean is greater than geometric mean and greater than harmonic mean. That is, **AM ≥ GM ≥ HM**

Example 1: The scores of four students is given as; 6, 8, 12 and 36. From the given scores, show that AM ≥ GM ≥ HM.

Solution

Step 1: Formula. $\overline{X} = AM = \frac{\Sigma x_i}{n}$, $GM = \text{Antilog}\left[\frac{\Sigma(\log x_i)}{n}\right]$, $HM = \frac{N}{\Sigma\left(\frac{1}{x_i}\right)}$. Where, N= n = 4

Step 2: Calculation of different mean

1) **Arithmetic Mean**

Arithmetic Mean(AM) $= \frac{6+8+12+36}{4}$ ....[Sum numerator values]

Arithmetic Mean(AM) $= \frac{62}{4}$ ....[Simplify equation fractions]

Arithmetic Mean(AM) $= \mathbf{15.5}$

2) **Geometric Mean**

Geometric Mean(GM) $= \text{Antilog}\left[\frac{\Sigma(\log x_i)}{n}\right]$....[Substitute data into the formula]

Geometric Mean(GM) $= \text{Antilog}\left[\frac{\log 6+\log 8+\log 12+\log 36}{4}\right]$ ....[Log values]

Geometric Mean(GM) $=$

$\text{Antilog}\left[\frac{0.77815125+0.903089987+1.079181246+1.556302501}{4}\right]$

Geometric Mean(GM) $= \text{Antilog}\left[\frac{4.316724984}{4}\right]$ .... .[Simplify bracket fraction]

Geometric Mean(GM) $= \text{Antilog}[1.079181246]$ ....[Antilog value]

Geometric Mean(GM) $= \mathbf{12}$

313

3) Harmonic Mean

$$\text{Harmonic Mean(HM)} = \frac{N}{\Sigma\left(\frac{1}{x_i}\right)}\dots\dots\dots\dots\text{[Substitute data into the formula]}$$

$$\text{Harmonic Mean(HM)} = \frac{4}{\frac{1}{6}+\frac{1}{8}+\frac{1}{12}+\frac{1}{36}}\dots\dots\dots\text{[Simplify denominator fractions]}$$

$$\text{Harmonic Mean(HM)} = \frac{4}{0.166666666+0.125+0.083333333+0.027777777}\text{[Sum values]}$$

$$\text{Harmonic Mean(HM)} = \frac{4}{0.402777777}\dots\dots\dots\dots\text{[Simplify equation fraction]}$$

$$\text{Harmonic Mean(HM)} = \mathbf{9.9}\dots\dots\dots\dots\dots\dots\text{[In one decimal places]}$$

Step 3: Justification. From our calculations, arithmetic mean (AM) is 15.5 , geometric mean (GM) is 12 and harmonic mean (HM) is 9.9. Therefore we see that, $AM(15.5) > GM(12) > HM(9.9)$.

Example 2: Use the frequency distribution table below and show that; AM $\geq$ GM $\geq$ HM.

Weight	$[30-35[$	$[35-40[$	$[40-45[$	$[45-50[$
No of students	4	7	10	19

<u>Solution</u>

Step 1: Formula. $\overline{X} = AM = \frac{\Sigma[f_ix_i]}{N}$, $GM = \text{Antilog}\left[\frac{\Sigma\{f_i(logx_i)\}}{N}\right]$, $HM = \frac{N}{\Sigma\left[\frac{f_i}{x_i}\right]}$, where, N =40

Step 2: Tabula derivation of unknowns. That is; $\Sigma[f_ix_i]$, $\Sigma\{f_i(logx_i)\}$ and $\Sigma\left[\frac{f_i}{x_i}\right]$

Original Table			Working Table			
			AM Unknown	GM Unknown		HM Unknown
Weight	$[f_i]$	$[x_i]$	f_ix_i	$logx_i$	$f_i(logx_i)$	$\frac{f_i}{x_i}$
30 - 35	4	32.5	130	1.511883361	6.047533444	0.123076923
35 - 40	7	37.5	262.5	1.574031268	11.01821887	0.186666666
40 - 45	10	42.5	425	1.62838893	16.2838893	0.235294117
45 - 50	19	47.5	902.5	1.67669361	31.85717858	0.4
	N=40		$\Sigma = 1720$		$\Sigma\{f_i(logx_i)\} = 65.2068202$	$\Sigma = 0.94503771$

Step 3: Data substitution and solving

1) **Arithmetic Mean**

$$\text{Arithmetic Mean}(\overline{X}) = AM = \frac{\Sigma[f_ix_i]}{N}\dots\dots\dots\text{[Substitute data into the formula]}$$

$$\text{Arithmetic Mean}(\overline{X}) = \frac{1720}{40}\dots\dots\dots\dots\dots\dots\text{[Simplify equation fraction]}$$

314

Arithmetic Mean($\overline{X}$) = **43**

2) **Geometric mean**

Geometric Mean(GM) = Antilog $\left[\frac{\Sigma\{f\,(\log x_i)\}}{N}\right]$.[Substitute data into the formula]

Geometric Mean(GM) = Antilog $\left[\frac{65.2068202}{40}\right]$ .... .[Simplify bracket fraction]

Geometric Mean(GM) = Antilog[1.630170505] ....[Antilog value]

Geometric Mean(GM) = **42. 7** ....[In one decimal place]

3) **Harmonic mean**

Harmonic Mean(HM) = $\frac{N}{\Sigma\left[\frac{f_i}{x_i}\right]}$ ....[Substitute data into the formula]

Harmonic Mean(HM) = $\frac{40}{0.94503771}$ ....[Simplify equation fraction]

Harmonic Mean(HM) = **42. 3** ....[In one decimal place]

Step 4: Justification. $AM(43) > GM(42.7) > HM(42.3)$.

Test Your Understanding

Exercise 11: The marks of students in G. B. H .S Jakiri in history is given as follows; 2, 10, 5, 6, 7, 8. Calculate the following;

1) Arithmetic Mean ....[Answer: **Arithmetic mean = 6.33**]
2) Harmonic Mean ....[Answer: **Harmonic mean = 4.86**]
3) Geometric Mean ....[Answer: **Geometric mean = 5.68**]
4) Quadratic Mean ....[Answer: **Quadratic mean = 6.81**]

Exercise 12: Using the frequency distribution tables below, calculate the values of; arithmetic mean (AM), harmonic mean (HM), geometric mean (GM) and quadratic mean (QM).

1) Frequency distribution table 1

Marks	$[10-20[$	$[20-30[$	$[30-40[$	$[40-50[$	$[50-60[$
Frequency	12	6	2	10	5

[Answer: **AM =32.14 , HM =24.82 , GM = 28.32 , and QM = 35.57**]

2) Frequency distribution table 2

Class interval	0-12	13-15	16-18	19-21
Frequency	4	12	20	1

[Answer: **AM = 16.64, HM = 16.14, GM = 16.40, and QM = 16.86**]

3) Frequency distribution table 3

Marks [x_i]	50	65	75	80	95	100
Frequency [f_i]	4	6	16	8	7	4

[Answer: AM = 77.67, HM = 75.08, GM = 76.42, and QM = 78.84]

8.2.2 Median

The median of a set of observation is defined as the middle value of the observation. Therefore, median is considered the mid-value as it is the value that divides the total observations or frequency into two equal parts. Generally, median is considered a better measure than mean in situation with open-ended frequency distribution table.

1) Algebraic method of calculating median

a) Calculation of median using discrete data without frequency

When dealing with ungroup data without frequency, median is the middle value of an arranged observations. The following should be noted,

1) When two observations are in the middle of an observation, to determine the median, we add the two values and divide it by 2. $\left(\text{Median} = \frac{\text{First middle number} + \text{Second middle number}}{2}\right)$

2) When only a value is in the middle of observations, the median is the value.

Example 1: Determine the median of the following observations; 2, 6, 8, 10, 12, 14 and 16.

Solution

Since our observations are orderly, we simply front and back count simultaneously. If we do this, we realize that 10 is the only middle number. Therefore, our median is 10.

Example 2: Compute the median of the following marks; 2, 3, 4, 5, 6, 7, 8 and 9.

Solution

Hint: Doing simultaneous front and back counting, we realized that 5 and 6 are in the middle of the observations.

Step 1: Formula. $\text{Median} = \frac{\text{First middle number} + \text{Second middle number}}{2}$

Where; First middle number = 5 and Second middle number = 6

Step 2: Data substation and solving

$\text{Median} = \frac{\text{First middle numner} + \text{Second middle number}}{2}$ ....[Enter data into the formula]

$\text{Medain} = \frac{5+6}{2}$ ....[Sum numerator values]

$$Medain = \frac{11}{2} \dots\dots\dots\dots\dots\dots\dots\dots\dots\dots\dots\dots\text{[Simplify equation fraction]}$$

$$Medain = \mathbf{5.5}$$

Example 3: Determine the value of median from the following marks; 10, 2, 17, 3, 14, 5, 6, 7, 18 and 9.

<u>Solution</u>

Hint: Looking at the marks, we realized that the marks are not in order. Therefore, we are required to first re-arrange the marks or make it orderly before any further solving.

Step 1: Re-arrangement of values in an orderly manner

Ascending order arrangement: 2, 3, 5. 6, 7, 9, 10, 14, 17 and 18

Descending order arrangement: 18, 17, 14, 10, 9. 7, 6, 5, 3 and 2

Step 2: Formula. $Median = \frac{\text{First middle number+Second middle number}}{2}$

Where; First middle number $= 7$ and Second middle number $= 9$

Step 3: Data substitution and solving

$$Median = \frac{\text{First middle number+Second middle number}}{2} \dots\text{[Substitute data into the formula]}$$

$$Median = \frac{7+9}{2} \dots\dots\dots\dots\dots\dots\dots\dots\dots\dots\dots\dots\dots\text{[sum numerator values]}$$

$$Median = \frac{16}{2} \dots\dots\dots\dots\dots\dots\dots\dots\dots\dots\dots\dots\text{[Simplify equation fraction]}$$

$$Median = \mathbf{8}$$

b) **Calculation of median using discrete data with frequency**

When using discrete data with frequency, the first cumulative frequency value that is greater than the value $\left[\frac{N+1}{2}\right]$ is considered the median. Suggested steps to follow in determining the median include;

1) Derive cumulative frequency from distribution.

2) Calculate the value of $\left[\frac{N+1}{2}\right]$, where N represents the sum of frequency. Identify from the values of cumulative frequencies, the first value that is greater than the value of $\left[\frac{N+1}{2}\right]$.

3) Derive the median. From the value identified in step (2) move horizontally towards it corresponding midpoint value and consider the value the median value.

Example 1: Find the median of the following distribution

Marks $[x_i]$	10	20	30	40	50	60
No of students $[f_i]$	7	12	17	19	21	24

Solution

Step 1: Derivation of Cumulative frequency

Original table		Working table	
$[x_i]$	$[f_i]$	Cumulative Frequency $[CF_i]$	
$x_1 = 10$	$f_1 = 7$	$CF_1 = f_1 + f_0 = 7 + 0 = 7$	We can see that $CF_4 = 55$ is the first CF_i value greater than the value of $\left[\frac{N+1}{2}\right] = 50.5$
$x_2 = 20$	$f_2 = 12$	$CF_2 = CF_1 + f_2 = 7 + 12 = 19$	
$x_3 = 30$	$f_3 = 17$	$CF_3 = CF_2 + f_3 = 19 + 17 = 36$	
$x_4 = 40$	$f_4 = 19$	$CF_4 = CF_3 + f_4 = 36 + 19 = 55$	
$x_5 = 50$	$f_5 = 21$	$CF_5 = CF_4 + f_5 = 55 + 21 = 76$	
$x_6 = 60$	$f_6 = 24$	$CF_6 = CF_5 + f_6 = 76 + 24 = 100$	
	$N = 100$		

Step 2: Calculation of $\left[\frac{N+1}{2}\right]$ in order to determine the median value.

1) Formula and data sorting. $\left[\frac{N+1}{2}\right]$, where $N = 100$

2) Data substitution and solving.

$$\frac{N+1}{2} = \frac{100+1}{2} \dots\dots\dots\dots\dots\dots\dots\dots\dots\dots.[\text{Sum numerator and divide by 2}]$$

$$\frac{N+1}{2} = 50.5$$

Note: Moving through the cumulative frequencies, we realized that 55 is the first value greater than 50.5. The midpoint value corresponding to 55 is 40. Therefore, our median is **40**.

c) Calculation of median using group data

When dealing with grouped data, the following formula is used in the determination of the median value.

$$Md = L_1 + \left[\frac{\frac{N}{2}-F}{F_m} \times C\right] \dots\dots\dots\dots\dots\dots\dots\dots\dots\dots\dots\dots(1)$$

$$Md = L_1 + C\left[\frac{\frac{N}{2}-\Sigma f_b}{F_M}\right] \dots\dots\dots\dots\dots\dots\dots\dots\dots\dots\dots(2)$$

Where;

Md = Median , L_1 = Lower limit of the median class,

F = Cummulative frequency before the median class , N =
Sum of frequency,

F_m = Frequency of the median class , C = Class size of the median class,

$\sum f_b$ = Sum of frequency before the median class .

The following steps are suggested in the determination of median when dealing with grouped data.

1) Examination of data. Identify the type of group (inclusive or exclusive) frequency and if the distribution is exclusive, convert to inclusive.

2) Derive median class. Establish the cumulative frequency of the distribution and obtain the value $\left[\frac{N}{2}\right]$, where N represents the sum of frequency. Identify from the values of cumulative frequencies, the first value that is greater than $\left[\frac{N}{2}\right]$. From the value identified move horizontally towards it corresponding class and consider the class to be the "median class".

3) Use the above formula and determine the median value

Example 1: Compute the median from the given frequency distribution table below.

Marks	5-10	10-15	15-20	20-25	25-30	30-35
Frequency	1	4	6	4	2	3

Solution

Step 1: Examination of data. The frequency distribution table is grouped and inclusive in nature. Therefore, we can continue to the next step.

Step 2: Derivation of the median class. Note that CF represents cumulative frequency

Origin Table		Working Table		
Marks	$[f_i]$	$\left[\frac{N}{2}\right]$	$[CF_i]$	
5-10	$f_1 =1$	Where,	$CF_1 = f_1 + f_0 = 1 + 0 = 1$	$CF_3 = 11$ is the first CF_i
10-15	$f_2 =4$	$N = 20$	$CF_2 = CF_1 + f_2 = 1 + 4 = 5$	that is greater than $\frac{N}{2} =$
15-20	$f_3 =6$		$CF_3 = CF_2 + f_3 = 5 + 6 = 11$	10.
20-25	$f_4 =4$		$CF_4 = CF_3 + f_4 = 11 + 4 = 15$	Looking at the mark
25-30	$f_5 =2$	$\frac{20}{2} = 10$	$CF_5 = CF_4 + f_5 = 15 + 2 = 17$	classes, CF_3 correspond
30-35	$f_6 =3$		$CF_6 = CF_4 + f_6 = 17 + 3 = 20$	to the class
	$N = 20$		Class width $[C] = L_2 - L_1 = 20 - 15 = 5$	15-20 [Median class].

Step 3: Stating of formula and data sorting plus solving

1) Formula. $\text{Median}(Md) = L_1 + \left[\frac{\frac{N}{2} - F}{F_m} \times C\right]$

Where; $L_1 = 15$, $F = 5$, $N = 20$, $F_m = 6$, $C=5$

2) Data substitution and solving.

$$\text{Median(Md)} = 15 + \left[\frac{\frac{20}{2}-5}{6} \times 5 \right] \quad\text{...................[Simplify numerator fraction]}$$

$$\text{Median(Md)} = 15 + \left[\frac{10-5}{6} \times 5 \right] \quad\text{...................[Simplify bracket numerator]}$$

$$\text{Median(Md)} = 15 + \left[\frac{5}{6} \times 5 \right] \quad\text{......[Simplify bracket fraction and multiply by 5]}$$

$$\text{Median(Md)} = 15 + 4.166666667 \quad\text{...............................[Sum values]}$$

$$\text{Median(Md)} = \mathbf{19.167} \quad\text{..............................[In three decimal places]}$$

Example 2: Determine the value of median from the frequency distribution table below

Class interval	40 - 49	50 - 59	60 - 69	70 - 79	80 - 89
Frequency	1	2	3	5	17

Solution

Step 1: Examination of data. The frequency distribution table is grouped and exclusive in nature. Therefore, we convert it to inclusive frequency table before any further working.

[original values]	40 - 49	50 - 59	60 - 69	70 - 79	80 - 89
[converted values]	39.5 - 49.5	49.5 - 59.5	59.5 - 69.5	69.5 - 79.5	79.5 - 89.5
Frequency	1	2	3	5	17

Step 2: Derivation of the median class. Note that CF represents cumulative frequency

Converted table		Working table		
Class intervals	$[f_i]$	$\left[\frac{N}{2}\right]$	Cumulative frequency $[CF_i]$	
39.5-49.5	$f_1 = 1$	Where,	$CF_1 = f_1 + f_0 = 1 + 0 = 1$	The first value of $[CF_i]$ greater than 14 is 28 $[CF_5]$
49.5-59.5	$f_2 = 2$	N=28	$CF_2 = CF_1 + f_2 = 1 + 2 = 3$	
59.5-69.5	$f_3 = 3$	$\frac{28}{2}=14$	$CF_3 = CF_2 + f_3 = 3 + 3 = 6$	
69.5-79.5	$f_4 = 5$		$CF_4 = CF_3 + f_4 = 6 + 5 = 11$	
79.5-89.5	$f_5 = 17$		$CF_5 = CF_4 + f_5 = 11 + 17 = 28$	Median class is
	$N = 28$			**79.5-89.5**

Step 3: Stating of formula and data sorting plus solving

1) Formula and data. $\text{Md} = L_1 + \left[\frac{\frac{N}{2}-F}{F_m} \times C \right]$, where; L_1=79.5, F =11, N=28, F_m=17,

C=10

2) Data substitution and solving.

$$\text{Median (Md)} = 79.5 + \left[\frac{\frac{28}{2}-11}{17} \times 10\right] \ldots\ldots\ldots\ldots\text{[Simplify numerator fraction]}$$

$$\text{Median (Md)} = 79.5 + \left[\frac{14-11}{17} \times 10\right] \ldots\ldots\ldots\ldots\text{[Simplify bracket numerator]}$$

$$\text{Median (Md)} = 79.5 + \left[\frac{3}{17} \times 10\right] \ldots\text{[Simplify bracket fraction and multiply by 10]}$$

$$\text{Median (Md)} = 79.5 + 1.764705882 \ldots\ldots\ldots\ldots\ldots\text{[Sum equation values]}$$

$$\text{Median (Md)} = \mathbf{81.26470588} \ldots\ldots\ldots\ldots\ldots\mathbf{[In\ three\ decimal\ places]}$$

2) Graphical method of calculating median

The determination of median graphically passes through three main steps as explained below.

1) Drawing of Ogives or Ogive. Here one is required to use the given frequency distribution table to draw a single ogive (greater than ogive or lesser than ogive) or both.

2) Marking. When dealing with both Ogives, we mark the point where they intersect. When dealing with a single ogive, we calculate the value $\left[\frac{N+1}{2}\right]$ or $\left[\frac{N}{2}\right]$ and mark the frequency that corresponds to the value when using discrete with frequency and grouped data respectively.

3) Determination of median. When using both Ogives, from the point of intersection, draw a downward vertical line to touch the class limit axis. The value on the vertical axis that the line touches to be the median. When dealing with an ogive, draw a horizontal line from the frequency marked point to touch the ogive. From the point it touches the ogive, draw a vertical line to touch the class limit axis and consider the value corresponding to the point it touches the axis the median.

Example 1: Compute the value of median graphically using the information of the frequency distribution table below.

Class	$[15-25[$	$[25-35[$	$[35-45[$	$[45-55[$	$[55-65[$	$[65-75[$
Frequency	4	11	19	14	0	2

<u>**Solution**</u>

Step 1: Draw a complete ogive. That is greater than ogive and lesser than ogive

Step 2: Derivation of median from the drawn Ogives	
Drawing of Ogives	**Explanation**
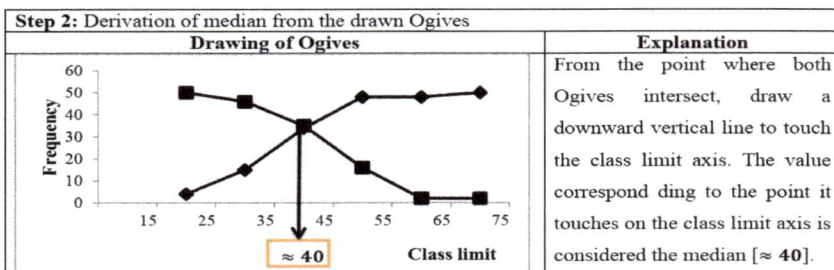 ≈ 40 **Class limit**	From the point where both Ogives intersect, draw a downward vertical line to touch the class limit axis. The value correspond ding to the point it touches on the class limit axis is considered the median [≈ **40**].

Test Your Understanding

Exercise 13: Compute the value of median using the following frequency distribution table below using the formula method.

1) Frequency distribution table 1

Marks	20	40	60	80
Frequency	15	2	10	4

[Answer: Median value = 40]

2) Frequency distribution table 2

Class interval	$[0 - 5[$	$[5 - 10[$	$[10 - 15[$	$[15 - 20[$
Frequenc	8	2	10	14

[Answer: Median value = 14]

3) Frequency distribution table 3

Classes	10 - 20	21 - 31	32 - 42	43 - 53	54 - 64
Frequency	4	6	10	7	3

[Answer: Median value = 37]

Exercise 14: Determine graphically the value of median from the frequency distribution table below

Class interval	$[10 - 20[$	$[20 - 30[$	$[30 - 40[$	$[40 - 50[$
Frequency	8	2	10	14

[Answer: Median value = 37]

8.2.3 Mode

The mode of a given observation or frequency distribution is the value that occur the maximum number of time or the value with the highest frequency. The following are worth knowing when dealing with mode;

a) When all element in the data set occurs the same number of time, we conclude by saying there is no mode.

b) When a given element in a data set occurs the highest time than the others, we say there is a Unimodal (single mode) situation.

c) When two values occurs the highest number of time in a set of values, both values are considered the mode. This situation is known as Bimodal

d) When more than two values occurs the highest number of times than others in a set of values, we consider all of them as the mode.

1) Algebraic method of calculating mode

a) When using discrete data without frequency

When using discrete data without frequency, the mode is simply the number or numbers that occur the highest time in a data set.

Example 1: Determine the mode the following distribution; 1, 2,2,2,3,4,5,6 and 6.

Solution

Note: Since our observations are in order, we simply identify the highest occurrence which is 2. Therefore our mode is 2.

Example 2: Compute the mode from the following data; 1, 7, 2, 6, 7, 8, 4, 4, 3, 5, 6, 4, 2 and 7.

Solution

Step 1: Re-arrangement of data or observations. 1, 2, 2, 3, 4, 4, 4, 5, 6, 6, 7, 7, 7 and 8.

Step 2: Compute the mode. Since 4 and 7 occurs the highest time than the others, it therefore means our mode is 4 and 7.

b) When using discrete data with frequency

When using discrete data with frequency, the mode is simply the number with the highest frequency. That is we examine the frequencies, identify the highest frequency and consider its corresponding value as the mode.

Example 1: From the frequency distribution table below, compute and illustrate graphically the mode score.

Scores	2	4	6	8	10	12	14
Frequency	3	10	15	4	6	15	2

Solution

Original table		Working table	Graphical illustration of mode
Scores	Frequency		
2	3	From the frequencies,	
4	10	we realize that the	
6	15	highest frequency is 15	
8	4	and corresponding	
10	6	scores having 15 as	
12	15	frequency are 6 and 12.	
14	2	Bimodal situation	
	N = 55		

c) When using grouped data

The first and most important step to take into consideration is to determine the modal class. The modal class is defined as the class with the highest frequency. The formula used to determine mode is given below.

$$\text{Mode}(M_O) = L_1 + \left[\frac{F_1 - F_0}{2F_1 - F_0 - F_2} \times C\right] \dots\dots\dots\dots\dots\dots\dots\dots\dots\dots\dots\dots\dots(1)$$

Where; M_O = Mode, L_1 = Lower limit of the modal class,

F_1 = Frequency of the modal class , F_0 = Frequency before the modal class

F_2 = Frequency after the modal class , C = class size of the modal class

$$\text{Mode}(M_O) = L_1 + \left[\frac{d_1}{d_1 + d_2} \times (L_2 - L_1)\right] \dots\dots\dots\dots\dots\dots\dots\dots\dots\dots\dots(2)$$

Where; M_O = Mode, L_1 = Lower limit of the modal class,

L_2 = Upper limit of the modal class

d_1 = Modal class frequency minus preceding class frquency

d_2 = Modal class frequency minus following class frequency

Note: It is worth knowing that in a situation where $2F_1 - F_0 - F_2$ gives the value zero, the formula appropriate in such situation is given as;

$$Mode(M_O) = \frac{f_1-f_0}{|f_1-f_0|+|f_1-f_2|} \times C \ldots\ldots\ldots\ldots\ldots\ldots\ldots\ldots\ldots\ldots\ldots\ldots\ldots\ldots(3)$$

The following suggested steps are worth using when determining the value of mode with grouped data.

1) Verification: Verify if the frequency distribution table is continuous or not in nature. If it is continuous, continue to the next step but if not, make it continuous.

2) Identification of modal class: Identify the class with the highest frequency and consider it the modal class.

Example 1: Calculate the mode of the students marks presented on the frequency distribution table below.

Marks	10 - 20	20 - 30	30 - 40	40 - 50	50 - 60	60 - 70
No of students	5	8	12	16	10	8

Solution

Step 1: Identification of modal class. Going through the frequencies, we see that $f_4 = 16$ is the highest and has as corresponding class 40-50. Therefore, our modal class is 40-50.

Marks	10 - 20	20 - 30	30 - 40	40 - 50	50 - 60	60 - 70
No of students	$f_1 = 5$	$f_2 = 8$	$f_3 = 12$	$f_4 = 16$	$f_5 = 10$	$f_6 = 8$

Step 3: Formula, data sorting and solving

Where: $L_1 = 40$, $L_2 = 50$, $F_1 = 16$, $F_0 = 12$, $F_2 = 10$, $d_1 = 16 - 12 = 4$, $d_2 = 16 - 10 = 6$, C $= 10$

Method 1

$$Mode(M_O) = L_1 + \left[\frac{F_1-F_0}{2F_1-F_0-F_2} \times C\right] \ldots\ldots\ldots\ldots\text{[Substitute data into the formula]}$$

$$Mode(M_O) = 40 + \left[\frac{16-12}{2(16)-12-10} \times 10\right] \ldots\ldots\ldots\ldots\text{[Expand denominator bracket]}$$

$$Mode(M_O) = 40 + \left[\frac{16-12}{32-12-10} \times 10\right]..\text{[Simplify fraction numerator and denominator]}$$

$$Mode(M_O) = 40 + \left[\frac{4}{10} \times 10\right] \ldots\ldots\ldots\ldots\text{[Simplify fraction and multiply by 10]}$$

$$Mode(M_O) = 40 + 4 \ldots\ldots\ldots\ldots\ldots\ldots\ldots\ldots\ldots\text{.......[Sum equation values]}$$

$$Mode(M_O) = \textbf{44}$$

Calculation of mode using formula 2

$$Mode(M_O)L_1 + \left[\frac{d_1}{d_1+d_2} \times (L_2 - L_1)\right] \ldots\ldots\ldots\ldots\text{[Substitute data into the formula]}$$

$Mode(M_O) = 40 + \left[\frac{4}{4+6} \times (50 - 40)\right]$.[Simplify denominator and inner bracket]

$Mode(M_O) = 40 + \left[\frac{4}{10} \times 10\right]$ ....[Simplify fraction]

$Mode(M_O) = 40 + [0.4 \times 10]$ ....[Simplify bracket]

$Mode(M_O) = 40 + 4$ ....[Sum equation values]

$Mode(M_O) = \mathbf{44}$

Example 2: Compute the mode from the frequency distribution below.

Marks	50 - 59	60 - 69	70 - 79	80 - 89	90 - 99	100 - 109
Frequency	5	20	40	50	30	6

<u>**Solution**</u>

Step 1: Verification: Frequency distribution table is not continuous in nature; therefore we convert it to continuous in nature.

Marks	50-59	60-69	70-79	80-89	90-99	100-109
Converted marks	49.5-59.5	59.5-69.5	69.5-79.5	79.5-89.5	89.5-99.5	99.5-109.5
Frequency	5	20	40	50	30	6

Step 2: Identification of modal class: From our given frequencies, $f_4 = 50$ is the highest frequency and our modal class is 79.5-89.5.

Marks	50-59	60-69	70-79	80-89	90-99	100-109
Converted marks	49.5-59.5	59.5-69.5	69.5-79.5	79.5-89.5	89.5-99.5	99.5-109.5
Frequency	$f_1 = 5$	$f_2 = 20$	$f_3 = 40$	$f_4 = 50$	$f_5 = 30$	$f_6 = 6$

Step 3: Formula, data sorting and solving

Method 1. $M_O = L_1 + \left[\frac{F_1 - F_0}{2F_1 - F_0 - F_2} \times C\right]$	Method 2. $M_O = L_1 + \left[\frac{d_1}{d_1 + d_2} \times (L_2 - L_1)\right]$
Where: $L_1 = 79.5$, $L_2 = 89.5$, $F_1 = 50$, $F_0 = 40$, $F_2 = 30$, $d_1 = 50 - 40 = 10$, $d_2 = 50 - 30 = 20$, $C = 89.5 - 79.5 = 10$	
$M_O = 79.5 + \left[\frac{50-40}{2(50)-40-30} \times 10\right]$ $M_O = 79.5 + \left[\frac{10}{100-40-30} \times 10\right]$ $M_O = 79.5 + \left[\frac{10}{30} \times 10\right]$ $M_O = 79.5 + 3.333333333$ $Mode(M_O) = \mathbf{82.83}$	$M_O = 79.5 + \left[\frac{10}{10+20} \times (89.5 - 79.5)\right]$ $M_O = 79.5 + \left[\frac{10}{30} \times 10\right]$ $M_O = 79.5 + [0.333333333 \times 10]$ $M_O = 79.5 + 3.333333333$ $Mode(M_O) = \mathbf{82.83}$

2) Graphical method of calculating mode

When determining mode graphically, the following steps are worth taken into consideration.

1) Verification: Make sure the frequency distribution table is continuous in nature.

2) Draw a histogram: Use the given frequency distribution table and draw a histogram

3) Crossing: Identify the tallest histogram bar. Draw from the angel of this bar a straight line touching the opposite histogram bar top angel.

4) Determination of mode: Mark the point where the lines intersect and from the mark, draw a vertical line to touch the class limit axis. Consider the value corresponding to the point it touches the class limit axis the mode.

Example 1: Compute the value of mode graphically

Class	$[0-10[$	$[10-20[$	$[20-30[$	$[30-40[$	$[40-50[$
Frequency	10	40	20	5	35

Solution

Step 1: Verification. Through observation we realize our table is continuous in nature.

Step 2: Drawing a histogram	Step 3: Crossing
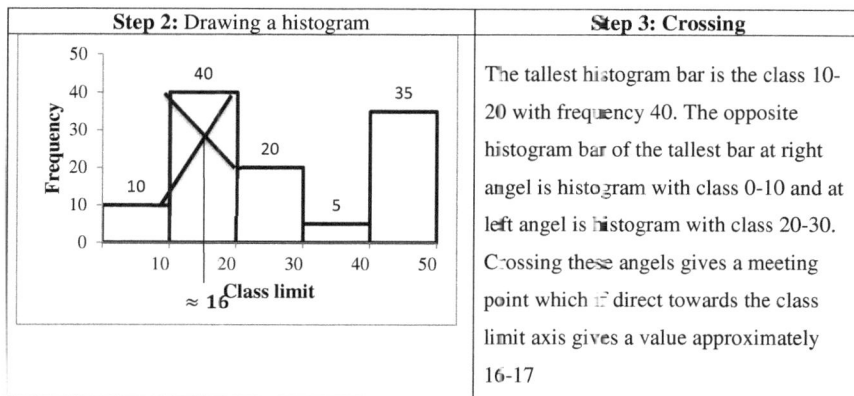 ≈ 16 Class limit	The tallest histogram bar is the class 10-20 with frequency 40. The opposite histogram bar of the tallest bar at right angel is histogram with class 0-10 and at left angel is histogram with class 20-30. Crossing these angels gives a meeting point which if direct towards the class limit axis gives a value approximately 16-17

Example 2: Graphically determine the mode value using the frequency distribution table below.

Scores	$[5-10[$	$[10-15[$	$[15-20[$	$[20-25[$
No of students	10	30	15	35

<u>**Solution**</u>

Step 1: Verification. Our frequency table is continuous in nature.

Step 2: Drawing of histogram	**Step 3**: Crossing
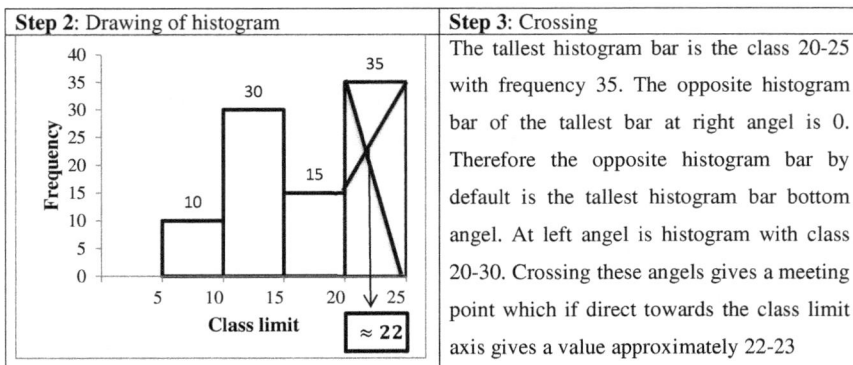	The tallest histogram bar is the class 20-25 with frequency 35. The opposite histogram bar of the tallest bar at right angel is 0. Therefore the opposite histogram bar by default is the tallest histogram bar bottom angel. At left angel is histogram with class 20-30. Crossing these angels gives a meeting point which if direct towards the class limit axis gives a value approximately 22-23

<u>**Test Your Understanding**</u>

Exercise 15: Compute the mode marks of students using the formula method from the frequency distribution tables below.

Marks	5	10	15	20	25	30
Frequency	12	7	6	2	20	19

[Answer: The value of mode = 25]

Class interval	0 - 5	5 - 10	10 - 15	15 - 20	20 - 25
Frequency	3	5	8	2	9

[Answer: The value of mode = 22]

Marks	10 - 19	20 - 29	30 - 39	40 - 49	50 - 59
No of students	10	6	9	3	4

[Answer: The value of mode = 17]

Note: It is worth knowing the empirical relationship between mean, median and mode is given as;

a) Mean $\geq$ Medain $\geq$ Mode [Round answer to the next integer]...................…...........(1)

   1) Mean = Median = Mode...................................…...................…................(1.1)

   2) Mean > $Median$ > $Mode$...................................…...............…..........……...(1.2)

b) Mean $-$ Mode = 3[Mean $-$ Medain]...................…..........…...….......…...........(2)

   1) Mean $= \dfrac{3\,\text{Median} - \text{Mode}}{2}$ ....…...............…...................…..........(2.1)

328

2) $\text{Median} = \frac{2\text{Mean} + \text{Mode}}{3}$ ....(2.2)

3) $\text{Mode} = 3\text{Median} - 2\text{Mean}$ ....(2.3)

Example 1: Use the numbers below and answer the questions that follow

70	66	60	55	61	63	72
68	60	60	63	60	75	68
59	71	53	76	64	64	52
64	64	68	64	66	67	63
64	70	69	68	63	59	57

1) Use the numbers above and construct a group frequency distribution table
2) Show that,

     c) Mean $\geq$ Medain $\geq$ Mode [Round answer to the next integer]

     d) Mean $-$ Mode $= 3[\text{Mean} - \text{Medain}]$

     e) Mode $= 3[\text{Medain}] - 2[\text{Mean}]$

Solution

1) Use the numbers above and construct a group frequency distribution table

Step 1: Orderly representation of values.

52	53	55	57	59	59	60
60	60	60	61	63	63	63
63	64	64	64	64	64	64
66	66	67	68	68	68	68
69	70	70	71	72	75	76

Step 2: Calculation of number of class interval and class size or magnitude

  a) Calculation of number of class interval

    Step 1: Formula. $K = 1 + 3.322[\text{Log n}]$, where n $= 35$

    Step 2: Data substitution and solving.

        $K = 1 + 3.322[\text{Log 35}]$ ....[Work log]

        $K = 1 + 3.322[1.544068044]$ ....[Solve]

        $K = 6$

  b) Calculation of class size or magnitude

    Step 1: Formula. $\text{Class Size}(S) = \frac{\text{Range}}{\text{Number of class interval}}$, where range $= 24$ [that is 76-52]

    Step 2: Data substitution and solving.

$$\text{Class Size(S)} = \frac{\text{Range}}{\text{Number of class interval}} \quad \dots\text{[Substitute data into the formula]}$$

$$\text{Class Size(S)} = \frac{24}{6} \dots\dots\dots\dots\dots\dots\dots\dots\dots\dots\dots\dots\dots\dots\dots\dots\text{[Simplify fraction]}$$

$$\text{Class Size(S)} = \mathbf{4}$$

Step 3: Construction of frequency distribution table

Class interval	52 - 56	57 - 61	62 - 66	67 - 71	72 - 76
Converted class interval	51.5 - 56.5	56.5 - 61.5	61.5 - 66.5	66.5 - 71.5	71.5 - 76.5
Frequency	3	8	12	9	3

2) **Show that,**

a) **Mean $\geq$ Medain $\geq$ Mode [Round answer to the next integer]**

Step 1: Formula. $\bar{X} = \frac{\Sigma[f_i x_i]}{N}$, $M_d = L_1 + \left[\frac{\frac{N}{2}-F}{F_m} \times C\right]$, $M_0 = L_1 + \left[\frac{F_1-F_0}{2F_1-F_0-F_2} \times C\right]$

Step 2: Tabula derivation of unknowns. That is, $[f_i x_i]$, median class and modal class

Original table			Working table			
		Mode Section	Mean section		Median section	
Converted class Interval	$[f_i]$		$[x_i]$	$[f_i(x_i)]$	$[CF_i]$	
51.5-56.5	$f_1 = 3$	Modal class 61.5-66.5	54	162	3	$\left[\frac{N}{2}\right] = \frac{35}{2} = 17.5$
56.5-61.5	$f_2 = 8$		59	472	11	
61.5-66.5	$f_3 = 12$		64	768	23	Median class
66.5-71.5	$f_4 = 9$		69	621	32	61.5-66.5
71.5-76.5	$f_5 = 3$		74	222	35	
	$N = 35$			$\Sigma = 2245$		

Step 3: Data substitution and solving

i. **Mean.**

$$\text{Mean } (\bar{X}) = \frac{\Sigma[f_i x_i]}{N} \dots\dots\dots\dots\dots\dots\dots\text{[Substitute data into the formula]}$$

$$\text{Mean } (\bar{X}) = \frac{2245}{35} \dots\dots\dots\dots\dots\dots\dots\dots\text{[Simplify equation fraction]}$$

$$\text{Mean } (\bar{X}) = \mathbf{64} \dots\dots\dots\dots\dots\dots\dots\dots\text{[Value of arithmetic mean]}$$

330

ii. Median.

$$\text{Median}(M_d) = L_1 + \left[\frac{\frac{N}{2}-F}{F_m} \times C\right] \ldots\ldots\ldots\text{[Substitute data into the formula]}$$

$$\text{Median}(M_d) = 61.5 + \left[\frac{\frac{35}{2}-11}{12} \times 5\right] \ldots\ldots\ldots\ldots\ldots\text{[Solve equation]}$$

$$\text{Median}(M_d) = \mathbf{64}$$

iii. Mode.

$$\text{Mode}(M_O) = L_1 + \left[\frac{F_1-F_0}{2F_1-F_0-F_2} \times C\right] \ldots\text{[Substitute data into the formula]}$$

$$\text{Mode}(M_O) = 61.5 + \left[\frac{12-8}{2(12)-8-9} \times 5\right] \ldots\ldots \ldots\ldots\ldots\text{[Solve equation]}$$

$$\text{Mode}(M_O) = \mathbf{64]}$$

Therefore, Mean = Medain = Mode making the given equation Mean $\geq$ Medain $\geq$ Mode true

b) **Mean − Mode = 3[Mean − Medain]**

$\quad$ 64 − 64 = 3[64 − 64] ……………..[Solve equation on both side separately]

$\quad$ **0 = 0** ………………………………………………………..**[The equation is valid]**

c) **Mode = 3[Medain] − 2[Mean]**

$\quad$ 64 = 3[64] − 2[64] …………..[Expand bracket on right hand side of equation]

$\quad$ 64 = 192 − 128 …………………...……..[Simplify right hand side of equation]

$\quad$ **64 = 64** ………………………...……...………..**[The equation is valid]**

Example 2: Fine the value of median and mode from the frequency distribution table below. Assume that the arithmetic mean $[\overline{X}]$ is 34. [Round answers to the next integer]

Marks	0-10	10-20	20-30	30-40	40-50	50-60
Frequency	5	15	20	f_4	20	10

Solution

Step 1: Frequency distribution table correction.

Original table		Working table		
Marks	$[f_i]$	$[x_i]$	$[f_i(x_1)]$	Calculation of the value of f_4
0 - 10	5	5	25	$\overline{X} = \frac{\sum[f_i(x_i)]}{N}$, where $\overline{X} = 34$
10 - 20	15	15	225	$34 = \frac{2200+35f_4}{70+f_4}$
20 - 30	20	25	500	
30 - 40	f_4	35	$35 + f_4$	$34(70 + f_4) = 2200 + 35f_4$
40 - 50	20	45	900	$2380 + 34f_4 = 2200 + 35f_4$
50 - 60	10	55	550	$2380 - 2200 = 35f_4 - 34f_4$

	$\Sigma = 70 + f_4$	$\Sigma = 2200 + 35f_4$	$f_4 = 180$

Step 2: Calculation of median and mode. It is worth knowing that our new frequency distribution table is given now as;

Marks	0 - 10	10 - 20	20 - 30	30 - 40	40 - 50	50 - 60
Frequency	5	15	20	180	20	10

a) Formula. $\text{Median}(M_d) = L_1 + \left[\frac{\frac{N}{2}-F}{F_m} \times C\right]$ and $\text{Mode}(M_0) = L_1 + \left[\frac{F_1-F_0}{2F_1-F_0-F_2} \times C\right]$

b) Tabula derivation of unknown

Original table		Working table			
		Mode section		Median section	
Marks	$[f_i]$		$[CF_i]$	$\frac{N}{2} = \frac{250}{2} = 125$	
0 - 10	5		5		
10 - 20	15	The modal class is	20		
20 - 30	20	30-40	40	The median class is	
30 - 40	180		220	30-40	
40 - 50	20		240		
50 - 60	10		250		
	N = 250				

c) Data substitution and solving.

Calculation of Median	Calculation 0f Mode
Formula. $M_d = L_1 + \left[\frac{\frac{N}{2}-F}{F_m} \times C\right]$	Formula. $M_0 = L_1 + \left[\frac{F_1-F_0}{2F_1-F_0-F_2} \times C\right]$
Where: $L_1 = 30$, $F = 40$, $F_m = 180$, $C = 10$, $F_1 = 180$, $F_0 = 20$, $F_2 = 20$, $N = 250$	
$M_d = 30 + \left[\frac{\frac{250}{2}-40}{180} \times 10\right]$ ....[Simplify] $M_d = 30 + [4.722222222]$ ....[Solve] $\mathbf{M_d = 35}$ ....[Value of median]	$M_0 = 30 + \left[\frac{180-20}{2(180)-20-20} \times 10\right]$ ......[Simplify] $M_0 = 30 + 5$ ....[Solve] $\mathbf{M_0 = 35}$ ....[Value of mode]

Example 3: What will be the mean marks and the mode marks of the students, considering that the median marks of 229 students in an examination is 46? Use the frequency table below in calculating the mean marks.

332

Class interval	10-20	20-30	30-40	40-50	50-70	70-80	80-90
Frequency	12	30	f_3	65	50	10	f_7

Solution

Hint: Class sizes are not all equal as [70-80] is having a class size of 20 which is different from normal class size of 10. All frequencies are not known [f_3 and f_7].

Step 1: Unequal class size Correction.

Class interval	10-20	20-30	30-40	40-50	50-60	60-70	70-80	80-90
Frequency	12	30	f_3	65	25	25	10	f_7

Step 2: Calculation of missing frequencies. Since total frequency and median are given, we use the median formula to obtain the missing frequencies values as seen below.

1) Formula and data. Median$(M_d) = L_1 + \left[\dfrac{\frac{N}{2}-F}{F_m} \times C\right]$, where; $M_d = 46$, $C = 10$, $N = 229$

2) Tabula derivation of unknown. That is, [F].

Original Table		Working Table	
Class Interval	[f_i]	Cumulative frequency	
10 - 20	12	12	
20 - 30	30	42	Since are median is
30 - 40	f_3	$42 + f_3$	46, it therefore mean
40 - 50	65	$107 + f_3$	it is within class
50 - 60	25	$132 + f_3$	interval of 40-50
60 - 70	25	$157 + f_3$	which is also known
70 - 80	10	$167 + f_3$	as the median class
80 - 90	f_7	$167 + f_3 + f_7$	
	$N = 167 + f_3 + f_7$		

3) Data substitution and solving.

Median$(M_d) = L_1 + \left[\dfrac{\frac{N}{2}-F}{F_m} \times C\right]$ ....[Substitute data into the formula]

$46 = 40 + \left[\dfrac{\frac{229}{2}-(42+f_3)}{65} \times 10\right]$ .....[Take 40 to the right and simplify the division]

$46 - 40 = \frac{114.5 - 42 - f_3}{65} \times 10$..[Divide both sides by 10 and multiple both side by 65]

$\frac{46-40}{10}(65) = 114.5 - 42 - f_3$ ...[Simplify equation and collect like teams together]

$39 = 72.5 - f_3$ ....…..........................…..…......[Make f_3 subject of formula]

$-f_3 = -33.5$ ....…..…......[Divide both sides of equation by -1]

$f_3 = 33.5 \cong 34$ ....…..[First missing value]

Note: Now that we know the value of (f_3) and total frequency (N), we apply the formula of sum of frequency to determine the value of (f_7) as seen below.

a) Formula. $N = \sum f_i$, where; f_i ranges from f_1 to f_7

b) Data substitution and solving

$229 = 167 + 34 + f_7$ ....…...[Simplify right hand side of the equation]

$229 = 201 + f_7$ ....…...[Make f_7 subject of formula and solve equation]

$f_7 = 28$ ....….[Second missing value]

Step 3: Calculation of mean

1) Formula and data. $\text{Mean}(\overline{X}) = \frac{\sum[f_i(x_i)]}{N}$, where; $N = 229$ and $\sum[f_i(x_i)] = ?$

2) Tabula derivation of unknown. That is $\sum[f_i(x_i)]$ and calculation of mean and mode

Original Table		Working Table	
Class interval	$[f_i]$	$[x_i]$	$[f_i(x_i)]$
10 - 20	12	$(10 + 20)/2 = 15$	$12(15) = 180$
20 - 30	30	$(20 + 30)/2 = 25$	$30(25) = 750$
30 - 40	34	$(30 + 40)/2 = 35$	$34(35) = 1,190$
40 - 50	65	$(40 + 50)/2 = 45$	$65(45) = 2,925$
50 - 60	25	$(50 + 60)/2 = 55$	$25(55) = 1,375$
60 - 70	25	$(60 + 70)/2 = 65$	$25(65) = 1,625$
70 - 80	10	$(70 + 80)/2 = 75$	$10(75) = 750$
80 - 90	28	$(80 + 90)/2 = 85$	$28(85) = 2,380$
	$N = 229$		$\sum = 11,175$

3) Data substitution and solving

$\text{Mean}(\overline{X}) = \frac{\sum[f_i(x_i)]}{N}$ ....[Substitute data into the formula]

$$\text{Mean}(\overline{X}) = \frac{11{,}175}{229} \ldots\ldots\ldots\ldots\ldots\ldots\ldots\ldots\ldots\ldots\ldots\ldots\text{[Solve equation]}$$

$$\text{Mean}(\overline{X}) = \mathbf{49} \ldots\ldots\ldots\ldots\ldots\ldots\ldots\ldots\ldots\ldots\ldots\ldots\textbf{[Rounded value]}$$

Step 4: Calculation of mode

a) Formula. $M_O = L_1 + \left[\frac{F_1 - F_0}{2(F_1) - F_0 - F_2} \times C\right]$

   Where; $L_1 = 40$, $F_1 = 65$, $F_0 = 34$, $F_2 = 25$, $C = 10$

b) Data substitution and solving

$$\text{Mode }(M_O) = L_1 + \left[\frac{F_1 - F_0}{2(F_1) - F_0 - F_2} \times C\right] \ldots\ldots\text{[Substitute data into the formula]}$$

$$\text{Mode }(M_O) = 40 + \left[\frac{65 - 34}{2(65) - 34 - 25} \times 10\right] \ldots\ldots\ldots\ldots\ldots\ldots\text{[Solve equation]}$$

$$\text{Mode }(M_O) = \mathbf{44} \ldots\ldots\ldots\ldots\ldots\ldots\ldots\ldots\ldots\ldots\ldots\ldots\textbf{[Rounded value]}$$

Example 4: The mode of the following distribution is 61. Determine the value of mean and median wages of works

Daily wages	a b	40-50	50-60	60-70	70-80	80-90
Midpoints	35					
No of workers	8	16	22	28	f_5	12

Solution

Note: The frequency table is not statistical in nature since all interval and frequencies are not given. Therefore, we are first required to complete the table.

Step 1: Calculation of unknown interval. [Being continuous in nature by default $(b = 40)$]

a) Formula and data. $\text{Midpoint} = \frac{\text{Lower limit} + \text{Upper limit}}{2}$, where; $\text{Upper limit} = \underline{b} = 40$ and $\text{Midpoint} = 35$.

b) Data substitution and solving.

$$35 = \frac{a + 40}{2} \ldots\ldots\ldots\ldots\ldots\ldots\ldots\ldots\ldots\ldots\text{[Multiply both sides of equation by 2]}$$

$$35(2) = \underline{a} + 40 \ldots\ldots\ldots\ldots\ldots\ldots\ldots\ldots\text{[Take 40 to the left side of equation]}$$

$$70 - 40 = \underline{a} \ldots\ldots\ldots\ldots\ldots\ldots\ldots\ldots\ldots\ldots\ldots\text{[Solve equation]}$$

$$a = \mathbf{30}$$

Step 2: Calculation of missing frequency

Hint: Since mode is given, it implies we can use the formula of mode in deriving the unknown frequency. It is worth knowing that our new frequency table is given as seen below.

Daily wages	30 - 40	40-50	50-60	60-70	70-80	80-90
No of workers	8	16	22	28	f_5	12

a) Formula and data. $M_O = L_1 + \left[\frac{F_1-F_0}{2F_1-F_0-F_2} \times C\right]$. Since mode is 61, therefore, our modal

class is [60-70]. Therefore, $L_1 = 60$, $F_1 = 28$, $F_0 = 22$, $C = 10$ and Mode$(M_O) = 66$

b) Data substitution and solving.

$61 = 60 + \left[\frac{28-22}{2(28)-22-f_5} \times 10\right]$ ......[Take 60 to the left and simplify the bracket]

$61 - 60 = \frac{6}{56-22-f_5} \times 10$ ....[Simplify denominator]

$1 = \frac{6}{34-f_5} \times 10$ ....[Multiply both sides of equation by $34 - f_5$]

$1(34 - f_5) = 60$ ....[Multiply bracket left side of equation by 6]

$34 - f_5 = 60$ ....[Collect like team together and solve equation]

$f_5 = 26$

Step 3: Calculation of mean

Note: It is worth knowing that our new frequency distribution table is given now as;

Daily wages	30 - 40	40 - 50	50 - 60	60 - 70	70 - 80	80 - 90
No of workers	8	16	22	28	26	12

1) Formula. $\bar{X} = \frac{\sum f_i x_i}{N}$, where; N = 112 and $\sum f_i x_i$ =?

2) Tabular determination of unknown. That is $\sum f_i x_i$

Original Table		Working Table	
Wages	Frequency	$[x_i = $ **(Upper Limit + Lower Limit)**$/2]$	$f_i x_i$
30 - 40	8	$(40 + 30)/2 = 70/2 = 35$	$8(35) = 280$
40 - 50	16	$(50 + 40)/2 = 90/2 = 45$	$16(45) = 720$
50 - 60	22	$(60 + 50)/2 = 110/2 = 55$	$22(55) = 1,210$
60 - 70	28	$(70 + 60)/2 = 130/2 = 65$	$28(65) = 1,820$
70 - 80	26	$(80 + 70)/2 = 150/2 = 75$	$26(75) = 1,950$
80 - 90	12	$(90 + 80)/2 = 170/2 = 85$	$12(85) = 1,020$
	N = 112		$\sum f_i x_i = 7000$

3) Data substitution and solving

Mean$(\bar{X}) = \frac{\sum f_i x_i}{N}$ ....[Substitute data into the formula]

Mean$(\bar{X}) = \frac{7,000}{112}$ ....[Simplify fraction]

$$\text{Mean}(\overline{X}) = 62.5$$

Step 4: Calculation of median

a) Formula. $\text{Median}(Md) = L_1 + \left[\dfrac{\frac{N}{2}-F}{F_m} \times C\right]$, where; $C = 10$

b) Determination of the median class. C. F represent cumulative frequency

Wages	30 – 40	40 – 50	50 – 60	60 – 70	70 – 80	80 – 90	$\dfrac{112}{2} = 56$
Frequency	8	16	22	28	26	12	Median class
C. F	8	24	46	74	100	112	**60 – 70**

c) Data substitution and solving

$$\text{Median (Md)} = 60 + \left[\dfrac{\frac{112}{2}-46}{74} \times 10\right] \ldots\ldots\ldots\ldots\ldots\ldots\ldots\text{[Solve equation]}$$

$$\text{Median (Md)} = \mathbf{61.4} \ldots\ldots\ldots\ldots\ldots\ldots\ldots\ldots\textbf{[In one decimal place]}$$

Example 5: Compute the value of mean, median and mode using information from the frequency distribution table below.

Class interval	[10 – 20[	[20 – 40[	[40 – 60[	[60 – 80[	[80 and above[
Frequency	7	10	12	5	8

Solution

Calculation of Mode

Hint: Since the class interval with the highest frequency is given, we can directly determine the value of mode. The modal class is therefore $[40 - 60[$

Step 1: Formula and data. $M_O = L_1 + \left[\dfrac{F_1-F_0}{2F_1-F_0-F_2} \times C\right]$, where; $L_1 = 40$, $F_1 = 12$, $F_0 = 10$, $C = 20$ and $f_2 = 5$

Step 2: Data substitution and solving.

$$\text{Mode}(M_O) = 40 + \left[\dfrac{12-10}{2(12)-10-5} \times 20\right]\ldots\text{[Simplify numerator and denominator]}$$

$$\text{Mode}(M_O) = 40 + \left[\dfrac{2}{9} \times 20\right] \ldots\ldots\ldots\ldots\text{[Simplify fraction and multiply by 20]}$$

$$\text{Mode}(M_O) = 40 + 4.444444444 \ldots\ldots\ldots\ldots\ldots\ldots\text{[Sum values]}$$

$$\text{Mode}(M_O) = \mathbf{44} \ldots\ldots\ldots\ldots\ldots\ldots\ldots\ldots\textbf{[Rounded value]}$$

Calculation of Median

Step 1: Formula. Median$(M_d) = L_1 + \left[\dfrac{\frac{N}{2}-F}{F_m} \times C\right]$, where; $C = 20$

Step 2: Determination of cumulative frequency [CF] and the median class

Class interval	Below 20	$[20 - 40[$	$[40 - 60[$	$[60 - 80[$	$[80$ and above$[$
Frequency	7	10	12	5	8
$[CF_i]$	7	17	29	34	42

Note: The value N is 42 and $\left[\dfrac{N}{2}\right] = \left[\dfrac{42}{2}\right] = 21$. The median class is; $[40 - 60[$. Therefore, $L_1 = 40, F = 17, F_m = 12, C = 20, N = 42$

Step 3: Data substitution and solving.

$$\text{Median}(M_d) = 40 + \left[\dfrac{\frac{42}{2}-17}{12} \times 20\right] \quad\dots\dots\dots\dots\dots\dots\dots\dots\text{[Simplify Numerator]}$$

$$\text{Median}(M_d) = 40 + \left[\dfrac{4}{12} \times 20\right] \quad\dots\dots\dots\dots\dots\text{[Simplify fraction and multiply by 20]}$$

$$\text{Median}(M_d) = 40 + 6.666666667 \quad\dots\dots\dots\dots\dots\dots\dots\dots\dots\dots\dots\text{[Sum values]}$$

$$\text{Median}(M_d) = \textbf{47} \quad\dots\dots\dots\dots\dots\dots\dots\dots\dots\dots\dots\dots\dots\textbf{[Rounded value]}$$

Calculation of Mean

Hint: Since the class size are equal, using the idea of mid-points or class size [20], the first lower class limit will be 0 and the last upper class limit will be 100.

Step 1: Formula and data. Mean $(\overline{X}) = \dfrac{\sum f_i x_i}{N}$, where; $N = 42$ and $\sum f_i x_i = ?$

Step 2: Tabular determination of unknown.

Class interval	$[0 - 20[$	$[20 - 40[$	$[40 - 60[$	$[60 - 80[$	$[80 - 100[$	
Frequency	7	10	12	5	8	
$[x_i]$	10	30	50	70	90	
$f_i x_i$	70	300	600	350	720	$\sum = 2,040$

Step 3: Data substitution and solving.

$$\text{Mean } (\overline{X}) = \dfrac{\sum f_i x_i}{N} \quad\dots\dots\dots\dots\dots\dots\dots\dots\dots\dots\text{[Substitute data into the formula]}$$

$$\text{Mean } (\overline{X}) = \dfrac{2,040}{42} \quad\dots\dots\dots\dots\dots\dots\dots\dots\dots\dots\dots\text{[Simplify fraction]}$$

$$\text{Mean } (\overline{X}) = \textbf{49} \quad\dots\dots\dots\dots\dots\dots\dots\dots\dots\dots\textbf{[Rounded value]}$$

Example 6: Determine the values of the missing frequencies, considering that the harmonic mean of 17 observations is 45.

Marks	20	40	80	100
Frequency	4	f_2	6	f_4

Solution

Hint: The value of harmonic mean makes it easy to derive the unknown frequencies through the harmonic mean formula.

Step 1: Calculating the values of unknown frequencies through harmonic mean.

1) Formula and data. Harmonic Mean(HM) $= \dfrac{N}{\sum f_i \left[\frac{1}{x_i}\right]}$, where; N = 17, HM = 45 and

$\sum f_i \left[\dfrac{1}{x_i}\right] = ?$

2) Tabular derivation of unknown. That is $\sum f_i \left[\dfrac{1}{x_i}\right]$.

Original table		Working table		Explanation
Mark	Frequency	$\dfrac{1}{x_i}$	$f_i \left[\dfrac{1}{x_i}\right]$	$10 + f_2 + f_4 = 17$
20	4	0.05	0.2	Making f_2 subject of formula, we get
40	f_2	0.025	$0.025f_2$	$f_2 = 7 - f_4$.This
80	6	0.0125	0.075	will be substituted
100	f_4	0.01	$0.01f_4$	in the place of f_2 in
	$\sum = 10 + f_2 + f_4$		$\sum = 0.275 + 0.025f_2 + 0.01f_4$	our working.

3) Data substitution and solving.

Harmonic Mean(HM) $= \dfrac{N}{\sum f_i \left[\frac{1}{x_i}\right]}$ ....[Substitute data into formula]

$45 = \dfrac{17}{0.275 + 0.025f_2 + 0.01f_4}$ ...[Transform the equation to have only one unknown]

$45 = \dfrac{17}{0.275 + 0.025(7 - f_4) + 0.01f_4}$ ....[Simplify denominator]

$45 = \dfrac{17}{0.275 + 0.175 - 0.025f_4 + 0.01f_4}$ .... Collect like terms in the denominator]

$45 = \dfrac{17}{0.45 - 0.015f_4}$ ....[Multiply both sides of equation by $0.45 - 0.015f_4$]

$20.25 - 0.675f_4 = 17$ ....[Collect like terms together and make f_4 subject]

$f_4 = 5$ ....**[Rounded value]**

Note: Since $10 + f_2 + f_4 = 17$ and $f_4 = 5$, substituting 5 in the place of (f_4) to get (f_2)

$10 + f_2 + 5 = 17$ ....…........[Collect like terms together]

$f_2 + 15 = 17$ ....[Take 15 to the right hand side of equation]

$f_2 = 17 - 15$ ....[Solve equation]

$f_2 = 2$

Test Your Understanding

Exercise 16: Fine the missing frequencies from the given frequency distribution table below, assuming the median to be 24 and total frequency of 100. [Approximate values to the nearest whole number]

Class interval	$[0 - 10[$	$[10 - 20[$	$[20 - 30[$	$[30 - 40[$	$[40 - 50[$
Frequency	14	f_2	27	f_4	15

[Answer: $f_2 = 35$ and $f_4 = 9$]

Exercise 17: Considering that the mode of a frequency distribution is 44, calculate the mean and median.

Marks	$[0 - 20[$	$[20 - 40[$	$[40 - 60[$	$[60 - 80[$	$[80 - 100[$
Frequency	5	18	f_3	12	5

[Answer: Mean = 48 and Median = 47]

Exercise 18: Compute the value of mean, median and mode using the frequency distribution table below. [Round up values to the nearest whole number]

Scores	$[0 - 10[$	$[10 - 20[$	$[20 - 30[$	$[30 - 40[$	$[40 - 50[$	$[50 - 60[$	$[60 - 70[$
Frequency	6	12	22	37	17	8	5

[Answer: Mean = 34, Median = 34 and Mode = 34]

Exercise 19: Given that the mean of 121 observations is 129, fine the values of the missing frequencies using the frequency distribution table below.

Marks	[below 50[	$[50 - 100[$	$[100 - 150[$	$[150 - 200[$	[200 and above[
Mid-points	25				225
Frequency	f_1	f_2	36	40	10

[Answer: $f_1 = 15$ and $f_2 = 20$]

Exercise 20: Calculate the value of median and mode from the given frequency distribution table, considering that arithmetic mean is 20.

Scores	10	15	20	25	30	Total
No of matches	3	f_2	18	f_4	10	56

[Answer: Median = 20 and Mode = 15]

340

CHAPTER NINE

MEASUREMENT OF DISPERSION

9.1 Meaning of measurement of dispersion

Dispersion measures the extents to which a set of observations are different from their respective mean. Major measures of dispersion include range, absolute mean deviation, median deviation, mode deviation, variance, standard deviation and their respective coefficients.

9.2. Types of measures of dispersion
9.2.1 Range

Range is simple the difference between the highest value (highest class limit) and the lowest value (lowest class limit). The formula for calculating range and coefficient of range is given as;

$$\text{Range} = \text{Highest value} - \text{lowest value} \dots\dots\dots(1)$$

$$\text{Coefficient of Range} = \frac{\text{Highest Value} - \text{Lowest Value}}{\text{Highest Value} + \text{Lowest Value}} \times 100 \dots\dots\dots(2)$$

$$\text{Range} = \text{Highest upper class limit} - \text{Lowest lower class limits} \dots\dots\dots(3)$$

$$\text{Coefficient of Range} = \frac{\text{Range}}{\text{Highest upper class limit} + \text{Lowest Lower class limit}} \times 100$$
$$\dots\dots\dots(4)$$

Note: Equation (1) and equation (2) are used when dealing with ungrouped data and equation (3) and (4) is used when dealing with grouped data. When dealing with ungrouped data, we are required to arranged the data in an increasing or decreasing order before solving.

Example 1: Calculate the range of the following scores; 2, 4, 8, 13, and 30

Solution

Hint: Data are well arranged in an increasing order. Therefore, we move to the next step.

Step 1: Formula and data. Range = Highest value − lowest value.

Where; Highest value = 30 and lowest value = 2

Step 2: Data substitution and solving.

Range = Highest value − lowest value ....[Substitute data into the formula]

Range = 30 − 2..[Solve equation]

Range = **28**

Example 2: Compute the range and coefficient of range using the data; 40, 24, 80, 15 and 10

Solution

Note: The values are not in order. Hence we are required to re-arrange them orderly.

Step 1: Re-arrangement. That is 10, 15, 24, 40 and 80

Step 2: Calculation of range

    a) Formula and data. Range = Highest value − lowest value.

            Where; Highest value = 80 and lowest value = 10

    b) Data substitution and solving.

            Range = Highest value − lowest value .[Substitute data into the formula]

            Range = 80 − 10...[Solve equation]

            Range = **70**

Step 3: Calculation of coefficient of range

    a) Formula: Coefficient of Range $= \frac{\text{Highest Value} - \text{Lowest Value}}{\text{Highest Value} + \text{Lowest Value}} \times 100$

    b) Data substitution and solving.

            Coefficient of Range $= \frac{80-10}{80+10} + 100$.....................[Simplify fraction]

            Coefficient of Range $= \frac{70}{90} \times 100$ ..[Simplify fraction and multiply by 100]

            Coefficient of Range = **78**

Example 3: Determine the range and coefficient of range of the following frequency distribution table below.

Scores	[0 − 5[	[5 − 10[	[10 − 15[	[15 − 20[	[20 − 25[
Frequency	4	6	2	10	23

Solution

a) Determination of Range

Step 1: Formula. Range = Highest upper class limit − Lowest lower class limits

  Where: Highest upper class limit(HUCL) = 25 and Lowest lower class limits(LLCL) = 0

Step 2: Data substitution and solving.

            Range = Range = HUCL − LLCL ....[Substitute data into the formula]

            Range = 25 − 0...[Solve equation]

            Range = **25**

b) Determination of Coefficient of Range

Step 1: Formula: Coefficient of Range $= \dfrac{Range}{Highest\ upper\ class\ limit + Lowest\ Lower\ class\ limit} \times 100$

Where: Highest uppercClass limit $= 25$ and Lowest lower class limits $= 0$

Step 2: Data substitution and solving.

Coefficient of Range $= \dfrac{25}{25+0} + 100$.... ....[Simplify denominator]

Coefficent of Range $= \dfrac{25}{25} \times 100$ ....[Simplify fraction and multiply by 100]

Coefficient of Range $= \mathbf{100}$

Example 4: Compute the value of range and coefficient of range using the frequency distribution table below.

Marks	[10 − 20[	[20 − 30[	]30 − 40[	[40 − 50[	[50 − 60[
Frequency	23	3	]0	4	12

Solution

a) Calculation of Range

Step 1: Formula. Range $=$ Highest upper class limit $-$ Lowest lower class limits

Where: Highest upper class limit(HUCL) $= 60$ and Lowest lower class limits(LLCL) $= 10$

Step 2: Data substitution and solving.

Range $=$ HUCL $-$ LLCL..........................[Substitute data into the formula]

Range $= 60 - 10$...…………..[Solve equation]

Range $= \mathbf{50}$

b) Calculation of Coefficient of Range

Step 1: Formula. Coefficient of Range $= \dfrac{Range}{Highest\ upper\ class\ limit + Lowest\ Lower\ class\ limit} \times 100$

Where: Highest uppercClass limit $= 60$ and Lowest lower class limits $= 10$

Step 2: Data substitution and solving.

Coefficient of Range $= \dfrac{50}{60+10} + 100$...................… ....[Simplify denominator]

Coefficient of Range $= \dfrac{50}{70} \times 100$ ....[Simplify fraction and multiply by 100]

Coefficient of Range $= \mathbf{71}$

Exercise 1: Compute the value of range and coefficient of range from the following marks of students realized during a math and an English test.

1) Math marks: 10, 20, 30, 40, 50, and 60.

[Answer: Range = 50 and coefficient of range = 71.43]

2) English marks: 30, 5, 70, 10, 60 and 20

[Answer: Range = 65 and coefficient of range = 86.67]

Exercise 2: Determine the value of range and coefficient of range using the frequency distribution tables below.

Marks	[0 − 20[	[20 − 40[	[40 − 60[	[60 − 80[	[80 − 100[
Frequency	23	3	10	4	12

[Answer: Range = 100 and coefficient of range = 100]

Marks	[25 − 50[	[50 − 75[	[75 − 100[	[100 − 125[	[125 − 150[
Frequency	20	15	10	40	62

[Answer: Range =125 and coefficient of range = 71.42]

9.2.2 Absolute Mean deviation

This measures how observations are different from their arithmetic mean. The formula for calculating absolute mean deviation for discrete data without frequency is given as;

$$\text{Mean Deviation } (\overline{X}D) = \frac{\Sigma|X_i - \overline{X}|}{n} = \frac{\Sigma|d|}{n} \text{ ....[Discrete Data without Frequency]}$$

$$\text{Mean Deviation } (\overline{X}D) = \frac{\Sigma[f_i|x_i - \overline{X}|]}{N} \text{ ....[Discrete Data with Frequency and group data]}$$

Example 1: Calculate the mean deviation of the following data; 2, 4, 6, 8 and 10.

Solution

Step 1: Formula and data. Mean Deviation $(\overline{X}D) = \frac{\Sigma|X_i - \overline{X}|}{n}$, where; n = 5 and $\Sigma|X_i - \overline{X}| =?$

Step 2: Calculation of mean $[\overline{X}]$.

a) Formula. Mean $(\overline{X}) = \frac{\Sigma X_i}{n}$, where; n = 5 and ΣX_i

b) Data substitution and solving.

$$\text{Mean } (\overline{X}) = \frac{\Sigma X_i}{n} \text{ ....[Substitute data into the formula]}$$

$$\text{Mean } (\overline{X}) = \frac{2+4+6+8+10}{5} \text{ ....[Sum numerator and divide by 5]}$$

344

Mean $(\overline{X}) = 6$

Step 3: Tabular derivation of unknown. That is $\sum|X_i - \overline{X}|$

X_i	2	4	6	8	10	
$|X_i - \overline{X}|$	$|2 - 6| = 4$	$|4 - 6| = 2$	$|6 - 6| = 0$	$|8 - 6| = 2$	$|10 - 6| = 4$	$\sum = 12$

Step 4: Data substitution and solving

Mean Deviation $(\overline{X}D) = \frac{\sum|X_i-\overline{X}|}{n}$ ....[Substitute data into the formula]

Mean Deviation $(\overline{X}D) = \frac{12}{6}$ ....[Solve equation]

Mean Deviation $(\overline{X}D) = 2$

Example 2: Compute the mean deviation of the following observations; 3, 4, 2, 10, 5, 20 and 50

Solution

Step 1: Formula and data. Mean Deviation $(\overline{X}D) = \frac{\sum|x_i-\overline{X}|}{n}$, where $n = 7$ and $\sum|X_i - \overline{X}| =?$

Step 2: Calculation of mean $[\overline{X}]$.

c) Formula. Mean $(\overline{X}) = \frac{\sum X_i}{n}$, where; $n = 7$ and $\sum X_i =?$

d) Data substitution and solving.

Mean $(\overline{X}) = \frac{\sum X_i}{n}$ ....[Substitute data into the formula]

Mean $(\overline{X}) = \frac{3+4+2+10+5+20+50}{7}$.................[Sum numerator and divide by 7]

Mean $(\overline{X}) = 13$ ....**[Rounded value]**

Step 3: Tabular determination of the unknown. That is $\sum|X_i - \overline{X}|$

X_i	2	3	4	5	10	20	50	
$|X_i - \overline{X}|$	11	10	9	8	3	7	37	$\sum =85$

Step 4: Data substitution and solving.

Mean Deviation $(\overline{X}D) = \frac{\sum|X_i-\overline{X}|}{n}$ ....[Substitute data into the formula]

Mean Deviation $(\overline{X}D) = \frac{85}{7}$............................ ....[Simplify fraction]

Mean Deviation $(\overline{X}D) = 12$...................................**[Rounded value]**

Example 3: Calculate the mean deviation using the frequency distribution table below

Scores	10	20	30	40	50	60
Frequency	15	5	20	8	12	4

Solution

Step 1: Formula. Mean Deviation $(\overline{X}D) = \frac{\Sigma[f_i|x_i - \overline{X}|]}{N}$, where; N = 64 and $\Sigma[f_i|x_i - \overline{X}|] =?$

Step 2: Calculation of mean.

   a) Formula and data. Mean$(\overline{X}) = \frac{\Sigma f_i x_i}{N}$, where; N = 64 and $\Sigma f_i x_i =?$

   b) Tabular determination of the unknown. That is $\Sigma f_i x_i$

Scores	10	20	30	40	50	60	
Frequency	15	5	20	8	12	4	
$f_i x_i$	150	100	600	320	600	240	$\Sigma = 2010$

   c) Data substitution and solving.

   $$\text{Mean}(\overline{X}) = \frac{\Sigma f_i x_i}{N} \dots\dots\dots\dots\dots\dots\dots\dots\dots\dots\text{[Substitute data into the formula]}$$

   $$\text{Mean}(\overline{X}) = \frac{2010}{64} \dots\dots\dots\dots\dots\dots\dots\dots\dots\dots\dots\dots\text{[Simplify fraction]}$$

   $$\text{Mean}(\overline{X}) = 31 \dots\dots\dots\dots\dots\dots\dots\dots\dots\dots\dots\dots\text{[Rounded value]}$$

Step 3: Tabular calculation of the unknowns. That is $|x_i - \overline{X}|$ and $f_i|x_i - \overline{X}|$

Original table		Working table	
Scores	Frequency	$|x_i - \overline{X}|$	$f_i|x_i - \overline{X}|$
10	15	$|10 - 31| = 21$	$15(21) = 315$
20	5	$|20 - 31| = 11$	$5(11) = 55$
30	20	$|30 - 31| = 1$	$20(1) = 20$
40	8	$|40 - 31| = 9$	$8(9) = 72$
50	12	$|50 - 31| = 19$	$12(19) = 228$
60	4	$|60 - 31| = 29$	$4(29) = 116$
	N = 64		$\Sigma f_i|x_i - \overline{X}| = 806$

Step 4: Data substitution and solving

   $$\text{Mean Deviation }(\overline{X}D) = \frac{\Sigma[f_i|x_i - \overline{X}|]}{N} \dots\dots\dots\dots\dots\text{[Substitute data into the formula]}$$

   $$\text{Mean Deviation }(\overline{X}D) = \frac{806}{64} \dots\dots\dots\dots\dots\dots\dots\dots\dots\text{[Simplify fraction]}$$

   $$\text{Mean Deviation }(\overline{X}D) = 13 \dots\dots\dots\dots\dots\dots\dots\dots\dots\dots\text{[Rounded value]}$$

Example 4: Compute the mean deviation using the frequency distribution table below

Students marks	$[10-20[$	$[20-30[$	$[30-40[$	$[50-60[$	$[60-70[$
No of students	10	16	8	5	15

Solution

Step 1: Formula and data. Mean Deviation $(\overline{X}D) = \frac{\sum[f_i|x_i-\overline{X}|]}{N}$, where; N = 54

Step 2: Determine the value of mean.

    a) Formula. Mean$(\overline{X}) = \frac{\sum f_ix_i}{N}$, where; N = 54 and $\sum f_ix_i$ =?

    b) Tabular determination of unknown $(\sum f_ix_i)$ and value of mean

Original table			Working table	Calculation of mean
Marks	Frequency	$[x_i]$	f_ix_i	Step 1: Formula. $\overline{X} = \frac{\sum f_ix_i}{N}$
$[10-20[$	10	15	150	Where; N = 54 and $\sum f_ix_i = 1880$
$[20-30[$	16	25	400	$\overline{X} = \frac{\sum f_ix_i}{N}$ ....[Substitute data]
$[30-40[$	8	35	280	
$[50-60[$	5	45	225	$\overline{X} = \frac{1880}{54}$ ....[Simplify fraction]
$[60-70[$	15	55	825	Mean$(\overline{X}) = 35$ ....[Rounded value]
	N = 54		$\sum = 1880$	

Step 3: Tabular determination of deviation. That is $\sum[f_i|x_i - \overline{X}|]$

Original Table		Working Table		
Marks	f_i	$[x_i]$	$|x_i - \overline{X}|$	$f_i|x_i - \overline{X}|$
$[10-20[$	10	15	$|15-35| = 20$	$10(20) = 200$
$[20-30[$	16	25	$|25-35| = 10$	$16(10) = 160$
$[30-40[$	8	35	$|35-35| = 0$	$8(0) = 0$
$[40-50[$	5	45	$|45-35| = 10$	$5(10) = 50$
$[60-70[$	15	55	$|55-35| = 20$	$15(20) = 300$
	N = 54			$\sum = 710$

Step 4: Data substitution and solving

        Mean Deviation $(\overline{X}D) = \frac{\sum[f_i|x_i-\overline{X}|]}{N}$ ....[Substitute data into the formula]

        Mean Deviation $(\overline{X}D) = \frac{710}{54}$ ....[Simplify fraction]

        Mean Deviation $(\overline{X}D) = 13$ ....[Rounded value]

Exercise 1: Calculate the value of mean deviation using the frequency distribution tables below

1) Frequency distribution table 1

Marks	1	2	3	4	5	6
Frequency	8	25	10	8	15	10

[Answer: Mean Deviation = 1.43 in two decimal places]

2) Frequency distribution table 2

Students marks	[0 − 20[	[20 − 40[	[40 − 60[	[60 − 80[	[80 − 100[
No of students	5	16	18	1	10

[Answer: Mean Deviation = 19.68 in two decimal places]

3) Frequency distribution table 3

Students marks	[1 − 9[	[10 − 18[	[19 − 27[	[28 − 36[	[37 − 45[
No of students	10	2	20	6	15

[Answer: Mean Deviation = 10.32 in two decimal places]

9.2.3 Median Deviation

Median deviation can also be considered as mean deviation about the median. The formula for calculating median deviation for discrete data without frequency is given as;

$$\text{Median Deviation(MD)} = \frac{\Sigma |X_i - Md|}{n} = \frac{\Sigma |d|}{n} \quad \ldots\ldots\ldots\ldots [\text{Discrete data without frequency}]$$

$$\text{Median Deviation (MD)} = \frac{\Sigma [f_i |x_i - Md|]}{N}. [\text{ Discrete data with frequency and grouped data}]$$

Example 1: Determine the median deviation of the following numbers; 2, 4, 6, 8, 10, and 12

Solution

Step 1: Formula and data. Median Deviation $= \frac{\Sigma |X_i - Md|}{n} = \frac{\Sigma |d|}{n}$, where; n = 6

Step 2: Calculation of median (Md).

a) Formula. $\text{Median(Md)} = \frac{\text{First middle numner} + \text{Second middle number}}{2}$

   Where; First middle numner = 6 and Second middle number = 8

b) Data substitution and solving.

$$\text{Median(Md)} = \frac{\text{First middle numner+Second middle number}}{2} \quad \ldots\ldots\ldots\text{[Substitute data]}$$

$$\text{Median(Md)} = \frac{6+8}{2} \ldots\ldots\ldots\ldots\ldots\ldots\ldots\ldots\text{[Sum numerator and divide by 2]}$$

$$\text{Median(Md)} = \mathbf{7}$$

Step 3: Calculation of deviation ($|X_i - Md|$) and median deviation

Numbers	Deviation	Calculation of median deviation
X_i	$|X_i - Md|$	Step 1: Median Deviation(MD) $= \frac{\Sigma|X_i - Md|}{n}$
2	$|2 - 7| = 5$	Where; n = 7 and $\Sigma|X_i - Md| = 18$
4	$|4 - 7| = 3$	Step 2: Data substitution and solving
6	$|6 - 7| = 1$	Median Deviation $= \frac{\Sigma|X_i - Md|}{n}$ $\ldots\ldots\ldots\ldots$[Substitute data]
8	$|8 - 7| = 1$	Median Deviation $= \frac{18}{7}$ $\ldots\ldots\ldots\ldots$[Simplify fraction]
10	$|10 - 7| = 3$	Median Deviation(MD) $= \mathbf{2.57}$.**[In two decimal places]**
12	$|12 - 7| = 5$	
	$\Sigma = \mathbf{18}$	

Example 2: Use the frequency distribution table below and determine the value of median deviation

Marks	1	3	5	6	8
Frequency	12	5	8	2	13

Solution

Step 1: Formula. Median deviation (MD) $= \frac{\Sigma[f_i|x_i - Md|]}{N}$, where; N = 40

Step 2: Calculation of median.

Marks	1	3	5	6	8
Frequency	12	5	8	2	13
Cumulative frequency	12	17	25	27	40
Note: (N + 1)/2 = 20.5. First cumulative frequency greater than 20.5 is 25. Therefore, the mean is **5**.					

Step 3: Calculation of deviation ($f_i|x_i - Md|$) and median deviation

Original table		Working table		Calculation of median deviation
Marks	Frequency	$|x_i - Md|$	$f_i|x_i - Md|$	Step 1: Median Deviation $= \frac{\sum[f_i|x_i - Md|]}{N}$
1	12	$|1 - 5| = 4$	$12(4) = 48$	Where; N = 40 and $f_i|x_i - Md| = 99$
3	5	$|3 - 5| = 2$	$5(2) = 10$	Step 2: Data substitution and solving
5	8	$|5 - 5| = 0$	$8(0) = 0$	$MD = \frac{99}{40}$ ....[Simplify fraction]
6	2	$|6 - 5| = 1$	$2(1) = 2$	Median Deviation(MD) = **2.475**
8	13	$|8 - 5| = 3$	$13(3) = 39$	
	N = 40		$\sum = 99$	

Example 3: Compute the median deviation from the frequency distribution table below

Marks	$[25 - 50[$	$[50 - 75[$	$[75 - 100[$	$[100 - 125[$	$[125 - 150[$
Frequency	10	23	8	5	15

Solution

Step 1: Formula. Median Deviation $= \frac{\sum[f_i|x_i - Md|]}{N}$, where; N = 61 and $\sum[f_i|x_i - Md|] = ?$

Step 2: Calculation of median.

a) Formula and data. $Median(Md) = L_1 + \left[\frac{\frac{N}{2} - F}{f_m} + C\right]$, where; N = 61

b) Derivation of cumulative frequency and median class and calculation of median

Original table			Working table		Calculation of median
Marks	$[f_i]$		$[CF_i]$	$\left[\frac{N}{2}\right]$	$Median(Md) = 50 + \left[\frac{\frac{61}{2} - 10}{23} \times 25\right]$
$[25 - 50[$	10		10	Where; N = 61	
$[50 - 75[$	23		33	$\frac{61}{2} = 30.5$	$Median(Md) = 50 + \left[\frac{30.5 - 10}{2} \times 25\right]$
$[75 - 100[$	8		41	Median class	$Median(Md) = 75 + 22.2826087$
$[100 - 125[$	5		46	$[50 - 75[$	$Median(Md) = $ **72**
$[125 - 150[$	15		61		
	N = 61				

Step 4: Calculation of deviation from the median.

Original Table			Working Table		
Marks	$[f_i]$	$[x_i]$	$|x_i - Md|$	$f_i|x_i - Md|$	
$[25 - 50[$	10	37.5	$|37.5 - 72| = 34.5$	$10(34.5) = 345$	
$[50 - 75[$	23	62.5	$|62.5 - 72| = 9.5$	$23(9.5) = 218.5$	
$[75 - 100[$	8	87.5	$|87.5 - 72| = 15.5$	$8(15.5) = 124$	
$[100 - 125[$	5	112.5	$|112.5 - 72| = 40.5$	$5(40.5) = 202.5$	
$[125 - 150[$	15	137.5	$|137.5 - 72| = 65.5$	$15(65.5) = 982.5$	
				$\Sigma = 1,872.5$	

Step 5: Calculation of median deviation.

$$\text{Median Deviation(MD)} = \frac{\Sigma[f_i|x_i - Md|]}{N} \ldots\ldots\ldots\ldots\text{[Substitute data into the formula]}$$

$$\text{Median Deviation(MD)} = \frac{1,872.5}{61} \ldots\ldots\ldots \ldots\ldots\ldots\ldots\ldots\text{[Simplify fraction]}$$

$$\text{Median Deviation(MD)} = 31 \ldots\ldots\ldots\ldots\ldots\ldots\ldots\ldots\ldots\text{[Rounded value]}$$

Test your understanding

Exercise 4: Calculate the median deviation using the frequency distribution tables below

1) Frequency distribution table 1

Marks	5	10	15	20	25	30
Frequency	15	2	10	18	5	10

[Answer: Median deviation = 7]

2) Frequency distribution table 2

Students marks	$[0 - 5[$	$[5 - 10[$	$[10 - 15[$	$[15 - 20[$	$[20 - 25[$
No of students	4	16	8	20	10

[Answer: Median deviation = 5.42 in two decimal places]

3) Frequency distribution table 3

Students marks	$[1 - 4[$	$[5 - 8[$	$[9 - 12[$	$[13 - 16[$	$[17 - 20[$
No of students	10	3	1	6	2

[Answer: Median deviation = 5.33 in two decimal places]

9.2.4: Modal deviation

Modal or mode deviation is also seen as mean deviation about the mode. The formula used for the calculation of mode deviation is given as seen below.

$$\text{Mode Deviation}(M_oD) = \frac{\sum|X_i - M_o|}{n} = \frac{\sum|d|}{n} \dots\dots\dots\dots[\text{Discrete data without frequency}]$$

$$\text{Mode Deviation}(M_oD) = \frac{\sum[f_i|x_i - M_o|]}{N} \dots[\text{Discrete data with frequency and grouped data}]$$

Example 1: Determine the mode deviation from the following values; 2, 3, 5, 6, 7 and 3.

Solution

Step 1: Formula. $\text{Mode Deviation}(M_oD) = \frac{\sum|X_i - M_o|}{n}$, where; $\sum|X_i - M_o| =?$

Step 2: Calculation of mode (M_o). Remember that with discrete data without frequency, the mode is the value that occur the highest number of time. If the values re-arranged, we will realized that 3 occurs the most (two times). Therefore, the mode is **3**.

Step 3: Determination of deviation. That is $\sum|X_i - M_o|$ and mode deviation

Original table	Working table	Calculation of mode deviation
Values	$|X_i - M_o|$	Step 1: Formula. $\text{Mode Deviation}(M_oD) = \frac{\sum|X_i - M_o|}{n}$
2	$|2 - 3| = 1$	Where; n = 6 and $\sum|X_i - M_o| = 10$
3	$|3 - 3| = 0$	Step 2: Data substitution and solving
3	$|3 - 3| = 0$	$\text{Mode Deviation}(M_oD) = \frac{\sum|X_i - M_o|}{n}$....[Substitute data]
5	$|5 - 3| = 2$	$\text{Mode Deviation}(M_oD) = \frac{10}{6}$ ....[Simplify fraction]
6	$|6 - 3| = 3$	$\text{Mode Deviation}(M_oD) = \mathbf{1.67}$.
7	$|7 - 3| = 4$	
	$\sum|X_i - M_o| = 10$	

Example 2: Compute the mode deviation of the frequency distribution table below.

Marks	2	4	8	10	12
Frequency	24	14	30	22	10

Solution

352

Step 1: Formula and data. Mode Deviation $(M_oD) = \frac{\Sigma[f_i|x_i - M_o|]}{N}$, where; $N = 100$

Step 2: Determination of mode. Remember that when dealing with discrete data with frequency and grouped data, the mode is simply the value or class with the highest frequency. Therefore, the value with the highest frequency (30) is 8.

Step 3: Determination of deviation $[|x_i - M_o|]$.

Original Table		Working Table	
Marks	Frequency	$|x_i - M_o|$	$f_i|x_i - M_o|$
2	24	$|2 - 8| = 6$	$24(6) = 144$
4	14	$|4 - 8| = 4$	$14(4) = 56$
8	30	$|8 - 8| = 0$	$30(0) = 0$
10	22	$|10 - 8| = 2$	$22(2) = 44$
12	10	$|12 - 8| = 4$	$10(4) = 40$
	N = 100		**$\Sigma = 284$**

Step 4: Data substitution and solving

$$\text{Mode Deviation}(M_oD) = \frac{284}{100} \ldots\ldots\ldots \ldots\ldots\ldots\ldots\text{[Simplify equation fraction]}$$

$$\text{Mode Deviation}(M_oD) = \mathbf{2.84}$$

Example 3: What is the value of mode deviation from the given frequency distribution table below.

Class interval	$[5 - 15[$	$[15 - 25[$	$[25 - 35[$	$[35 - 45[$	$[45 - 55[$
Frequency	12	8	18	32	10

Solution

Step 1: Formula. Mode Deviation $(M_oD) = \frac{\Sigma[f_i|x_i - M_o|]}{N}$, where; $N = 80$

Step 2: Calculation of mode.

a) Formula. $\text{Mode}(M_o) = L_1 + \left[\frac{f_1 - f_0}{2f_1 - f_0 - f_2} \times C\right]$ and modal is $[35 - 45[$

b) Data substitution and solving.

$$\text{Mode}(M_o) = 35 + \left[\frac{32 - 18}{2(32) - 18 - 10} \times 10\right] \ldots\ldots\ldots\ldots\ldots\ldots\text{[Solve equation]}$$

$$\text{Mode}(M_o) = \mathbf{39} \ldots\ldots\ldots\ldots\ldots\ldots\ldots\ldots\ldots\ldots\text{[Rounded value]}$$

Step 3: Calculation of unknown. That is; $f_i|x_i - M_o|$

Original Table		Working Table		
Class Interval	$[f_i]$	x_i	$|x_i - M_o|$	$f_i|x_i - M_o|$
$[5 - 15[$	12	10	$|10 - 39| = 29$	$12(29) = 348$
$[15 - 25[$	8	20	$|20 - 39| = 19$	$8(19) = 152$
$[25 - 35[$	18	30	$|30 - 39| = 9$	$18(9) = 162$
$[35 - 45[$	32	40	$|40 - 39| = 1$	$32(1) = 32$
$[45 - 55[$	10	50	$|50 - 39| = 11$	$10(11) = 110$
	$N = 80$			$\Sigma = 804$

Step 4: Data substitution and solving of mode deviation.

$$\text{Mode Deviation}(M_oD) = \frac{\Sigma[f_i|x_i - M_o|]}{N} \quad\ldots\ldots\ldots\text{[Substitute data into the formula]}$$

$$\text{Mode Deviation}(M_oD) = \frac{804}{80} \quad\ldots\ldots\ldots\ldots\ldots\ldots\ldots\ldots\ldots\ldots\text{[Simplify fraction]}$$

$$\text{Mode Deviation}(M_oD) = \mathbf{10.05}$$

Test your understanding

Exercise 5: Calculate the value of modal deviation using the frequency distribution tables below

a) Frequency distribution table 1

Marks	11	21	31	41	51	61
Frequency	5	25	10	8	15	10

[Answer: Modal Deviation = 15.89 in two decimal places]

b) Frequency distribution table 2

Students marks	$[0 - 20[$	$[20 - 40[$	$[40 - 60[$	$[60 - 80[$	$[80 - 100[$
No of students	15	6	20		10

[Answer: Modal deviation = 21.87 in two decimal places]

c) Frequency distribution table 3

Students marks	$[10 - 19[$	$[20 - 29[$	$[30 - 39[$	$[40 - 49[$	$[50 - 59[$
No of students	10	20	30	20	10

[Answer: Modal deviation = 8.89 in two decimal places]

9.2.5 Variance

Variance measure how far each element in the set is from the mean. Generally, variance is calculated by taking the difference between each number in the data set and the mean, squaring the difference and finally dividing the sum of the squares by the number or sum of frequency. The value of a variance is always positive and a zero variance shows that a given element is not different from the mean.

$$\text{Variance}(\sigma^2) = \frac{\sum(X_i - \bar{X})^2}{N} \quad \dots\dots\dots\dots\dots\dots\dots\dots\dots\dots\dots\dots\dots\dots\dots\dots\dots\dots\dots(1)$$

$$\text{Variance}(\sigma^2) = \frac{\sum[f_i(X_i - \bar{X})^2]}{N} \quad \dots\dots\dots\dots\dots\dots\dots\dots\dots\dots\dots\dots\dots\dots\dots\dots\dots(2)$$

$$\text{Variance}(\sigma^2) = \frac{\sum[f_i(x_i^2)] - \frac{[\sum f_i(x_i)]^2}{N}}{N} \quad \dots\dots\dots\dots\dots\dots\dots\dots\dots\dots\dots\dots\dots\dots\dots(3)$$

$$\text{Variance}(\sigma^2) = \frac{\sum f_i(x_i^2)}{\sum f_i} - \left[\frac{\sum f_i(x_i)}{\sum f_i}\right]^2 \quad \dots\dots\dots\dots\dots\dots\dots\dots\dots\dots\dots\dots\dots(4)$$

Note: Equation (1) is used when dealing with discrete data without frequency and equation (2), (3) and (4) are used when dealing with discrete data without frequency and grouped data.

Example 1: Determine the variance of the following data; 6, 11, 14, 10, 8, 11 and 9

Solution

Step 1: Formula. $\text{Variance}(\sigma^2) = \frac{\sum(X_i - \bar{X})^2}{N}$, where; N = 7 and $\bar{X}$ = 9.86[In two decimal places]

Step 2: Determine of $[\sum(X_i - \bar{X})^2]$ and calculation of variance

Values (X_i)	$(X_i - \bar{X})^2$	Calculation of variance
6	$(6 - 9.86)^2 = 14.8996$	Step 1: Formula. $\text{Variance}(\sigma^2) = \frac{\sum(X_i - \bar{X})^2}{N}$
11	$(11 - 9.86)^2 = 1.2996$	Where; $\sum(X_i - \bar{X})^2 = 38.8572$ and N = 7
14	$(14 - 9.86)^2 = 17.1396$	Step 2: Data substitution and solving
10	$(10 - 9.86)^2 = 0.0196$	$\text{Variance}(\sigma^2) = \frac{38.8572}{7}$ ....[Solve equation]
8	$(8 - 9.86)^2 = 3.4596$	$\text{Variance}(\sigma^2) = \mathbf{5.55}$
11	$(11 - 9.86)^2 = 1.2996$	
9	$(9 - 9.86)^2 = 0.7396$	

Example 2: Determine the variance using the frequency distribution table below.

Marks	$[50 - 52[$	$[52 - 54[$	$[54 - 56[$	$[56 - 58[$	$[58 - 60[$
No of students	17	35	28	15	5

Solution

Step 1: Formula. $\text{Variance}(\sigma^2) = \frac{\sum[f_i(x_i-\bar{X})^2]}{N}$ or $\text{Variance}(\sigma^2) = \frac{\sum f_i(x_i^2)}{\sum f_i} - \left[\frac{\sum f_i(x_i)}{\sum f_i}\right]^2$

Step 2: Tabula derivation of unknown. That is; $\bar{X}$, $\sum[f_i(X_i - \bar{X})^2]$, $\sum f_i(x_i^2)$ and $\sum f_i(x_i)$

Original Table		Mean Section		Variance Section			
Class	$[f_i]$	$[x_i]$	$f_i(x_i)$	$(X_i - \bar{X})^2$	$f_i(x_i - \bar{X})^2$	(x_i^2)	$f_i(x_i^2)$
$[50-52[$	17	51	867	9.7344	165.4848	2,601	44,217
$[52-54[$	35	53	1,855	1.2544	43.904	2,809	98,315
$[54-56[$	28	55	1,540	0.7744	21.6832	3,025	84,700
$[56-58[$	15	57	855	8.2944	124.416	3,249	48,735
$[58-60[$	5	59	295	23.8144	119.072	3,481	17,405
	$N = 100$		$\sum = 5,412$		$\sum = 474.56$		$\sum = 293,372$

a) Calculating of mean ($\bar{X}$).

$\text{Mean}(\bar{X}) = \frac{\sum f_i x_i}{N}$..................................[Data substitution and solving]

$\text{Mean}(\bar{X}) = \frac{5,412}{100}$ ....[Simplify fraction]

$\text{Mean}(\bar{X}) = \mathbf{54}$ ....**[Rounded value]**

Step 3: Data substitution and solving.

Formula 1: $\text{Variance}(\sigma^2) = \frac{\sum f_i(x_i^2)}{\sum f_i} - \left[\frac{\sum f_i(x_i)}{\sum f_i}\right]^2$ .....[Substitute data into the formula]

$\text{Variance}(\sigma^2) = \frac{293,372}{100} - \left[\frac{5,412}{100}\right]^2$ ....[Simplify fractions]

$\text{Variance}(\sigma^2) = 2,933.72 - [54.12]^2$.[Work exponent and expand bracket]

$\text{Variance}(\sigma^2) = 2,933.72 - 2,928.9744$ ....[Solve equation]

$\text{Variance}(\sigma^2) = \mathbf{5}$ ....**[Rounded value]**

Formula 2: $\text{Variance}(\sigma^2) = \frac{\sum[f_i(x_i^2)]-\frac{[\sum f_i(x_i)]^2}{N}}{N}$ ......[Substitute data into the formula]

$\text{Variance}(\sigma^2) = \frac{293,372-\frac{[5,412]^2}{100}}{100}$ ....[Simplify exponent and divide fraction]

$\text{Variance}(\sigma^2) = \frac{293,372-292,897.44}{100}$[Simplify numerator and divide by 100]

$\text{Variance}(\sigma^2) = \mathbf{5}$ ....**[Rounded value]**

Formula 3: $\text{Variance}(\sigma^2) = \frac{\sum[f_i(x_i-\bar{X})^2]}{N}$ ....[Substitute data into the formula]

$$\text{Variance}(\sigma^2) = \frac{474.56}{100} \ldots\ldots\ldots\ldots\ldots\ldots\ldots\ldots\ldots\ldots\ldots\ldots\text{[Solve equation]}$$

$$\text{Variance}(\sigma^2) = 5 \ldots\ldots\ldots\ldots\ldots\ldots\ldots\ldots\ldots\ldots\ldots\text{[Rounded value]}$$

Example 2: From the frequency distribution table below, determine the variance

Class	$[4-12[$	$[12-20[$	$[20-28[$	$[28-36[$	$[36-44[$
Frequency	21	14	4	10	1

Solution

Step 1: Formula. $\text{Variance}(\sigma^2) = \frac{\Sigma[f_i(X_i-\overline{X})^2]}{N}$ or $\text{Variance}(\sigma^2) = \frac{\Sigma[f_i(x_i^2)]-\frac{[\Sigma f_i(x_i)]^2}{N}}{N}$

Where; N = 50

Step 2: Tabula determination of unknown. That is; $\overline{X}$, $f_i(X_i - \overline{X})^2]$ and $[f_i(x_i^2)]$

Original table		Mean		Variance			
Class	$[f_i]$	$[x_i]$	$f_i(x_i)$	$(X_i - \overline{X})^2$	$[f_i(X_i - \overline{X})^2]$	(x_i^2)	$[f_i(x_i^2)]$
$[4-12[$	21	8	168	80.2816	1,685.9136	64	1,344
$[12-20[$	14	16	224	0.9216	12.9024	256	3,584
$[20-28[$	4	24	96	49.5616	198.2464	576	2,304
$[28-36[$	10	32	320	226.2016	2,262.016	1,024	10,240
$[36-44[$	1	40	40	530.8416	530.8416	1,600	1,600
	N = 50		$\Sigma = 848$		$\Sigma = 4,689.92$		$\Sigma = 19,072$

a) Calculation of mean

$$\text{Mean}(\overline{X}) = \frac{\Sigma f_i x_i}{N} \ldots\ldots\ldots\ldots\ldots\ldots\ldots\ldots\ldots\text{[Substitute data into the formula]}$$

$$\text{Mean}(\overline{X}) = \frac{848}{50} \ldots\ldots\ldots\ldots\ldots\ldots\ldots\ldots\ldots\ldots\ldots\ldots\text{[Solve equation]}$$

$$\text{Mean}(\overline{X}) = 17 \ldots\ldots\ldots\ldots\ldots\ldots\ldots\ldots\ldots\ldots\ldots\text{[Rounded value]}$$

Step 3: Data substitution and solving.

Method 1	Method 2	Method 3
$\sigma^2 = \frac{\Sigma[f_i(X_i-\overline{X})^2]}{N}$	$\sigma^2 = \frac{\Sigma[f_i(x_i^2)]-\frac{[\Sigma f_i(x_i)]^2}{N}}{N}$	$\sigma^2 = \frac{\Sigma f_i(x_i^2)}{\Sigma f_i} - \left[\frac{\Sigma f_i(x_i)}{\Sigma f_i}\right]^2$
$\sigma^2 = \frac{4,689.92}{50}$	$\sigma^2 = \frac{19,072-\frac{[848]^2}{50}}{50}$	$\sigma^2 = \frac{19,072}{50} - \left[\frac{848}{50}\right]^2$
$\text{Variance}(\sigma^2) = 93.7984$	$\sigma^2 = \frac{19,072-14,382,08}{50}$	$\sigma^2 = 381.44 - 287.6416$
	$\text{Variance}(\sigma^2) = 93.7984$	$\text{Variance}(\sigma^2) = 93.7984$

Exercise 6: Compute the variance of the following data; 148, 153, 156, 157 and 160.

[Answer: Variance$(\sigma^2) = 16.56$]

Exercise 7: Calculate the value of variance using the frequency distribution tables below

1) Frequency distribution table 1

Marks	3	6	9	12	15	17
Fre uency	5	2	10	4	1	3

[Answer: Variance$(\sigma^2) = 18.4224$]

2) Frequency distribution table 2

Students marks	[0 − 10[	[10 − 20[	[20 − 30[	[30 − 40[	[40 − 50[
N of students	15	13	2	5	10

[Answer: Variance$(\sigma^2) = 246.22$ in two decimal places]

3) Frequency distribution table 3

Students marks	[10 − 19[	[20 − 29[	[30 − 39[	[40 − 49[	[50 − 59[
No of students	0	2	7	20	8

[Answer: Variance$(\sigma^2) = 191.13$ in two decimal places]

9.2.6 Standard Deviation

Standard deviation measures the dispersion of a data set relative to it mean and it is calculated as the square root of variance. Standard deviation value is always positive and sensitive to outliers. The formula used for the calculation of standard deviation is given as seen below.

$$\text{Standard Deviation}(\sigma) = \sqrt{\frac{\sum(X_i - \bar{X})^2}{n}} \quad \dots \dots \dots (1)$$

$$\text{Standard Deviation}(\sigma) = \sqrt{\frac{\sum X_i^2}{n} - \left(\frac{\sum X_i}{n}\right)^2} \quad \dots \dots (2)$$

$$\text{Standard Deviation}(\sigma) = \sqrt{\frac{\sum[f_i(X_i - \bar{X})^2]}{N}} \quad \dots \dots (3)$$

$$\text{Standard Deviation}(\sigma) = \sqrt{\frac{\sum[f_i(X_i^2)]}{N} - \left(\frac{\sum[f_i(X_i)]}{N}\right)^2} \quad \dots \dots (5)$$

$$\text{Standard Deviation}(\sigma) = \sqrt{\frac{\sum[f_i(X_i^2)]}{N} - \bar{X}^2} \quad \dots \dots (6)$$

Note: Equation (1) and (2) are used when dealing with discrete data without frequency. Equation (3), (4) and (5) are used when dealing with discrete data with frequency and grouped data.

Example 1: Determine the standard deviation from the following scores; 3, 6, 8, 10, 12 and 14.

<u>Solution</u>

Step 1: Formula. Standard Deviation$(\sigma) = \sqrt{\frac{\Sigma(X_i - \overline{X})^2}{n}}$ or Standard Deviation$(\sigma) =$

$\sqrt{\frac{\Sigma X_i^2}{n} - \left(\frac{\Sigma X_i}{n}\right)^2}$

Where; $n = 6$, $\Sigma(X_i - \overline{X})^2 = ?$, $\Sigma X_i^2 = ?$ And $\left(\frac{\Sigma X_i}{n}\right)^2 = ?$

Step 2: Calculation of mean.

a) Formula. Mean$(\overline{X}) = \frac{\Sigma X_i}{n}$, where; $n = 6$ and $\Sigma X_i = ?$

b) Data substitution and solving.

$\quad$ Mean$(\overline{X}) = \frac{\Sigma X_i}{n}$ ....[Substitute data into the formula]

$\quad$ Mean$(\overline{X}) = \frac{3+6+8+10+12+14}{6}$...................[Sum numerator and divide by 6]

$\quad$ Mean$(\overline{X}) = 9$...**[Rounded value]**

Step 3: Tabula determination of unknown. That is $\Sigma(X_i - \overline{X})^2$, and ΣX_i^2

Original Table	Working Table		
[X_i]	[X_i^2]	$(X_i - \overline{X})$	$(X_i - \overline{X})^2$
3	$3^2 = 9$	$3 - 9 = -6$	$-6^2 = 36$
6	$6^2 = 36$	$6 - 9 = -3$	$-3^2 = 9$
8	$8^2 = 64$	$8 - 9 = -1$	$-1^2 = 1$
10	$10^2 = 100$	$10 - 9 = 1$	$1^2 = 1$
12	$12^2 = 144$	$12 - 9 = 3$	$3^2 = 9$
14	$14^2 = 196$	$14 - 9 = 5$	$5^2 = 25$
$\Sigma = 53$	$\Sigma = 549$		$\Sigma = 81$

Step 4: Data substitution and solving.

$\quad$ **Formula 1:** Standard Deviation$(\sigma) = \sqrt{\frac{\Sigma X_i^2}{n} - \left(\frac{\Sigma X_i}{n}\right)^2}$ ....[Substitute data]

$\quad$ Standard Deviation$(\sigma) = \sqrt{\frac{549}{6} - \left(\frac{53}{6}\right)^2}$ [Simplify fraction and exponent]

Standard Deviation$(\sigma) = \sqrt{91.5 - 78.02777778}$.[Simplify root values]

Standard Deviation$(\sigma) = \sqrt{13.47222222}$ ....[Work root]

Standard Deviation$(\sigma) = 4$ ....**[Rounded value]**

Formula 2: Standard Deviation$(\sigma) = \sqrt{\frac{\Sigma(x_i - \overline{X})^2}{n}}$ ....[Substitute data]

Standard Deviation$(\sigma) = \sqrt{\frac{81}{6}}$ ....[Simplify fraction]

Standard Deviation$(\sigma) = \sqrt{13.5}$ ....[Work root]

Standard Deviation$(\sigma) = 4$ ....**[Rounded value]**

Example 2: Determine the standard deviation using the frequency distribution table below showing scores of students during an examination. [Round answers to no decimal places]

Scores	$[0-5[$	$[5-10[$	$[10-15[$	$[15-20[$	$[20-25[$
No of students	6	12	5	1	18

Solution

Step 1: Formula. Standard Deviation$(\sigma) = \sqrt{\frac{\Sigma[f_i(x_i - \overline{X})^2]}{N}}$ ....(1)

Standard Deviation$(\sigma) = \sqrt{\frac{\Sigma[f_i(x_i{}^2)]}{N} - \left(\frac{\Sigma[f_i(x_i)]}{N}\right)^2}$ ....(2)

Where; N = 42

Step 2: Tabula determination of unknown. That is $\overline{X}$, $f_i(x_i - \overline{X})^2$, $\Sigma[f_i(x_i{}^2)]$ and $\Sigma[f_i(x_i)]$.

Original Table		Calculation of Mean		Standard Deviation			
		$\overline{X} = \frac{\Sigma f_i x_i}{N}$		Using deviation values		Using midpoint values	
	f_i	x_i	$f_i x_i$	$(x_i - \overline{X})^2$	$[f_i(x_i - \overline{X})^2]$	$[x^2]$	$[f_i(x_i{}^2)]$
$[0-5[$	6	2.5	15	132.25	793.5	6.25	37.5
$[5-10[$	12	7.5	90	42.25	507	56.25	675
$[10-15[$	5	12.5	62.5	2.25	11.25	156.25	781.25
$[15-20[$	1	17.5	17.5	12.25	12.25	306.25	306.25
$[20-25[$	18	22.5	405	72.25	1,300.5	506.25	9,112.5
	N = 42		**$\Sigma = 590$**		**$\Sigma = 2,624.5$**		**$\Sigma = 10,912.5$**

a) Calculation of mean.

$$\text{Mean}(\overline{X}) = \frac{\sum f_i x_i}{N} \quad \dots\dots\dots\dots\dots \quad \dots\dots\dots[\text{Substitute data into the formula}]$$

$$\text{Mean}(\overline{X}) = \frac{590}{42} \quad \dots\dots\dots\dots\dots\dots\dots\dots\dots\dots\dots\dots\dots\dots..\dots\dots..[\text{Solve equation}]$$

$$\text{Mean}(\overline{X}) = 14 \quad \dots\dots\dots\dots\dots\dots\dots\dots\dots\dots\dots\dots\dots..[\textbf{Rounded value}]$$

Step 3: Calculation of standard deviation

Formula 1: Standard Deviation$(\sigma) = \sqrt{\frac{\sum[f_i(x_i{}^2)]}{M} - \left(\frac{\sum[f_i(x_i)]}{N}\right)^2}$ $\dots\dots..[\text{Substitute data}]$

Standard Deviation$(\sigma) = \sqrt{\frac{10{,}912.5}{42} - \left(\frac{590}{42}\right)^2}$.[Simplify   fraction   and exponents]

Standard Deviation$(\sigma) = \sqrt{259.8214286 - 197.3356009}$ ..[Simplify value]

Standard Deviation$(\sigma) = \sqrt{62.48582767}$ $\dots\dots$ $\dots\dots\dots\dots\dots.$[Work root]

Standard Deviation$(\sigma) = 8$ $\dots\dots\dots\dots\dots\dots\dots\dots\dots$..[**Rounded value**]

Formula 2: Standard Deviation$(\sigma) = \sqrt{\frac{\sum[f_i(x_i - \overline{X})^2]}{N}}$ $\dots\dots\dots\dots\dots.\dots\dots$[Substitute data]

Standard Deviation$(\sigma) = \sqrt{\frac{2{,}624.5}{42}}$ $\dots\dots\dots\dots\dots\dots..\dots\dots$[Simplify fraction]

Standard Deviation$(\sigma) = \sqrt{62.48809524}$ $\dots\dots\dots\dots$.........[Work root]

Standard Deviation$(\sigma) = 8$ $\dots\dots\dots\dots\dots\dots\dots\dots$.......[**Rounded value**]

Test your understanding

Exercise 8: Calculate the value of standard deviation using the frequency distribution tables below

1) Frequency distribution table 1

Marks	25	50	75	100	125	150
Frequency	19	2	10	25	8	3

[**Answer: Standard Deviation $(\sigma) = 38.69$ in two decimal places**]

2) Frequency distribution table 2

Students marks	[0 − 10[	[10 − 20[	[20 − 30[	[30 − 40[	[40 − 50[
No of students	9	3	2	5	1

[**Answer: Standard deviation $(\sigma) = 13.82$ in two decimal places**]

3) Frequency distribution table 3

Students marks	[10 − 19[	[20 − 29[	[30 − 39[	[40 − 49[	[50 − 59[
No of students	4	2	6	10	8

[**Answer: Standard deviation $(\sigma) = 13.10$ in two decimal places**]

9.2.7 Coefficient of Variation

The coefficient of variation also called relative standard deviation is a measure of relative variability. It is the ratio of the standard deviation to the mean. It is considered as a standardized measure of dispersion of a probability or frequency distribution and is often expressed as a percentage.

$$\text{Coefficient of Variation} = \frac{\text{Standard Deviation}}{\text{Arithmetic Meam}} \times 100 \dots\dots\dots\dots\dots\dots\dots\dots\dots\dots\dots\dots\dots\dots(1)$$

$$\text{Coefficient of Variation} = \frac{\sqrt{\frac{\Sigma(x_i-\bar{x})^2}{n}}}{\frac{\Sigma x_i}{n}} \times 100 \equiv \frac{\sqrt{\frac{\Sigma x_i^2}{n}-\left(\frac{\Sigma x_i}{n}\right)^2}}{\frac{\Sigma x_i}{n}} \times 100 \dots\dots\dots\dots\dots\dots(2)$$

$$\text{Coefficient of Variation} = \frac{\sqrt{\frac{\Sigma[f_i(x_i-\bar{x})^2]}{N}}}{\frac{\Sigma f_ix_i}{N}} \times 100 \equiv \frac{\sqrt{\frac{\Sigma[f_i(x_i^2)]}{N}-\left(\frac{\Sigma[f_i(x_i)]}{N}\right)^2}}{\frac{\Sigma f_ix_i}{N}} \times 100 \dots\dots\dots\dots(3)$$

Note: Equation (1) is the general equation and is mostly used when standard deviation and mean are given. Equation (2) is used when dealing with discrete data without frequency and equation (3) is used when dealing with data with frequency.

Example 1: Compute the value of standard deviation and coefficient of variation from the following scores; 13, 60, 5, 10, 2, 30 and 14.

Solution

1) **Calculation of Standard Deviation (σ)**

Step 1: Formula. Standard Deviation$(\sigma) = \sqrt{\frac{\Sigma x_i^2}{n}-\left(\frac{\Sigma x_i}{n}\right)^2}$, where; n = 7

Step 2: Tabular determination of unknown. That is $[x_i^2]$ and calculation of standard deviation

$[x_i]$	x_i^2	Calculation of Standard Deviation
2	$2^2 = 4$	Formula. Standard Deviation$(\sigma) = \sqrt{\frac{\Sigma x_i^2}{n}-\left(\frac{\Sigma x_i}{n}\right)^2}$
5	$5^2 = 25$	
10	$10^2 = 100$	$\sigma = \sqrt{\frac{4994}{7}-\left(\frac{134}{7}\right)^2}$ ......[Work exponent and divide fractions]
13	$13^2 = 169$	$\sigma = \sqrt{713.4285714 - 366.4489796}$ ....[Simplify]
14	$14^2 = 196$	$\sigma = \sqrt{346.9795918}$ ....[Root the value]
30	$30^2 = 900$	Standard Deviation$(\sigma) = \mathbf{18.62738822}$
60	$60^2 = 3600$	
$\Sigma x_i = \mathbf{134}$	$\Sigma x_i^2 = \mathbf{4994}$	

2) Calculation of Coefficient of Variation

Step 1: Formula. $\text{CV} = \frac{\text{Standard Deviation}}{\text{Arithmetic Meam}} \times 100$ or $\text{CV} = \frac{\sqrt{\frac{\sum x_i^2}{n} - \left(\frac{\sum x_i}{n}\right)^2}}{\frac{\sum x_i}{n}} \times 100$, where; n =7

Step 2: Calculation of arithmetic mean.

a) Formula. $\text{Mean}(\overline{X}) = \frac{\sum x_i}{n}$, where $n = 7$ and $\sum x_i = 134$ (as seen above)

b) Data substitution and solving.

$\text{Mean}(\overline{X}) = \frac{\sum x_i}{n}$ ………………….…..…[Substitute data into the formula]

$\text{Mean}(\overline{X}) = \frac{134}{7}$…………………………………..…[Simplify fraction]

$\text{Mean}(\overline{X}) = \mathbf{19.14285714}$ ……………….…………[Value of mean]

Step 3: Data substitution and solving

Method 1

a) Formula. $\text{Coefficient of Variation(CV)} = \frac{\text{Standard Deviation}}{\text{Arithmetic Meam}} \times 100$

Where;   Standard Divation = 18.62738822   and   Arithmetic Meam
19.14285714

b) Data substitution and solving

$\text{Coefficient of Variation(CV)} = \frac{18.62738822}{19.14285714} \times 100$ ……[Solve equation]

$\text{Coefficient of Variation(CV)} = 97$ ……………..……..…[Rounded value]

Method 2

a) Using formula 1: $\text{Coefficient of Variation(CV)} = \frac{\sqrt{\frac{\sum x_i^2}{n} - \left(\frac{\sum x_i}{n}\right)^2}}{\frac{\sum x_i}{n}} \times 100$

$\text{Coefficient of Variation(CV)} = \frac{\sqrt{\frac{4994}{7} - \left(\frac{134}{7}\right)^2}}{\frac{134}{7}} \times 100$...[Work   fraction   and exponent]

$\text{Coefficient of Variation(CV)} = \frac{\sqrt{713.4285714 - 366.4489796}}{19.14285714} \times 100$ ……..[Simplify]

$\text{Coefficient of Variation(CV)} = \frac{\sqrt{346.9795918}}{19.14285714} \times 100$ ……..[Work root and solve]

$\text{Coefficient of Variation(CV)} = 97$ ……..……………..…..……[Rounded value]

b) Using formula 2: $\text{Coefficient of Variation} = \frac{\sqrt{\frac{\sum(x_i - \overline{X})^2}{n}}}{\frac{\sum x_i}{n}} \times 100$, where; $\overline{X} = 19$, $n = 7$

i) Tabula determination of unknown. That is; $\sum(X_i - \overline{X})^2$

$[x_i]$	$X_i - \overline{X}$	$(X_i - \overline{X})^2$
2	$2 - 19 = -17$	$-17^2 = -17 \times -17 = 289$
5	$5 - 19 = -14$	$-14^2 = -14 \times -14 = 196$
10	$10 - 19 = -9$	$-9^2 = -9 \times -9 = 81$
13	$13 - 19 = -6$	$-6^2 = -6 \times -6 = 36$
14	$14 - 19 = -5$	$-5^2 = -5 \times -5 = 25$
30	$30 - 19 = 11$	$11^2 = 11 \times 11 = 121$
60	$60 - 19 = 41$	$41^2 = 41 \times 41 = 1,681$
$\sum x_i = 134$		$\sum(X_i - \overline{X})^2 = 2,428$

ii) Data substitution and solving

$$\text{Coefficient of Variation(CV)} = \frac{\sqrt{\frac{2,428}{7}}}{\frac{134}{7}} \times 100 \ \ldots\ldots\ldots\ldots\ldots[\text{Simplify fractions}]$$

$$\text{Coefficient of Variation(CV)} = \frac{\sqrt{346.8571429}}{19.14285714} \times 100 \ \ldots\ldots\ldots[\text{Simplify numerator}]$$

$$\text{Coefficient of Variation(CV)} = \frac{18.62410113}{19.14285714} \times 100 \ \ldots\ldots\ldots\ldots[\text{Solve equation}]$$

$$\text{Coefficient of Variation(CV)} = \mathbf{97} \ \ldots\ldots\ldots\ldots\ldots\ldots\ldots\ldots[\textbf{Rounded value}]$$

Example 2: Calculate the value of standard deviation and coefficient of variation using the frequency distribution table below

Marks	5	10	15	20
Frequency	3	12	5	10

Solution

a. Calculation of standard deviation

Step 1: Formula. Standard Deviation$(\sigma) = \sqrt{\frac{\sum[f_i(x_i - \overline{X})^2]}{N}}$ where; $N = 30$ and $\sum[f_i(x_i - \overline{X})^2]$

Step 2: Tabula determination of first unknown. That is mean $\left[\overline{X} = \frac{\sum f_i x_i}{N}\right]$

Original table		Working table	Calculation of mean
Marks	Frequency	$f_i x_i$	Step 1: Formula. Mean$(\overline{X}) = \frac{\sum f_i x_i}{N}$
5	3	$3(5) = 15$	Where; $\sum f_i x_i = 410$ and $N = 30$
10	12	$12(10) = 120$	Step 2: Data substitution and solving
15	5	$5(15) = 75$	Mean$(\overline{X}) = \frac{410}{30}$ ....[Simplify fraction]
20	10	$10(20) = 200$	Mean$(\overline{X}) = \mathbf{14}$ ....[**Rounded value**]
	N = 30	$\mathbf{\sum = 410}$	

Step 3: Calculation of second unknown. That is; $\sum[f_i(x_i - \overline{X})^2]$

Original Table		Working Table	
Marks	Frequency	$(x_i - \overline{X})^2$	$f_i(x_i - \overline{X})^2$
5	3	$(5 - 14)^2 = 81$	$3(81) = 243$
10	12	$(10 - 14)^2 = 16$	$12(16) = 192$
15	5	$(15 - 14)^2 = 1$	$5(1) = 5$
20	10	$(20 - 14)^2 = 36$	$10(36) = 360$
	N = 30		$\mathbf{\sum f_i(x_i - \overline{X})^2 = 800}$

Step 4: Data substitution and solving

$$\text{Standard Deviation}(\sigma) = \sqrt{\frac{\sum[f_i(x_i - \overline{X})^2]}{N}} \quad ....[\text{Substitute data into the formula}]$$

$$\text{Standard Deviation}(\sigma) = \sqrt{\frac{800}{30}} \quad ....[\text{Simplify fraction}]$$

$$\text{Standard Deviation}(\sigma) = \sqrt{26.66666667} \quad ....[\text{Work root}]$$

$$\text{Standard Deviation}(\sigma) = \mathbf{5} \quad ....[\textbf{Rounded value}]$$

b. Calculation of coefficient of variation

Step 1: Formula. Coefficient of Variation $= \frac{\text{Standard Diviation}}{\text{Arithmetic Meam}} \times 100$, where; SD $= 5$, $\overline{X} = 14$

Step 2: Data substitution and solving.

$$\text{Coefficient of variation} = \frac{\text{Standard Diviation}}{\text{Arithmetic Mean}} \times 100 \quad ....[\text{Substitute data}]$$

$$\text{Coefficient of variation} = \frac{5}{14} \times 100 ....[\text{Solve equation}]$$

$$\text{Coefficient of variation} = \mathbf{36} ....[\textbf{Rounded value}]$$

Example 3: Compute the standard deviation and coefficient of variation using the frequency distribution table below

Class interval	[10 − 20[	[20 − 30[	[30 − 40[	[40 − 50[	[50 − 60[
Frequency	12	4	10	8	16

Solution

a. Calculation of standard deviation

Step 1: Formula. Standard Deviation$(\sigma) = \sqrt{\frac{\Sigma[f_i(x_i-\overline{X})^2]}{N}}$

$\qquad$ Standard Deviation$(\sigma) = \sqrt{\frac{\Sigma[f_i(x_i^2)]}{N}} - \overline{X}^2$, where; N = 50

Step 2: Tabula determination of unknown. That is; $\overline{X}$ and $[f_i(x_i - \overline{X})^2]$ and $\Sigma[f_i(x_i^2)]$

Original table		Calculation of mean		Standard deviation			
		$\overline{X} = \frac{\Sigma f_i x_i}{N}$		Using midpoint		Using deviation	
Class	f_i	x_i	$f_i(x_i)$	x_i^2	$f_i(x_i^2)$	$(x_i - \overline{X})^2$	$f_i(x_i - \overline{X})^2$
[10 − 20[	12	15	180	225	2,700	501.76	6,021.12
[20 − 30[	4	25	100	625	2,500	153.76	615.04
[30 − 40[	10	35	350	1,225	12,250	5.76	57.6
[40 − 50[	8	45	360	2,025	16,200	57.76	462.08
[50 − 60[	16	55	880	3,025	48,400	309.76	4,956.16
	N = 50		$\Sigma = 1,870$		$\Sigma = 82,050$		$\Sigma = 12,112$

a) Calculation of mean

$\qquad$ Mean$(\overline{X}) = \frac{\Sigma f_i x_i}{N}$ ....[Substitute data into the formula]

$\qquad$ Mean$(\overline{X}) = \frac{1,870}{50}$ ....[Simplify fraction]

$\qquad$ Mean$(\overline{X}) = \mathbf{37.4}$

Step 3: Calculation of standard deviation

Method 1	Method 2
Formula: $\sigma = \sqrt{\frac{\sum[f_i(x_i-\overline{X})^2]}{N}}$	Formula: $\sigma = \sqrt{\frac{\sum[f_i(x_i^2)]}{N}} - \overline{X}^2$
$\sigma = \sqrt{\frac{12,112}{50}}$ ....(Divide values)	$\sigma = \sqrt{\frac{82,050}{50} - 37.4^2}$ ...(Divide and square value)
$\sigma = \sqrt{242.24}$ ......(Square root value)	$\sigma = \sqrt{1,641 - 1,398.76}$ ....(Subtract values)
$\sigma = 15.56406117$	$\sigma = \sqrt{242.24}$ ....(Square root value)
	$\sigma = 15.56406117$

b. Calculation of coefficient of variation

Method 1

Step 1: Formula and data. $CV = \frac{Standard\ Deviation}{Arithmetic\ Meam} \times 100$

Where; $SD = 15.56406117$ and $\overline{X} = 37.4$

Step 2: Data substitution and solving.

Coefficient of Variation(CV) $= \frac{Standard\ Deviation}{Arithmetic\ Meam} \times 100$................[Substitute data]

Coefficient of Variation(CV) $= \frac{15.56406117}{37.4} \times 100$.......................[Solve equation]

Coefficient of Variation(CV) $= 42$ ....[Rounded value]

Method 2

Step 1: Coefficient of Variation(CV) $= \frac{\sqrt{\frac{\sum[f_i(x_i-\overline{X})^2]}{N}}}{\frac{\sum f_i x_i}{N}} \times 100 \equiv \frac{\sqrt{\frac{\sum[f_i(x_i^2)]}{N} - \left(\frac{\sum[f_i(x_i)]}{N}\right)^2}}{\frac{\sum f_i x_i}{N}} \times 100$

Step 2: Data substitution and solving. See formula component values in the table above

a) **Using formula 1**

Coefficient of Variation(CV) $= \frac{\sqrt{\frac{\sum[f_i(x_i-\overline{X})^2]}{N}}}{\frac{\sum f_i x_i}{N}} \times 100$ ....[Substitute data]

Coefficient of Variation(CV) $= \frac{\sqrt{\frac{12,112}{50}}}{\frac{1,870}{50}} \times 100$ ....[Simplify fractions]

Coefficient of Variation(CV) $= \frac{\sqrt{242.24}}{37.4} \times 100$ ....[Solve equation]

Coefficient of Variation(CV) $= 42$ ....[Rounded value]

b) **Using formula 2**

$$\text{Coefficient of Variation(CV)} = \frac{\sqrt{\frac{\Sigma[f_i(x_i{}^2)]}{N} - \left(\frac{\Sigma[f_i(x_i)]}{N}\right)^2}}{\frac{\Sigma f_i x_i}{N}} \times 100 \dots\dots\dots\dots[\text{Substitute data}]$$

$$\text{Coefficient of Variation(CV)} = \frac{\sqrt{\frac{82,050}{50} - \left(\frac{1,870}{50}\right)^2}}{\frac{1,870}{50}} \times 100 \quad [\text{Simplify fractions and}$$

exponent]

$$\text{Coefficient of Variation(CV)} = \frac{\sqrt{1641 - 1398.76}}{37.4} \times 100 \dots\dots\dots\dots[\text{Simplify numerator}]$$

$$\text{Coefficient of Variation(CV)} = \frac{\sqrt{242.24}}{37.4} \times 100 \dots\dots\dots\dots\dots\dots[\text{Work root}]$$

$$\text{Coefficient of Variation(CV)} = \frac{15.56406117}{37.4} \times 100 \dots\dots\dots\dots\dots\dots[\text{Solve}]$$

$$\text{Coefficient of Variation(CV)} = \mathbf{42} \dots\dots\dots\dots\dots\dots\dots[\textbf{Rounded value}]$$

Test your understanding

Exercise 9: Calculate the value of coefficient of variation using the frequency distribution tables below. [Leave answer to no decimal place]

1) Frequency distribution table 1

Marks	3	6	9	12	15	17
Frequency	14	2	10	11	8	3

[Answer: Coefficient of variation = 51]

2) Frequency distribution table 2

Students marks	$[0-10[$	$[10-20[$	$[20-30[$	$[30-40[$	$[40-50[$
No of students	4	1	6	5	2

[Answer: Coefficient of variation = 52]

3) Frequency distribution table 3

Students marks	$[10-19[$	$[20-29[$	$[30-39[$	$[40-49[$	$[50-59[$
No of students	10	17	9	20	8

[Answer: Coefficient of variation = 38]

9.2.8 Quartile deviation [Quartile coefficient of dispersion]

Quartile divides a given data set into three; first quartile, second quartile and third quartile. Quartile deviation or semi-interquartile range is one-half the difference between the first quartile and the third quartile. Quartile deviation allows one to compare dispersion for two or more set of data.

$$\text{Quartile Deviation(Qd)} = \frac{Q_3 - Q_1}{2} \quad \dots\dots\dots\dots\dots\dots\dots\dots\dots\dots\dots\dots\dots\dots\dots(1)$$

$$\text{Inter} - \text{Quartile Range(IQR)} = Q_3 - Q_2 \quad \dots\dots\dots\dots\dots\dots\dots\dots\dots\dots\dots\dots(2)$$

$$\text{Coefficient of Quartile Deviation} = \frac{Q_3 - Q_1}{Q_3 + Q_1} \times 100 \quad \dots\dots\dots\dots\dots\dots\dots\dots\dots(3)$$

Where; Q_1, Q_2 and Q_3 are considered the real values of each respective quartile. $Q_1 = \frac{1(n+1)}{4}$ item , $Q_2 = \frac{2(n+1)}{n} = Q_3 - Q_2$ and $Q_3 = \frac{3(n+1)}{4}$

1) Situation of Discrete Data without Frequency

Here, the data should be in order, that is should be in an increasing or decreasing order. In order to determine the quartile deviation, we first compute the position item and real value. The real value of a given quartile (Q_n) is calculated using the formula below.

$$\text{Real value of } Q_n = PV + DPV[VPV - PV] \quad \dots\dots\dots\dots\dots\dots\dots\dots\dots\dots\dots\dots\dots(1)$$

Where; PV = Position value,

DPV = Decimal of position,

VPV = Value after position value

Example 1: Compute the value of quartile deviation and inter-quartile range using the data below; 2, 4, 5, 6, 7, 10, 23 and 32

Solution

Step 1: Formula. $\text{Quartile Deviation(Qd)} = \frac{Q_3 - Q_1}{2}$ $\dots\dots\dots\dots\dots\dots\dots\dots\dots\dots\dots\dots(1)$

$\text{Inter} - \text{Quartile Range(IQR)} = Q_3 - Q_1 \dots\dots\dots\dots\dots\dots\dots\dots\dots\dots\dots(2)$

Step 2: Determination of position item.

a) First quartile position item.

   i) Formula. First Quartile$(Q_1) = \frac{1(n+1)}{4}$, where; n = 8

   ii) Data substitution and solving

   First Quartile$(Q_1) = \frac{1(8+1)}{4}$ $\dots\dots\dots\dots\dots\dots\dots\dots$[Simplify numerator]

   First Quartile$(Q_1) = \frac{9}{4}$ $\dots\dots\dots\dots\dots\dots\dots\dots\dots$[Simplify fraction]

   First Quartile$(Q_1) = 2.25\dots$[2 indicate the second item in the data. That is 4]

b) Third quartile position item.

   i) Formula. Third Quartile$(Q_3) = \frac{3(n+1)}{4}$, where; n = 8

ii) Data substitution and solving

Third Quartile$(Q_3) = \frac{3(8+1)}{4}$..............................[Simplify numerator]

Third Quartile$(Q_3) = \frac{27}{4}$..............................[Simplify fraction]

Third Quartile$(Q_3) = \mathbf{6.75}$ .....[6 indicate the eight item in the data. That is 10]

Step 3: Calculation of real values.

a) **First Quartile (Q_1) Real Value**

i) Formula. $Q_1 = PV + DPV[PPV - PV]$, where; PV = 4, DPV = 0.25, PPV = 5

ii) Data substitution and solving

First Quartile $(Q_1) = 4 + 0.25[5 - 4]$ ....[Expand bracket]

First Quartile $(Q_1) = 4 + 0.25$ ....[Sum values]

First Quartile $(Q_1) = \mathbf{4.25}$ ....**[Real value of Q_1]**

b) **Third Quartile (Q_3) Real Value**

i) Formula. $Q_3 = PV + DPV[PPV - PV]$, where; PV = 10, DPV = 0.75, PPV = 23

ii) Data substitution and solving

Third Quartile $(Q_3) = 23 + 0.75[23 - 10]$ ....[Expand bracket]

Third Quartile $(Q_3) = 23 + 9.75$ ....[Sum values]

Third Quartile $(Q_3) = \mathbf{32.75}$ ....**[Real value of Q_3]**

Step 4: Calculation of quartile deviation and inter-quartile range

Calculation of quartile deviation	Calculation of inter-quartile range
Step 1: Formula. $Qd = \frac{Q_3 - Q_1}{2}$	Step 1: Formula. $IQR = Q_3 - Q_1$
Where; $Q_3 = 32.75$ and $Q_1 = 4.25$	
$Qd = \frac{Q_3 - Q_1}{2}$ ....[Substitute data]	$IQR = Q_3 - Q_1$ ....[Substitute data]
$Qd = \frac{32.75 - 4.25}{2}$ ......[Simplify numerator]	$IQR = 32.75 - 4.25$ ....[Simplify equation]
$Qd = \frac{28.5}{2}$ ....[Simplify fraction]	$Inter - Quartile\ Range(IQR) = \mathbf{28.5}$
Quartile Deviation$(Qd) = \mathbf{14.25}$	

Example 2: Use the given data and answer the following questions; 25, 18, 30, 8, 15, 5, 10, 35, 40 and 45. Determine;

1) The first quartile [Q_1], second quartile [Q_2] and third quartile [Q_3]

2) The quartile deviation

3) Inter-quartile range

4) Coefficient of quartile deviation

Solution

Note: From observation, we realized that our values are not in order. Therefore, we are required to re-arrange this data in an increasing order. That is; 5, 8, 10, 15, 18, 25, 30, 35, 40 and 45

1) The first quartile $[Q_1]$, second quartile $[Q_2]$ and third quartile $[Q_3]$

Step 1: General Formula and data. $Q_n = PV + DPV[VPV - PV]$

Step 2: Determination of position value

a) Formula and data. $Q_1 = \frac{1(n+1)}{4}$, $Q_1 = \frac{2(n+1)}{4}$ and $Q_1 = \frac{3(n+1)}{4}$, where; n = 10

b) Data substitution and solving.

i. $Q_1 = \frac{1(10+1)}{4} \rightarrow Q_1 = \frac{11}{4} \rightarrow Q_1 = 2.75$....[PV = 8, DPV = 0.75 and VPV = 10]

ii. $Q_2 = \frac{2(10+1)}{4} \rightarrow Q_2 = \frac{22}{4} \rightarrow Q_2 = 5.5$...[PV = 18, DPV = 0.5 and VPV = 25]

iii. $Q_3 = \frac{3(10+1)}{4} \rightarrow Q_3 = \frac{33}{5} \rightarrow Q_3 = 8.25$...[PV = 35, DPV = 0.25 and VPV = 40]

Step 3: Data substitution and solving of real values of quartiles. That is $[Q_1, Q_2$ and $Q_3]$

a) $Q_1 = 8 + 0.75[10 - 8] \rightarrow Q_1 = 8 + 1.5 \rightarrow Q_1 = 9.5 \cong 10$..........[Real value of Q_1]

b) $Q_2 = 18 + 0.5[25 - 18] \rightarrow Q_2 = 18 + 3.5 \rightarrow Q_2 = 21.5 \cong 22$......[Real value of Q_2]

c) $Q_3 = 35 + 0.25[40 - 35] \rightarrow Q_3 = 35 + 1.25 \rightarrow Q_3 = 36.25 \cong 36$[Real value of Q_1]

2) The quartile deviation

Step 1: Formula. Quartile Deviation$(Qd) = \frac{Q_3 - Q_1}{2}$, where; $Q_1 = 9.5$ and $Q_3 = 36.25$

Step 2: Data substitution and solving.

Quartile Deviation$(Qd) = \frac{36.25 - 9.5}{2}$....... ....[Simplify numerator]

Quartile Deviation$(Qd) = \frac{26.75}{2}$ ....[Simplify fraction]

Quartile Deviation$(Qd) = 13.375$

3) Inter-quartile range

Step 1: Formula. Inter $-$ Quartile Range(IQR) $= Q_3 - Q_1$. where; $Q_1 = 9.5$ and $Q_3 = 36.25$

Step 2: Data substitution and solving.

Inter $-$ Quartile Range(IQR) $= Q_3 - Q_1$ ....[Substitute data into the formula]

Inter $-$ Quartile Range(IQR) $= 36.25 - 9.5$......... ....[Simplify equation]

Inter $-$ Quartile Range(IQR) $= 26.75$

4) Coefficient of quartile deviation

Step 1: Formula. Coefficient of Quarter Deviation$(CQD) = \frac{Q_3 - Q_1}{Q_3 + Q_1} \times 100$

Where; $Q_1 = 9.5$ and $Q_3 = 36.25$

Step 2: Data substitution and solving.

Coefficient of Quarter Deviation$(CQD) = \frac{Q_3 - Q_1}{Q_3 + Q_1} \times 100$ ....[Substitute data]

Coefficient of Quarter Deviation$(CQD) = \frac{36.25 - 9.5}{36.25 + 9.5} \times 100$ ....[Simplify equation]

Coefficient of Quarter Deviation$(CQD) = \frac{26.75}{45.75} \times 100$ ....[Solve equation]

Coefficient of Quarter Deviation$(CQD) = \textbf{58}$ ....**[Rounded value]**

2) Situation of Discrete Data with Frequency

It is worth knowing that the procedures used above are different with discrete data with frequency, the following suggested steps are used in determining quartile deviation with ungrouped data with frequency

1) Derivation of cumulative frequency [CF]. It is highly recommended to look for cumulative frequency greater than.

2) Determination of real values of Q_1 and Q_3. Calculate the value $\left[\frac{1N+1}{4}\right]$ and $\left[\frac{3N+1}{4}\right]$ for the first and third quartile respectively. Identify from the cumulative frequency the first value greater than $\left[\frac{1N+1}{4}\right]$ and $\left[\frac{3N+1}{4}\right]$ and consider it the real value Q_1 and Q_3 respectively.

3) Final solving. Substitute the values of Q_1 and Q_3 in the general equation $\left[\frac{Q_3 - Q_2}{2}\right]$ to determine the quartile deviation and/or in the equation $[Q_3 - Q_1]$ to determine the value of inter-quartile range.

Example 1: Compute the value of quartile deviation and inter-quartile deviation using the frequency distribution table below.

Marks	5	8	12	15	19
Frequency	4	8	9	13	18

Solution

Step 1: Formula. Quartile Deviation$(Qd) = \frac{Q_3 - Q_1}{2}$ and $IQR = Q_3 - Q_1$

Step 2: Tabula determination of Q_3 and Q_1 real values.

Marks	5	8	12	15	19	
Frequency	4	8	9	13	18	$\sum = N = 52$
Cumulative frequency	4	12	21	34	52	

a) **Calculation of Q_1 real value**

   i) Formula. First Quartile$(Q_1) = \frac{1(N)+1}{4}$, where; N = 52

   ii) Data substitution and solving

   First Quartile$(Q_1) = \frac{1(52)+1}{4}$ ....[Simplify numerator]

   First Quartile$(Q_1) = \frac{53}{4}$ ....[Simplify fraction]

   First Quartile$(Q_1) = \mathbf{13.25 \cong 13}$ ....…....……**[Real value of Q_1]**

b) **Calculation of Q_3 real value**

   i) Formula. Third Quartile$(Q_3) = \frac{1(N)+1}{4}$, where; N = 52

   ii) Data substitution and solving

   Third Quartile$(Q_3) = \frac{3(52)+1}{4}$ ....…....[Simplify numerator]

   Third Quartile$(Q_3) = \frac{157}{4}$ ....[Simplify fraction]

   Third Quartile$(Q_3) = \mathbf{39.25 \cong 39}$ ....**[Real value of Q_1]**

Step 3: Calculation of quartile deviation and inter-quartile range.

Calculation of quartile deviation	Calculation of inter-quartile range
Step 1: Formula. Qd $= \frac{Q_3 - Q_1}{2}$	Step 1: Formula. IQR $= Q_3 - Q_1$
Where: $Q_1 = 12$ and $Q_3 = 19$	
Qd $= \frac{Q_3 - Q_1}{2}$ ....[Substitute data] Qd $= \frac{19-12}{2}$ ....[Solve equation] Quartile Deviation(Qd) $= \mathbf{3.5}$	IQR $= Q_3 - Q_1$ ....[Substitute data] IQR $= 19 - 12$ . ...[Simplify equation] Inter $-$ Quarter Range(IQR) $= \mathbf{7}$

Example 2: From the frequency distribution table below, answer the questions that follows

Scores	5	10	15	20	25
Frequency	20	15	5	13	7

Determine the following;
   1) The first quartile, second quartile and third quartile
   2) The quartile deviation

3) The inter-quartile range

4) The coefficient of quartile deviation.

<u>Solution</u>

1) The first quartile, second quartile and third quartile

Step 1: General formula. Remember that when dealing with discrete data with frequency, the real values of quartiles are the values whose cumulative frequency is the first to be greater than the value of $Q_n = \left[\frac{n[N+1]}{4}\right]$. Where n represent the quartiles. That is $[Q_1, n = 1$ and $Q_2, n = 2$ and $Q_3, n = 3]$.

Step 2: Tabular determination of quartile real values

Original Table			Working Table		
Marks	f_i	CF_i	Calculation of Q_1	Calculation of Q_2	Calculation of Q_3
5	20	20	$Q_1 = \frac{1(N)+1}{4}$	$Q_2 = \frac{2(N)+1}{4}$	$Q_3 = \frac{3(N)+1}{4}$
10	15	35	$Q_1 = \frac{1(60)+1}{4}$	$Q_2 = \frac{2(60)+1}{4}$	$Q_3 = \frac{3(60)+1}{4}$
15	5	40			
20	13	53	$Q_1 = 15.25$	$Q_2 = 30.25$	$Q_3 = 45.25$
25	7	60	Real value = 5	Real value = 10	Real value = 20
	N = 60		$Q_1 = 5$	$Q_2 = 10$	$Q_3 = 20$

2) The quartile deviation

Step 1: Formula. Quartile Deviation$(Qd) = \frac{Q_3 - Q_1}{2}$, where; $Q_3 = 20$ and $Q_1 = 5$

Step 2: Data substitution and solving.

Quartile Deviation$(Qd) = \frac{Q_3 - Q_1}{2}$ ....[Substitute data]

Quartile Deviation$(Qd) = \frac{20-5}{2}$..............[Simplify numerator and divide by 2]

Quartile Deviation$(Qd) = 7.5 \cong 8$

3) The inter-quartile range

Step 1: Formula. Inter $-$ Quartile Range$(IQR) = Q_3 - Q_1$, where; $Q_3 = 20$ and $Q_1 = 5$

Step 2: Data substitution and solving.

Inter $-$ Quartile Range$(IQR) = Q_3 - Q_1$ ....[Substitute data]

Inter $-$ Quartile Range$(IQR) = 20 - 5$...............................[Solve equation]

Inter $-$ Quartile Range$(IQR) = 15$

4) The coefficient of quartile deviation.

Step 1: Formula and data. $CQD = \frac{Q_3 - Q_1}{Q_3 + Q_1} \times 100$, where; $Q_3 = 20$ and $Q_1 = 5$

Step 2: Data substitution and solving.

$$CQD = \frac{20-5}{20+5} \times 100 \dots\dots\dots\dots\dots\dots\dots\dots\text{[Simplify numerator and denominator]}$$

$$CQD = \frac{15}{25} \times 100 \dots\dots\dots\dots\dots\dots\dots\dots\text{[Divide fraction and multiply by 100]}$$

Coefficient of Quartile Deviation(CQD) = **60**

3) Situation of with Grouped Data

In situation with grouped data with frequency, the formula used for the calculation of quartiles real values differ with that of discrete data with and without frequency seen above. Here,

$$Q_1 = LQ_1 + \left[\frac{\frac{1N}{4}-F}{fQ_1} \times C\right], \quad Q_1 = LQ_2 + \left[\frac{\frac{2N}{4}-F}{fQ_2} \times C\right] \text{ and } Q_3 = LQ_3 + \left[\frac{\frac{3N}{4}-F}{fQ_3} \times C\right]$$

Where;

1) LQ_1, LQ_2 and LQ_3 are lower limits of first, second and third quartile class respectively

2) fQ_1, fQ_2 and fQ_3 are the frequency of the first, second and third quartile class respectively

3) F is the cumulative frequency before the quartile class

4) C is the class size of the quartile class

The following suggested steps are recommended in determining quartile deviation with grouped data.

1) Derivation of cumulative frequency [CF]. It is highly recommended to look for cumulative frequency greater than.

2) Determination of the quartile class. Calculate the value of $\left[\frac{1N}{4}\right]$ and $\left[\frac{3N}{4}\right]$ for the first and third quartile respectively. Identify from the cumulative frequency the first value greater than the values of $\left[\frac{1N}{4}\right]$ and $\left[\frac{3N}{4}\right]$ and consider it the first quartile class and third quartile class respectively.

3) Determination of quartile real values. Use the formula $Q_1 = LQ_1 + \left[\frac{\frac{1N}{4}-F}{fQ_1} \times C\right]$ to calculate the real value of first quartile and $Q_3 = LQ_3 + \left[\frac{\frac{3N}{4}-F}{fQ_3} \times C\right]$ for the real value of third quartile.

4) Final calculation. Substitute the real values of Q_3 and Q_1 in the formula $\frac{Q_3-Q_1}{2}$ to obtain the value of quartile deviation.

Example 1: Determine the quartile deviation and inter-quartile range of the frequency distribution table below.

Marks	$[0-10[$	$[10-20[$	$[20-30[$	$[30-40[$	$[40-50[$
Frequency	11	18	25	28	30

Solution

Step 1: Formula. Quartile Deviation $= \frac{Q_3-Q_1}{2}$ and Inter $-$ Quartile Range $= Q_3 - Q_2$

Step 2: Tabula determination of median class for Q_3 and Q_1. $[CF_i]$ represent cumulative frequency.

Original table			Working table	
			Quartile class for Q_1	Quartile class for Q_3
Marks	$[f_i]$	$[CF_i]$	Formula. $Q_1 = \left[\frac{1N}{4}\right]$	Formula. $Q_3 = \left[\frac{3N}{4}\right]$
$[0-10[$	11	11	$Q_1 = \frac{1(112)}{4} = 28$	$Q_3 = \frac{3(112)}{4} = 84$
$[10-20[$	18	29	Looking through $[CF_i]$, we	Looking through $[CF_i]$, we
$[20-30[$	25	54	realized that 54 is the first	realized that 112 is the first
$[30-40[$	28	82	value greater than 28,	value greater than 84,
$[40-50[$	30	112	therefore, the quartile class is	therefore, the quartile class
	N = 112		$[10-20[$	is $[40-50[$

Step 3: Calculation of the real value of Q_3 and Q_1.

a) Calculation of the real value of Q_3

i) Formula. Third Quartile$(Q_3) = LQ_3 + \left[\frac{\frac{3N}{4}-F}{fQ_3} \times C\right]$

Where; $LQ_3 = 40$, $fQ_3 = 30$, $F = 82$, $N = 112$, $C = 10$

ii) Data substitution and solving

Third Quartile$(Q_3) = 40 + \left[\frac{\frac{3(112)}{4}-82}{30} \times 10\right]$ ....[Simplify numerator]

Third Quartile$(Q_3) = 40 + \left[\frac{84-82}{30} \times 10\right]$ ....[Simplify bracket]

Third Quartile$(Q_3) = 40 + 0.666666666$ ....[Sum values]

Third Quartile$(Q_3) = \mathbf{40.66666667}$ ....[Real value of Q_3]

b) Calculation of the real value of Q_1

i) Formula. First Quartile$(Q_1) = LQ_1 + \left[\frac{\frac{1N}{4}-F}{fQ_1} \times C\right]$

Where; $LQ_1 = 10$, $fQ_1 = 18$, $F = 11$, $N = 112$, $C = 10$

ii) Data substitution and solving

$$\text{First Quartile}(Q_1) = 10 + \left[\frac{\frac{1(112)}{4}-11}{18} \times 10\right] \text{ ...[Simplify bracket numerator]}$$

$$\text{First Quartile}(Q_1) = 10 + \left[\frac{28-11}{18} \times 10\right] \text{ ....[Simplify bracket]}$$

$$\text{First Quartile}(Q_1) = 10 + 9.444444444 \text{ ....[Sum values]}$$

$$\text{First Quartile}(Q_1) = \mathbf{19.44444444} \text{ ....[Real value of } \mathbf{Q_1]}$$

Step 4: Calculation of quartile deviation and inter-quartile range

a) **Calculation of Quartile deviation [Qd]**

   i) Formula. $Qd = \frac{Q_3-Q_1}{2}$, where; $Q_3 = 40.66666667$ and $Q_1 = 19.44444444$

   ii) Data substitution and solving.

   $$\text{Quartile Deviation}(Qd) = \frac{40.66666667-19.44444444}{2} \qquad \text{......[Simplify}$$

   numerator]

   $$\text{Quartile Deviation}(Qd) = \frac{21.22222223}{2} \text{ ....[Simplify fraction]}$$

   $$\text{Quartile Deviation}(Qd) = \mathbf{11} \text{ ....[Rounded value]}$$

b) **Calculation of inter-quartile range [IQR]**

   i) Formula. $IQR = Q_3 - Q_1$, where; $Q_3 = 40.66666667$ and $Q_1 = 19.44444444$

   ii) Data substitution and solving.

   $$\text{Inter} - \text{Quartile Range}(IQR) = Q_3 - Q_1 \text{ ....[Substitute data]}$$

   $$\text{Inter} - \text{Quartile Range}(IQR) = 40.66666667 - 19.44444444 \text{ ....[Solve]}$$

   $$\text{Inter} - \text{Quartile Range}(IQR) = \mathbf{21} \text{ ....[Rounded value]}$$

Example 2: Use the frequency distribution table below and answer the following questions.

Marks	$[1-9[$	$[10-19[$	$[20-29[$	$[30-39[$	$[40-49[$
No of students	4	12	9	2	13

Calculate the values of the following;

1) First quartile $[Q_1]$, second quartile $[Q_2]$ and third quartile $[Q_3]$
2) Quartile deviation
3) Inter-quartile range
4) Coefficient of quartile deviation.

378

Solution

Hint: The table is not continuous and hence we are required to make it continuous before any further working. Remember that to change non-continuous table to continuous table, we subtract 0.5 from the lower limits and add 0.5 to the upper limits.

Marks	$[0.5 - 9.5[$	$[9.5 - 19.5[$	$[19.5 - 29.5[$	$[29.5 - 39.5[$	$[39.5 - 49.5[$
No of students	4	12	9	2	13

1) **First quartile $[Q_1]$, second quartile $[Q_2]$ and third quartile $[Q_3]$**

Step 1: General formula and data. $Q_n = LQ_n + \left[\frac{\frac{nN}{4}-F}{fQ_n} \times C\right]$, where; $n = 1, 2,$ and 3

Step 3: Tabula determination of quartile class.

Original Table			Working Table		
			Q_1	Q_2	Q_3
Marks	$[f_i]$	$[CF_i]$	$Q_1 = \left[\frac{1N}{4}\right]$	$Q_1 = \left[\frac{2N}{4}\right]$	$Q_3 = \left[\frac{3N}{4}\right]$
$[0.5 - 9.5[$	4	4	$Q_1 = \frac{1(40)}{4}$	$Q_2 = \frac{2(40)}{4}$	$Q_3 = \frac{3(40)}{4}$
$[9.5 - 19.5[$		16 (Q_1)	$Q_1 = 10$	$Q_2 = 20$	$Q_3 = 30$
$[19.5 - 29.5[$		25 (Q_2)	Quartile class	Quartile class	Quartile class
$[29.5 - 39.9[$	12	27	$[9.5 - 19.5[$	$[19.5 - 29.5[$	$[39.9 - 49.5[$
$[39.5 - 49.5[$	9	40 (Q_3)			
	N = 40				

Step 4: Calculation of the real value of Q_3, Q_1 and Q_2

a) **Calculation of the real value of Q_3**

    i) Formula. $Q_3 = LQ_3 + \left[\frac{\frac{3N}{4}-F}{fQ_3} \times C\right]$

        Where; $LQ_3 = 39.5$, $fQ_3 = 13$, $F = 27$, $C = 10$, $N=40$

    ii) Data substitution and solving

$$\text{Third Quartile}(Q_3) = 39.5 + \left[\frac{\frac{3(40)}{4}-27}{13} \times 10\right] \text{ [Simplify bracket numerator]}$$

$$\text{Third Quartile}(Q_3) = 39.5 + \left[\frac{30-27}{13} \times 10\right] \text{..............[Simplify bracket]}$$

$$\text{Third Quartile}(Q_3) = 39.5 + 2.307692308 \text{...................[Sum values]}$$

$$\text{Third Quartile}(Q_3) = \mathbf{41.80769231} \text{................[Real value of } Q_3\text{]}$$

379

b) Calculation of the real value of Q_2

i) Formula. $Q_2 = LQ_2 + \left[\dfrac{\frac{2N}{4}-F}{fQ_2} \times C\right]$

Where; $LQ_2 = 19.5$, $fQ_2 = 9$, $F = 16$, $C = 10$, $N=40$

ii) Data substitution and solving.

Second Quartile$(Q_2) = 19.5 + \left[\dfrac{\frac{2(40)}{4}-16}{9} \times 10\right]$..[Simplify bracket numerator]

Second Quartile$(Q_2) = 19.5 + \left[\dfrac{20-16}{9} \times 10\right]$ ....[Simplify bracket]

Second Quartile$(Q_2) = 19.9 + 4.444444444$ ....[Sum values]

Second Quartile$(Q_2) = \mathbf{23.94444444}$ ....**[Real value of Q_2]**

c) Calculation of the real value of Q_1

i) Formula. $Q_1 = LQ_1 + \left[\dfrac{\frac{1N}{4}-F}{fQ_1}(C)\right]$

Where; $LQ_1 = 9.5$, $fQ_1 = 12$, $F = 4$, $N = 40$, $C = 10$

ii) Data substitution and solving.

First Quartile$(Q_1) = 9.5 + \left[\dfrac{\frac{1(40)}{4}-4}{12} \times 10\right]$ ....[Simplify bracket fraction]

First Quartile$(Q_1) = 9.5 + \left[\dfrac{10-4}{12} \times 10\right]$ ....[Simplify bracket]

First Quartile$(Q_1) = 9.5 + 5$ ....[Sum values]

First Quartile$(Q_1) = \mathbf{14.5}$ ....**[Real value of Q_1]**

2) Quartile deviation

Step 1: Formula. Quartile Deviation$(Qd) = \dfrac{Q_3-Q_1}{2}$

Where: $Q_3 = 41.80769231$ and $Q_1 = 14.5$

Step 2: Data substitution and solving.

Quartile Deviation$(Qd) = \dfrac{Q_3-Q_1}{2}$ ....[Substitute data]

Quartile Deviation$(Qd) = \dfrac{41.80769231-14.5}{2}$ ....[Solve]

Quartile Deviation$(Qd) = \mathbf{14}$ ....**[Rounded value]**

3) Inter-quartile range

Step 1: Formula. IQR $= Q_3 - Q_2$, where; $Q_3 = 41.80769231$ and $Q_1 = 14.5$

Step 2: Data substitution and solving.

Inter − Quartile Range(IQR) $= Q_3 − Q_2$ ....[Substitute data]

Inter − Quartile Range(IQR) $= 41.80769231 − 14.5$.........................[Solve]

Inter − Quartile Range(IQR) $= \textbf{27}$ ....[Rounded value]

4) Coefficient of quartile deviation.

Step 1: Formula and data. $CQD = \frac{Q_3 − Q_1}{Q_3 + Q_1} \times 100$, where; $Q_3 = 41.80769231$ and $Q_1 = 14.5$

Step 2: Data substitution and solving.

$CQD = \frac{41.80769231 − 14.5}{41.80769231 + 14.5} \times 100$.............[Simplify numerator and denominator]

$CQD = \frac{27.30769231}{56.30769231} \times 100$ ....[Simplify fraction and multiply by 100]

$CQD = \textbf{48}$ ....[Rounded value]

Application exercise

Exercise 10: Use the frequency distribution table below and answer the questions that follows

Marks	10	20	15	19	11	18	13	17	12	14	16
Frequency	2	1	16	4	4	6	8	9	5	10	10

a) Construct a grouped frequency distribution table using the class interval of; 0 – 5, 5 – 10, 10 – 15, 15 – 20 and 20 – 25. [**Answer:** ....**See appendix**]

b) Calculate the value of the following

    i) Range ....[**Answer: 25**]

    ii) Variance ....[**Answer: 6.52 in two decimal places**]

    iii)   Interquartile range ....[**Answer: 2.358 in three decimal places**]

    iv) Geometric mean ....[**Answer: 15.417 in three decimal places**]

    v) Coefficient of variation ....[**Answer: 16.328 in three decimal places**]

Exercise 11: Use the grouped frequency distribution table below and answer the following questions

X_i	50-100	100-150	150-200	200-250	250-300	300-350	350-500	500-700
f_i	3	14	25	20	18	26	10	4

1) Calculate the value of the following

    a) Mode ....[**Answer: 306**]

    b) Median ....[**Answer: 650**]

    c) 1st and 3rd quartile ....[**Answer: $Q_1 = 176$ and $Q_3 = 319$ rounded value**]

2) How many stock items cost;

Exercise 12: Answer the following questions using the table below. Leave your answers in two decimal places.

Class Interval	0 – 10	10 – 20	20 – 30	30 – 40	40 – 50
Frequency	6	8	5	5	2

1) Calculate the value of mean deviation about the mean ....[Answer: **10.85**]

2) Calculate the value of mean deviation about the median ....[Answer: **10.62**]

3) Calculate the value of mean deviation about the mode ....[Answer: **10.92**]

9.3 Symmetry and Skewness

9.3.1 The concept of Symmetry

A symmetrical distribution is a distribution in which the mean$(\overline{X})$, median $(M_d)$and mode $(M_o)$are equal$[\overline{X} = M_d = M_o]$. A symmetrical distribution is illustrated graphically as seen below.

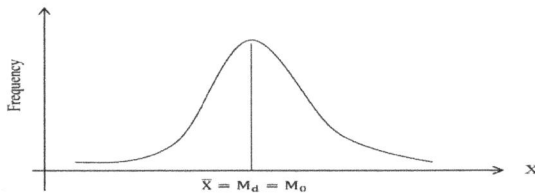

$$\overline{X} = M_d = M_o$$

Example 1: Determine the skewness of the distribution below using Karl Pearson coefficient of skewness.

Class	54	59	64	69	74
Frequency	3	8	12	9	3

Solution

Step 1: Determination of mean, median and mode values...[Cum.F = for cumulative frequency]

1) Formula.

$$\text{Mean}(\overline{X}) = \frac{\sum f_i x_i}{N}, \text{Median}(M_d) =$$

First Class with Cum. F greater than $\left[\frac{N+1}{2}\right]$

$$\text{Mode}(M_o) = \text{Class with Highest Frequency}$$

2) Tabular determination of unknowns and solving

Original Table			Mean	Median					Mode		
Marks	Frequency	$f_i x_i$		Marks	Frequency	Cum. F	$\frac{N+1}{2} = \frac{35+1}{2} = 18$		Marks	Frequency	
54	3	162		54	3	3			54	3	
59	8	472		59	8	11			59	8	
64	12	768		64	12	23			64	12	$M_o = 64$
69	9	621		69	9	32	$M_d = 64$		69	9	
74	3	222		74	3	35			74	3	
	$\Sigma = 35$	$\Sigma = 2,245$			$\Sigma = 35$					$\Sigma = 35$	
		$\overline{X} = \frac{2,245}{35} = 64$									

Step 2: Plotting and interpretation..................... [Mean(64) = Median(64) = Mode(64)]

Illustration	Explanation
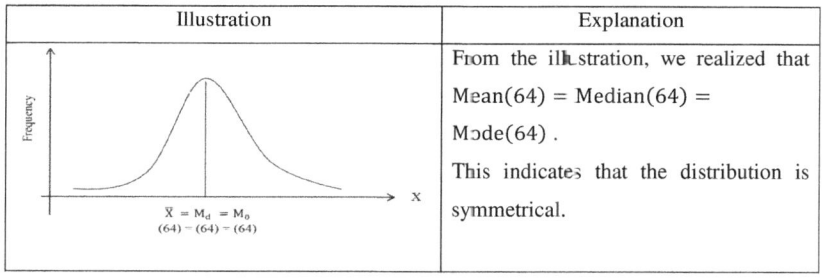
$\overline{X} = M_d = M_o$
$(64) - (64) - (64)$	From the illustration, we realized that Mean(64) = Median(64) = Mode(64) . This indicates that the distribution is symmetrical.

9.3.2 The concept of Skewness

Skewness is defined as the absence of symmetry. The skewness (asymmetry) of a distribution as well as the symmetry of a distribution can be determined using the Karl Pearson coefficient of skewness(S_k).

$$\text{Karl Pearson Coefficient of Skewness }(S_k) = \frac{\text{Mean}(\overline{X}) - \text{Mode}(M_o)}{\text{Standard Deviation}(\sigma)} \quad\dots\dots\dots\dots\dots\dots(1)$$

$$\text{Karl Pearson Coefficient of Skewness }(S_k) = \frac{3[\text{Mean}(\overline{X}) - \text{Median}(M_d)]}{\text{Standard Deviation}(\sigma)} \quad\dots\dots\dots\dots\dots(2)$$

N/B: The Karl Pearson coefficient of skewness ($\pm S_k$) shows the direction of skewness and its value gives the extent of skewness. The possible values of Karl Pearson coefficient of skewness and interpretation is given below.

Karl Pearson Coefficient of Skewness $(S_k) > 0$ ....[Positive Skewness]

Karl Pearson Coefficient of Skewness $(S_k) = 0$ ....[No Skewness or Symmetry]

Karl Pearson Coefficient of Skewness $(S_k) < 0$ ....[Negative Skewness]

Note: Skewness can also be determined graphically. The graphical illustration of positive skewness and negative skewness is represented below.

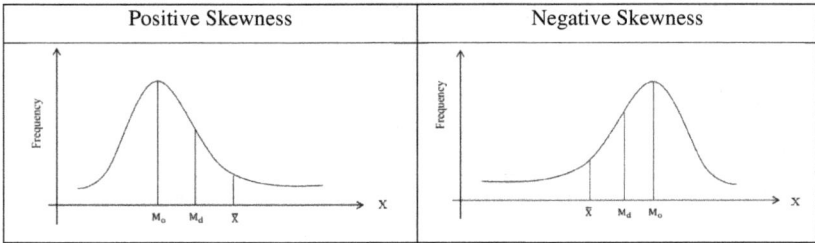

Positive Skewness	Negative Skewness

N/B: In a positive skewed situation, the mean is greater than the median and the median I greater than the mode. In a negative skewed situation, the mean is less than the median and the median is less than the mode.

Example 1: The mean of a distribution is 61.4 and the mode of the distribution is 61.13. Given that the standard deviation of the distribution is 1.76, determine the skewness of the distribution.

<u>**Solution**</u>

Step 1: Formula. Karl Pearson Coefficient of Skewness $(S_k) = \frac{\text{Mean}(\overline{X}) - \text{Mode}(M_o)}{\text{Standard Deviation}(\sigma)}$

Where; $\text{Mean}(\overline{X}) = 61.4$, $\text{Mode}(M_o) = 61.13$ and $\text{Standard Deviation}(\sigma) = 1.76$

Step 2: Data substitution and solving

Karl Pearson Coefficient of Skewness $(S_k) = \frac{61.4 - 61.13}{1.76}$ ....[Simplify numerator]

Karl Pearson Coefficient of Skewness $(S_k) = \frac{0.27}{1.76}$ ....[Solve equation]

Karl Pearson Coefficient of Skewness $(S_k) = \mathbf{0.1534}$ ....[In four decimal places]

Step 3: Interpretation: The Karl Pearson coefficient of skewness value is greater than zero$[S_k = 0.1534, \therefore S_k > 0]$. This means that the distribution is positively skewed.

Example 2: Given that $N = 10$, $\sum X = 450$, $\sum X^2 = 24,250$ and mode equal to 43. Compute the Karl Pearson coefficient of skewness.

<u>**Solution**</u>

Step 1: Formula. Karl Pearson Coefficient of Skewness $(S_k) = \frac{\text{Mean}(\overline{X}) - \text{Mode}(M_o)}{\text{Standard Deviation}(\sigma)}$

Where; $\text{Mean}(\overline{X}) = \frac{\sum X}{N} = \frac{450}{10} = 45$, $\text{Standard Deviation}(\sigma) = \sqrt{\frac{\sum X^2}{N} - \overline{X}^2} = \sqrt{\frac{24,250}{10} - 45^2} = 20$

Step 2: Data substitution and solving

$$\text{Karl Pearson Coefficient of Skewness } (S_k) = \frac{45-43}{20} \dots \dots \dots \text{[Simplify numerator]}$$

$$\text{Karl Pearson Coefficient of Skewness } (S_k) = \frac{2}{20} \dots \dots \dots \dots \dots \text{[Solve equation]}$$

$$\text{Karl Pearson Coefficient of Skewness } (S_k) = \mathbf{0.1} \dots \dots [\ S_k > 0, \text{ Positively Skewed}]$$

Example 3: Determine the skewness of the distribution below

Class	$1-5$	$6-10$	$11-15$	$16-20$	$21-25$	$26-30$	$31-35$
Frequency	20	27	29	38	48	53	70

Solution

a) Graphical method

Step 1: Determination of mean, median and mode values…[Cum.F stands for cumulative frequency]

1) Formula. $\text{Mean}(\overline{X}) = \frac{\sum f_i x_i}{N}$ …………………………………………………………..………(1)

$$\text{Median}(M_d) = L_1 + \left[\frac{\frac{N}{2}-F}{F_m} \times C\right] \dots \dots \dots \dots \dots \dots \dots \dots \dots(2)$$

$$\text{Mode}(M_o) = L_1 + \left[\frac{F_1-F_0}{2F_1-F_0-F_2} \times c\right] \dots \dots \dots \dots \dots \dots \dots(3)$$

3) Tabular determination of unknowns

Working Table				Mean		Median		Mode		
Class	Frequency	x_i	$f_i x_i$	Class	Frequency	Cum. F		Class	Frequency	Modal Class
0.5 – 5.5	20	3	60	0.5 – 5.5	20	20	$\frac{N}{2}=\frac{285}{2}=142.5$	0.5 – 5.5	20	30.5 – 35.5
5.5 – 10.5	27	8	216	5.5 – 10.5	27	47	Median Class	5.5 – 10.5	27	Data
10.5 – 15.5	29	13	377	10.5 – 15.5	29	76	[20.5 – 25.5]	10.5 – 15.5	29	$L_1=30.5$
15.5 – 20.5	38	18	684	15.5 – 20.5	38	114	Data	15.5 – 20.5	38	$F_1=70$
20.5 – 25.5	48	23	1,104	20.5 – 25.5	48	162	$L_1=20.5$	20.5 – 25.5	48	$F_0=53$
25.5 – 30.5	53	28	1,484	25.5 – 30.5	53	215	$F_m=48$	25.5 – 30.5	53	$F_2=0$
30.5 – 35.5	70	33	2,310	30.5 – 35.5	70	285	$F=114$	30.5 – 35.5	70	$C=5$
	$\sum=285$		$\sum=6,235$		$\sum=285$		$C=5$			

4) Data substitution and solving

Mean	Median	Mode
$\text{Mean}(\overline{X}) = \frac{6,235}{285}$ $\text{Mean}(\overline{X}) = 22$	$\text{Median}(M_d) = 20.5 +$ $\left[\frac{\frac{285}{2}-114}{48} \times 5\right]$ $\text{Median}(M_d) = 20.5 + \left[\frac{28.5}{48} \times 5\right]$	$\text{Mode}(M_o) = 30.5 +$ $\left[\frac{70-53}{2(70)-53-0} \times 5\right]$ $\text{Mode}(M_o) = 30.5 + \left[\frac{17}{87} \times 5\right]$ $\text{Mode}(M_o) = 31$

	Median(M_d) = 23	
	Mean(22) < $Median$(23) < $Mode$(31)	

Step 2: Plotting and interpretation

Illustration	Explanation
	From the illustration we realized that $Mean(22) < Median(23) < Mode(31)$. This indicates that the distribution is negatively skewed.

2) Algebraic method

Step 1: Formula. Karl Pearson Coefficient of Skewness $(S_k) = \frac{Mean(\overline{X}) - Mode(M_o)}{Standard\ Deviation(\sigma)}$

Where; Mean$(\overline{X}) = 22$ and Mode$(M_o) = 31$

Step 2: Determination of unknown.....................................[Standard Deviation(σ)]

1) Formula. Standard Deviation$(\sigma) = \sqrt{\frac{\sum f_i(x_i - \overline{X})^2}{N}}$, where; $N = 285$ and $\overline{X} = 22$

2) Tabular determination of unknown...................................[$\sum f_i(x_i - \overline{X})^2$]

Class	Frequency	x_i	$(x_i - \overline{X})^2$	$f_i(x_i - \overline{X})^2$
0.5 − 5.5	20	3	361	7,220
5.5 − 10.5	27	8	196	5,292
10.5 − 15.5	29	13	81	2,349
15.5 − 20.5	38	18	16	608
20.5 − 25.5	48	23	1	48
25.5 − 30.5	53	28	36	1,908
30.5 − 35.5	70	33	121	8,470
	$\sum = 285$			$\sum = 25,895$

3) Data substitution and solving

Standard Deviation$(\sigma) = \sqrt{\frac{25,895}{285}}$ ....[Simplify fraction]

Standard Deviation$(\sigma) = \sqrt{90.85964912}$ ....[Work square root]

Standard Deviation$(\sigma) = \mathbf{9.532}$ ....[In three decimal places]

Step 3: Data substitution and solving.

Karl Pearson Coefficient of Skewness $(S_k) = \frac{22-31}{9.532}$ ....[Simplify numerator]

Karl Pearson Coefficient of Skewness $(S_k) = \frac{-9}{9.532}$ ....[Solve equation]

Karl Pearson Coefficient of Skewness $(S_k) = \mathbf{-0.944}$ ....[$S_k < 0$,   Negatively skewed]

Example 4: Determine the skewness of the distribution below

Class	$0-20$	$20-40$	$40-60$	$60-80$	$80-100$
Frequency	13	25	27	19	16

<u>Solution</u>

a) Graphical method

Step 1: Determination of mean, median and mode values…[Cum.F stands for cumulative frequency]

1) Formula. $\text{Mean}(\overline{X}) = \frac{\sum f_i x_i}{N}$ ....(1)

$$\text{Median}(M_d) = L_1 + \left[\frac{\frac{N}{2}-F}{F_m} \times C\right] ....(2)$$

$$\text{Mode}(M_o) = L_1 + \left[\frac{F_1-F_0}{2F_1-F_0-F_2} \times c\right] ....(3)$$

2) Tabular determination of unknowns

Working Table				Mean	Median				Mode		
Class	Frequency	x_i	$f_i x_i$	Class	Frequency	Cum. F	$\frac{N}{2}=\frac{100}{2}=50$	Class	Frequency	Modal Class	
$0-20$	13	10	130	$0-20$	13	13		$0-20$	13	$40-60$	
$20-40$	25	30	750	$20-40$	25	38	Median Class	$20-40$	25	Data	
$40-60$	27	50	1,350	$40-60$	27	65	$[40-60]$	$40-60$	27	$L_1=40, F_1=27$	
$60-80$	19	70	1,330	$60-80$	19	84	Data	$60-80$	19	$F_0=25, F_2=19$	
$80-100$	16	90	1,440	$80-100$	16	100	$L_1=40, F_m=27$	$80-100$	16	$C=20$	
	$\sum=100$		$\sum=5,000$		$\sum=100$		$F=38$ and $C=20$		$\sum=100$		

3) Data substitution and solving

Mean	Median	Mode
$\text{Mean}(\overline{X}) = \frac{5,000}{100}$ $\text{Mean}(\overline{X}) = 50$	$\text{Median}(M_d) = 40 + \left[\frac{\frac{100}{2}-38}{27} \times 20\right]$ $\text{Median}(M_d) = 40 + \left[\frac{12}{27} \times 20\right]$ $\text{Median}(M_d) = 49$	$\text{Mode}(M_o) = 40 + \left[\frac{27-25}{2(27)-25-19} \times 20\right]$ $\text{Mode}(M_o) = 40 + \left[\frac{2}{10} \times 20\right]$ $\text{Mode}(M_o) = 44$
	$\text{Mean}(50) > \text{Median}(49) > \text{Mode}(44)$	

Step 2: Plotting and interpretation

Illustration	Explanation
M_o (44) M_d (49) $\bar{x}$ (50)	From the illustration we realized that Mean(50) > $Median(49)$ > $Mode(44)$. This indicates that the distribution is positively skewed.

b) Algebraic method

Step 1: Formula. Karl Pearson Coefficient of Skewness $(S_k) = \frac{Mean(\bar{X})-Mode(M_o)}{Standard\ Deviation(\sigma)}$

Where; Mean$(\bar{X})$ = 50 and Mode(M_o) = 44

Step 2: Determination of unknown...[Standard Deviation(σ)]

1) Formula. Standard Deviation$(\sigma) = \sqrt{\frac{\sum f_i(x_i-\bar{X})^2}{N}}$, where; N = 100 and $\bar{X}$ = 50

2) Tabular determination of unknown.....................................[$\sum f_i(x_i - \bar{X})^2$]

Class	Frequency	x_i	$(x_i - \bar{X})^2$	$f_i(x_i - \bar{X})^2$
0 − 20	13	10	1,600	20,800
20 − 40	25	30	400	10,000
40 − 60	27	50	0	0
60 − 80	19	70	400	7,600
80 − 100	16	90	1,600	25,600
	$\sum = 100$			$\sum = 64,000$

3) Data substitution and solving

$\qquad$ Standard Deviation$(\sigma) = \sqrt{\frac{64,000}{100}}$ ....[Simplify fraction]

$\qquad$ Standard Deviation$(\sigma) = \sqrt{640}$ ....[Work square root]

$\qquad$ Standard Deviation$(\sigma) = 25.298$ ....[In three decimal places]

Step 3: Data substitution and solving.

$\qquad$ Karl Pearson Coefficient of Skewness $(S_k) = \frac{50-44}{25.298}$ ....[Simplify numerator]

$\qquad$ Karl Pearson Coefficient of Skewness $(S_k) = \frac{6}{25.298}$ ....[Solve equation]

$\qquad$ Karl Pearson Coefficient of Skewness $(S_k) = 0.2372..$[S_k > 0, Positively skewed]

Example 5: Given that Karl Pearson coefficient of skewness of a distribution (S_k) is -0.25, the mean ($\bar{X}$) is 50 and the coefficient of variation (CV) is 35%, fine the value of median and mode.

Solution

1) Determination of mode

Step 1: Formula. Karl Pearson Coefficient of Skewness (S_k) $= \frac{\text{Mean}(\bar{X}) - \text{Mode}(M_o)}{\text{Standard Deviation}(\sigma)}$

Where; $S_k = -0.25$, $\text{Mean}(\bar{X}) = 50$ and Standard Deviation$(\sigma) = \frac{CV \times \bar{X}}{100} = \frac{35 \times 50}{100} = \frac{1,750}{100} = 17.5$

Step 2: Data substitution and solving

$-0.25 = \frac{50 - \text{Mode}(M_o)}{17.5}$ ....[Linearized equation]

$-4.375 = 50 - \text{Mode}(M_o)$ ....[Collect like terms together]

$\text{Mode}(M_o) = 50 + 4.375$ ....[Solve equation]

$\text{Mode}(M_o) = \mathbf{54.375}$

2) Determination of median

Step 1: Formula. Karl Pearson Coefficient of Skewness (S_k) $= \frac{3[\text{Mean}(\bar{X}) - \text{Median}(M_d)]}{\text{Standard Deviation}(\sigma)}$

Where; $S_k = -0.25$, $\text{Mean}(\bar{X}) = 50$ and Standard Deviation$(\sigma) = 17.5$

Step 2: Data substitution and solving

$-0.25 = \frac{3[50 - \text{Median}(M_d)]}{17.5}$ ....[Simplify numerator and linearized equation]

$-4.375 = 150 - 3\text{Median}(M_d)$ ....[Collect like terms together]

$3\text{Median}(M_d) = 154.375$ ....[Divide both sides of equation by (3)]

$\text{Median}(M_d) = \mathbf{51.458}$ ....[In three decimal places]

Test Your Understanding

Exercise 12: The mean of a distribution is 4 and its median is 5. Given that the standard deviation is 2.61, find the Karl Pearson coefficient of skewness(S_k) .[Answer: $\mathbf{S_k = -1.1494}$]

Exercise 13: From the table below, determine the skewness of the distribution using Karl Pearson coefficient of skewness.

Class	$0-10$	$10-20$	$20-30$	$30-40$	$40-50$	$50-60$	$60-70$
Frequency	6	12	22	24	16	12	8

[Answer: Karl Pearson Coefficient of Skewness (S_k) = 0.1856. Positively Skewed]

390

Exercise 14: Using the table below, determine the skewness of the distribution using Karl Pearson coefficient of skewness.

Marks	5	10	15	20	25	30	35
Frequency	8	15	20	32	23	17	5

[**Answer: Karl Pearson Coefficient of Skewness (S_k) = 0. Symmetry**]

Exercise 15: Given that the Karl Pearson coefficient of skewness of a distribution (S_k) is -0.93, the mean $(\overline{X})$ is 10 and the coefficient of variation (CV) is 32.1%, fine the value of median and mode...................[**Answer: Median(M_d) = 11 and Mode(M_0) = 12.98**]

Note: Apart from the Karl Pearson coefficient of skewness, we also have other measures such as; Bowley's coefficient of skewness, Kelly's Coefficient of Skewness and Coefficient measure of skewness based on moments.

1) Bowley's Coefficient of Skewness

The Bowley's coefficient of skewness (Sk_b) is based on the quarters of a distribution. It is calculated using the formula below.

$$\text{Bowley's Coefficient of Skewness}(Sk_b) = \frac{Q_3 + Q_1 - 2M_d}{Q_3 - Q_1} \quad \ldots\ldots\ldots\ldots\ldots\ldots\ldots(1)$$

$$\text{Bowley's Coefficient of Skewness}(Sk_b) = \frac{(Q_3 - Q_2) - (Q_2 - Q_1)}{(Q_3 - Q_2) + (Q_2 - Q_1)} \quad \ldots\ldots\ldots\ldots\ldots\ldots(2)$$

$$\text{Bowley's Coefficient of Skewness}(Sk_b) = \frac{(Q_3 - M) - (M - Q_1)}{Q_3 - Q_1} \quad \ldots\ldots\ldots\ldots\ldots\ldots(3)$$

N/B: The interpretation of the Bowley's coefficient of skewness remains the same as that of Karl Pearson coefficient of skewness.

Example 1: Calculate the Bowley's coefficient of skewness given that $Q_3 = 56.6$, $Q_1 = 44.1$ and $M_d = 55.35$.

Solution

Step 1: Formula. Bowley's Coefficient of Skewness$(Sk_b) = \frac{Q_3 - Q_1 - 2M_d}{Q_3 - Q_1}$

Where; $Q_3 = 56.6$, $Q_1 = 44.1$ and $M_d = 55.35$.

Step 2: Data substitution and solving

$$\text{Bowley's Coefficient of Skewness}(Sk_b) = \frac{56.6 + 44.1 - 2(55.35)}{56.6 - 44.1} \quad \ldots\ldots\ldots[\text{Simplify equation}]$$

$$\text{Bowley's Coefficient of Skewness}(Sk_b) = \frac{-10}{12.5} \quad \ldots\ldots\ldots\ldots\ldots\ldots\ldots\ldots[\text{Solve equation}]$$

Bowley's Coefficient of Skewness$(Sk_b) = -0.8$....[Answer: $Sk_b < 0, Negatively\ Skewed$]

Example 2: Compute the Bowley's coefficient of skewness given that $Q_3 = 70$, $Q_1 = 30$ and $M_d = 38$.

<u>**Solution**</u>

Step 1: Formula. Bowley's Coefficient of Skewness$(Sk_b) = \frac{Q_3+Q_1-2M_d}{Q_3-Q_1}$

Where; $Q_3 = 70$, $Q_1 = 30$ and $M_d = 38$.

Step 2: Data substitution and solving

Bowley's Coefficient of Skewness$(Sk_b) = \frac{70+30-2(38)}{70-30}$ ....[Simplify equation]

Bowley's Coefficient of Skewness$(Sk_b) = \frac{24}{40}$ ....[Solve equation]

Bowley's Coefficient of Skewness$(Sk_b) = 0.6$ ...[Answer: $Sk_b > 0, Positively\ Skewed$]

<u>Test Your Understanding</u>

Exercise 16: Determine the Bowley's coefficient of skewness given that $Q_3 = 115.7$, $Q_1 = 105.8$ and $M_d = 110.7$..................[**Answer: 0.01, $Sk_b > 0, Positively\ Skewed$**]

Exercise 17: Calculate the Bowley's coefficient of skewness given that $Q_3 - Q_2 = 100$ and $Q_2 - Q_1 = 120$..............................[**Answer: -0.09, $Sk_b < 0, Negatively\ Skewed$**]

---

2) Kelly's Coefficient of Skewness

Kelly's coefficient of skewness (Sk_D) is based on the deciles (D_i) of the given set of observations. It is calculated using the formula below.

Kelly's Coefficient of Skewness$(Sk_D) = \frac{D_9-D_1-2D_5}{D_9-D_1}$ ....(1)

Where; (D_9) represent the ninth deciles of the given set of observation

(D_1) represent the first deciles of the given set of observation

(D_5) represent the fifth deciles of the given set of observation (Median)

3) Coefficient measure of skewness based on moments

Coefficient measure of skewness based on moments (β_1) is based on the central moments of observations. It is calculated using the formula below.

Coefficient Measure of Skewness based on Moments$(\beta_1) = \frac{\mu_3{}^2}{\mu_2{}^3}$ ....(1)

Where; (μ_3) represent the third central moments of the observations

(μ_2) represent the second central moments of the observations

CHAPTER TEN

INDEX NUMBER

10.1 Meaning of Index Numbers

An index number is a statistical measure or device designed to measure changes in a variable or group of related variables with respect to time, geographical location or other characteristics such as; income, profession etc. It is worth knowing that an index number is a number that expresses the relative changes not absolute changes in variables, with comparison from the base (reference) year.

Base year period is the period in time in the past against or from which all comparison are made. It is statistically accepted that the base year or the reference year if not specified is by default 100. The magnitude or degree of relative change between periods or geographical location using the concept of index number can be obtained through two main methods.

1) Index Point method

This refers to the percentage difference between the indexes value of two periods (current and base). Algebraically, it is expressed as;

$$Index\ Point = \frac{Current\ period\ index - Base\ period\ index}{Base\ period\ index} \times 100$$
............................(1)

A positive index point indicate increase, a negative index point indicate decrease and a zero index point indicate no change or equilibrium situation.

2) Reference Method

Here, we compare the base index (by default =100) to the calculated index value (current index). If the current index value is greater than the base index, it signifies an increase and if the current index value is lesser than base index value, it indicates a decrease and if they are equal, then it shows no change. Algebraically the magnitude of change using this method is expressed as;

$$Reference = Current\ index\ value\ (n) - base\ index\ value(100)$$
..................(2)

10.2 Major Concepts of Index Number

10.2.1 Price Relative (Price Index)

It is the ratio of price of a single commodity of a period (current period) from a previous period (base period). Price relative can be transform to percentage when price relative is being multiplied by 100. Algebraically, it is expressed as;

$$Relative\ Price\ (RP) = \frac{Current\ Price(P_1)}{Base\ Price\ (P_0)} \times 100$$

....(1)

Example 1: The price of Garri per bag was 10,000 FRS in the year 2000. In the year 2005, it increases to 15,000 FRS. Assuming that the year 2000 is the base year, calculate the relative price and interpret your result.

<u>Solution</u>

Step 1: Formulas and data. $Relative\ Price\ (RP) = \frac{P_1}{P_0} \times 100$

Where; $P_1 = Price\ of\ the\ year\ 2005 = 15{,}000$, $P_0 = Price\ of\ the\ year\ 2000 = 10{,}000$

Step 2: Data substitution and solving

$Relative\ Price\ (RP) = \frac{P_1}{P_0} \times 100$ ....[Substitute data]

$Relative\ Price\ (RP) = \frac{15{,}000}{10{,}000} \times 100$ ....[Simplify fraction]

$Relative\ Price\ (RP) = 1.5 \times 100$

....[Solve]

$Relative\ Price\ (RP) = \mathbf{150\%}$

Step 3: Interpretation of result.

1) **Index Point Method**

   a) Formula. $Index\ Point = \frac{Current\ period\ index - Base\ period\ index}{Base\ period\ index} \times 100$

   Where; $Current\ period\ index = 150$ and $Base\ period\ index = 100$

   b) Data substitution and solving

   $Index\ Point = \frac{Current\ period\ index - Base\ period\ index}{Base\ period\ index} \times 100$ ....[Substitute data]

   $Index\ Point = \frac{150-100}{100} \times 100$ ....[Simplify fraction and multiply by 100]

   $Index\ Point = \mathbf{50\%}$

2) **Reference Method**
   a) Formula. $Reference = Current\ index\ value - base\ index\ value$
      Where; $Current\ period\ index = 150$ and $Base\ period\ index = 100$
   b) Data substitution and solving
      $Reference = Current\ index\ value - base\ index\ value$.....[Substitute data]
      $Reference = 150 - 100$ ....[Simplify equation by subtraction]
      $Reference = 50\%$

Conclusion: Garri in the year 2005 is 150 times expensive or 50 times more expensive than was in the year 2000.

10.2.2 Relative Quantity (Quantity index)

Relative quantity measures the changes in the volume quantity of commodities between two periods. It answers the question of "by how much or in what quantity has quantity change over time? When multiplied by 100, it is expresses the relative quantity in percentages.

$$Relative\ Quantity(RQ) = \frac{Current\ Quantity\ (Q_1)}{Base\ Quantity\ (Q_0)} \times 100$$
....(2)

Example 1: The quantity of cocoa production in Cameroon in the years 2003 is considered to be 500 Million tons. Assuming that the national institute forecasted that Cameroon will produce 800 Million tons in 2006, determine the relative quantity of cocoa production in Cameroon and interpret your answer.

Solution

Step 1: Formula. $Relative\ Quantity(RQ) = \frac{Current\ Quantity\ (Q_1)}{Base\ Quantity\ (Q_0)} \times 100$

Where; $Q_1 = Quantity\ of\ 2006 = 800,000,000$ and $Q_0 = Quantity\ of\ 2003 = 500,000,000$

Step 2: Data substitution and solving

$Relative\ Quantity(RQ) = \frac{Q_1}{Q_0} \times 100$ ....[Substitute data]

$Relative\ Quantity(RQ) = \frac{800,000,000}{500,000,000} \times 100$ ....[Simplify equation fraction]

$Relative\ Quantity(RQ) = 1.6 \times 100$ ....[Solve equation]

$$Relative\ Quantity(RQ) = \mathbf{160\%}$$

Step 3: Interpretation of result

1) **Index Point Method**

    a) Formula. $Index\ Point = \frac{Current\ period\ index - Base\ period\ index}{Base\ period\ index} \times 100$

        Where; $Current\ period\ index = 160$ and $Base\ period\ index = 100$

    b) Data substitution and solving

        $Index\ Point = \frac{Current\ period\ index - Base\ period\ index}{Base\ period\ index} \times 100$ ....[Substitute data]

        $Index\ Point = \frac{160-100}{100} \times 100$ ....[Simplify fraction and multiply by 100]

        $Index\ Point = \mathbf{60\%}$

2) **Reference Method**

    a) Formula. $Reference = Current\ index\ value - base\ index\ value$

        Where; $Current\ period\ index = 160$ and $Base\ period\ index = 100$

    b) Data substitution and solving

        $Reference = Current\ index\ value - base\ index\ value$ ..[Substitute data]

        $Reference = 160 - 100$ ....[Simplify equation by subtraction]

        $Reference = \mathbf{60\%}$

Conclusion: Cameroons cocoa production in the year 2005 will increase by 60%.

10.2.3 Relative Value (Value Index)

Value index measures or compare the changes in the monetary value (import, export, production and consumption) of commodities over time. It is worth knowing that the value is a function of price (cost) and quantity. Relative value can also be expressed in percentage as seen be;

$$Relative\ Value = \frac{Current\ year\ price_{(P_1)} \times Quantity\ of\ current\ year_{(Q_1)}}{Base\ year\ price_{(P_1)} \times Quantity\ of\ base\ year_{(Q_0)}} \times 100$$

..................(3)

Example 1: During school resumption in 2016, Mr. Tabi a teacher in G.T.H.S Jakiri purchased 50 Engineering books to give to outstanding students. Considering that the price per book is 10.000 FRS for which in 2015, it costed 15,000 FRS. Compute the relative value of this transaction.

<u>**Solution**</u>

Step 1: $Relative\ Value = \dfrac{Current\ year\ price_{(P_1)} \times Quantity\ of\ current\ year_{(Q_1)}}{Base\ year\ price_{(P_0)} \times Quantity\ of\ base\ year_{(Q_0)}} \times 100$

Where; $P_1 = Price\ for\ 2016 = 10,000,\ Q_1 = Quantity\ for\ 2015 = 50$

$P_0 = Price\ for\ 2015 = 15,000,\ Q_0 = Quantity\ of\ 2015 = 50$

Step 2: Data substitution and solving

$Relative\ Value = \dfrac{P_1 Q_1}{P_0 Q_0} \times 100$ ....[Substitute data]

$Relative\ Value = \dfrac{10,000 \times 50}{15,000 \times 50} \times 100$ ....[Simplify numerator and denominator]

$Relative\ Value = \dfrac{500,000}{750,000} \times 100$ ....[Simplify fraction]

$Relative\ Value = 0.666666666 \times 100$ ....[Solve equation]

$Relative\ Value = \mathbf{67\%}$ ....**[Rounded value]**

Step 3: Interpretation of result

1) **Index Point Method**

   a) Formula. $Index\ Point = \dfrac{Current\ period\ index - Base\ period\ index}{Base\ period\ index} \times 100$

   Where; $Current\ period\ index = 67$ and $Base\ period\ index = 100$

   b) Data substitution and solving

   $Index\ Point = \dfrac{Current\ period\ index - Base\ period\ index}{Base\ period\ index} \times 100$[Substitute data]

   $Index\ Point = \dfrac{67 - 100}{100} \times 100$ ......[Simplify fraction and multiply by 100]

   $Index\ Point = \mathbf{-33\%}$

2) **Reference Method**

   a) Formula. $Reference = Current\ index\ value - base\ index\ value$

Where; $Current\ period\ index = 67$ and $Base\ period\ index = 100$

b) Data substitution and solving

$Reference = Current\ index\ value - base\ index\ value$ [Substitute data]

$Reference = 67 - 100$ ....[Simplify equation by 100]

$Reference = \mathbf{-33\%}$

Conclusion: The monetary value of the book has reduced by 33%

Example 2: The price and quantity demanded of rice in 1980, 1990 and 2000 are presented in the table below.

Item	Years					
	1980		1990		2000	
	Price	Quantity	Price	Quantity	Price	Quantity
Rice	150	50	200	25	100	75

Work Required

1) Calculate the relative price and relative quantity of rice between 1980 and 1990, using 1980 as the base year.

2) Calculate the relative price and relative quantity between 1990 and 2000, using 1990 as the reference year

3) Calculate the relative price and relative quantity of 1990 and 2000 using 1980 as the base year.

4) Calculate the relative value between the period 1980 and 2000, using 1980 as the base year.

Solution

1) **Calculate the relative price and relative quantity of rice between 1980 and 1990, using 1980 as the base year.**

Step 1: Formula. $Relative\ Price = \frac{P_1}{P_0} \times 100$, where; $P_1 = 200$ $P_0 = 150$

Step 2: Data substitution and solving.

$Relative\ price = \frac{200}{150} \times 100$........................ ....[Simplify fraction]

$Relative\ price = 1.333333333 \times 100$

....[Solve]

$Relative\ price = \mathbf{133}$ %   ....[Rounded value]

Step 3: Interpretation. Price has increased by 33 % in 1990. Economically, this increase in price will reduce demand and increase cost of living.

2) **Calculate the relative price and relative quantity between 1990 and 2000, using 1990 as the reference year**

Step 1: Formula and data. $Relative\ Quantity = \frac{Q_1}{Q_0} \times 100$, where; $Q_0 = 25$ $Q_1 = 75$

Step 2: Data substitution and solving.

$Relative\ Quantity = \frac{75}{25} \times 100$ ....[Simplify fraction]

$Relative\ Quantity = 3 \times 100$

....[Solve]

$Relative\ Quantity = \mathbf{300\%}$

Step 3: Interpretation. Quantity demanded for rice has increased by 200%. Economically, quantity demand increment is mainly caused by reduction in price.

3) **Calculate the relative price and relative quantity of 1990 and 2000 using 1980 as the base year.**

   i) **Calculate the relative price and relative quantity of 1990 using 1980 as the base year.**

Step 1: Formulas. $Relative\ Price = \frac{P_1}{P_0} \times 100$ and $Relative\ Quantity = \frac{Q_1}{Q_0} \times 100$

Step 2: Data sorting, data substitution and solving

a) $Relative\ Price = \frac{P_1}{P_0} \times 100$, where; ; $P_1 = 200$, $P_0 = 150$

    $Relative\ Price = \frac{200}{150} \times 100$ ....…..…..........[Simplify fraction]

    $Relative\ Price = 1.333333333 \times 100$

    ....…........................…[Solve]

    $Relative\ Price =$ **133** % ....[That is 33 % increase in price]

b) $Relative\ Quantity = \frac{Q_1}{Q_0} \times 100$, where; $Q_1 = 25$, $Q_0 = 50$

    $Relative\ Quantity = \frac{25}{50} \times 100$ ....[Simplify fraction]

    $Relative\ Quantity = 0.5 \times 100$

    ....…...…..............[Solve]

    $Relative\ Quantity =$ **50%** ....[That is 50% decrease in quantity]

  ii) Calculate the relative price and relative quantity of 2000 using 1980 as the base year.

Step 1: Formulas. $Relative\ Price = \frac{P_1}{P_0} \times 100$ and $Relative\ Quantity = \frac{Q_1}{Q_0} \times 100$

Step 2: Data sorting, data substitution and solving

a) $Relative\ Price = \frac{P_1}{P_0} \times 100$, where; $P_1 = 100$, $P_0 = 150$

    $Relative\ Price = \frac{100}{150} \times 100$ ....…......[Simplify fraction]

    $Relative\ Price = 0.666666666 \times 100$

    ....…......................…....[Solve]

    $Relative\ Price =$ **67%** ....…...[That is 33 % decrease in price]

b) $Relative\ Quantity = \frac{Q_1}{Q_0} \times 100$, where; ; $Q_1 = 75$, $Q_0 = 50$

    $Relative\ Quantity = \frac{75}{50} \times 100$ ....…..............…...[Simplify fraction]

    $Relative\ Quantity = 1.5 \times 100$

    ....…....................................[Solve]

$Relative\ Quantity = 150\%$ ....[That is 50% increase in **quantity**]

4) **Calculate the relative value between the period 1980 and 2000, using 1980 as the base year.**

Step 1: Formula and data. $Relative\ Value = \frac{P_1 Q_1}{P_0 Q_0} \times 100$

Where; $P_1 = 100$, $Q_1 = 75$, $P_0 = 150$, $Q_0 = 50$

Step 2: Data substitution and solving

$Relative\ Value = \frac{100 \times 75}{150 \times 50} \times 100$ ....[Simplify numerator and denominator]

$Relative\ Value = \frac{7,500}{7,500} \times 100$ ....[Simplify fraction]

$Relative\ Value = 1 \times 100$

....[Solve]

$Relative\ Value = \mathbf{100\%}$

Step 3: Interpretation. From the relative value obtain, it is statistically clear that there has being no change in consumers expenditure in the year 2000.

10.3 Types of Index Number

The major types of index numbers are represented on the graph below

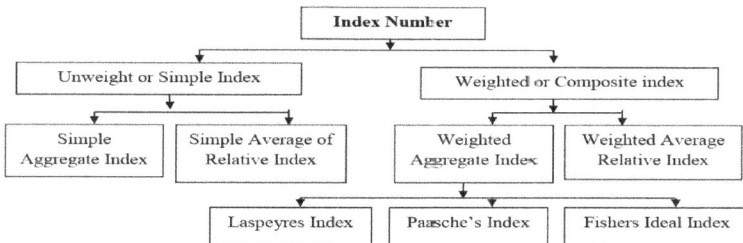

10.3.1 Simple or Unweight Indices

Simple index is also called unweighted index because no item is given more or less importance in relation to other items. There are two main methods of calculating simple indices; simple aggregative method and simple average of price relative method.

a) **Simple Aggregative Method**

The formula used for the calculation of price index and quantity their respective procedures are given below.

Current Price Index $= \frac{\Sigma P_1}{\Sigma P_0} \times 100$ ....(1)

Current Quanitity Index $= \frac{\Sigma Q_1}{\Sigma Q_0} \times$

100...(2)

Example 1: Find the simple aggregative index number using the table below for the year 2014, using 2010 as the base year.

Commodities	Prices	
	2010	2014
A	200	400
B	100	200
C	50	60

Solution

Step 1: Formula and data. Current Price Index $= \frac{\sum P_1}{\sum P_0} \times 100$

Where; P_1 = price of 2014 P_0 = price of 2010

Step 2: Tabular derivation of formula components

Commodities	Prices		Working	
	$2010 = P_0$	$2014 = P_1$	P_0	P_1
A	200	400	200	400
B	100	200	100	200
C	50	60	50	60
			$\sum P_0 = 350$	$\sum P_1 = 660$

Step 3: Data substitution and solving.

Current Price Index $= \frac{\sum P_1}{\sum P_0} \times 100$ ....[Substitute data]

Current Price Index $= \frac{660}{350} \times 100$........ ....[Simplify fraction]

Current Price Index $= 1.885714286 \times 100$ ....[Solve]

Current Price Index $= \textbf{189}$ ....**[Rounded value]**

Step 4: Interpretation of result: There is an increase in price by 89 % in 2014 as against 2010.

Example 2: Determine the index number for the following year; 1980, 1990, 2000, and 2010, using 1980 as the base year from the table below.

Years	1980	1990	2000	2010
Price	120	150	200	400

Solution

Hint: When more than two years are given, we calculate the index number using the successive segmentation method. That is we calculate the price index for 1980 and its base year (which in this case is still 1980), 1990 with base year 1980, 2000 with base year 1990, and 2010 with base year 2000.

Step 1: Formula and data. Current Price Index $= \frac{\sum P_1}{\sum P_0} \times 100$

Where; P_0 = Price for 1980, and P_1 = prices of 1980,1990,2000 and 2010 respectively

Step 2: Data substitution and solving.

Method 1 [Tabula method. (PI = Price Index)]		Method 2 [Algebraic method]
Years	Respective current price index	**Current price index for 1980**
1980	Current PI $= \frac{120}{120} \times 100 = \textbf{100}$	Current Price Index $= \frac{120}{120} \times 100 = \textbf{100}$
1990	Current PI $= \frac{150}{120} \times 100 = \textbf{125}$	**Current price index for 1990**
		Current Price Index $= \frac{150}{120} \times 100 = \textbf{125}$
2000	Current PI $= \frac{200}{120} + 100 = \textbf{166.6}^+$	**Current price index for 2000**
		Current Price Index $= \frac{200}{120} \times 100 = \textbf{166.6}^+$
2010	Current PI $= \frac{400}{120} \times 100 = \textbf{333.3}^+$	**Current price index for 2010**
		Current Price Index $= \frac{400}{120} \times 100 = \textbf{333.3}^+$

Step 3: Interpretation of results: The price index for 1980 shows there is a zero increase in price, 1990 witness a 25% increase in price from 1980, followed by a 67% increase in price in 2000 from 1980 and finally 233 % increase in price in 2010 from 1980. Therefore, the result shows a successive increase in the price from 1980 to 2010.

Example 3: Compute the simple aggregative index number for the year 2005 and 2009 using 2001 as the base year from the statistics presented below.

Years	Commodities				
	A	B	C	D	E
2001	20	10	15	30	50
2005	10	10	50	50	45
2009	50	41	12	4	10

Solution

Step 1: Formula and data. Current Price Index(CPI) $= \frac{\Sigma P_1}{\Sigma P_0} \times 100$

Where; P_0 = Price of 2001 and P_1 = Price of 2005 ..[CPI of 2005 using 2001 as base]

P_0 = Price of 2001 and P_1 = Price of 2009 ..[CPI of 2009 using 2001 as base]

Step 2: Tabula determination of formula component

Commodities	Years		
	$2001 = P_0$	$2005 = P_1$	$2009 = P_1$
A	20	10	50
B	10	10	41
C	15	50	12
D	30	50	4
E	50	45	10
	$\Sigma P_0 = \textbf{125}$	$\Sigma P_1 = \textbf{165}$	$\Sigma P_1 = \textbf{117}$

Step 3: Data substitution and solving

a) **Current price index for 2005 using 2001 as base year**

    i) Formula: Current Price Index(CPI) $= \frac{\sum P_1}{\sum P_0} \times 100$

        Where; $\sum P_1 = 165$ and $\sum P_0 = 125$

    ii) Data substitution and solving

        Current Price Index(CPI) $= \frac{\sum P_1}{\sum P_0} \times 100$ ....[Substitute data]

        Current Price Index $= \frac{165}{125} \times 100$ ....…......[Simplify fraction]

        Current Price Index $= 1.32 \times 100$ ....…...............[Solve]

        Current Price Index $= \mathbf{132\%}$ ....[32% increase in price from 2001 to 2005]

b) **Current price index for 2009 using 2001 as base year**

    i) Formula: Current Price Index(CPI) $= \frac{\sum P_1}{\sum P_0} \times 100$

        Where; $\sum P_1 = 117$ and $\sum P_0 = 125$

    ii) Data substitution and solving

        Current Price Index $= \frac{117}{125} \times 100$ ....[Simplify fraction]

        Current Price Index $= 0.936 \times 100$ ....[Solve]

        Current Price Index $= \mathbf{93.6\%}$ ....[6.4% reduction in price from 2001 to 2009]

Example 4: Using the table below, determine the price index of the following years (2001, 2003, 2006, 2007 and 2010) using 2006 as the base year.

Years	2001	2003	2006	2007	2010
Price	263	268	279	288	296

Solution

Step 1: Formula and data. Current Price Index $= \frac{\sum P_1}{\sum P_0} \times 100$

    Where; $P_0 = $ Price of 2006, $P_1 = $ prices of 2001, 2003, 2006, 2007, 2010, N = 5

Step 2: Data substitution and solving

Years	Price	Calculation of individual current price index
		General formula: $\left[\frac{P_1}{P_0} \times 100\right]$
2001	263	$\frac{263}{279} \times 100 = \mathbf{94.26523297}$ ....[2001 and 2006]
2003	268	$\frac{268}{279} \times 100 = \mathbf{96.05734767}$ ....[2003 and 2006]

2006 = P$_0$	279	$\frac{279}{279} \times 100 = \mathbf{100}$ ....[2006 and 2006]	
2007	288	$\frac{288}{279} \times 100 = \mathbf{103.2258065}$ ....[2007 and 2006]	
2010	296	$\frac{296}{279} \times 100 = \mathbf{106.09319}$ ....[2010 and 2006]	

b) Simple Average of Price Relative method

Here, the price of each item in the current year is taken as a percentage of the price of corresponding item of the base year. To obtain the index, we divide the percentages by the number of items.

$$\text{Current Price Index} = \frac{\sum \frac{P_1}{P_0} \times 100}{N} \equiv \frac{1}{N}\sum\left(\frac{P_1}{P_0} \times 100\right)...(1)$$

$$\text{Current Qunaity Index} = \frac{\sum \frac{Q_1}{Q_0} \times 100}{N} \equiv \frac{1}{N}\sum\left(\frac{Q_1}{Q_0} \times 100\right) ....(2)$$

$$\text{Current Price Index} = \text{Antilog}\left[\frac{1}{N}\sum \text{Log}\left(\frac{P_1}{P_0} \times 100\right)\right] ....(3)$$

$$\text{Current Quantity Index} = \text{Antilog}\left[\frac{1}{N}\sum \text{Log}\left(\frac{Q_1}{Q_0} \times 100\right)\right] ....(4)$$

Example 1: Compute by simple average of price relative method the price index of 2004 taking 1990 as the base year from the data presented in the table below.

Commodities	Price	
	2001	2006
A	60	80
B	40	60
C	30	70

Solution

a) Using Arithmetic mean concept

Step 1: Formula. $\text{Current Price Index} = \frac{\sum \frac{P_1}{P_0} \times 100}{N}$ OR $\text{Current Price Index} = \frac{1}{N}\sum\left(\frac{P_1}{P_0} \times 100\right)$

Where; $P_1 = $ Prices of 2006, $P_0 = $ Prices of 2001 and N = Number of items (3)

Step 2: Tabular determination of formula component

Commodities	Price		Working
	$2001 = P_0$	$2006 = P_1$	$\left[\frac{P_1}{P_0} \times 100\right]$
A	60	80	$\frac{P_1}{P_0} \times 100 \rightarrow \frac{80}{60} \times 100 = 133.3333333$
B	40	60	$\frac{P_1}{P_0} \times 100 \rightarrow \frac{60}{40} \times 100 = 150$
C	30	70	$\frac{P_1}{P_0} \times 100 \rightarrow \frac{70}{30} \times 100 = 233.3333333$
			$\sum \frac{P_1}{P_0} \times 100 = 517$ ....**[Rounded value]**

Step 3: Data substitution and solving

Method 1: Current Price Index $= \frac{\sum \frac{P_1}{P_0} \times 100}{N}$	Method 2: Current Price Index $= \frac{1}{N} \sum \left(\frac{P_1}{P_0} \times 100\right)$
Where; $\sum \frac{P_1}{P_0} \times 100 = 517$, N = 3	
Current Price Index $= \frac{517}{3}$ ....[Solve]	Current Price Index $= \frac{1}{3}(517)$ ..[Solve]
Current Price Index $= \mathbf{172}$	Current Price Index $= \mathbf{172}$

b) Using Geometric mean concept

Step 1: Formula. Current PI $= \text{Antilog}\left[\frac{\sum \text{Log}\left(\frac{P_1}{P_0} \times 100\right)}{N}\right]$

....…….....(1)

Current PI $= \text{Antilog}\left[\frac{1}{N} \sum \text{Log}\left(\frac{P_1}{P_0} \times 100\right)\right]$ ....(2)

Step 2: Tabular determination of formula component

Commodities	Price		Working	
	$2001 = P_0$	$2006 = P_1$	$\left[\frac{P_1}{P_0} \times 100\right]$	$\left[\text{Log}\left(\frac{P_1}{P_0} \times 100\right)\right]$
A	60	80	133.3333333	2.124938737
B	40	60	150	2.176091259
C	30	70	233.3333333	2.367976785
				$\sum \text{Log}\left(\frac{P_1}{P_0} \times 100\right) =$ 6.669006781

Step 3: Data substitution and solving

Current PI = Antilog $\left[\dfrac{\sum \text{Log}\left(\frac{P_1}{P_0}\times 100\right)}{N}\right]$	Current PI = Antilog $\left[\dfrac{1}{N}\sum \text{Log}\left(\dfrac{P_1}{P_0}\times 100\right)\right]$
Current PI = Antilog $\left[\dfrac{6.669006781}{3}\right]$.....[Simplify] Current PI = Antilog [2.22300226] Current PI = **167**	Current PI = Antilog $\left[\dfrac{1}{3}(6.669006781)\right]$ Current PI = Antilog[2.223002256] Current PI = **167**

Example 2: The table below reports the price of several food items from 2001 to 2006.

Food Items	Prices in 2001	Prices in 2006
Bread (Loaf)	200	250
Egg (Dozen)	210	100
Milk (Littre)	40	60
Orange Juice (Bottle)	120	30

Calculate the price index of price for 2006 using 2001 as the base year price using,

1) Simple aggregative method
2) Simple Average of Price Relative method

Solution

1) Simple aggregative method

Step 1: Formula and data. Current Price Index $= \dfrac{\sum P_1}{\sum P_0} \times 100$

Where $P_1 = 2006$ and $P_0 = 2001$

Step 2: Data substitution and solving

Tabula solving			Calculation of current price index
Food Items	P_0	P_1	Formula. Current Price Index $= \dfrac{\sum P_1}{\sum P_0} \times 100$
Bread (Loaf)	200	250	Where; $\sum = 570$ and $\sum = 440$
Egg (Dozen)	210	100	Current Price Index $= \dfrac{440}{570} \times 100$
Milk (Littre)	40	60	Current PI $= 0.771929824 \times 100$
Juice (Bottle)	120	30	Current PI $= \textbf{77}$
	$\sum = \textbf{570}$	$\sum = \textbf{440}$	

2) Simple Average of Price Relative method

Step 1: Formula and data. Current Price Index $= \dfrac{\sum \frac{P_1}{P_0} \times 100}{N}$, where; $P_1 = 2006$, $P_0 = 2001$, N

=4

Step 2: Data Substitution and solving

Tabular method				Calculation of current price index
Food Items	P_0	P_1	Working $\left[\frac{P_1}{P_0} \times 100\right]$	Formula. Current PI $= \dfrac{\sum \frac{P_1}{P_0} \times 100}{N}$
Bread (Loaf)	200	250	125	Where; $\sum \frac{P_1}{P_0} = 347.6190476$
Egg (Dozen)	210	100	47.61904762	Current PI $= \dfrac{347.6190476}{4}$
Milk (Littre)	40	60	150	Current PI $= \mathbf{87}$
Juice (Bottle)	120	30	25	
			347.6190476	

10.3.2 Unweighted Indices [Weighted Aggregate Index]

1) Laspeyre's method

This method was developed by Ernet Louis Etienne Laspeyres in 1871 for determining price increase. Laspeyres method assumes that the base period quantity and price are still realistic in the next period. That is the quantity and price used for an item is the same in both base and current period.

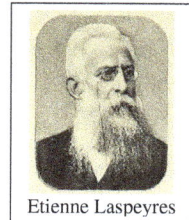

Etienne Laspeyres

Laspeyres Price Index (LPI): LPI $= \frac{\sum P_1 Q_0}{\sum P_0 Q_0} \times 1$ .... (1)

Laspeyres Quantity Index (LQI): LQI $= \frac{\sum Q_1 P_0}{\sum Q_0 P_0} \times 100$ .... (2)

Example 1: Use the table below and calculate the Laspeyres Price Index and quantity index of three commodities.

Commodities	Price		Quantity	
	2001	2002	2001	2002
Rice	500	520	20	50
Cocoyam	400	450	60	32
Garri	200	360	30	15

Solution [Separate Table Technique]

1. Calculation of Laspeyres price index

Step 1: Formula and data. Laspeyres price index $= \frac{\sum P_1 Q_0}{\sum P_0 Q_0} \times 100$

Where; P_1 = Prices of 2002, P_0 = Price of 2001, and Q_0 = Quantities of 2001

Step 2: Tabular determination of formula components

Commodities	Prices		Quantities	Working	
	P_0	P_1	Q_0	(P_1Q_0)	(P_0Q_0)
Rice	500	520	20	$520 \times 20 = 10,400$	$500 \times 20 = 10,000$
Cocoyam	400	450	60	$450 \times 60 = 27,000$	$400 \times 60 = 24,000$
Garri	200	360	30	$360 \times 30 = 10,800$	$200 \times 30 = 6,000$
				$\sum P_1Q_0 = 48,200$	$\sum P_0Q_0 = 40,000$

Step 3: Data substitution and solving

$$\text{Laspeyres price index} = \frac{48,200}{40,000} \times 100 \dots\dots\dots\dots\dots\dots\text{[Simplify fraction]}$$

$$\text{Laspeyres price index} = 1.205 \times 100 \dots\dots\dots\dots\dots\dots\dots\text{[Solve]}$$

$$\text{Laspeyres price index} = \mathbf{120.5}$$

2. Calculation of Laspeyres quantity index

Step 1: Formula and data. Laspeyres quantity index $= \frac{\sum Q_1P_0}{\sum Q_0P_0} \times 100$

Where; $Q_1 =$ Quantity of 2002, $Q_0 =$ Quantity of 2001 $P_0 =$ Price of 2001

Step 2: Tabular determination of formula component

Commodities	Price	Quantity		Working	
	P_0	Q_0	Q_1	Q_1P_0	Q_0P_0
Rice	500	20	50	25,000	10,000
Cocoyam	400	60	32	12,800	24,000
Garri	200	30	15	3,000	6,000
				$\sum Q_1P_0 = 40,800$	$\sum Q_0P_0 = 40,000$

Step 3: Data substitution and solving

$$\text{Laspeyres quantity index} = \frac{40,800}{40,000} \times 100 \dots\dots\dots\dots\dots\text{[Simplify fraction]}$$

$$\text{Laspeyres quantity index} = 1.02 \times 100 \dots\dots\dots\dots\dots\dots\text{[Solve]}$$

$$\text{Laspeyres quantity index} = \mathbf{102}$$

Example 2: Compute the price and quantity index number for the following data below using 1980 as the base year, using the laspeyres price index

Years	Item A		Item B		Item C	
	Price	Quantity	Price	Quantity	Price	Quantity
1990	4	50	3	10	2	5
2010	10	45	6	8	3	4

<u>Solution [Combine Table Technique]</u>

Step 1: Formulas: Laspeyres quanity index $= \frac{\sum Q_1 P_0}{\sum Q_0 P_0} \times 100$

Laspeyres price index $= \frac{\sum P_1 Q_0}{\sum P_0 Q_0} \times 100$

Where; Q_1 = Quantity of 2010 , P_0 = Price of 1990, Q_0 = Quantity of 1990, P_1 = Price of 2010

Step 2: Tabular determination of formula components

Items	Original Table				Working			
					Laspeyres price index		Laspeyres quanity index	
	P_0	Q_0	P_1	Q_1	$P_1 Q_0$	$F_0 Q_0$	$Q_1 P_0$	$Q_0 P_0$
A	4	50	10	45	500	200	180	200
B	3	10	6	8	60	30	24	30
C	2	5	3	4	15	10	8	10
					$\sum = 575$	$\sum = 240$	$\sum = 212$	$\sum = 240$

Step 3: Data substitution and solving

a) **Calculation of Laspeyres Price Index**

i) Formula. Laspeyres quanity index $= \frac{\sum Q_1 P_0}{\sum Q_0 P_0} \times 100$

Where; $\sum Q_1 P_0 = 212, \sum Q_0 P_0 = 240$

ii) Data substitution and solving

Laspeyres quanity index $= \frac{212}{240} \times 100$ ....[Simplify fraction]

Laspeyres quanity index $= 0.883333333 \times 100$

...........................[Solve]

Laspeyres quanity index $= $ **88** ....[Rounded value]

b) **Calculation of Laspeyres Quantity Index**

i) Formula. Laspeyres price index $= \frac{\sum P_1 Q_0}{\sum P_0 Q_0} \times 100$

Where; $\sum P_1 Q_0 = 575, \sum P_0 Q_0 = 240$

ii) Data substitution and solving

Laspeyres price index $= \frac{575}{240} \times 100$ ....[Simplify fraction]

Laspeyres price index $= 2.39583333333 \times 100$ ....[Solve]

Laspeyres price index $= $ **240** ....[Rounded value]

411

Example 3: Use the table below and test Laspeyres method of calculating price and index number.

Items	2000		2005		2010	
	Price	Quantity	Price	Quantity	Price	Quantity
A	12	42	14	40	20	30
B	14	52	13	50	14	45
C	16	32	20	30	15	40

Work Required

1) Time Reversal Test (Using the period 2000 to 2005)
2) Factor Reversal Test (Using the period 2005 to 2012, with 2005 being the base period)

<u>Solution</u>

1) Time Reversal Test (Using the period 2000 to 2005) [Test is based on condition]

Step 1: Condition. Time Reversal Test $= \frac{\sum P_1 Q_0}{\sum P_0 Q_0} \times \frac{\sum P_0 Q_1}{\sum P_1 Q_1} = 1$

Step 2: Tabular determination of formula components

Items	Original Table				Working			
	P_0	Q_0	P_1	Q_1	$P_1 Q_0$	$P_0 Q_0$	$P_0 Q_1$	$P_1 Q_1$
A	12	42	14	40	588	504	480	560
B	14	52	13	50	676	728	700	650
C	16	32	20	30	640	512	480	600
					$\sum = 1,940$	$\sum = 1,744$	$\sum = 1,660$	$\sum = 1,810$

Step 3: Data substitution and solving

Time Reversal Test $= \frac{1,940}{1,744} \times \frac{1,660}{1,810}$ ....[Simplify fractions]

Time Reversal Test $= 1.112385321 \times 0.917127071$ ....[Solve]

Time Reversal Test $= \mathbf{1.020198692}$ ....[Test Reversal Test $\neq 1$]

Step 4: Interpretation. From the result, since both sides are not equal to one, we can conclude that Laspeyres index method fail to satisfy the time reversal test.

412

2) Factor Reversal Test (Using the period 2005 to 2012, with 2005 being the base period)

Step 1: Condition. Factor Reversal Test $= \frac{\sum P_1 Q_0}{\sum P_0 Q_0} \times \frac{\sum Q_1 P_0}{\sum Q_0 P_0} = \frac{\sum P_1 Q_1}{\sum P_0 Q_0}$

Step 2: Tabular determination of formula components

Items	Original Table				Working			
	P_0	Q_0	P_1	Q_1	$P_1 Q_0$	$P_0 Q_0$	$Q_1 P_0$	$P_1 Q_1$
A	14	40	20	30	800	560	420	600
B	13	50	14	45	700	650	585	630
C	20	30	15	40	450	600	800	600
					$\sum = 1,950$	$\sum = 1,810$	$\sum = 1,805$	$\sum = 1,830$

Step 3: Data substitution and solving

Factor Reversal Test $= \frac{1,950}{1,810} \times \frac{1,805}{1,810} = \frac{1,830}{1,810}$ . ...[Simplify fractions]

Factor Reversal Test $= 1.077348066 \times 0.997237569 = 1.011049724$ .....[Solve]

Factor Reversal Test $= \mathbf{1.0744 \neq 1.0110}$ ....[In four decimal places]

Step 4: Interpretation. The result shows that both sides are not equal, therefore Laspeyres method fail to satisfy the factor reversal test.

2) Paasche's Index

This method was developed by Herman Paasche's and is also called Current weighted Index. It is a weighted harmonic average or mean of the relative index that uses the current period as weight. Note that this method fails to satisfy the time reversal test and factor reversal test. The main difference between Paasche's and Laspeyres method is only in the period used for weight, as Paasche's uses the current year weight for the denominator of the weighted index.

Herman Passche

Paasche's Price Index (PPI) $= \frac{\sum P_1 Q_1}{\sum P_0 Q_1} \times 100$ ....(1)

Paasche's Quantity Index (PQI) $= \frac{\sum Q_1 P_1}{\sum Q_0 P_1} \times 100$ ....(2)

Example 1: The price and quantity changes of three goods, A, B and C from 2000 to 2005 are given in the table below. Calculate the Paasche's price and quantity index of these goods

413

Goods	Prices		Quantities	
	2000	2005	2000	2005
A	4	12	50	30
B	8	10	10	5
C	5	15	40	15

Solution [Separate Table Technique]

1. Calculation of Paasche's price index

Step 1: Formula and data. Paasche's price index $= \frac{\sum P_1 Q_1}{\sum P_0 Q_1} \times 100$

Where; P_1 = Price of 2005, P_0 = Prices of 2000 , Q_1 = Quantities of 2005

Step 2: Tabula determination of formula components

Commodities	Prices		Quantities	Working	
	P_0	P_1	Q_1	$(P_1 Q_1)$	$(P_0 Q_1)$
A	4	12	30	$12 \times 30 = 360$	$4 \times 30 = 120$
B	8	10	5	$10 \times 5 = 50$	$8 \times 5 = 40$
C	5	15	15	$15 \times 15 = 225$	$5 \times 15 = 75$
				$\sum P_1 Q_1 = 635$	$\sum P_0 Q_1 = 235$

Step 3: Data substitution and solving

Paasche's price index $= \frac{635}{235} \times 100$ ....[Simplify fraction]

Paasche's price index $= 2.70212766 \times 100$ ....[Solve]

Paasche's price index $= \mathbf{270}$ ....**[Rounded value]**

2. Calculation of Paasche's quantity index

Step 1: Formula and data. Paasche's quantity index $= \frac{\sum Q_1 P_1}{\sum Q_0 P_1} \times 100$

Where; Q_1 = Quantity of 2005, Q_0 = Quantity of 2000 and P_1 = Price of 2005

Step 2: Tabular determination of formula components

Goods	Prices	Quantities		Working	
	P_1	Q_0	Q_1	$Q_1 P_1$	$Q_0 P_1$
A	12	50	30	360	600
B	10	10	5	50	100
C	15	40	15	225	600
				$\sum = 635$	$\sum = 1,200$

Step 3: Data substitution and solving

$$\text{Paasche's quantity index} = \frac{635}{1,200} \times 100 \ldots\ldots\ldots\ldots\ldots\ldots\ldots\text{[Simplify fraction]}$$

$$\text{Paasche's quantity index} = 0.529166666 \times 100 \ldots\ldots\ldots\ldots\ldots\ldots\ldots\ldots\text{[Solve]}$$

$$\text{Paasche's quantity index} = \mathbf{53} \ldots\ldots\ldots\ldots\ldots\ldots\ldots\ldots\ldots\text{[\textbf{Rounded value}]}$$

Example 2: From the table below, determine the price index and quantity index of these three goods using the Paasche's method.

Goods	2010		2014	
	Price	Quantity	Price	Quantity
Sugar	8	20	10	5
Flour	4	40	22	25
Milk	16	10	8	5

Solution [Combine Table Technique]

Step 1: Formulas and data. Paasche's price index $= \frac{\sum P_1 Q_1}{\sum P_0 Q_1} \times 100$

$$\text{Paasche's quantity index} = \frac{\sum Q_1 P_1}{\sum Q_0 P_1} \times 100$$

Where; P_1 = Price of 2014, Q_1 = Quantity of 2014 P_0 = Price of 2010, Q_0 = Quanitity of 2010

Step 2: Tabular determination of formula components

Goods	Original Table				Working			
					Paasche's Price Index		Paasche's Quantity Index	
	P_0	Q_0	P_1	Q_1	$P_1 Q_1$	$P_0 Q_1$	$Q_1 P_1$	$Q_0 P_1$
Sugar	8	20	10	5	50	40	50	200
Flour	4	40	22	25	550	100	550	880
Milk	16	10	8	5	40	80	40	80
					$\sum = 640$	$\sum = 220$	$\sum = 640$	$\sum = 1,160$

Step 3: Data substitution and solving

a) **Calculation of Paasche's price index**

  i)   Paasche's price index $= \frac{\sum P_1 Q_1}{\sum P_0 Q_1} \times 100$, where; $\sum P_1 Q_1 = 640, \sum P_0 Q_1 = 220$

  ii) Data substitution and solving

$$\text{Paasche's price index} = \frac{640}{220} \times 100 \ldots\ldots\ldots \ldots\ldots\ldots\text{[Simplify fraction]}$$

Paasche's price index = 2.909090909 × 100 ....[Solve]

Paasche's price index = **291** ....**[Rounded value]**

b) **Calculation of Paasche's quantity index**

Paasche's quantity index = $\frac{640}{1,160}$ × 100 ....[Simplify fraction]

Paasche's quantity index = 0.551724137 × 100 ....[Solve]

Paasche's quantity index = **55** ....[Rounded value]

Example 3: The following are prices and consumption quantities for three goods in 1995 and 2009. Verify if the Paasche's method of calculating index number respect the time reversal test and factor reversal test.

Items	Years			
	1995		2009	
	Price	Quantity	Price	Quantity
A	2	20	3	21
B	18	3	36	2
C	3	18	4	23

Solution

1) **Time reversal test**

Step 1: Condition. Time reversal test = $\frac{\sum P_1 Q_1}{\sum P_0 Q_1} \times \frac{\sum P_0 Q_0}{\sum P_1 Q_0} = 1$

Step 2: Tabular determination of formula components

Items	Original Table				Working			
	P_0	Q_0	P_1	Q_1	$P_1 Q_1$	$P_0 Q_1$	$P_0 Q_0$	$P_1 Q_0$
A	2	20	3	21	63	42	40	60
B	18	3	36	2	72	36	54	108
C	3	18	4	23	92	69	54	72
					$\sum = 227$	$\sum = 147$	$\sum = 148$	$\sum = 240$

Step 3: Data substitution and solving

Time reversal test = $\frac{227}{147} \times \frac{148}{240}$ ....[Simplify fractions]

Time reversal test = 1.544217687 × 0.616666666 ....[Solve]

Time reversal test = **0.952267573** ....**[Time reversal test ≠ 1]**

416

Step 4: Interpretation. Paasche's method fail to satisfy the time reversal test.

2) Factor reversal test

Step 1: Condition. Factor reversal test $= \frac{\sum P_1 Q_1}{\sum P_0 Q_1} \times \frac{\sum Q_1 P_1}{\sum Q_0 P_1} = \frac{\sum P_1 Q_1}{\sum P_0 Q_0}$

Step 2: Tabular determination of formula components

Items	Original Table				Working			
	P_0	Q_0	P_1	Q_1	$P_1 Q_1$	$P_0 Q_1$	$Q_c P_1$	$P_0 Q_0$
A	2	20	3	21	63	42	60	40
B	18	3	36	2	72	36	108	54
C	3	18	4	23	92	69	72	54
					$\Sigma = 227$	$\Sigma = 147$	$\Sigma = 240$	$\Sigma = 148$

Step 3: Data substitution and solving

Factor reversal test $= \frac{227}{147} \times \frac{227}{240} = \frac{227}{148}$ ....[Simplify fractions]

Factor reversal test $= 1.544217687 \times 0.945833333 = 1.533783784$

........[Solve]

Factor reversal test $= \mathbf{1.4606 \neq 1.5338}$ ....[In four decimal places]

Step 4: Interpretation. From the statistical result, it's seen that both sides are not equal; hence fail the factor reversal test

3. Fisher's ideal method

Fishers price index (FPI) number is given by the geometric mean of the Lasperey's and Paasche's index numbers $\left\{ \sqrt{[\text{Laspeyres index}] [\text{Paasches index}]} \right\}$. The fisher's method uses both base period and current period as weight and hence avoid the statistical bias associated with Laspeyres and Paasche's method. This satisfies both the time reversal test and factor reversal test but fails to satisfy the circular test.

Ronald A. Fisher

$$\text{FP I} = \sqrt{\left[\frac{\sum P_1 Q_1}{\sum P_0 Q_0} \times 100 \right] \left[\frac{\sum P_1 Q_1}{\sum P_0 Q_1} \times 100 \right]} \equiv \text{FPI} \sqrt{\left\{ \left[\frac{\sum P_1 Q_1}{\sum P_0 Q_0} \right] \left[\frac{\sum P_1 Q_1}{\sum P_0 Q_1} \right] \times 10,000 \right\}} \quad \dots\dots\dots\dots(1)$$

$$\text{F Q I} = \sqrt{\left[\frac{\sum Q_1 P_0}{\sum Q_0 P_0} \times 100 \right] \left[\frac{\sum Q_1 P_1}{\sum Q_0 P_1} \times 100 \right]} \equiv \text{FOI} \sqrt{\left\{ \left[\frac{\sum Q_1 P_0}{\sum Q_0 P_0} \right] \left[\frac{\sum Q_1 P_1}{\sum Q_0 P_1} \right] \times 10,000 \right\}} \quad \dots\dots\dots(2)$$

Example 1: From the table below, determine the fishers ideal price and quantity index using 1998 as the base year

417

	1998		2005	
Item	Price	Quantity	Price	Quantity
A	3	6	3	6
B	4	4	4	5
C	2	3	4	3

Solution [Separate Table Technique]

1. **Calculation of fishers price index**

Step 1: Formula and data. Fishers price index $= \sqrt{\left[\frac{\sum P_1 Q_1}{\sum P_0 Q_0} \times 100\right]\left[\frac{\sum P_1 Q_1}{\sum P_0 Q_1} \times 100\right]}$

Where; P_1 = Price of 2005, P_0 = Price of 1998, Q_1 = Quantity of 2005, Q_0 = Quantity of 1998

Step 2: Tabula determination of formula component

True table					Working			
						Laspeyres index		Paasches index
Items	P_0	Q_0	P_1	Q_1	$P_1 Q_1$	$P_0 Q_0$	$P_1 Q_1$	$P_0 Q_1$
A	3	6	3	6	18	18	18	18
B	4	4	4	5	20	16	20	20
C	2	3	4	3	12	6	12	6
					$\sum = 50$	$\sum = 40$	$\sum = 50$	$\sum = 44$

Step 3: Data substitution and solving

a) **Using formula 1**

   i. Formula. Fishers price index $= \sqrt{\left[\frac{\sum P_1 Q_1}{\sum P_0 Q_0} \times 100\right]\left[\frac{\sum P_1 Q_1}{\sum P_0 Q_1} \times 100\right]}$

   ii. Data substitution and solving

   Fishers price index $= \sqrt{\left[\frac{50}{40} \times 100\right]\left[\frac{50}{44} \times 100\right]}$ ....[Simplify fractions]

   Fishers price index $= \sqrt{[1.25 \times 100][1.136363636 \times 100]}$.[Simplify brackets]

   Fishers price index $= \sqrt{[125][113.6363636]}$ ....[Expand brackets]

   Fishers price index $= \sqrt{14204.54545}$ ....…....[Work root]

   Fishers price index $= \textbf{119}$ ....…......**[Rounded value]**

b) Using formula 2

    i.   Formula. Fishers price index $= \sqrt{\left\{\left[\frac{\sum P_1 Q_1}{\sum P_0 Q_0}\right]\left[\frac{\sum P_1 Q_1}{\sum P_0 Q_1}\right] \times 10,000\right\}}$

    ii. Data substitution and solving

        Fishers price index $= \sqrt{\left\{\left[\frac{50}{40}\right]\left[\frac{50}{44}\right] \times 10,000\right\}}$ ……….....[Simplify fraction]

        Fishers price index $= \sqrt{\{[1.25][1.13636363636] \times 10,000\}}$.[Expand bracket]

        Fishers price index $= \sqrt{\{1.420454545 \times 10,000\}}$ …….[Multiply values]

        Fishers price index $= \sqrt{14204.54545}$ ……. ……….....…….[Work root]

        Fishers price index $= \mathbf{119}$ ………………………….…..[Rounded value]

2. Calculation of fisher quantity index

Step 1: Formula and data. Fisher quantity index $= \sqrt{\left[\frac{\sum Q_1 P_0}{\sum Q_0 P_0} \times 100\right]\left[\frac{\sum Q_1 P_1}{\sum Q_0 P_1} \times 100\right]}$

        Where; $Q_1 =$ Quantity of 2005, $Q_0 =$ Quantity of 1998 , $P_0 =$ Price of 1998 and $P_1 =$ Price of 2005

Step 2: Tabular determination of formula components

Original table					Working		
					Laspeyres index		Paasches index	
Item	P_0	Q_0	P_1	Q_1	$Q_1 P_0$	$Q_0 P_0$	$Q_1 P_1$	$Q_0 P_1$
A	3	6	3	6	18	18	18	18
B	4	4	4	5	20	16	20	16
C	2	3	4	3	6	6	12	12
					$\Sigma = 44$	$\Sigma = 40$	$\Sigma = 50$	$\Sigma = 46$

Step 3: Data substitution and solving

a) Using formula 1

    i.   Formula. Fisher quantity index $= \sqrt{\left[\frac{\sum Q_1 P_0}{\sum Q_0 P_0} \times 100\right]\left[\frac{\sum Q_1 P_1}{\sum Q_0 P_1} \times 100\right]}$

    ii. Data substitution and solving

        Fisher quantity index $= \sqrt{\left[\frac{44}{40} \times 100\right]\left[\frac{50}{46} \times 100\right]}$ ……..........[Simplify fractions]

        Fisher quantity index $= \sqrt{[1.1 \times 100][1.086956522 \times 100]}$ [Simplify brackets]

        Fisher quantity index $= \sqrt{[110][108.6956522]}$ ………....…….[Expand brackets]

        Fisher quantity index $= \sqrt{11,956.52174}$ …………….…….....…….[Work root]

Fisher quantity index = **109** ....[Rounded value]

b. Using formula 2

i. Formula. Fisher quantity index = $\sqrt{\left\{\left[\frac{\sum Q_1 P_0}{\sum Q_0 P_0}\right]\left[\frac{\sum Q_1 P_1}{\sum Q_0 P_1}\right] \times 10,000\right\}}$

ii. Data substitution and solving.

Fisher quantity index = $\sqrt{\left\{\left[\frac{44}{40}\right]\left[\frac{50}{46}\right] \times 10,000\right\}}$ ....[Simplify fraction]

Fisher quantity index = $\sqrt{\{[1.1][1.086956522] \times 10,000\}}$ .....[Simplify bracket]

Fisher quantity index = $\sqrt{\{1.195652174 \times 10,000\}}$ ....[Multiply values]

Fisher quantity index = $\sqrt{11956.52174}$ ....[Work root]

Fisher quantity index = **109** ....[Rounded value]

Example 2: Calculate the fishers ideal price index and quantity index from the following data presented in the table below

Items	2003		2007	
	Price	Quantity	Price	Quantity
A	20	80	24	90
B	22	100	11	104
C	28	60	34	60
D	16	56	20	58

Solution [Combine Table Technique]

Step 1: Formulas FPI $= \sqrt{\left[\frac{\sum P_1 Q_1}{\sum P_0 Q_0} \times 100\right]\left[\frac{\sum P_1 Q_1}{\sum P_0 Q_1} \times 100\right]}$, FQI $=$

$\sqrt{\left[\frac{\sum Q_1 P_0}{\sum Q_0 P_0} \times 100\right]\left[\frac{\sum Q_1 P_1}{\sum Q_0 P_1} \times 100\right]}$.

Where: P_1 = Price of 2007, Q_1 = Quanitity of 2007, P_0 = Price of 2003 and Q_0 = Quantity of 2003

420

Step 2: Tabular determination of formula component

Items	Original Table				Working						
					Fishers Price Index			Fishers Quantity Index			
	P_0	Q_0	P_1	Q_1	P_1Q_1	P_0Q_0	P_0Q_1	Q_1P_0	Q_0P_0	Q_1P_1	Q_0P_1
A	20	80	24	90	2,160	1,600	1,800	1,800	1,600	2,160	1,920
B	22	100	11	104	1,144	2,200	2,288	2,288	2,200	1,144	1,100
C	28	60	34	60	2,040	1,680	1,680	1,680	1,680	2,040	2,040
D	16	56	20	58	1,160	896	928	928	896	1,160	1,120
				$\Sigma =$	6,540	6,376	6,696	6,696	6,376	6,540	6,180

Step 3: Data substitution and solving

Fishers Price Index	Fishers Quantity Index
$FPI = \sqrt{\left[\frac{6,540}{6,376} \times 100\right]\left[\frac{6,540}{6,696} \times 100\right]}$	$FQI = \sqrt{\left[\frac{6,696}{6,376} \times 100\right]\left[\frac{6,540}{6,180} \times 100\right]}$
$FPI = \sqrt{[102.5721455][97.6702509]}$	$FQI = \sqrt{[105.0188206][105.8252427]}$
$FPI = \sqrt{10,018.24719}$	$FQI = \sqrt{11,113.54218}$
Fisher Price Index $= \mathbf{100}$	Fisher Quantity Index $= \mathbf{105}$

Example 3: From the table below prove that the fisher idea method of calculating index number satisfy the time reversal test and factor reversal test.

Items	1990		2000	
	Price	Quantity	Price	Quantity
A	50	100	20	140
B	30	80	40	50

Solution

1. Time Reversal Test (TRT) Calculation

Step 1: Condition. Time Reversal Test (TRT) $= \left\{\left[\sqrt{\left(\frac{\sum P_1Q_0}{\sum P_0Q_0}\right)\left(\frac{\sum P_1Q_1}{\sum P_0Q_1}\right)}\right]\left[\sqrt{\left(\frac{\sum P_0Q_1}{\sum P_1Q_1}\right)\left(\frac{\sum P_0Q_0}{\sum P_1Q_1}\right)}\right]\right\} = 1$

Where; P_1 = Price of 2000, Q_1 = Quantity of 2000, P_0 = Price of 1990, Q_0 = Qunatity of 1990

Step 2: Tabular determination of formula components

Items	Original Table				Working			
	P_0	Q_0	P_1	Q_1	P_1Q_0	P_0Q_0	P_1Q_1	P_0Q_1
A	50	100	20	140	2,000	5,000	2,800	7,000
B	30	80	40	50	3,200	2,400	2,000	1,500
					$\Sigma = 5,200$	$\Sigma = 7,400$	$\Sigma = 4,800$	$\Sigma = 8,500$

Step 3: Data substitution and solving.

$$\text{Time Reversal Test (TRT)} = \left\{ \left[\sqrt{\left(\frac{5,200}{7,400}\right)\left(\frac{4,800}{8,500}\right)} \right]\left[\sqrt{\left(\frac{8,500}{4,800}\right)\left(\frac{7,400}{4,800}\right)} \right] \right\}$$

$$\text{TRT} = \left\{ \left[\sqrt{(0.702702702)(0.564705882)} \right]\left[\sqrt{(1.770833333)(1.541666667)} \right] \right\}$$

$$\text{Time Reversal Test (TRT)} = \left\{ \left[\sqrt{0.396820349} \right]\left[\sqrt{2.730034722} \right] \right\}$$

$$\text{Time Reversal Test (TRT)} = \{ [0.629936781][1.652281672] \}$$

$$\text{Time Reversal Test (TRT)} = 1 \dots\dots\dots\dots\dots\dots\dots\text{[Time Reversal Test = 1]}$$

Step 4: From our result, we realize that both sides are equal to 1, meaning fishers ideal method of calculating index number satisfy the time reversal test.

2. Time Reversal Test (TRT) Calculation

Step 1: Formula. Time Reversal Test $(\text{TRT}) = \sqrt{\frac{(\Sigma P_1 Q_1)^2}{(\Sigma P_0 Q_0)^2}} = \frac{\Sigma P_1 Q_1}{\Sigma P_0 Q_0}$

Where; $\quad P_1 = $ Price of 2000, $\quad Q_1 = $ Quantity of 2000, $\quad P_0 = $ Price of 1990, $Q_0 = $ Qunatity of 1990

Step 2: Tabular determination of formula components

Items	Original Table				Working	
	P_0	Q_0	P_1	Q_1	P_0Q_0	P_1Q_1
A	50	100	20	140	5,000	2,800
B	30	80	40	50	2,400	2,000
					$\Sigma = 7,400$	$\Sigma = 4,800$

Step 3: Data substitution and solving

$$\text{Time Reversal Test (TRT)} = \sqrt{\frac{(4,800)^2}{(7,400)^2}} = \frac{4,800}{7,400} \quad \dots\dots\dots\dots\text{[Simplify exponent and}$$
fraction]

$$\text{Time Reversal Test (TRT)} = \sqrt{\frac{23,040,000}{54,760,000}} = 0.648648648$$

$$\text{Time Reversal Test (TRT)} = \sqrt{0.420745069} = 0.648648648 \dots\dots\dots\dots\text{[Work root]}$$

422

Time Reversal Test (TRT) = **0.6486 = 0.6436** ………......[In four decimal places]

Step 4: Seeing from the statistics, we realize that both sides are equal, hence we can conclude that the fisher ideal method satisfy the factor reversal test.

Example 4: Use the table below and answer the following questions

Items	January 2001		January 2006	
	Price	Quantity	Price	Quantity
Books	100	40	50	60
Pen	200	23	250	10
Ruler	300	42	130	70

Calculate the price index for 2006 using 2001 as the base year using the following methods

1) Laspeyres method
2) Paasche's method
3) Fisher ideal index method

Solution [Separate Table Technique]

1) Laspeyres method

Step 1: Formula. Laspeyres Price Index(LPI) $= \frac{\sum P_1 Q_0}{\sum P_0 Q_0} \times 100$,

Where; P_1 = Price of 2006, P_0 = Price of 2001 and Q_0 = Quantity of 2001

Step 2: Data substitution and solving

Tabular solving				Working		Algebraic solving
Items	P_0	Q_0	P_1	$P_1 Q_0$	$P_0 Q_0$	$LP = \frac{\sum P_1 Q_0}{\sum P_0 Q_0} \times 100$
Books	100	40	50	2,000	4,000	$LP = \frac{13,210}{21,200} \times 100$
Pen	200	23	250	5,750	4,600	
Ruler	300	42	130	5,460	12,600	$LP = 0.623113207 \times 100$
				$\sum = 13,210$	$\sum = 21,200$	$LP = 62$ ......[Rounded value]

2) Paasche's method

Step 1: Formula. Paasche Price Index (PPI) $= \frac{\sum P_1 Q_1}{\sum P_0 Q_1} \times 100$

Where; P_1 = Price of 2006, P_0 = Price of 2001 and Q_1 = Quantity of 2006

423

Step 2: Data substitution and solving

Tabular solving				Working		Algebraic solving
Items	P_0	P_1	Q_1	P_1Q_1	P_0Q_1	$PPI = \frac{\sum P_1Q_1}{\sum P_0Q_1} \times 100$
Books	100	50	60	3,000	6,000	
Pen	200	250	10	2,500	2,000	$PPI = \frac{14,600}{29,000} \times 100$
Ruler	300	130	70	9,100	21,000	$PPI = 0.503448275 \times 100$
				$\sum = 14,600$	$\sum = 29,000$	$PPI = \mathbf{50}$ ....**[Rounded]**

3) Fisher ideal index method

Step 1: Formula. Fisher Price Index(FPI) $= \sqrt{\text{[Laspeyres index] [Paasches index]}}$

Where; Laspeyres index $= 62$ and Paasches index $= 50$

Step 2: Data substitution and solving.

Fisher Price Index(FPI) $= \sqrt{[62][50]}$ ....…..............[Expand bracket]

Fisher Price Index(FPI) $= \sqrt{3,100}$ ....[Work root]

Fisher Price Index(FPI) $= \mathbf{56}$ ....**[Rounded value]**

Test Your Understanding

Exercise 1: The prices and quantity of fish, cow meat and chicken used by trader from 2011 to 2013 are summarized in the table below.

product	2011		2012		2013	
	Price	Quantity	Price	Quantity	Price	Quantity
Meat	150	40	300	34	400	25
Fish	300	30	800	25	900	20
Chicken	200	50	250	45	300	20

1) Calculate the price index and quantity index of these products for 2011, using 2010 as the base year using the Laspeyres method. (Assume 2011 quantity to be 20% more than 2010 quantity)

2) Statisticians forecast a 30% increase in the prices of meat and fish only and also said this price increase will lead to a 10% reduction in the quantity of meat and fish and a 5% increase in the quantity of chicken from 2013. Determine the value of price and quantity index for 2014, using 2011 as the base year using the Paasche's method.

[Answer: ....…...................**See appendix]**

Exercise 2: Use the table below and answer the following questions [Leave answer in four decimal places]

	Items					
	A [Measured in bags]		B [Measured in Kilograms]		C [Measured in litters]	
Years	Price	Quantity	Price	Quantity	Price	Quantity
2005	8	25	3	15	4	6
2016	20	50	9	10	6	4

Calculate the price index and quantity index for 2016 using 2005 as the base year using the following methods

1) Laspeyres Method ....[**Answer: LPI = 91.4498 and LQI= 165.7993**]

2) Paasche's Method ....[**Answer: PPI = 59.1928 and PQI = 107.3171**]

3) Fisher Ideal Index Method ....[**Answer: FPI = 76.2185 and FQI = 133.3907**]

PART THREE

CALCULUS

CHAPTER ELEVEN

DIFFERENTIATION

Differential calculus deals with differentiation. Differentiation is the process of finding the derivative of a function. The derivative of a function is the rate of change of that function with respect to a variable. That is, differentiation measures how one variable changes with respect to another. The derivative of a function (y) with respect to variable (x) is written as;

$$y' = f'(x) = \frac{dy}{dx} \dots\dots\dots\dots\dots\dots\dots\dots\dots\dots\dots\dots\dots\dots\dots\dots\dots\dots\dots\dots\dots(1)$$

11.1 Laws of Differentiation

The laws of differentiation are standard formulas or expression that helps in determining of a derivative of a function.

11.1.1 Power Rule

The power rule or power law is used to determine the derivative of a function with exponent. The exponent can be a positive or negative number and the law is expressed as seen below.

$$\text{If } Y = x^n, \text{ then } \frac{dY}{dx} = n(x)^{n-1} \dots\dots\dots\dots\dots[\text{When exponent is a positive number}]$$

$$\text{If } Y = \frac{1}{x^n} = x^{-n}, \text{ then } \frac{dY}{dx} = -n(x)^{-n-1} \dots[\text{When exponent is a negative number}]$$

Example 1: Determine the derivative of the following

1) $Y = x^7$
2) $Y = x^{-2}$

Solution

1) $Y = x^7$

Step 1: Formula. If $Y = x^{\pm n}$, then $\frac{dY}{dx} = \pm n(x)^{\pm n-1}$, where; n = 7

Step 2: Data substitution and solving

$$\text{If } Y = x^7, \text{ then } \frac{dY}{dx} = 7(x^{7-1}) \dots\dots\dots\dots\dots\dots\dots\dots\dots\dots \dots\dots\dots[\text{Simplify exponent}]$$

$$\frac{dY}{dx} = 7(x^6) \dots\dots\dots\dots\dots\dots\dots\dots\dots\dots\dots \dots\dots\dots \dots\dots\dots\dots[\text{Open bracket}]$$

$$\frac{dY}{dx} = 7x^6$$

2) $Y = x^{-2}$ .... $\left[Y = x^{-2} = Y = \frac{1}{x^2} \right]$

Step 1: Formula. If $Y = x^{\pm n}$, then $\frac{dY}{dx} = \pm n(x)^{\pm n-1}$, where; $n = -2$

Step 2: Data substitution and solving

If $Y = x^{-2}$, then $\frac{dY}{dx} = -2(x^{-2-1})$ ....[Simplify exponent]

$\frac{dY}{dx} = -2(x^{-3}) = -2 \times \frac{1}{x^3}$ ....[Multiply terms]

$\frac{dY}{dx} = -\frac{2}{x^3}$

Example 2: Determine the derivative of the following

1) $Y = \sqrt{x^3}$

2) $Y = \sqrt[4]{x}$

<u>**Solution**</u>

1) $Y = \sqrt{x^3}$ .... $\left[Y = \sqrt{x^3} = (x^3)^{\frac{1}{2}} = x^{\frac{3}{2}} \right]$

Step 1: Formula. If $Y = x^{\pm n}$, then $\frac{dY}{dx} = \pm n(x)^{\pm n-1}$, where: $n = x^{\frac{3}{2}}$

Step 2: Data substitution and solving

If $Y = x^{\frac{3}{2}}$, then $\frac{dY}{dx} = \frac{3}{2}\left(x^{\frac{3}{2}-1}\right)$ ....[Simplify exponent]

$\frac{dY}{dx} = \frac{3}{2}x^{\frac{1}{2}} = \frac{3}{2}\sqrt{x} = \frac{3\sqrt{x}}{2}$

2) $Y = \sqrt[4]{x}$ .... $\left[Y = \sqrt[4]{x} = (x)^{\frac{1}{4}} \right]$

Step 1: Formula. If $Y = x^{\pm n}$, then $\frac{dY}{dx} = \pm n(x)^{\pm n-1}$, where: $n = x^{\frac{3}{2}}$

Step 2: Data substitution and solving

If $Y = (x)^{\frac{1}{4}}$, then $\frac{dY}{dx} = \frac{1}{4}\left(x^{\frac{1}{4}-1}\right)$ ....[Simplify exponent]

$\frac{dY}{dx} = \frac{1}{4}x^{-\frac{3}{4}} = \frac{1}{4} \times \frac{1}{x^{\frac{3}{4}}}$ ....[Multiply terms]

$\frac{dY}{dx} = \frac{1}{4} \times \frac{1}{\sqrt[4]{x^3}} = \frac{1}{4\sqrt[4]{x^3}}$

<u>**Test Your Understanding**</u>

Exercise 1: Differentiate the following functions with respect to variable x

1) $Y = x^4$ ....[Answer: $4x^3$]

2) $Y = x^{-3}$ ....[Answer: $-\frac{3}{x^4}$]

3) $Y = \sqrt{x}$ ....[Answer: $\frac{1}{2}x^{-\frac{1}{2}} = \frac{1}{2\sqrt{x}}$]

---

4) $Y = \sqrt[3]{x^2}$ ....[**Answer:** $\frac{2}{3\sqrt[3]{x}}$]

11.1.2 Complementary Rules

These are rules that facilitate the application of power rule when power functions above are modified.

If $Y = C$, then $\frac{dY}{dx} = 0$ ....,,..........[Constant Rule]

If $Y = x$, then $\frac{dY}{dx} = 1$ ....[Identity Function Rule]

If $Y = Kf(x)$, then $\frac{d}{dx}[Kf(x)] = K\frac{d}{dx}[f(x)]$ ....[Constant Multiple Rule]

Example 1: Differentiate the following equations

1) $Y = 2x$
2) $Y = 8x^2$

Solution

1) $Y = 2x$...[$Y = 2x = 2x^1$]

Step 1: Formula. If $Y = Kf(x)$, then $\frac{d}{dx}[Kf(x)] = K\frac{d}{dx}[f(x)]$, where; $K = 2$ and $f(x) = x$

Step 2: Data substitution and solving

If $Y = 2x$, then $\frac{d}{dx}[2x] = 2\frac{d}{dx}[x]$ ....[Apply identify function rule]

$\frac{d}{dx}[2x] = 2[1] = \mathbf{2}$

2) $Y = 8x^2$

Step 1: Formula. If $Y = Kf(x)$, then $\frac{d}{dx}[Kf(x)] = K\frac{d}{dx}[f(x)]$, where; $K = 8$ and $f(x) = x^2$

Step 2: Data substitution and solving

If $Y = 8x^2$, then $\frac{d}{dx}[8x^2] = 8\frac{d}{dx}[x^2]$ ....[Apply power rule]

$\frac{d}{dx}[8x^2] = 8[2(x^{2-1})]$ ....[Simplify power]

$\frac{d}{dx}[8x^2] = 8[2(x^1)] = 8[2(x)]$ ....[Open bracket]

$\frac{d}{dx}[8x^2] = 8[2(x^1)] = \mathbf{16x}$

Example 2: Differentiate the following equations

1) $Y = 4\sqrt{x}$
2) $Y = 3\sqrt[3]{x^2}$

1) $Y = 4\sqrt{x}$ ....$\left[Y = 4\sqrt{x} = Y = 4\left(x^{\frac{1}{2}}\right)\right]$

Step 1: Formula. If $Y = Kf(x)$, then $\frac{d}{dx}[Kf(x)] = K\frac{d}{dx}[f(x)]$, where; $K = 4$ and $f(x) = x^{\frac{1}{2}}$

Step 2: Data substitution and solving

$\quad$ If $Y = 4\left(x^{\frac{1}{2}}\right)$, then $\frac{d}{dx}\left[4\left(x^{\frac{1}{2}}\right)\right] = 4\frac{d}{dx}\left[x^{\frac{1}{2}}\right]$ ....[Apply power rule]

$\quad \frac{d}{dx}\left[4\left(x^{\frac{1}{2}}\right)\right] = 4\left[\frac{1}{2}\left(x^{\frac{1}{2}-1}\right)\right]$ ....[Simplify exponent]

$\quad \frac{d}{dx}\left[4\left(x^{\frac{1}{2}}\right)\right] = 4\left[\frac{1}{2}\left(x^{-\frac{1}{2}}\right)\right]$ ....[Open brackets]

$\quad \frac{d}{dx}\left[4\left(x^{\frac{1}{2}}\right)\right] = \frac{4}{2}\left(x^{-\frac{1}{2}}\right) = 2 \times \frac{1}{x^{\frac{1}{2}}} = \frac{2}{x^{\frac{1}{2}}} = \frac{2}{\sqrt{x}}$

2) $Y = 3\sqrt[3]{x^2}$ ....$\left[Y = 3\sqrt[3]{x^2} = 3\left(x^{\frac{2}{3}}\right)\right]$

Step 1: Formula. If $Y = Kf(x)$, then $\frac{d}{dx}[Kf(x)] = K\frac{d}{dx}[f(x)]$, where; $K = 3$ and $f(x) = x^{\frac{2}{3}}$

Step 2: Data substitution and solving

$\quad$ If $Y = 3\left(x^{\frac{2}{3}}\right)$, then $\frac{d}{dx}\left[3\left(x^{\frac{2}{3}}\right)\right] = 3\frac{d}{dx}\left[x^{\frac{2}{3}}\right]$ ....[Apply power rule]

$\quad \frac{d}{dx}\left[3\left(x^{\frac{2}{3}}\right)\right] = 3\left[\frac{2}{3}\left(x^{\frac{2}{3}-1}\right)\right]$ ....[Simplify exponent]

$\quad \frac{d}{dx}\left[3\left(x^{\frac{2}{3}}\right)\right] = 3\left[\frac{2}{3}\left(x^{-\frac{1}{3}}\right)\right]$ ....[Open bracket]

$\quad \frac{d}{dx}\left[3\left(x^{\frac{2}{3}}\right)\right] = \frac{6}{3}\left(x^{-\frac{1}{3}}\right) = 2 \times \frac{1}{x^{\frac{1}{3}}}$ ....[Multiply terms]

$\quad \frac{d}{dx}\left[3\left(x^{\frac{2}{3}}\right)\right] = \frac{2}{x^{\frac{1}{3}}} = \frac{2}{\sqrt[3]{x}}$

Test Your Understanding

Exercise 2: Differentiate the following equations

1) $f(x) = 1,000$ ....[**Answer: 0**]

2) $g(x) = x$ ....[**Answer: 1**]

3) $f(x) = 4x^2$ ....[**Answer: 8x**]

4) $g(x) = 10\sqrt{x}$ ....[**Answer: $5x^{-\frac{1}{2}} = \frac{5}{\sqrt{x}}$**]

11.1.3 Sum and Difference Rule

The addition and subtraction rule is used when adding and subtracting two functions.

$\quad$ If $Y = f(x) + g(x)$, then $\frac{dY}{dx} = \frac{d}{dx}[f(x)] + \frac{d}{dx}[g(x)]$ ....[Sum Rule]

$\quad$ If $Y = f(x) - g(x)$, then $\frac{dY}{dx} = \frac{d}{dx}[f(x)] - \frac{d}{dx}[g(x)]$ ....[Difference Rule]

Example1: Differentiate the following functions with respect to variable x

1) $Y = 2x^2 + 1$

2) $Q = x^3 - 3x$

<u>Solution</u>

1) $Y = 2x^2 + 1$

Step 1: Formula. If $Y = f(x) + g(x)$, then $\frac{dY}{dx} = \frac{d}{dx}[f(x)] + \frac{d}{dx}[g(x)]$

Where; $f(x) = 2x^2$ and $g(x) = 1$

Step 2: Data substitution and solving

If $Y = 2x^2 + 1$, then $\frac{dY}{dx} = \frac{d}{dx}[2x^2] + \frac{d}{dx}[1]$ ....[Apply power and constant rule]

$\frac{dY}{dx} = 2(2x^{2-1}) + 0$...[Work power and bracket]

$\frac{dY}{dx} = 4x^1 + 0$ ....[Solve equation]

$\frac{dY}{dx} = 4x$

2) $Q = x^3 - 3x$.. ....$[Q = x^3 - 3x = x^3 - 3x^1]$

Step 1: Formula. If $Q = f(x) - g(x)$, then $\frac{dQ}{dx} = \frac{d}{dx}[f(x)] - \frac{d}{dx}[g(x)]$

Where; $f(x) = x^3$ and $g(x) = 3x$

Step 2: Data substitution and solving

If $Y = x^3 - 3x$, then $\frac{dQ}{dx} = \frac{d}{dx}[x^3] - \frac{d}{dx}[3x]$ ....[Apply common multiple rule]

$\frac{dQ}{dx} = \frac{d}{dx}[x^3] - 3\frac{d}{dx}[x]$ ....[Apply power rule]

$\frac{dQ}{dx} = 3(x^{3-1}) - 3[1(x^{1-1})]$ ....[Simplify exponents]

$\frac{dQ}{dx} = 3(x^2) - 3[1(x^0)]$ ....[Open bracket. Remember; $(n^0 = 1)$]

$\frac{dQ}{dx} = 3x^2 - 3$

Example 2: Differentiate the following functions

1) $Y = x + x - 2x$

2) $Y = 3x^2 - 2x + 4$

1) $Y = x + x - 2x$ ....[$Y = x + x - 2x = x^1 + x^1 - 2x$]

$\frac{d}{dx}(x + x - 2x) = \frac{d}{dx}(x) + \frac{d}{dx}(x) - \frac{d}{dx}(2x)$ ....[Apply constant multiple rule]

$\frac{d}{dx}(x + x - 2x) = \frac{d}{dx}(x) + \frac{d}{dx}(x) - 2\frac{d}{dx}(x)$ ....[Apply identify function rule]

$\frac{d}{dx}(x + x - 2x) = 1 + 1 - 2(1)$ ....[Open bracket and solve]

$\frac{d}{dx}(x + x - 2x) = 0$

2) $Y = 3x^2 - 2x + 4$ ....[$Y = 3x^2 - 2x + 4 = 3x^2 - 2x^1 + 4$]

$\frac{d}{dx}(3x^2 - 2x + 4) = \frac{d}{dx}(3x^2) - \frac{d}{dx}(2x) + \frac{d}{dx}(4)$ ...[Apply constant multiple rule]

$\frac{d}{dx}(3x^2 - 2x + 4) = 3\frac{d}{dx}(x^2) - 2\frac{d}{dx}(x) + \frac{d}{dx}(4)$

$\frac{d}{dx}(3x^2 - 2x + 4) = 3[2x] - 2[1] + 0$ ....[Open brackets and solve]

$\frac{d}{dx}(3x^2 - 2x + 4) = 6x - 2$

Example 3: Find the derivative of the following functions

1) $Y = 3|X| - \sqrt{X}$

2) $Y = \frac{1}{X} + \frac{1}{X^3}$

1) $Y = 3|x| - \sqrt{x}$ ....$\left[Y = 3|x| - \sqrt{x} = Y = 3|x| - x^{\frac{1}{2}}\right]$

Step 1: Formula. If $Y = f(x) - g(x)$, then $\frac{dY}{dx} = \frac{d}{dx}[f(x)] - \frac{d}{dx}[g(x)]$

Where; $f(x) = 3|x|$ and $g(x) = x^{\frac{1}{2}}$

Step 2: Data substitution and solving

$\frac{dY}{dx} = 3.\frac{d}{dx}|x| - \frac{d}{dx}(\sqrt{x}) = 3.\frac{d}{dx}|x| - \frac{d}{dx}\left(x^{\frac{1}{2}}\right)$ ....[Apply power rule]

$\frac{dY}{dx} = 3.\frac{|x|}{x} - \frac{1}{2}\left(x^{\frac{1}{2}-1}\right)$ ....[Simplify exponent]

$\frac{dY}{dx} = 3.\frac{|x|}{x} - \frac{1}{2}x^{-\frac{1}{2}} = 3.\frac{|x|}{x} - \frac{1}{2} \times \frac{1}{x^{\frac{1}{2}}}$ ....[Simplify equation]

$\frac{dY}{dx} = \frac{3|x|}{x} - \frac{1}{2\sqrt{x}}$

2) $Y = \frac{1}{X} + \frac{1}{X^3}$ ....$\left[Y = \frac{1}{X} + \frac{1}{X^3} = x^{-1} + x^{-3}\right]$

Step 1: Formula. If $Y = f(x) + g(x)$, then $\frac{dY}{dx} = \frac{d}{dx}[f(x)] + \frac{d}{dx}[g(x)]$

Where; $f(x) = x^{-1}$ and $g(x) = x^{-3}$

Step 2: Data substitution and solving

$\frac{dY}{dx} = \frac{d}{dx}[x^{-1}] + \frac{d}{dx}[x^{-3}]$ ....[Apply power rule]

$\frac{dY}{dx} = -1[x^{-1-1}] + (-3)[x^{-3-1}]$ ....[Simplify exponents]

$\frac{dY}{dx} = -1[x^{-2}] - 3[x^{-4}] = \left[-1 \times \frac{1}{x^2}\right] + \left[-3 \times \frac{1}{x^4}\right]$ ....[Open bracket]

$\frac{dY}{dx} = -\frac{1}{x^2} - \frac{3}{x^4}$

Test Your Understanding

Exercise 3: Differentiate the following functions with respect to variable x

1) $Y = x^3 + 5$...[Answer: $3x^2$]

2) $Y = 5x^2 + 3x$...[Answer: $10x + 3$]

3) $Y = x^2 - 3$ ....[Answer: $2x$]

4) $Y = x^3 - 2x^4$ ....[Answer: $3x^2 - 8x^3$]

5) $Y = -6x^3 + 4x^2 - 2x^4$ ....[Answer; $-18x^2 + 8x - 8x^3$]

6) $Y = 2x^3 + 3x - 17$...[Answer: $6x^2 + 3$]

11.1.4 Product and Quotient Rule

The product rule is used to find the product of two functions and the quotient rule is used to find the quotient of two functions.

$$\frac{d}{dx}(UV) = V\frac{dU}{dx} + U\frac{dV}{dx}$$ ....[Product Rule]

$$\frac{d}{dx}\left(\frac{U}{V}\right) = \frac{V\frac{dU}{dx} - U\frac{dV}{dx}}{V^2}$$ ....[Quotient Rule]

N/B: The product rule can be explain theoretically as; hold the (V) function constant and differentiate the (U) function with respect to (x) plus hold the (U) function constant and differentiate the (V) function with respect to (x).

Example 1: Differentiate; $(x + 1)(x^2 + 3)$

Solution

Step 1: Formula. $\frac{d}{dx}(UV) = V\frac{dU}{dx} + U\frac{dV}{dx}$, where; $U = (x + 1)$ and $V = (x^2 + 3)$

433

Step 2: Determination of $\left(\frac{dU}{dx}\right)$ and $\left(\frac{dV}{dx}\right)$

$\left(\frac{dU}{dx}\right)$ ....[U = x + 1]	$\left(\frac{dV}{dx}\right)$ ....[V = x² + 3]
$\frac{dU}{dx}$ = x + 1 ....[Identity and constant rule]	$\frac{dV}{dx}$ = x² + 3 ....[Power and constant rule]
$\frac{dU}{dx}$ = 1 + 0 ....[Add values]	$\frac{dV}{dx}$ = 2(x²⁻¹) + 0 .[Work power and bracket]
$\frac{dU}{dx}$ = **1**	$\frac{dV}{dx}$ = 2x¹ + 0 ....[Solve. Remember (a¹ = a)]
	$\frac{dV}{dx}$ = **2x**

Step 3: Data substitution and solving.......................................[$\frac{dU}{dx}$ = 1 and $\frac{dV}{dx}$ = 2x]

$\frac{d}{dx}(UV) = [(x^2 + 3)1] + [(x + 1)2x]$ ....[Open inner brackets]

$\frac{d}{dx}(UV) = [x^2 + 3] + [2x^2 + 2x]$ ....[Open main brackets]

$\frac{d}{dx}(UV) = x^2 + 3 + 2x^2 + 2x$ ....[Collect like terms]

$\frac{d}{dx}(UV) = \mathbf{3x^2 + 2x + 3}$

Example 2: Differentiate the function; $Y = (2x^2 + 4x)(x^2 - 4x^2)$

Solution

Step 1: Formula. $\frac{d}{dx}(UV) = V\frac{dU}{dx} + U\frac{dV}{dx}$, where; U = (2x² + 4x) and V = (x² − 4x²)

Step 2: Data substitution and solving

$\frac{d}{dx}(UV) = \left[(x^2 - 4x^2)\frac{d}{dx}(2x^2 + 4x)\right] + \left[(2x^2 + 4x)\frac{d}{dx}(x^2 - 4x^2)\right]$ ..[Apply power rule]

$\frac{d}{dx}(UV) = [(x^2 - 4x^2)(4x + 4)] + [(2x^2 + 4x)(2x - 8x)]$ ....[Open inner brackets]

$\frac{d}{dx}(UV) = [4x^3 + 4x^2 - 8x^3 - 8x^2] + [4x^3 - 16x^3 + 8x^2 - 32x^2]$ ..[Simplify brackets]

$\frac{d}{dx}(UV) = [-4x^3 - 4x^2] + [-12x^3 - 24x^2]$ ....[Open brackets]

$\frac{d}{dx}(UV) = -4x^3 - 4x^2 - 12x^3 - 24x^2$ ....[Solve]

$\frac{d}{dx}(UV) = \mathbf{-16x^3 - 28x^2}$

Example 3: Differentiate the function; $Y = \frac{x^2+3x-4}{2x+1}$

Solution

Step 1: Formula. $\frac{d}{dx}\left(\frac{U}{V}\right) = \frac{V\frac{dU}{dx} - U\frac{dV}{dx}}{V^2}$, where; U = x² + 3x − 4 and V = 2x + 1

Step 2: Determination of $\left(\frac{dU}{dx}\right)$ and $\left(\frac{dV}{dx}\right)$

$\left(\frac{dU}{dx}\right)$ .....[$U = x^2 + 3x - 4 = x^2 + 3x^1 - 4$]	$\left(\frac{dV}{dx}\right)$ ....[$V = 2x + 1 = 2x^1 + 1$]
$\frac{dU}{dx} = x^2 + 3x^1 - 4$ ...[Power and constant rule]	$\frac{dV}{dx} = 2x^1 + 1$ .....[Power and constant rule]
$\frac{dU}{dx} = 2(x^{2-1}) + 1(3x^{1-1}) - 0$ ..[Work power]	$\frac{dV}{dx} = 1(2x^{1-1}) + 0$ ....[Simplify power]
$\frac{dU}{dx} = 2(x^1) + 1(3x^0)$ ....[$(a^1 = a), a^0 = 1$]	$\frac{dV}{dx} = 1(2x^0)$ ....[Remember; $a^0 = 1$]
$\frac{dU}{dx} = 2(x) + 1(3)$ ....[Open brackets]	$\frac{dV}{dx} = 1(2)$ ....[Open bracket]
$\frac{dU}{dx} = \mathbf{2x + 3}$	$\frac{dV}{dx} = \mathbf{2}$

Step 3: Data substitution and solving..................... ....[$\left(\frac{dU}{dx} = 2x + 3\right)$ and $\left(\frac{dV}{dx} = 2\right)$]

$\frac{d}{dx}\left(\frac{U}{V}\right) = \frac{[(2x+1)(2x+3)]-[(x^2+3x-4)(2)]}{(2x+1)^2}$ ....[Open inner brackets]

$\frac{d}{dx}\left(\frac{U}{V}\right) = \frac{[4x^2+8x+3]-[2x^2+6x-8]}{(2x+1)^2}$ ....[Open main brackets]

$\frac{d}{dx}\left(\frac{U}{V}\right) = \frac{4x^2+8x+3-2x^2-6x+8}{(2x+1)^2}$ ....…....[Simplify numerator]

$\frac{d}{dx}\left(\frac{U}{V}\right) = \frac{2x^2+2x+11}{(2x+1)^2}$

Example 4: Differentiate; $Y = \frac{(x+1)(x+2)}{(x-1)(x-2)}$

Solution

Step 1: Formula. $\frac{d}{dx}\left(\frac{U}{V}\right) = \frac{V\frac{dU}{dx}-U\frac{dV}{dx}}{V^2}$......................…....[$y = \frac{(x+1)(x+2)}{(x-1)(x-2)} = \frac{x^2+2x+x+2}{x^2-2x-x+2} = \frac{x^2+3x+2}{x^2-3x+2}$]

Where; $U = x^2 + 3x + 2$ and $V = x^2 - 3x + 2$

Step 2: Data substitution and solving

$\frac{d}{dx}\left(\frac{U}{V}\right) = \frac{[(x^2-3x+2)\frac{d}{dx}(x^2+3x+2)]-[(x^2+3x+2)\frac{d}{dx}(x^2-3x+2)]}{(x^2-3x+2)^2}$...[Apply power and constant rules]

$\frac{d}{dx}\left(\frac{U}{V}\right) = \frac{[(x^2-3x+2)(2x+3)]-[(x^2+3x+2)(2x-3)]}{(x^2-3x+2)^2}$ ....…....[Open inner brackets]

$\frac{d}{dx}\left(\frac{U}{V}\right) = \frac{[2x^3-6x^2+4x+3x^2-9x+6]-[2x^3+6x^2+4x-3x^2-9x-6]}{(x^2-3x+2)^2}$[Simplify and open main brackets]

$\frac{d}{dx}\left(\frac{U}{V}\right) = \frac{[2x^3-3x^2-5x+6]-[2x^3+3x^2-5x-6]}{(x^2-3x+2)^2} = \frac{2x^3-3x^2-5x+6-2x^3-3x^2+5x+6}{(x^2-3x+2)^2}$..........[Simplify]

$\frac{d}{dx}\left(\frac{U}{V}\right) = \frac{-6x^2+12}{(x^2-3x+2)^2}$

Exercise 4: Differentiate the following functions

1) $Y = (2x + 3)(3x - 2)$...[Answer: $12x + 5$]

2) $Y = (1 - 4x^3)(3x^2 - 5x + 2)$ ....[Answer: $-60x^4 + 80x^3 - 24x^2 + 6x - 5$]

3) $Y = \frac{x}{x+1}$ ....[Answer: $\frac{1}{(x+1)^2}$]

4) $Y = \frac{2x+1}{x-2}$ ....[Answer: $\frac{-5}{(x-2)^2}$]

11.2 Chain Rule

To apply the chain rule, we let the function within the bracket to be U and the new function obtained can be called given transform function.

$$\frac{dy}{dx} = \frac{dy}{dU} \times \frac{dU}{dx} \dots\dots\dots\dots\dots\dots\dots\dots\dots\dots\dots\dots\dots\dots\dots\dots\dots\dots\dots\dots\dots\dots\dots\dots(1)$$

N/B: The theoretical explanation of the chain rule is given as; differentiate the given transform function with respect to (U) and multiply by the derivative of the function (U) with respect to (x).

Example 1: Differentiate $y = (3x + 2)^3$

Solution

Step 1: Formula. $\frac{dy}{dx} = \frac{dy}{dU} \times \frac{dU}{dx}$

Step 2: Determination of $\left(\frac{dy}{dU}\right)$...[Let $U = (3x + 2)$]

$$[y = (3x + 2)^3 \Rightarrow y = (U)^3 \Rightarrow y = U^3]$$

$y = U^3$ ....[Apply power function rule]

$\frac{dy}{dU} = 3[U^{3-1}]$ ....[Simplify power and open bracket]

$\frac{dy}{dU} = 3U^2$ ....[Since $[3U^2]$, substituting $[U]$ gives us $[3(3x + 2)^2]$]

Step 3: Determination of $\left(\frac{dU}{dx}\right)$............................[Remember $U = (3x + 2) = (3x^1 + 2)$]

$U = 3x + 2$ ....[Apply power function rule and constant rule]

$\frac{dU}{dx} = 1(3x^{1-1}) + 0$ ....[Simplify power and open bracket]

$\frac{dU}{dx} = 3x^0 + 0$ ....[Apply zero power principle. Remember; $(n^0 = 1)$]

$\frac{dU}{dx} = 3$

Step 4: Data substitution and solving................................$[\frac{dy}{dU} = 3U^2$ and $\frac{dU}{dx} = 3]$

$\frac{dy}{dx} = 3U^2 \times 3$ ....…..…....[Multiply values]

$\frac{dy}{dx} = 9U^2$ ....…..…................…....[Substitute the value of (U)]

$\frac{dy}{dx} = 9(3x + 2)^2$

Example 2: Differentiate; $y = \frac{1}{(1-2x)^4}$

Solution

Step 1: Formula. $\frac{dy}{dx} = \frac{dy}{dU} \times \frac{dU}{dx}$...…....….[$y = \frac{1}{(1-2x)^4} = (1 - 2x)^{-4}$]

Step 2: Determination of $\left(\frac{dy}{dU}\right)$..........................… ....… ..….......[Let $U = (1 - 2x)$]

$$[y = (1 - 2x)^{-4} \Rightarrow y = (U)^{-4} \Rightarrow y = U^{-4}]$$

$y = U^{-4}$ ....…… ….....…...[Apply power function rule]

$\frac{dy}{dU} = -4[U^{-4-1}]$ ....…...[Simplify power and open bracket]

$\frac{dy}{dU} = -4U^{-5}$ ....…..[Since $[-4U^{-5}]$, substituting U] gives us $[-4(1 - 2x)^{-5}]$]

Step 3: Determination of $\left(\frac{dU}{dx}\right)$..........................…..….[Remember $U = (1 - 2x) = (1 - 2x^1)$]

$U = 1 - 2x$ ....…...[Apply power function rule and constant rule]

$\frac{dU}{dx} = 0 - 1(2x^{1-1})$ ....…....[Simplify power and open bracket]

$\frac{dU}{dx} = 0 - 2x^0$ ....…...[Apply zero power principle. Remember; $(n^0 = 1)$]

$\frac{dU}{dx} = -2$

Step 4: Data substitution and solving...................... ….....…...... $\frac{dy}{dU} = -4U^{-5}$ and $\frac{dU}{dx} = -2$]

$\frac{dy}{dx} = -4U^{-5} \times -2$ ....…........[Multiply values]

$\frac{dy}{dx} = 8U^{-5}$ ....…...............….........[Substitute the value of (U)]

$\frac{dy}{dx} = 8(1 - 2x)^{-5}$

Example 3: Differentiate; $y = \frac{1}{\sqrt{x^2+1}}$

Solution

Step 1: Formula. $\frac{dy}{dx} = \frac{dy}{dU} \times \frac{dU}{dx}$..............................….......[$y = \frac{1}{\sqrt{x^2+1}} = \frac{1}{(x^2+1)^{\frac{1}{2}}} = (x^2 + 1)^{-\frac{1}{2}}$]

Step 2: Determination of $\left(\frac{dy}{dU}\right)$...[Let $U = (x^2 + 1)$]

$$\left[y = (x^2 + 1)^{-\frac{1}{2}} \Rightarrow y = (U)^{-\frac{1}{2}} \Rightarrow y = U^{-\frac{1}{2}}\right]$$

$y = U^{-\frac{1}{2}}$ ....[Apply power function rule]

$\frac{dy}{dU} = -\frac{1}{2}\left[U^{-\frac{1}{2}-1}\right]$ ....[Simplify power and open bracket]

$\frac{dy}{dU} = -\frac{1}{2}U^{-\frac{3}{2}}$ ....[Since $\left[-\frac{1}{2}U^{-\frac{3}{2}}\right]$, substituting [U] gives us $\left[-\frac{1}{2}(x^2 + 1)^{-\frac{3}{2}}\right]$]

Step 3: Determination of $\left(\frac{dU}{dx}\right)$..[Remember $U = (x^2 + 1)$]

$U = x^2 + 1$ ....[Apply power function rule and constant rule]

$\frac{dU}{dx} = 2(x^{2-1}) + 0$ ....[Simplify power and open bracket]

$\frac{dU}{dx} = 2x^1 + 0 = 2x$

Step 4: Data substitution and solving...............................[$\frac{dy}{dU} = -\frac{1}{2}U^{-\frac{3}{2}}$ and $\frac{dU}{dx} = 2x$]

$\frac{dy}{dx} = -\frac{1}{2}U^{-\frac{3}{2}} \times 2x$ ....[Multiply values]

$\frac{dy}{dx} = -\frac{2x}{2}U^{-\frac{3}{2}} = -xU^{-\frac{3}{2}}$..............................[Substitute the value of (U)]

$\frac{dy}{dx} = -x(x^2 + 1)^{-\frac{3}{2}}$ .....[Re-write equation. Remember; $\left(a^{-n} = \frac{1}{a^n}\right)$ and $(a)^{\frac{n}{m}} = \sqrt[m]{(a)^n}$]

$\frac{dy}{dx} = \frac{-x}{(x^2+1)^{\frac{3}{2}}} = \frac{-x}{\sqrt[2]{(x^2+1)^3}} = \frac{-x}{\sqrt{(x^2+1)^3}}$

Example 4: Differentiate; $y = 3(3x + 1)^{\frac{2}{3}}$

Solution

Step 1: Formula. $\frac{dy}{dx} = \frac{dy}{dU} \times \frac{dU}{dx}$

Step 2: Determination of $\left(\frac{dy}{dU}\right)$...[Let $U = (3x + 1)$]

$$\left[y = 3(3x + 1)^{\frac{2}{3}} \Rightarrow y = 3(U)^{\frac{2}{3}} \Rightarrow y = 3U^{\frac{2}{3}}\right]$$

$y = 3U^{\frac{2}{3}}$ ....[Apply constant multiple rule. $\left(\frac{d}{dx}kf(x) = k\frac{d}{dx}[f(x)]\right)$]

$\frac{dy}{dU} = 3U^{\frac{2}{3}} = 3 \times \frac{d}{dU}\left(U^{\frac{2}{3}}\right)$ ....[Apply power rule]

$\frac{dy}{dU} = 3 \times \frac{2}{3}\left(U^{\frac{2}{3}-1}\right) = 3 \times \frac{2}{3}U^{-\frac{1}{3}}$ ....[Multiply values and simplify fraction]

$\frac{dy}{dU} = \frac{6}{3}U^{-\frac{1}{3}} = 2U^{-\frac{1}{3}}$.................[Since $\left[2U^{-\frac{1}{3}}\right]$, substituting [U] gives us $\left[2(3x + 1)^{-\frac{1}{3}}\right]$]

Step 3: Determination of $\left(\frac{dU}{dx}\right)$..............................[Remember $U = (3x + 1) = (3x^1 + 1)$]

$U = 3x + 1$ ....[Apply power function rule and constant rule]

$\frac{dU}{dx} = 1(3x^{1-1}) + 0$ ....[Simplify power and open bracket]

$\frac{dU}{dx} = 3x^0 + 0$ ....[Solve. Remember; $(n^0 = 1)$]

$\frac{dU}{dx} = 3(1) = 3$

Step 4: Data substitution and solving.............................. .....$[\frac{dy}{dU} = 2U^{-\frac{1}{3}}$ and $\frac{dU}{dx} = 3]$

$\frac{dy}{dx} = 2U^{-\frac{1}{3}} \times 3$ ....[Multiply values]

$\frac{dy}{dx} = 6U^{-\frac{1}{3}}$..[Substitute the value of (U)]

$\frac{dy}{dx} = 6(3x + 1)^{-\frac{1}{3}}$ ......[Re-write equation. Remember; $\left(a^{-n} = \frac{1}{a^n}\right)$ and $(a)^{\frac{n}{m}} = \sqrt[m]{(a)^m}]$

$\frac{dy}{dx} = 6(3x + 1)^{-\frac{1}{3}} = \frac{6}{(3x+1)^{\frac{1}{3}}} = \frac{6}{\sqrt[3]{(3x+1)^1}} = \frac{6}{\sqrt[3]{3x+1}}$

Test Your Understanding

Exercise 5: Differentiate the following functions

1) $Y = (1 + x^2)^3$ ....[**Answer: $6x(1 + x^2)^2$**]

2) $Y = (2 - 5x)^4$.. .[**Answer: $-20(2 - 5x)^3$**]

3) $Y = \frac{1}{(2x+1)^3}$... ....[**Answer: $-\frac{6}{(2x+1)^4}$**]

4) $Y = (7 - 2x^3)^{-4}$...[**Answer: $\frac{24x^2}{(7-2x^3)^5}$**]

11.3 Partial Differentiation

Partial differentiation is used when the function to be differentiated is made up of more than one variable. To differentiate a function with many variables, we differentiate the function with each of the variable. Consider the function with three variables $[f(x, y, z)]$, we have to differentiate the function with respect to;

1) $(x) = f_x(x, y, z) = \frac{d}{dx}(x, y, z)$....................[y and z terms are considered constant]

2) $(y) = f_y(x, y, z) = \frac{d}{dy}(x, y, z)$.. ....[x and z terms are considered constant]

3) $(z) = f_z(x, y, z) = \frac{d}{dz}(x, y, z)$.. ....[x and y terms are considered constant]

N/B: When a term is a product of two variables$[(x$ and $y)$ or $(x$ and $z)$ or $(y$ and $z)]$, we differentiate only the term element under differentiation. A function can be differentiated many times, such as; first differentiation(y'), second differentiation (y'') and n^{th} differentiation $\left(y^{n^{th}}\right)$.

Example 1: Differentiate the following functions

1) $f(x,y) = x^2 + 4y^2$
2) $f(x,y) = x^2 - 4xy + 2y^2$

Solution

1) $f(x,y) = x^2 + 4y^2$

a) Differentiation with respect to variable (x).$[\, f(x,y) = x^2 + 4y^2 = f_x(x,y) = x^2 + 0]$

$f_x(x,y) = \frac{d}{dx}(xy) = 2(x^{2-1}) + 0$ ....[Simplify power]

$f_x(x,y) = \frac{d}{dx}(xy) = 2(x^1) = 2(x)$ ....[Open bracket]

$f_x(x,y) = \frac{d}{dx}(xy) = 2x$

b) Differentiation with respect to variable (y).$\left[\, f(x,y) = x^2 + 4y^2 = f_y(x,y) = 0 + 4y^2\right]$

$f_y(x,y) = \frac{d}{dy}(x,y) = 0 + 2(4y^{2-1})$ ....[Simplify power]

$f_y(x,y) = \frac{d}{dy}(x,y) = 2(4y^1) = 2(4y)$ ....[Open bracket]

$f_y(x,y) = \frac{d}{dy}(x,y) = 8y$

2) $f(x,y) = x^2 - 4xy + 2y^2$

a) Respect to variable (x).........$[\, f(x,y) = x^2 - 4xy + 2y^2 = f_x(x,y) = x^2 - 4xy + 0]$

$f_x(x,y) = \frac{d}{dx}(x,y) = 2(x^{2-1}) - 1(4x^{1-1})y + 0$ ....[Simplify exponents]

$f_x(x,y) = \frac{d}{dx}(x,y) = 2(x^1) - 1(4x^0)y = 2(x) - 1(4)y$ ....[Open brackets]

$f_x(x,y) = \frac{d}{dx}(x,y) = 2x - 4y$

b) Respect to variable (y)........$\left[\, f(x,y) = x^2 - 4xy + 2y^2 = f_y(x,y) = 0 - 4xy + 2y^2\right]$

$f_y(x,y) = \frac{d}{dy}(x,y) = 0 - 4x.1(y^{1-1}) + 2(2y^{2-1})$ ....[Simplify exponents]

$f_y(x,y) = \frac{d}{dy}(x,y) = 0 - 4x.1(y^0) + 2(2y^1) = -4x.1(1) + 2(2y)$.[Open brackets]

$f_y(x,y) = \frac{d}{dy}(x,y) = -4x + 4y$

Example 2: Differentiate the following functions

1) $f(x,y) = x^4 - x^3y + x^2y^2 - xy^3 + y^4$
2) $f(x,y,z) = x^2y^3z^4$
3) $f(x,y) = \frac{x+y}{x-y}$

440

1) $f(x, y) = x^4 - x^3y + x^2y^2 - xy^3 + y^4$

a) Respect to (x)..$f_x(x, y) = x^4 - x^3y + x^2y^2 - xy^3 + 0]$

$f_x(x, y) = \frac{d}{dx}(x, y) = 4(x^{4-1}) - 3(x^{3-1})y + 2(x^{2-1})y^2 - 1(x^{1-1})y^3 + 0[\text{Simplify}]$

$f_x(x, y) = \frac{d}{dx}(x, y) = 4(x^3) - 3(x^2)y + 2(x^1)y^2 - 1(x^0)y^3 \ldots\ldots[\text{Open brackets}]$

$f_x(x, y) = \frac{d}{dx}(x, y) = \mathbf{4x^3 - 3x^2y + 2xy^2 - y^3}$

b) Respect to (y)..$f_y(x, y) = 0 - x^3y + x^2y^2 - xy^3 + y^4]$

$f_y(x, y) = \frac{d}{dy}(x, y) = 0 - x^3.1(y^{1-1}) + x^2.2(y^{2-1}) - x.3(y^{3-1}) + 4(y^{4-1})$

.[Simplify]

$f_y(x, y) = \frac{d}{dy}(x, y) = -x^3.1(y^0) + x^2.2(y^1) - x.3(y^2) + 4(y^3) \ldots[\text{Open brackets}]$

$f_y(x, y) = \frac{d}{dy}(x, y) = -x^3 + x^2 2y - x3y^2 + 4y^3 = \mathbf{-x^3 + 2x^2y - 3xy^2 + 4y^3}$

2) $f(x, y, z) = x^2y^3z^4$

$f_x(x, y, z) = \frac{d}{dx}(x, y, z) = 2(x^{2-1})y^3z^4 = 2(x^1)y^3z^4 = \mathbf{2xy^3z^4}$

$f_y(x, y, z) = \frac{d}{dy}(x, y, z) = x^2.3(y^{3-1})z^4 = x^2.3(y^2)z^4 = x^2.3y^2z^4 = \mathbf{3x^2y^2z^4}$

$f_z(x, y, z) = \frac{d}{dz}(x, y, z) = x^2y^3.4(z^{4-1}) = x^2y^3.4(z^3) = x^2y^3.4z^3 = \mathbf{4x^2y^3z^3}$

3) $f(x, y) = \frac{x+y}{x-y}$

a) Respect of variable (x)....$\left[\frac{d}{dx}\left(\frac{U}{V}\right) = \frac{V\frac{dU}{dx} - U\frac{dV}{dx}}{V^2}\right]$. where; $(U = x + y)$ and $(V = x - y)$

$f(x, y) = \frac{x+y}{x-y}$ ....[Apply the quotient rule]

$f_x(x, y) = \frac{d}{dx}\left(\frac{U}{V}\right) = \frac{[(x-y)(1+0)]-[(x+y)(1-0)]}{(x-y)^2}$ ....[Simplify brackets]

$f_x(x, y) = \frac{d}{dx}\left(\frac{U}{V}\right) = \frac{[(x-y)(1)]-[(x+y)(1)]}{(x-y)^2}$ ....[Open inner brackets]

$f_x(x, y) = \frac{d}{dx}\left(\frac{U}{V}\right) = \frac{[x-y]-[x+y]}{(x-y)^2} = \frac{x-y-x-y}{(x-y)^2} = \frac{-2y}{(x-y)^2}$

b) Respect to variable (y)......$\left[\frac{d}{dy}\left(\frac{U}{V}\right) = \frac{V\frac{dU}{dx} - U\frac{dV}{dx}}{V^2}\right]$, where; $(U = x + y)$ and $(V = x - y)$

$f(x, y) = \frac{x+y}{x-y}$ ....[Apply the quotient rule]

$f_y(x, y) = \frac{d}{dy}\left(\frac{U}{V}\right) = \frac{[(x-y)(0+1)]-[(x+y)(0-1)]}{(x-y)^2}$ ....[Simplify brackets]

$$f_y(x, y) = \frac{d}{dy}\left(\frac{U}{V}\right) = \frac{[(x-y)(1)]-[(x+y)(-1)]}{(x-y)^2} \quad ....\text{[Open inner brackets]}$$

$$f_y(x, y) = \frac{d}{dy}\left(\frac{U}{V}\right) = \frac{[x-y]-[-x-y]}{(x-y)^2} \quad ....\text{[Open main brackets]}$$

$$f_y(x, y) = \frac{d}{dy}\left(\frac{U}{V}\right) = \frac{x-y+x+y}{(x-y)^2} = \frac{2x}{(x-y)^2}$$

Example 3: Differentiate the following functions

1) $f(x, y) = (5x + 3)(6x + 2y)$

2) $f(x, y) = (x^2 + 3y^3)^4$

<u>Solution</u>

1) $f(x, y) = (5x + 3)(6x + 2y)$

a) Respect to variable (x)... $\left[\frac{d}{dx}(UV) = V\frac{dU}{dx} + U\frac{dV}{dx}\right]$, where; $U = (5x + 3)$, $V =$

$(6x + 2y)$

$$\frac{d}{dx}(UV) = \left[(6x + 2y)\frac{d}{dx}(5x + 3)\right] + \left[(5x + 3)\frac{d}{dx}(6x + 2y)\right] \quad ....\text{[Differentiate]}$$

$$\frac{d}{dx}(UV) = [(6x + 2y)(5 + 0)] + [(5x + 3)(6 + 0)] \quad ....\text{[Open inner brackets]}$$

$$\frac{d}{dx}(UV) = [30x + 10y] + [30x + 18] \quad ....\text{[Open main brackets]}$$

$$\frac{d}{dx}(UV) = 30x + 10y + 30x + 18 \quad ....\text{[Solve]}$$

$$\frac{d}{dx}(UV) = \mathbf{60x + 10y + 18}$$

b) Respect to variable (y)... $\left[\frac{d}{dy}(UV) = V\frac{dU}{dy} + U\frac{dV}{dy}\right]$, where; $U = (5x + 3)$, $V =$

$(6x + 2y)$

$$\frac{d}{dy}(UV) = \left[(6x + 2y)\frac{d}{dy}(5x + 3)\right] + \left[(5x + 3)\frac{d}{dy}(6x + 2y)\right] \quad ....\text{[Differentiate]}$$

$$\frac{d}{dy}(UV) = [(6x + 2y)(0)] + [(5x + 3)(0 + 2)] \quad ....\text{[Open inner brackets]}$$

$$\frac{d}{dy}(UV) = [0] + [10x + 6] \quad ....\text{[Open main brackets]}$$

$$\frac{d}{dy}(UV) = 0 + 10x + 6 \quad ....\text{[Solve]}$$

$$\frac{d}{dy}(UV) = \mathbf{10x + 6}$$

2) $f(x, y) = (x^2 + 3y^3)^4$

a) Respect to variable (x)............$\left[\frac{df}{dx} = \frac{df}{dU} \times \frac{dU}{dx}\right]$, where; $(U = x^2 + 3y^3)$ and $df = U^4$

$$\frac{df}{dx} = \frac{d}{dU}(U^4) \times \frac{d}{dx}(x^2 + 3y^3) \quad ....\text{[Differentiate]}$$

442

$\frac{df}{dx} = 4U^3 \times (2x + 0)$ ....…[Multiply terms]

$\frac{df}{dx} = 8xU^3$ ....…..[Substitute U]

$\frac{df}{dx} = \mathbf{8x(x^2 + 3y^3)^3}$

b) Respect to variable (y)……..…… $\left[\frac{df}{dy} = \frac{df}{dU} \times \frac{dU}{dy}\right]$, where; $(U = x^2 + 3y^3)$ and $df = U^4$

$\frac{df}{dy} = \frac{d}{dU}(U^4) \times \frac{d}{dy}(x^2 + 3y^3)$ ....…............….…………[Differentiate]

$\frac{df}{dy} = 4U^3 \times (0 + 9y^2)$ ....[Multiple terms]

$\frac{df}{dy} = 36y^2U^3$ ....…...…………..……..[Substitute U]

$\frac{df}{dy} = \mathbf{36y^2(x^2 + 3y^3)^3}$

Test Your Understanding

Exercise 6: Differentiate the following functions

1) $f(x, y) = 3x^3 - 4y^2 + 5$...............................[**Answer:** $\frac{df}{dx} = \mathbf{9x^2}$ and $\frac{df}{dy} = \mathbf{-8y}$]

2) $f(x, y, z) = x^2 + y^3 + z^4$................[**Answer:** $\frac{df}{dx} = \mathbf{2x}$, $\frac{df}{dy} = \mathbf{3y^2}$, and $\frac{df}{dz} = \mathbf{4z^3}$]

3) $f(x, y) = \sqrt{x} + \sqrt{y}$ ....[**Answer:** $\frac{df}{dx} = \frac{1}{2\sqrt{x}}$ and $\frac{df}{dy} = \frac{1}{2\sqrt{y}}$]

Exercise 7: Differentiate the following functions

1) $f(x, y) = x^2y^3$...................................[**Answer:** $\frac{df}{dx} = \mathbf{2xy^3}$ and $\frac{df}{dy} = \mathbf{3x^2y^2}$]

2) $f(x, y) = x^2y^3 - 2y^2$ ....[**Answer:** $\frac{df}{dx} = \mathbf{2xy^3}$ and $\frac{df}{dy} = \mathbf{3x^2y^2 - 4y}$]

3) $f(x, y) = In(xy)$...[**Answer:** $\frac{df}{dx} = \frac{1}{x}$ and $\frac{df}{dy} = \frac{1}{y}$]

Exercise 8: Differentiate the following functions

1) $f(x, y) = (2x^2 + 4y)(x^2 - 4y^2)$

[**Answer:** $\frac{df}{dx} = \mathbf{4x^3 + 4x^2 - 16xy^2 + 8xy}$ and $\frac{df}{dy} = \mathbf{4x^2 - 48y^2 - 16x^2y}$]

2) $f(x, y) = \frac{(6x+7y)}{(5x+3y)}$............................[**Answer:** $\frac{df}{dx} = \frac{-17y}{(5x+3y)^2}$ and $\frac{df}{dy} = \frac{17x}{(5x+3y)^2}$]

3) $f(x, y) = (3x^3 - 4y^2)^2$

[**Answer:** $\frac{df}{dx} = \mathbf{18x^2(3x^3 - 4y^2)}$ and $\frac{df}{dy} = \mathbf{-16y(3x^3 - 4y^2)}$]

CHAPTER TWELVE

INTEGRATION

Integral calculus deals with integration. Integration is the process of finding a function when its derivative is known. Integral calculus is sub-divided into indefinite integration and definite integration.

12.1 Indefinite Integration

An indefinite integration is an integration function without integration limits. An indefinite integration is given as seen below.

$$\int f(x)\, dx \ldots\ldots\ldots\ldots\ldots\ldots\ldots\ldots\ldots\ldots\ldots\ldots\ldots\ldots\ldots\ldots\ldots\ldots\ldots(1)$$

Integration is also considered as anti-derivation. That is the derivative of an integral is equal to the original function.

$$\frac{d}{dx}\int f(x)\, dx = f(x) \ldots\ldots\ldots\ldots\ldots\ldots\ldots\ldots\ldots\ldots\ldots\ldots\ldots\ldots\ldots(2)$$

The calculation of the indefinite integration is made easy thanks to the following basic rules of integration seen below.

12.1.1 The Power Rule

The power rule of an indefinite integration depends on the exponent value and the exponent value sign.

$$\int x^n\, dx = \frac{x^{n+1}}{n+1} + c \ldots\ldots\ldots\ldots\ldots\ldots\ldots\ldots[\text{Power rule when } (n \neq -1)]$$
$$\int x^{-1}\, dx = \ln|x| + c \ldots\ldots\ldots\ldots\ldots\ldots\ldots\ldots[\text{Power rule when } (n = -1)]$$

N/B: It is worth knowing that (n) represents negative values (except negative one) and positive values.

Example 1: Compute the integral of $f(x) = x^4$

<u>**Solution**</u>

Step 1: Formula. $\int x^n\, dx = \frac{x^{n+1}}{n+1} + c$, where; $n = 4$

Step 2: Data substitution and solving

$$\int x^n\, dx = \frac{x^{4+1}}{4+1} + c \ldots\ldots\ldots\ldots\ldots\ldots[\text{Simplify numerator power and denominator}]$$
$$\int x^n\, dx = \frac{x^5}{5} + c = \frac{1}{5}x^5 + \mathbf{c}$$

Example 2: Determine the integral of the following functions

1) $\int x^{-2}\,dx$

2) $\int \frac{1}{\sqrt{x}}\,dx$

Solution

1) $\int x^{-2}\,dx$

Step 1: Formula. $\int x^n\,dx = \frac{x^{n+1}}{n+1} + c$, where; $n = -2$

Step 2: Data substitution and solving

$\int x^{-2}\,dx = \frac{x^{-2+1}}{-2+1} + c$ ....[Simplify numerator and denominator]

$\int x^{-2}\,dx = \frac{x^{-1}}{-1} + c$ ....[Re-expressed equation]

$\int x^{-2}\,dx = -\frac{1}{1}x^{-1} + c = -\frac{1}{x} + \mathbf{c}$

2) $\int \frac{1}{\sqrt{x}}\,dx$.....................................$\left\{ \int \frac{1}{\sqrt{x}}\,dx = \int \sqrt{x}^{-1}\,dx = \int \left(x^{\frac{1}{2}} \right)^{-1}\,dx = \int x^{-\frac{1}{2}}\,dx \right\}$

Step 1: Formula. $\int x^n\,dx = \frac{x^{n+1}}{n+1} + c$, where; $n = -\frac{1}{2}$

Step 2: Data substitution and solving

$\int x^{-\frac{1}{2}}\,dx = \frac{x^{-\frac{1}{2}+1}}{-\frac{1}{2}+1} + c$ ....[Simplify numerator and denominator]

$\int x^{-\frac{1}{2}}\,dx = \frac{x^{\frac{1}{2}}}{\frac{1}{2}} + c$ ....[Re-expressed equation]

$\int x^{-\frac{1}{2}}\,dx = 2x^{\frac{1}{2}} + c = 2\sqrt{x} + \mathbf{c}$

Test Your Understanding

Exercise 17: Compute the integral of the following

1) $\int x^5\,dx$..[**Answer:** $\frac{1}{6}x^6 + \mathbf{c}$]

2) $\int \sqrt[3]{x}\,dx$..[**Answer:** $\frac{3}{4}x^{\frac{4}{3}} + \mathbf{c}$]

3) $\int x^{-1}\,dx$ ....[**Answer:** $\ln|x| + \mathbf{c}$]

12.1.2 Complementary integration rules

Apart of the power rule seen above, other basic rules of integration when dealing with one variable function is given below.

$\int Kf(x)\,dx = K\int f(x)\,dx = K * \frac{x^{n+1}}{n+1} + c$ ....[Constant multiple rule]

$\int K\,dx = Kx + c$ ....[Non-zero constant rule]

$\int 0 \, dx = c$ ....[Zero constant rule]

Example 1: Calculate the integral of $f(x) = 2x^2$

Solution

Step 1: Formulas. $\int Kf(x) \, dx = K \int f(x) \, dx = K * \frac{x^{n+1}}{n+1} + c$, where; $n = 2$

Step 2: Data substitution and solving

$\int 2x^2 \, dx = 2 \int x^2 \, dx = 2 * \frac{x^{2+1}}{2+1} + c$ ....[Simplify numerator power and denominator]

$\int 2x^2 \, dx = 2 * \frac{x^3}{3} + c$ ....[Multiply terms]

$\int 2x^2 \, dx = \frac{2x^3}{3} + \mathbf{c}$

Example 2: Given that; $f(x) = 3x^2$, shows that; $\frac{d}{dx} \int f(x) \, dx = f(x)$

Solution

Step 1: Determination of integral of $[f(x) = 3x^2]$

$\int 3x^2 \, dx = 3 \int x^2 \, dx = 3 * \frac{x^{2+1}}{2+1} + c$ ....[Simplify numerator power and denominator]

$\int 3x^2 \, dx = 3 * \frac{x^3}{3} + c$ ....[Multiply terms]

$\int 3x^2 \, dx = \frac{3x^3}{3} + \mathbf{c} = \mathbf{x^3 + c}$

Step 2: Determination of derivative of $[\int 3x^2 \, dx = x^3 + c]$

$\frac{d}{dx} \int f(x) \, dx = x^3 + c$ ....[Apply power and constant rule]

$\frac{d}{dx} \int f(x) \, dx = 3(x^{3-1}) + 0$ ....[Simplify exponent]

$\frac{d}{dx} \int f(x) \, dx = 3(x^2) + 0$ ....[Open bracket and solve]

$\frac{d}{dx} \int f(x) \, dx = \mathbf{3x^2 = f(x)}$

Test Your Understanding

Exercise 18: Compute the integral of the following

1) $\int \frac{1}{2} x^5 \, dx$..[Answer: $\frac{1}{12} x^6 + \mathbf{c}$]

2) $\int 6x^2 \, dx$..[Answer: $\mathbf{2x^3 + c}$]

3) $\int 4 \sqrt[3]{x} \, dx$..[Answer: $\mathbf{3x^{\frac{4}{3}} + c}$]

4) $\int 10 \, dx$..[Answer: $\mathbf{10x + c}$]

446

12.1.3 Addition and subtraction rule of integration

The addition and subtraction of two functions is calculated using the formulas below.

$$\int [f(x) + g(x)]\, dx = \int f(x)\, dx + \int g(x)\, dx \dots \dots (1)$$

$$\int [f(x) - g(x)]\, dx = \int f(x)\, dx - \int g(x)\, dx \dots \dots (2)$$

Example 1: Find the integral of the following

1) $\int [2x^3 + 4x^2]\, dx$

2) $\int [\sqrt{x} + 4]\, dx$

Solution

1) $\int [2x^3 + 4x^2]\, dx$

Step 1: Formula. $\int [f(x) + g(x)]\, dx = \int f(x)\, dx + \int g(x)\, dx$, where; $f(x) = 2x^3$ and $g(x) = 4x^2$

Step 2: Data substitution and solving

$\int [2x^3 + 4x^2]\, dx = \int 2x^3\, dx + \int 4x^2\, dx \dots \dots$[Apply constant multiple rule]

$\int [2x^3 + 4x^2]\, dx = 2 \int x^3\, dx + 4 \int x^2\, dx \dots \dots$[Apply power rule]

$\int [2x^3 + 4x^2]\, dx = 2 * \left[\frac{x^{3+1}}{3+1} + c\right] + 4 * \left[\frac{x^{2+1}}{2+1} + c\right]$ [Simplify exponents and open brackets]

$\int [2x^3 + 4x^2]\, dx = \left[\frac{2}{4}x^4 + c\right] + \left[\frac{4}{3}x^3 + c\right]$ ...[Solve. Consider only one integral constant]

$\int [2x^3 + 4x^2]\, dx = \frac{2}{4}x^4 + \frac{4}{3}x^3 + c$

2) $\int [\sqrt{x} + 4]\, dx \dots \dots \left[\int [\sqrt{x} + 4]\, dx = \int \left[x^{\frac{1}{2}} + 4\right] dx\right]$

Step 1: Formula. $\int [f(x) + g(x)]\, dx = \int f(x)\, dx + \int g(x)\, dx$, where; $f(x) = x^{\frac{1}{2}}$ and $g(x) = 4$

Step 2: Data substitution and solving

$\int \left[x^{\frac{1}{2}} + 4\right] dx = \int x^{\frac{1}{2}}\, dx + \int 4\, dx \dots \dots$[Apply power and constant rules]

$\int \left[x^{\frac{1}{2}} + 4\right] dx = \left[\frac{x^{\frac{1}{2}+1}}{\frac{1}{2}+1} + c\right] + [4x + c] \dots \dots$[Simplify power and denominator]

$\int \left[x^{\frac{1}{2}} + 4\right] dx = \left[\frac{x^{\frac{3}{2}}}{\frac{3}{2}} + c\right] + [4x + c] \dots \dots$[Open bracket and solve. Rem (a) = 1(a)]

$\int \left[x^{\frac{1}{2}} + 4\right] dx = \frac{2}{3}x^{\frac{3}{2}} + 4x + c$

Example 2: Derive the integral of the following

1) $\int [20 - 5x]\, dx$

2) $\int \left[x - \frac{1}{x}\right] dx$

<u>Solution</u>

1) $\int [20 - 5x]\, dx$

Step 1: Formula. $\int [f(x) - g(x)]\, dx = \int f(x)\, dx - \int g(x)\, dx$, where; $f(x) = 20$ and $g(x) = 5x$

Step 2: Data substitution and solving

$\int [20 - 5x]\, dx = \int 20\, dx - \int 5x\, dx$ ....[Apply constant multiple rule]

$\int [20 - 5x]\, dx = \int 20\, dx - 5 \int x\, dx$ ....[Apply constant and power rule]

$\int [20 - 5x]\, dx = [20x + c] - 5\left[\frac{x^{1+1}}{1+1} + c\right]$ ......[Simplify numerator and denominator]

$\int [20 - 5x]\, dx = [20x + c] - 5\left[\frac{x^2}{2} + c\right]$ ....[Open brackets and solve]

$\int [20 - 5x]\, dx = 20x - \frac{5}{2}x^2 + c$

2) $\int \left[x - \frac{1}{x}\right] dx$..$\left[\int \left[x - \frac{1}{x}\right] dx = \int [x - x^{-1}]\, dx\right]$

Step 1: Formula. $\int [f(x) - g(x)]\, dx = \int f(x)\, dx - \int g(x)\, dx$, where; $f(x) = x$ and $g(x) = x^{-1}$

Step 2: Data substitution and solving

$\int [x - x^{-1}]\, dx = \int x\, dx - \int x^{-1}\, dx$ ....[Apply power rule]

$\int [x - x^{-1}]\, dx = \left[\frac{x^{1+1}}{1+1} + c\right] - [|x| + c]$ ....[Simplify numerator and denominator]

$\int [x - x^{-1}]\, dx = \left[\frac{x^2}{2} + c\right] - [|x| + c]$ ....[Open brackets and solve]

$\int [x - x^{-1}]\, dx = \frac{1}{2}x^2 - |x| + c$

Example 3: Compute the integral of the following

1) $\int [8x + 4x^2 - 6x^2]\, dx$

2) $\int [-3x^2 + x - 5]\, dx$

<u>Solution</u>

1) $\int [8x + 4x^2 - 6x^2]\, dx$

$\int [8x + 4x^2 - 6x^2]\, dx$ ....[Apply constant multiple rule]

$\int [8x + 4x^2 - 6x^2]\, dx = 8 \int x\, dx + 4 \int x^2\, dx - 6 \int x^2\, dx$ ......[Apply power rule]

$\int [8x + 4x^2 - 6x^2]\, dx = 8\left[\frac{x^{1+1}}{1+1} + c\right] + 4\left[\frac{x^{2+1}}{2+1} + c\right] - 6\left[\frac{x^{2+1}}{2+1} + c\right]$ ......[Simplify]

$\int [8x + 4x^2 - 6x^2]\, dx = 8\left[\frac{x^2}{2} + c\right] + 4\left[\frac{x^3}{3} + c\right] - 6\left[\frac{x^3}{3} + c\right]$ ....[Open brackets]

$\int [8x + 4x^2 - 6x^2]\, dx = \frac{8}{2}x^2 + \frac{4}{3}x^3 - \frac{6}{3}x^3 + c$ ....[Simplify fraction]

448

$$\int [8x + 4x^2 - 6x^2]\, dx = 4x^2 + \frac{4}{3}x^3 - 2x^3 + c$$

2) $\int [-3x^2 + x - 5]\, dx$

$\int [-3x^2 + x - 5]\, dx$ ....[Apply constant multiple rule]

$\int [-3x^2 + x - 5]\, dx = -3 \int x^2\, dx + \int x\, dx - \int 5\, dx$ ..[Power and constant rule]

$\int [-3x^2 + x - 5]\, dx = -3 \left[\frac{x^{2+1}}{2+1} + c\right] + \left[\frac{x^{1+1}}{1+1} + c\right] - [5x + c]$ ..[Simply brackets]

$\int [-3x^2 + x - 5]\, dx = -3 \left[\frac{x^3}{3} + c\right] + \left[\frac{x^2}{2} + c\right] - [5x + c]$ ....[Open brackets]

$\int [-3x^2 + x - 5]\, dx = \frac{-3}{3}x^3 + \frac{1}{2}x^2 - 5x + c$ ....[Simplify fraction]

$\int [-3x^2 + x - 5]\, dx = -x^3 + \frac{1}{2}x^2 - 5x + c$

Example 4: Find the integral of the following functions

1) $\int \left[\left(x - \frac{1}{x}\right)^2\right] dx$

2) $\int \left[\sqrt{x}(x^2 + x + 1)\right] dx$

Solution

1) $\int \left[\left(x - \frac{1}{x}\right)^2\right] dx$........ $\left\{\int \left[\left(x - \frac{1}{x}\right)\left(x - \frac{1}{x}\right)\right] dx = \int \left[x^2 - 2 - \frac{1}{x^2}\right] dx = \int [x^2 - 2 + x^{-2}]\, dx\right\}$

$\int \left[x^2 - 2 + \frac{1}{x^2}\right] dx = \int x^2\, dx - \int 2\, dx + \int x^{-2}\, dx$ ....[Power and constant rule]

$\int \left[x^2 - 2 + \frac{1}{x^2}\right] dx = \left[\frac{x^{2+1}}{2+1} + c\right] - [2x + c] + \left[\frac{x^{-2+1}}{-2+1} + c\right]$ ......[Simplify brackets]

$\int \left[x^2 - 2 + \frac{1}{x^2}\right] dx = \left[\frac{x^3}{3} + c\right] - [2x + c] + \left[\frac{x^{-1}}{-1} + c\right]$ ....[Open brackets]

$\int \left[x^2 - 2 + \frac{1}{x^2}\right] dx = \frac{1}{3}x^3 - 2x - x^{-1} + c$ ....[Re-expressed equation]

$\int \left[\left(x - \frac{1}{x}\right)^2\right] dx = \int \left[x^2 - 2 + \frac{1}{x^2}\right] dx = \frac{1}{3}x^3 - 2x - \frac{1}{x^1} + c$

2) $\int \left[\sqrt{x}(x^2 + x + 1)\right] dx$........................ $\left\{\int \left[x^{\frac{1}{2}}(x^2 + x - 1)\right] dx = \int \left[x^{\frac{5}{2}} + x^{\frac{3}{2}} + x^{\frac{1}{2}}\right] dx\right\}$

$\int \left[x^{\frac{5}{2}} + x^{\frac{3}{2}} + x^{\frac{1}{2}}\right] dx = \int x^{\frac{5}{2}}\, dx + \int x^{\frac{3}{2}}\, dx + \int x^{\frac{1}{2}}\, dx$ ....[Apply power rule]

$\int \left[x^{\frac{5}{2}} + x^{\frac{3}{2}} + x^{\frac{1}{2}}\right] dx = \left[\frac{x^{\frac{5}{2}+1}}{\frac{5}{2}+1} + c\right] + \left[\frac{x^{\frac{3}{2}+1}}{\frac{3}{2}+1} + c\right] + \left[\frac{x^{\frac{1}{2}+1}}{\frac{1}{2}+\_} + c\right]$ [Simplify brackets]

$\int \left[x^{\frac{5}{2}} + x^{\frac{3}{2}} + x^{\frac{1}{2}}\right] dx = \left[\frac{x^{\frac{7}{2}}}{\frac{7}{2}} + c\right] + \left[\frac{x^{\frac{5}{2}}}{\frac{5}{2}} + \_\right] + \left[\frac{x^{\frac{3}{2}}}{\frac{3}{2}} + c\right]$ ....[Open brackets]

$\int \left[\sqrt{x}(x^2 + x + 1)\right] dx = \int \left[x^{\frac{5}{2}} + x^{\frac{3}{2}} + x^{\frac{1}{2}}\right] dx = \frac{2}{7}x^{\frac{7}{2}} + \frac{2}{5}x^{\frac{5}{2}} + \frac{2}{3}x^{\frac{3}{2}} + c$

449

Example 5: determine the integral of the following functions

1) $\int \left[\frac{2x^4+x^2}{x}\right] dx$

2) $\int [8x^2 + 3x^{-4} + x^{-8}] dx$

Solution

1) $\int \left[\frac{2x^4+x^2}{x}\right] dx$.................................$\left\{\int \left[\frac{2x^4+x^2}{x}\right] dx = \int \left[\frac{2x^4}{x} + \frac{x^2}{x}\right] dx = \int [2x^3 + x] dx\right\}$

$\int [2x^3 + x] dx = \int 2x^3 dx + \int x dx = 2\int x^3 dx + \int x dx$ ....[Apply power rule]

$\int [2x^3 + x] dx = 2 \left[\frac{x^{3+1}}{3+1} + c\right] + \left[\frac{x^{1+1}}{1+1} + c\right]$ ....[Simplify brackets]

$\int [2x^3 + x] dx = 2 \left[\frac{x^4}{4} + c\right] + \left[\frac{x^2}{2} + c\right]$ ....[Open brackets and solve]

$\int [2x^3 + x] dx = \frac{1}{2}x^4 + \frac{1}{2}x^2 + c$

2) $\int [8x^2 + 3x^{-4} + x^{-8}] dx$

$\int [8x^2 + 3x^{-4} + x^{-8}] dx = \int 8x^2 dx + \int 3x^{-4} dx + \int x^{-8} dx$[Constant multiple rule]

$\int [8x^2 + 3x^{-4} + x^{-8}] dx = 8 \int x^2 dx + 3 \int x^{-4} dx + \int x^{-8} dx$.....[Apply power rule]

$\int [8x^2 + 3x^{-4} + x^{-8}] dx = 8 \left[\frac{x^{2+1}}{2+1} + c\right] + 3 \left[\frac{x^{-4+1}}{-4+1} + c\right] + \left[\frac{x^{-8+1}}{-8+1} + c\right]$ ....[Simplify]

$\int [8x^2 + 3x^{-4} + x^{-8}] dx = 8 \left[\frac{x^3}{3} + c\right] + 3 \left[\frac{x^{-3}}{-3} + c\right] + \left[\frac{x^{-7}}{-7} + c\right]$ ....[Open brackets]

$\int [8x^2 + 3x^{-4} + x^{-8}] dx = \frac{8}{3}x^3 - \frac{1}{x^3} - \frac{1}{7x^7} + c$

Test Your Understanding

Exercise 19: Determine the integral of the following

1) $\int [4x^3 + 4x] dx$...[Answer: $x^4 + 2x^2 + c$]

2) $\int [6x^2 - 4x + 2] dx$ ....[Answer: $2x^3 - 2x^2 + 2x + c$]

3) $\int \left[\frac{3x^3+x^2}{x}\right] dx$ ....[Answer: $x^3 + \frac{1}{2}x^2 + c$]

4) $\int [x^3 + 3x^2 + x + 4] dx$ ....[Answer: $\frac{1}{4}x^4 + x^3 + \frac{1}{2}x^2 + 4x + c$]

5) $\int [1 - 10x + 9x^2] dx$.....................................[Answer: $x - 5x^2 + 3x^3 + c$]

6) $\int [3x^2 + 2x - x^{-3}] dx$ ....[Answer: $x^3 + x^2 + \frac{1}{2x^2} + c$]

Note: Other types of integration techniques include the integration by substitution and integration by part which is analyzed in this document. It is worth knowing that there are many indefinite integration rules apart of the above mentioned. Other major rules are seen below

$\int e^x dx = e^x + c$ ....(1)

450

$$\int e^{Kx} \, dx = \frac{1}{K} e^{Kx} + c \, \text{................................} \quad \text{..(2)}$$

$$\int \sin(x) \, dx = -\cos(x) + c \, \text{..(3)}$$

$$\int \cos(x) \, dx = \sin(x) + c \, \text{..(4)}$$

12.2 Definite Integration

Definite integration has a start value called lower limit or bound and end value called upper limit. The general form of a definite integration is given below.

$$\int_a^b f(x) \, dx \, \text{..(1)}$$

It is worth knowing that (b) represents the upper limit and (a) represents the lower limit. The rules of definite integration are given below.

12.2.1 The Power Rule

The power rules as with the case of indefinite integration also depends on the exponent as seen below.

$$\int_a^b x^n \, dx = \frac{b^{n+1} - a^{n+1}}{n+1} \, \text{...}\text{Power rule when } (n \neq -1)]$$

$$\int_a^b x^{-1} \, dx = \ln|b| - \ln|a| = \ln \left| \frac{b}{a} \right| \, \text{.......................}\text{[Power rule when } (n = -1)]$$

Example 1: Calculate the integral of the following

1) $\int_0^4 x^3 \, dx$

2) $\int_1^2 \frac{1}{x} \, dx$

Solution

1) $\int_0^4 x^3 \, dx$

Step 1: Formula. $\int_a^b x^n \, dx = \frac{b^{n+1} - a^{n+1}}{n+1}$, where; $x^n = x^3$, $b = 4$, $a = 0$ and $n = 3$

Step 2: Data substitution and solving

$$\int_0^4 x^3 \, dx = \frac{4^{3+1} - 0^{3+1}}{3+1} \, \text{..}\text{[Simplify equation]}$$

$$\int_0^4 x^3 \, dx = \frac{4^4 - 0^4}{4} = \frac{4^4}{4} \, \text{................................}\text{[Work exponent and simplify fraction]}$$

$$\int_0^4 x^3 \, dx = \frac{256}{4} = \mathbf{64}$$

2) $\int_1^2 \frac{1}{x} \, \mathbf{dx} \text{...} \left[\int_1^2 \frac{1}{x} \, dx = \int_1^2 x^{-1} \, dx \right]$

Step 1: Formula. $\int_a^b x^{-1} \, dx = \ln|b| - \ln|a| = \ln\left|\frac{b}{a}\right|$, where; b = 2 and a = 1

Step 2: Data substitution and solving

$\int_1^2 x^{-1} \, dx = \ln|2| - \ln|1| = \ln\left|\frac{2}{1}\right|$ ....[Simplify fraction]

$\int_1^2 x^{-1} \, dx = \mathbf{\ln|2|}$

12.2.2 Complementary Rules

Other basic rules that complement the power rule include the following.

$\int_a^b K \, dx = K(b - a)$ ....[Constant rule]

$\int_a^a f(x) \, dx = 0$ ....[Identical limits rule]

$\int_a^b Kf(x) \, dx = K \int_a^b f(x) \, dx$ ....[Constant multiplication rule]

$\int_a^b f(x) \, dx = - \int_b^a f(x) \, dx$ ....[Backward integration rule]

Example 1: Calculate the integral of the following

1) $\int_1^2 10 \, dx$

2) $\int_{-1}^1 x^3 \, dx$

<u>Solution</u>

1) $\int_1^2 \mathbf{10 \, dx}$

Step 1: Formula. $\int_a^b K \, dx = K(b - a)$, where; K = 10, b = 2 and a = 1

Step 2: Data substitution and solving

$\int_1^2 10 \, dx = 10(2 - 1)$ ....[Simplify bracket]

$\int_1^2 10 \, dx = 10(1)$ ....[Open bracket]

$\int_1^2 10 \, dx = \mathbf{10}$

2) $\int_{-1}^1 \mathbf{x^3 \, dx}$

Step 1: Formula. $\int_a^b x^n \, dx = \frac{b^{n+1} - a^{n+1}}{n+1}$, where; $x^n = x^3$, b = 1, a = −1 and n = 3

Step 2: Data substitution and solving

$\int_{-1}^1 x^3 \, dx = \frac{1^{3+1} - (-1)^{3+1}}{3+1}$ ....[Simplify equation]

$\int_{-1}^1 x^3 \, dx = \frac{1^4 + 1^4}{4}$ ....[Solve equation]

$\int_{-1}^1 x^3 \, dx = \frac{2}{4} = \frac{1}{2}$

Example 2: Determine the integral of the following

1) $\int_1^2 5x^4\,dx$

2) $\int_{-4}^{-2} 6x^2\,dx$

<u>Solution</u>

1) $\int_1^2 5x^4\,dx$

Step 1: Formula. $\int_a^b Kf(x)\,dx = K \int_a^b f(x)\,dx$, where; $K = 5$, $f(x) = x^4$, $b = 2$ and $a = 1$

Step 2: Data substitution and solving

$\int_1^2 5x^4\,dx = 5\int_1^2 x^4\,dx$ ....[Apply power rule]

$\int_1^2 5x^4\,dx = 5\left[\frac{2^{4+1}-1^{4+1}}{4+1}\right]$ ....[Simplify bracket]

$\int_1^2 5x^4\,dx = 5\left[\frac{2^5-1^5}{5}\right] = 5\left[\frac{31}{5}\right]$ ....[Open bracket and solve]

$\int_1^2 5x^4\,dx = \mathbf{31}$

2) $\int_{-4}^{-2} 6x^2\,dx$

Step 1: Formula. $\int_a^b Kf(x)\,dx = K \int_a^b f(x)\,dx$, where; $K = 6$, $f(x) = x$, $b = -2$ and $a = -4$

Step 2: Data substitution and solving

$\int_{-4}^{-2} 6x^2\,dx = 6\int_{-4}^{-2} x^2\,dx$ ....[Apply power rule]

$\int_{-4}^{-2} 6x^2\,dx = 6\left[\frac{-2^{2+1}-(-4)^{2+1}}{2+1}\right]$ ....[Simplify bracket]

$\int_{-4}^{-2} 6x^2\,dx = 6\left[\frac{-2^3+4^3}{3}\right] = 6\left[\frac{56}{3}\right]$ ....[Open bracket and solve]

$\int_{-4}^{-2} 6x^2\,dx = \mathbf{112}$

Test Your Understanding

Exercise 20: Determine the integral of the following

1) $\int_0^3 x\,dx$ ....[Answer: 9/2]

2) $\int_2^4 5\,dx$ ....[Answer: 10]

3) $\int_2^6 x^{-1}\,dx$..[Answer: In|3|]

4) $\int_{-2}^{-1} 5x^4\,dx$...[Answer: 31]

5) $\int_4^9 3\sqrt{x}\,dx$ ....[Answer: 38]

6) $\int_1^2 3x^{-1}\,dx$ ....[Answer: 3In|2|]

12.2.3 Addition and Subtraction Rule

The rules of addition and subtraction of two functions are given below.

$$\int_a^b [f(x) + g(x)]\, dx = \int_a^b f(x)\, dx + \int_a^b g(x)\, dx \ \ldots\ldots\ldots\ldots\ldots\ldots\ldots\ldots\text{[Addition rule]}$$

$$\int_a^b [f(x) - g(x)]\, dx = \int_a^b f(x)\, dx - \int_a^b g(x)\, dx \ \ldots\ldots\ldots\ldots\ldots\ldots\ldots\text{[Subtraction rule]}$$

Example 1: Determine the integral of the following equations

1) $\int_0^4 [x + 2]\, dx$

2) $\int_0^5 [30 - 4x]\, dx$

<u>Solution</u>

1) $\int_0^4 [x + 2]\, dx$

Step 1: Formula. $\int_a^b [f(x) + g(x)]\, dx = \int_a^b f(x)\, dx + \int_a^b g(x)\, dx$, where; $f(x) = x$ and $g(x) = 2$

Step 2: Data substitution and solving

$$\int_0^4 [x + 2]\, dx = \int_0^4 x\, dx + \int_0^4 2\, dx \ \ldots\ldots\ldots\ldots\ldots\ldots\ldots\text{[Apply power and constant rule]}$$

$$\int_0^4 [x + 2]\, dx = \left[\frac{4^{1+1} - 0^{1+1}}{1+1}\right] + [2(4 - 0)] \ \ldots\ldots\ldots\ldots\ldots\ldots\ldots\ldots\text{[Simplify brackets]}$$

$$\int_0^4 [x + 2]\, dx = \left[\frac{4^2 - 0^2}{2}\right] + [2(4)] = \left[\frac{8}{2}\right] + [8] \ \ldots\ldots\ldots\ldots\ldots\text{[Open brackets and solve]}$$

$$\int_0^4 [x + 2]\, dx = \left[\frac{4^2 - 0^2}{2}\right] + [2(4)] = \mathbf{16}$$

2) $\int_0^5 [30 - 4x]\, dx$

Step 1: Formula. $\int_a^b [f(x) - g(x)]\, dx = \int_a^b f(x)\, dx - \int_a^b g(x)\, dx$

Where; $f(x) = 30$ and $g(x) = 4x$

Step 2: Data substitution and solving

$$\int_0^5 [30 - 4x]\, dx = \int_0^5 30\, dx - \int_0^5 4x\, dx \ \ldots\ldots\ldots\ldots\ldots\ldots\text{[Apply constant multiple rule]}$$

$$\int_0^5 [30 - 4x]\, dx = \int_0^5 30\, dx - 4 \int_0^5 x\, dx \ \ldots\ldots\ldots\ldots\ldots\text{[Apply constant and power rules]}$$

$$\int_0^5 [30 - 4x]\, dx = [30(5 - 0)] - 4 \left[\frac{5^{1+1} - 0^{1+1}}{1+1}\right] \ \ldots\ldots\ldots\ldots\ldots\ldots\text{[Simplify brackets]}$$

$$\int_0^5 [30 - 4x]\, dx = [30(5)] - 4 \left[\frac{5^2 - 0^2}{2}\right] = [30(5)] - 4 \left[\frac{25}{2}\right] \ \ldots\text{[Open brackets and solve]}$$

$$\int_0^5 [30 - 4x]\, dx = 150 - 50 = \mathbf{100}$$

Example 2: Find the integral of the following functions

1) $\int_1^2 (x^3 - 2x + 5)\, dx$

2) $\int_0^1 x(x^2 - 1)\, dx$

Solution

1) $\int_1^2 (x^3 - 2x + 5)\, dx$

$\int_1^2 (x^3 - 2x + 5)\, dx = \int_1^2 x^3\, dx - \int_1^2 2x\, dx + \int_1^2 5\, dx$ [Apply constant multiple rule]

$\int_1^2 (x^3 - 2x + 5)\, dx = \int_1^2 x^3\, dx - 2\int_1^2 x\, dx + \int_1^2 5\, dx$ ....[Apply power rule]

$\int_1^2 (x^3 - 2x + 5)\, dx = \left[\frac{2^{3+1} - 1^{3+1}}{3+1}\right] - 2\left[\frac{2^{1+1} - 1^{1+1}}{1+1}\right] + [5(2-1)]$[Simplify brackets]

$\int_1^2 (x^3 - 2x + 5)\, dx = \left[\frac{2^4 - 1^4}{4}\right] - 2\left[\frac{2^2 - 1^2}{2}\right] + [5(1)]$ ....[Open brackets and solve]

$\int_1^2 (x^3 - 2x + 5)\, dx = \frac{23}{4} = 5\frac{3}{4} = 5.75$

2) $\int_0^1 x(x^2 - 1)\, dx$..........$\left[\int_0^1 x(x^2 - 1)\, dx = \int_0^1 [x * x^2 - 1 * x]\, dx = \int_0^1 [x^3 - x]\, dx\right]$

$\int_0^1 [x^3 - x]\, dx = \int_0^1 x^3\, dx - \int_0^1 x\, dx$ ....[Apply Power rule]

$\int_0^1 [x^3 - x]\, dx = \left[\frac{1^{3+1} - 0^{3+1}}{3+1}\right] - \left[\frac{1^{1+1} - 0^{1+1}}{1+1}\right]$ ....[Simplify brackets]

$\int_0^1 [x^3 - x]\, dx = \left[\frac{1^4 - 0^4}{4}\right] - \left[\frac{1^2 - 0^2}{2}\right] = \left[\frac{1}{4}\right] - \left[\frac{1}{2}\right]$ ....[Open brackets and solve]

$\int_0^1 [x^3 - x]\, dx = -\frac{1}{4}$

Test Your Understanding

Exercise 21: Find the integral of the following equations

1) $\int_0^{50} [50 - x]\, dx$...[Answer: 1,250]

2) $\int_0^4 [-2x + 14]\, dx$...[Answer: 40]

3) $\int_1^3 [3x + 3]\, dx$ ....[Answer: 18]

4) $\int_{-1}^1 [x^2 + 4x - 3]\, dx$...[Answer: $-\frac{10}{3}$]

N/B: Other major technique of integration includes; integration by substitution (apply to reverse the chain rule), integration by part (apply to reverse the product rule), integration by successive reduction and integration by partial fraction.

APPENDIX [Answers of Exercises]

Chapter Two: [Financial, Geometric and Weight Measurement]

Exercise 13

Triangle 1	Triangle 2
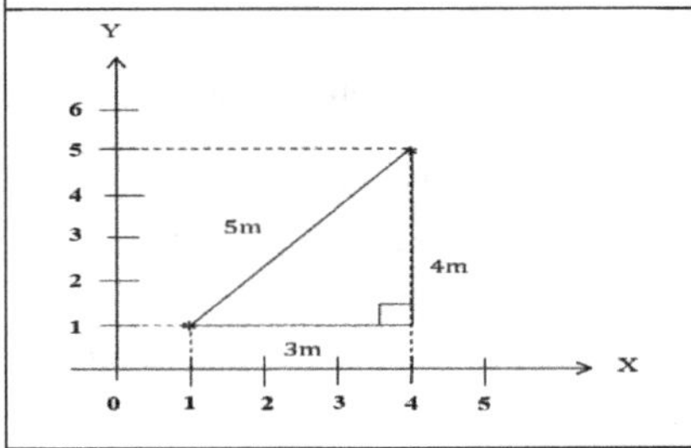	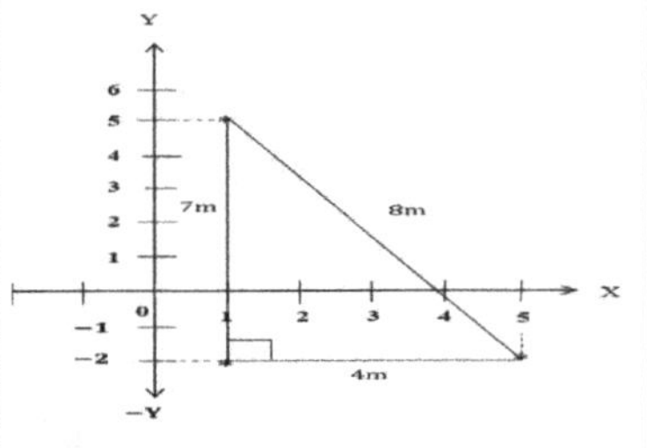
Perimeter = 5m + 3m + 4m = **12m**	Perimeter = 7m + 4m + 8m = **19m**

Exercise 14

Triangle 1	Triangle 2	Triangle 3	Triangle 4
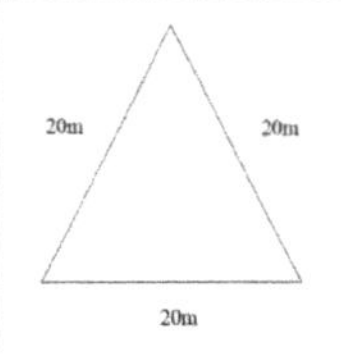	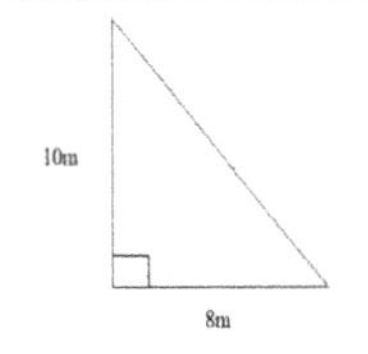	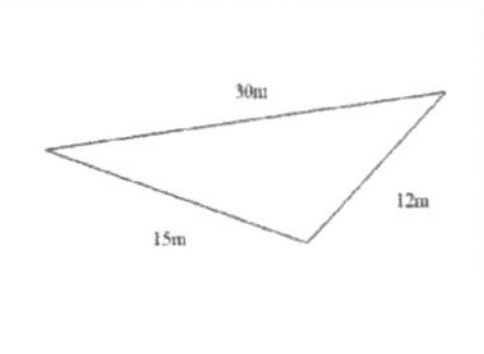	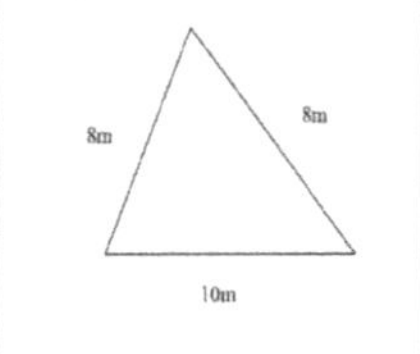
60 m	**31m**	**57m**	**26m**

Exercise 19

Rectangle	Triangle	Circles	Square
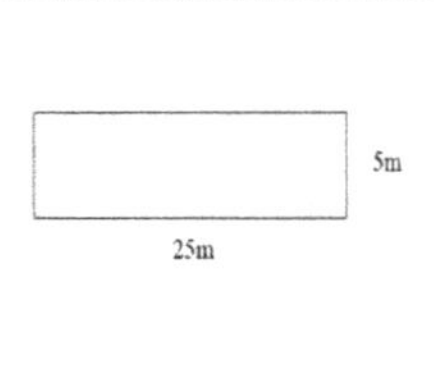	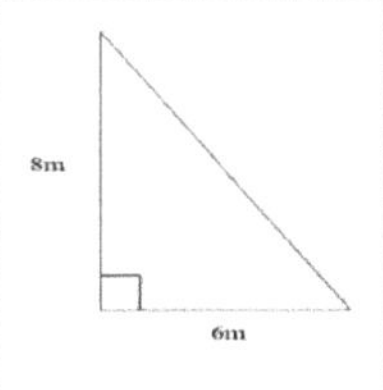	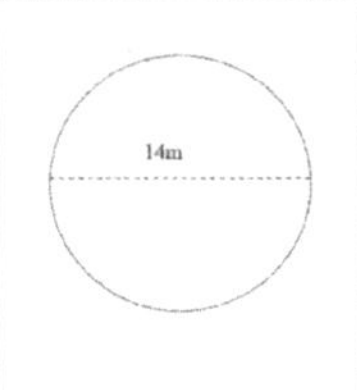	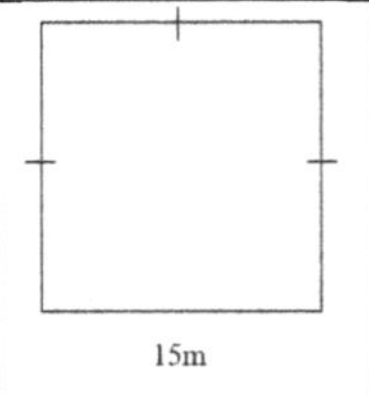
60m	**24m**	**44m**	**60m**

456

Exercise 25

a) (2,1), (4,0) and (5,7)	**b)** (−2,1), (3,4) and (3,1)
(graph)	*(graph)*

N/B: Land is the largest land 2 $(7.5m^2)$ and land 1 area is $4m^2$. The son should choose land 2

Chapter Three: [Mathematical Equations]

Exercise 2

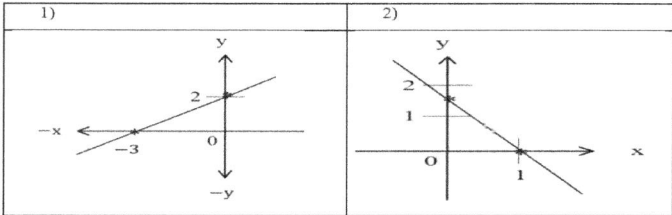

Exercise 3

1)	x	−2	−1	0	1	2
	y	$-2(-2)-1=3$	$-2(-1)-1=1$	$-2(0)-1=-1$	$-2(1)-1=-3$	$-2(2)-1=-5$

2)	x	−2	−1	0	1	2
	y	−2	−2	−2	−2	−2

3)	x	−2	−1	0	1	2
	y	−2	−1	0	1	2

Exercise 4

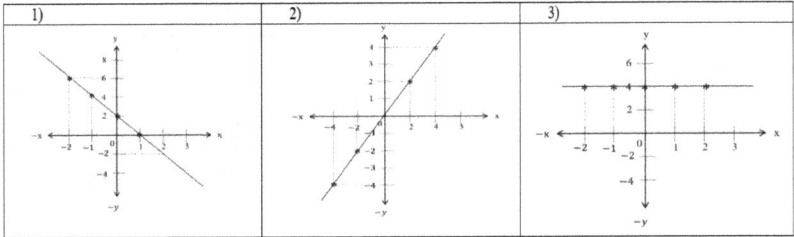

Exercise 15

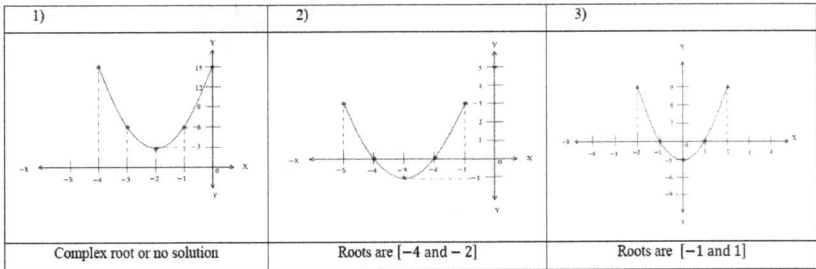

1)	2)	3)
Complex root or no solution	Roots are [−4 and − 2]	Roots are [−1 and 1]

Exercise 18

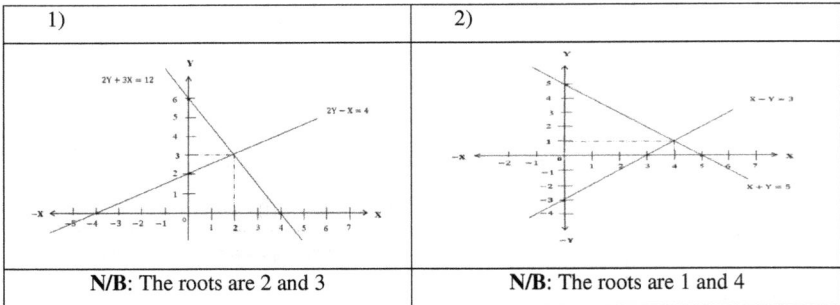

1)	2)
N/B: The roots are 2 and 3	**N/B**: The roots are 1 and 4

Exercise 21

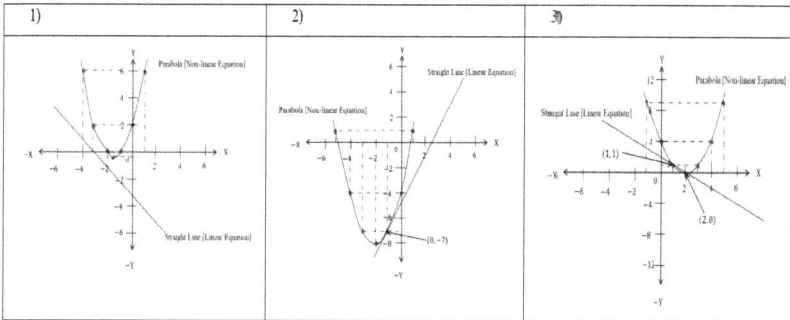

Exercise 22

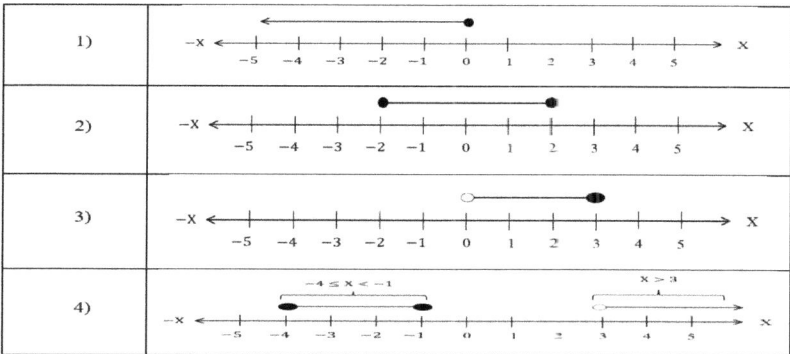

Exercise 26

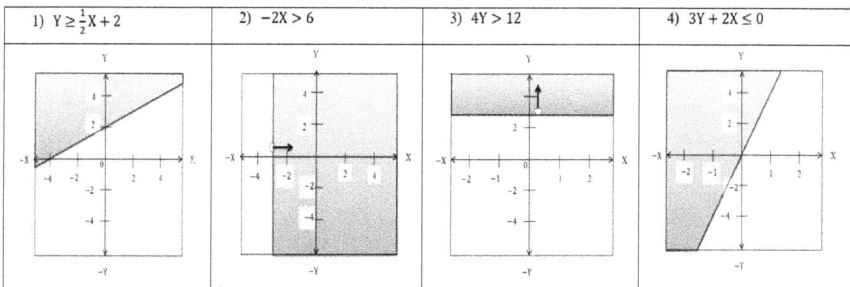

459

Exercice 27

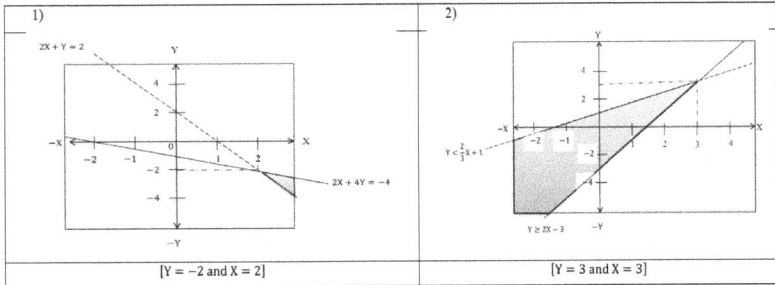

1)	2)
[Y = −2 and X = 2]	[Y = 3 and X = 3]

Exercice 28

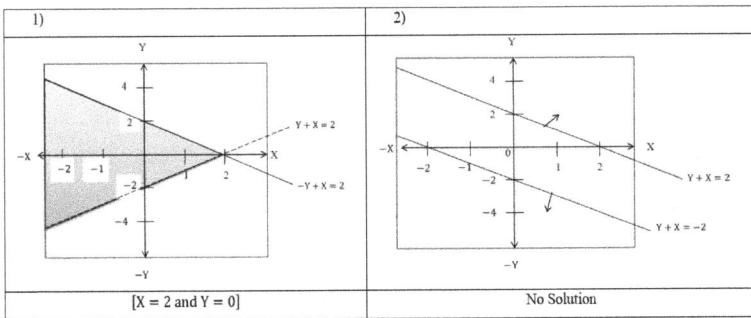

1)	2)
[X = 2 and Y = 0]	No Solution

Chapter Four: [Matrix and Equation]

Exercise 1

1)	2)	3)
$A = \begin{bmatrix} 2 & -4 \\ -8 & 10 \end{bmatrix}$	$A = \begin{bmatrix} 2 & 6 \\ -8 & -12 \end{bmatrix}$	$A = \begin{bmatrix} -4 & 6 \\ 10 & -12 \end{bmatrix}$

Exercise 2

1)	2)	3)	4)
$A = \begin{bmatrix} \frac{1}{2} & -0.5 \\ -1 & -8 \\ 0 & -\frac{5}{2} \end{bmatrix}$	$N = \begin{bmatrix} 5 & \frac{1}{2} & -0.5 \\ 1 & 0 & -\frac{5}{2} \end{bmatrix}$	$D = \begin{bmatrix} 5 & \frac{1}{2} \\ 1 & 0 \end{bmatrix}$	$P = \begin{bmatrix} 5 & 0 & 1 \\ \frac{1}{2} & -1 & 0 \\ -0.5 & -8 & -\frac{5}{2} \end{bmatrix}$

Exercise 3

1)	2)
$A + B = \begin{bmatrix} 5 & 6 \\ 2 & 5 \end{bmatrix}$	$B - A = \begin{bmatrix} -1 & 2 \\ 6 & -5 \end{bmatrix}$

Exercise 4

1)	2)
$A + B \begin{bmatrix} 6 & 5 & -6 \\ -1 & 0 & -7 \\ 3 & 4 & 7 \end{bmatrix}$	$B - A = \begin{bmatrix} 0 & -3 & -6 \\ 3 & 0 & 1 \\ 2 & 2 & 1 \end{bmatrix}$

Exercise 5

1)	2)	3)
$AB = \begin{bmatrix} 12 \\ 6 \end{bmatrix}$	$DC = \begin{bmatrix} -\frac{3}{2} \\ -\frac{3}{2} \\ 3 \end{bmatrix}$	$EF = \begin{bmatrix} -5 & 5 \\ 10 & 10 \\ 2 & 16 \\ 5 & 5 \end{bmatrix}$

Exercise 6

1)	2)
$\lambda I - A = \begin{bmatrix} \lambda - 10 & 8 \\ -4 & \lambda + 2 \end{bmatrix}$	$\alpha B + \beta C = \begin{bmatrix} \alpha + 2\beta & 2\alpha + \beta \\ 4\alpha + 3\beta & \alpha + \beta \end{bmatrix}$

Exercise 12

1)	2)	3)
$Adj(A) = \begin{bmatrix} 5 & -6 \\ -7 & 10 \end{bmatrix}$	$Adj(D) = \begin{bmatrix} 1 & -4 \\ 3 & 1 \end{bmatrix}$	$Adj(M) = \begin{bmatrix} 5 & 1 \\ 0 & 2 \end{bmatrix}$

Exercise 13: Minor Matrix $= \begin{bmatrix} 2 & 2 & 2 \\ -2 & 3 & 3 \\ 0 & -10 & 0 \end{bmatrix}$

Exercise 14

1)	2)
$Co - Factor = \begin{bmatrix} 0 & -26 & 0 \\ 0 & 28 & 0 \\ 0 & -22 & 0 \end{bmatrix}$	$Co - Factor = \begin{bmatrix} -8 & 6 & -1 \\ 0 & 0 & 0 \\ 0 & -6 & 1 \end{bmatrix}$

Exericse 15

1)	2)
$\text{Adj(M)} = \begin{bmatrix} 2 & -11 & 25 \\ 2 & 4 & -10 \\ -2 & 6 & -10 \end{bmatrix}$	$\text{Adj(H)} = \begin{bmatrix} 2 & -1 & -2 \\ 2 & -2 & -2 \\ 6 & -5 & -8 \end{bmatrix}$

Exercise 17: $M^{-1} = \begin{bmatrix} -\frac{2}{16} & -\frac{6}{16} & \frac{6}{16} \\ -\frac{6}{16} & -\frac{14}{16} & \frac{5}{16} \\ \frac{12}{16} & -\frac{12}{16} & \frac{4}{16} \end{bmatrix}$

Exercise 18

First System	Second System	Third System
$\begin{bmatrix} 3 & -4 \\ -2 & 1 \end{bmatrix}\begin{bmatrix} X \\ Y \end{bmatrix} = \begin{bmatrix} 9 \\ -3 \end{bmatrix}$	$\begin{bmatrix} 0 & 2 \\ -5 & -3 \\ 6 & 0 \end{bmatrix}\begin{bmatrix} X \\ Y \end{bmatrix} = \begin{bmatrix} 2 \\ 1 \\ 6 \end{bmatrix}$	$\begin{bmatrix} \frac{1}{2} & -3 & 1 \\ 0 & 1 & -4 \end{bmatrix}\begin{bmatrix} X \\ Y \\ Z \end{bmatrix} = \begin{bmatrix} 2 \\ 3 \\ 5 \end{bmatrix}$

Exercise 19

First System	Second System	Third System
$\begin{bmatrix} 1 & 2 & 1 \\ 2 & 1 & 1 \\ 1 & 1 & 3 \end{bmatrix}\begin{bmatrix} X \\ Y \\ Z \end{bmatrix} = \begin{bmatrix} 9 \\ 7 \\ 10 \end{bmatrix}$	$\begin{bmatrix} 1 & 3 & 1 \\ 2 & -1 & 1 \\ -2 & 2 & -1 \end{bmatrix}\begin{bmatrix} X \\ Y \\ Z \end{bmatrix} = \begin{bmatrix} 1 \\ 5 \\ -8 \end{bmatrix}$	$\begin{bmatrix} 1 & -2 & 4 \\ 0 & 8 & -14 \\ 3 & -9 & 0 \end{bmatrix}\begin{bmatrix} X \\ Y \\ Z \end{bmatrix} = \begin{bmatrix} 44 \\ -33 \\ 61 \end{bmatrix}$

Exercise 20

Equation 1	Equation 2	Equation 3
$-7Y = 2$ ....(1) $2X + Y = 5$ ....(2)	$2X + Y + Z = 3$ ....(1) $X + 3Y - Z = 7$ ....(2) $X + Y + Z = 1$ ....(3)	$2X + 3Y + Z = 2$ ....(1) $X + 2Y + 3Z = 4$ ....(2)

Exercise 21

	Equation 1	Equation 2	Equation 3
Coefficient Matrix	$\begin{bmatrix} 1 & 2 \\ 3 & -4 \end{bmatrix}$	$\begin{bmatrix} 0 & 1 & -3 \\ 2 & 0 & 4 \\ 3 & 1 & 7 \end{bmatrix}$	$\begin{bmatrix} 5 & -3 & 0 \\ 0 & 0 & 1 \end{bmatrix}$
Augmented Matrix	$\begin{bmatrix} 1 & 2 & 3 \\ 3 & -4 & 1 \end{bmatrix}$	$\begin{bmatrix} 0 & 1 & -3 & 5 \\ 2 & 0 & 4 & 9 \\ 3 & 1 & 7 & -3 \end{bmatrix}$	$\begin{bmatrix} 5 & -3 & 0 & 8 \\ 0 & 0 & 1 & -3 \end{bmatrix}$

Chapter Seven: [The Concept of Statistics]

Exercise 1

Class interval	Frequency		Class interval	Frequency
$[46-56[$	2		$[46-55]$	2
$[56-66[$	2		$[56-65]$	2
$[66-76[$	6		$[66-75]$	6
$[76-86[$	9		$[76-85]$	9
$[86-96[$	9		$[86-95]$	9
$[96-106[$	4		$[96-105]$	4
	$N=32$			$N=32$

Exercise 2

	Exercise 2.1								
Marks	5	8	12	15	19	24	30	42	Total
Frequency	4	4	2	5	5	1	3	1	24
	Exercise 2.2								
Marks	5 - 12	12 - 19	19 - 26	26 - 33	33 - 40	40 - 47			
Frequency	8	6	6	3	0	1	$N=24$		

Exercise 2.3

Relative frequency		Percentage Relative frequency		Cumulative frequency		Relative cumulative frequency	
Mark	f_i	Mark	f_i	Mark	f_i	Mark	f_i
5 - 12	0.333333333	5 - 12	33.33333333	5 - 12	8	5 - 12	0.071428571
12-19	0.25	12-19	25	12-19	14	12-19	0.125
19-26	0.25	19-26	25	19-26	20	19-26	0.178571428
26-33	0.125	26-33	12.5	26-33	23	26-33	0.205357142
33-40	0	33-40	0	33-40	23	33-40	0.205357142
40-47	0.041666666	40-47	4.166666667	40-47	24	40-47	0.214285714
	$N=1$		$N=100$		$N=112$		$N=1$

Exercise 3

Class interval	$[10-20[$	$[20-30[$	$[30-40[$	$[40-50[$	$[50-60[$	$[60-70[$
Frequency	18	12	5	25	25	25

Marks	$[5-10[$	$[10-15[$	$[15-20[$	$[20-25[$	$[25-30[$
Midpoints	7.5	12.5	17.5	22.5	27.5
No of students	10	25	12	20	14

Exercise 4.1

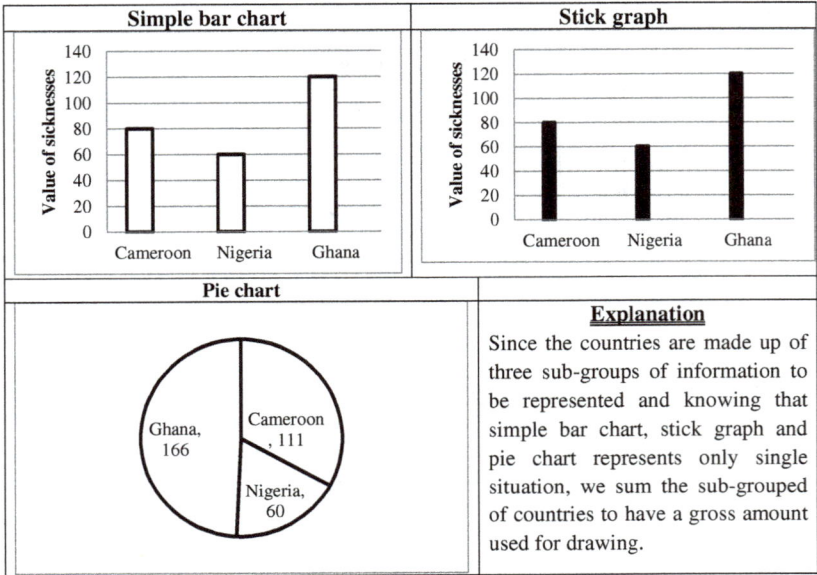

Exercise 4.2

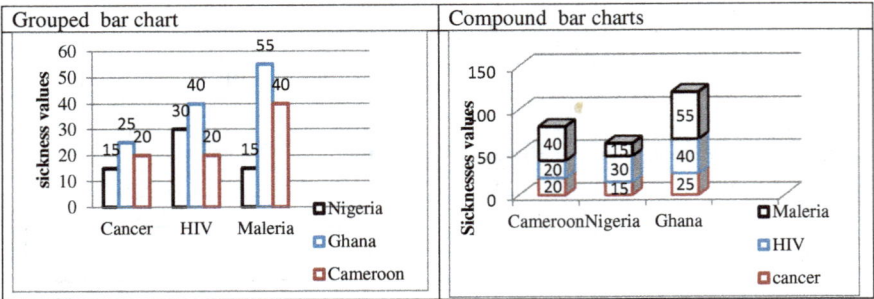

464

Exercise 5

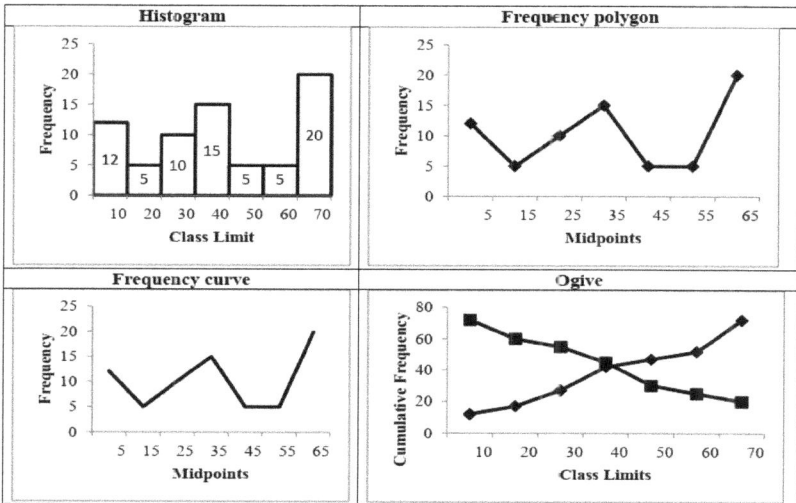

Exercise 6

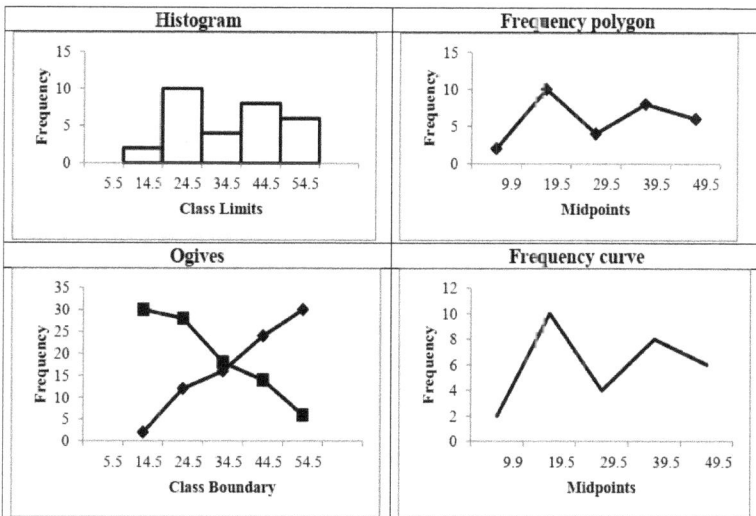

REFERENCE

Bayne C. S. (2010). Business Statistics. JMSB, Concordia University. Cam.Ed john Wiley and sons, Canada Ltd.

Black, K., & Eldredge, D. L. (2002). *Business and economic statistics using Microsoft Excel*. South-Western Pub.

Davis, H. T., & Nelson, W. F. C. (1935). Elements of statistics with applications to economic data.

Dean, S., & Illowsky, B. (2013). *Principles of business statistics*. Connexions, Rice University.

Del D. Business Statistics; Frequency distribution and graphical representation. Faculty of business administration, university of New Brunswick.

Demir, M., & Demir, Ş. Ş. The Book of ICSSER Abstracts.

Duman E. Y. Introduction to probability and statistics; measure of central tendency and sampling distribution. Istanbul kultur university , faculty of engineering.

El-Taha M. Introduction to business statistics. Department of mathematics and statistics. University of southern Maine, 96 falmouth street Portland, ME 04104-9300.

Floyed J. E. (2010): Statistics for Economist; A beginning. University of Toronto July 2, 2010.

Francis A. (2008). Business mathematics and statistics. British library cataloguing-in-publication data.

Gilmartin, K., & Rex, K. (1999). *Student Toolkit: Working with Charts, Graphs and Tables*. Open University.

Kelley, J. T. (2002). Using Graphs and Visuals to Present Financial Information. *Prieiga per internetq:< http://home. xnet. com/~ jkelley/Publications/Using_Graphs. pdf>,(prisijungta 2010-02-17)*.

Smith, M. J. (2014). Statistical Analysis Handbook A Comprehensive Handbook of Statistical Concepts. *Techniques and Software Tool http://www. statsref. com/HTML/index. html*.

466

The university of reading statistical services center (2001). Informative presentation of tables, graphs and statistics. Biometrics advisory and support services to DFID UK.

Tiemann T. K. (2010). Introductory Business Statistics. Global text project funded by the Jacobs foundation, Zurich, Switzerland.

Varalakshmi V., Suseela N., sundaram T. G. G., Ezh larasi S., & Indrani B. (2004). Statistics. Tamilnada textbook and educational service corporation. College road, channai-600-006.

www.ingramcontent.com/pod-product-compliance
Lightning Source LLC
Chambersburg PA
CBHW060425220326
41598CB00021BA/2288